Fast and Effective Embedded Systems Design

Fast and Effective Embedded Systems Design

From Bits and Bytes to IoT, with the Arm Mbed

Third Edition

Rob Toulson
Tim Wilmshurst
Tom Spink

Amsterdam • Boston • Heidelberg • London • New York • Oxford • Paris
San Diego • San Francisco • Singapore • Sydney • Tokyo
Newnes is an imprint of Elsevier

ELSEVIER

Newnes
An imprint of Elsevier

Academic Press is an imprint of Elsevier
125 London Wall, London EC2Y 5AS, United Kingdom
525 B Street, Suite 1650, San Diego, CA 92101, United States
50 Hampshire Street, 5th Floor, Cambridge, MA 02139, United States
The Boulevard, Langford Lane, Kidlington, Oxford OX5 1GB, United Kingdom

ISBN: 978-0-323-95197-5

For information on all Elsevier publications visit our website at
https://www.elsevier.com/books-and-journals

Publisher: Katey Birtcher
Acquisition Editor: Stephen R. Merken
Editorial Project Manager: Sara Valentino
Production Project Manager: Nandhini Thanga Alagu
Cover Designer: Mark Rogers

Typeset by TNQ Technologies

Contents

Introduction

It's now eleven years since the first edition of this book was published, and seven years since the second. Microprocessors are of course still everywhere, providing "intelligence" in cars, mobile phones, household and office equipment, TVs and entertainment systems, medical products, aircraft—the list is endless. Those everyday products, where a little computer is hidden inside to add intelligence, are called *embedded systems*.

It is still not so long ago that designers of embedded systems had to be electronics experts, or software experts, or both. Nowadays, with user-friendly and sophisticated building blocks available for our use, both the specialist and the beginner can quickly engage in successful embedded system construction and design. One such building block was the *Mbed*, a self-contained microcontroller development board, launched in 2009 by the renowned computer giant Arm. The Mbed was the central theme of the first edition of this book, and through it all the main topics of embedded system design were introduced.

Technology has continued its onward gallop since those earlier editions, and there is already much from the second edition which needs updating, rewriting, or replacing. The single Mbed device that we based the first book on has spawned an extended family of "Mbed-enabled" devices. An "ecosystem" of products, development tools, and community support has emerged, and the phrase "internet of things" (IoT) is now on everyone's lips. The creators of the Mbed, Arm, repositioned the Mbed concept, notably by placing it at the heart of their IoT developments. The original Mbed LPC1768 device, however, is still there, widely used and well established, in industry, among hobbyists, and in colleges and universities worldwide.

Alongside hardware developments, there has been stunning growth in software techniques. As part of the Mbed ecosystem, Arm have crafted a sophisticated operating system, an essential framework through which program development takes place. While in earlier editions we tried to restrict ourselves more or less to use of the C programming language, this is now no longer possible. The Mbed operating system requires some use of C++, as more advanced applications are encountered.

In light of all these changes, we as authors made a number of decisions regarding this new edition. We have retained the Mbed LPC1768, but recognize that it's approaching the end of its working life. Therefore, in parallel, we've introduced an alternative board, the Nucleo F401RE. You can work through the first 10 chapters of the book using either of these. As if two boards aren't enough, we introduce a third, the ST IoT discovery kit. We use this primarily for Chapters 11–13, the IoT part of the book.

In terms of content, we have dropped the audio and internet chapters to make way for a pair of explicitly IoT-focused chapters. There is also a completely new chapter on using the real time operating system, or RTOS.

Broadly, the book divides into three parts. Chapters 1 to 10 provide a wide-ranging introduction to embedded systems. These chapters aim to give full support to the reader, moving you through a carefully constructed series of concepts and exercises. They start from basic principles and simple projects, and move on to more advanced system design. The next three chapters, 11–13, sit clearly in the realm of wireless and the IoT. They start with the essential wireless techniques which underlie IoT capability, before building up a complex IoT example, linking sensors through intermediate devices right up to the cloud and back! Programming here also includes some Java and JavaScript. The final two chapters, mainly based on the Mbed LPC1768 but not depending on it, look inside the microcontroller to cover more advanced or specialist topics. They explain how some of the most fundamental microcontroller activities are undertaken and how they can be controlled—for example, setting clock speed or optimizing power consumption.

All this book asks of you at the very beginning is a basic grasp of electrical/electronic theory. The book adopts a "learning through doing" approach. To get started, you will need either the LPC1768 or the F401RE development board, an internet-connected computer, and at least some of the various additional electronic components identified. You won't need every single one of these if you choose not to do a certain experiment or book section. You'll also need a digital voltmeter, and ideally use of an oscilloscope.

Each chapter is based on a major topic in embedded systems. Each has some theoretical introduction, and may have more theory within the chapter. Most chapters then proceed as series of practical experiments. Have your board ready to connect up the next circuit, and download and compile the next example program. As your confidence grows, so will your creativity and originality; you will start to turn your own ideas into working projects.

You will find that this book rapidly helps you to

- Understand and apply the key aspects of embedded systems,
- Learn from scratch, or develop your skills in embedded C/C++ programming,
- Understand and apply the key aspects of the Arm Mbed operating system,

- Develop your understanding of electronic components and configurations,
- Produce designs and innovations you never thought you were capable of!

If you are a university or college instructor, then this book offers a "complete solution" for your embedded systems course. All three authors are experienced university lecturers, and had your students in mind when writing this book. The book contains a structured sequence of practical and theoretical learning activity. Ideally you should equip every student or student pair with an Mbed development board, prototyping breadboard, and component kit. These are highly portable, so development work is not confined to the college or university lab. Later in the course students will start networking their devices together. PowerPoint presentations for each chapter are available to instructors via the book web site, as well as answers to quiz questions.

Because the need for electronic theory is limited, this book is accessible to disciplines which would not normally aim to take embedded systems. The book is meant to be accessible to Year 1 undergraduates, though we expect it will more often be used in the years which follow. Students are likely to be studying one of the branches of engineering, physics, or computer science. The book will also be of interest to the practicing professional and the hobbyist.

The first ideas for this book came from Rob Toulson. When at Anglia Ruskin University in Cambridge, he started teaching with those new-fangled Mbed devices, working in close cooperation with the Mbed design team at Arm. Tim Wilmshurst—then at the University of Derby—joined to form a writing duo, and together Rob and Tim brought the first two editions to life. Tim then proposed, and has since led, development of the third edition. A new and welcome co-author, Tom Spink, joins the team for this edition. A lecturer in computer science at the University of St Andrews, Tom brings particular expertise in IoT and the software end of things. He and Tim are also authors of Arm Education MOOCs (massive open online courses) on embedded systems and IoT. Rob meanwhile has moved on from academia to become an active entrepreneur in digital music technologies (https://www.rt60.uk/), but has continued to play a supporting role in the new edition.

Because Tim has written several books on embedded systems in the past, a few background sections and diagrams have been taken from these and adapted for inclusion in this book. There seemed no point in "reinventing the wheel" where introductory explanations were needed.

Acknowledgments

The authors would like to thank staff from Arm, including Donatien Garnier, Robert Iannello, David MacKenzie, and Andy Powers, for support in the development of this edition, and in earlier work on the MOOCs. Thanks also to those who develop and maintain content on the Mbed OS web site (https://os.Mbed.com/). We have made repeated use of this site and benefited from the many example programs therein.

We would further like to thank most warmly the following for their comments on draft chapters. They read and evaluated with care and attention, making valued suggestions for improvements to content or code, exposing lurking typos or technical inaccuracy, and brought their shared teaching expertise to how ideas should be presented and should flow. In alphabetical order they are Hammam Alsafrjalani of the University of Miami, Haitham Abu Ghazaleh of Tarleton State University, John Larkin of Whitworth University, Erik Petrich of the University of Oklahoma, and Atoussa Tehrani of Florida International University. Despite the value of these inputs, any errors or weaknesses in the book remain the authors' responsibility!

Thanks to Liz A of sweetclipart.com for the use of the car image in Figure 1.3.

Companion Web Site

www.embedded-knowhow.co.uk

PART 1

Essentials of Embedded Systems, and Developing with Mbed

CHAPTER 1

Embedded Systems, Microcontrollers, and Arm

1.1 Introducing Embedded Systems

1.1.1 What is an Embedded System?

We are all familiar with the idea of a desktop or laptop computer, and the amazing processing that they can do. These computers are general-purpose; we can get them to do different things at different times, depending on the application or program we run on them. At the very heart of such computers, we would find one or more *microprocessors*, tiny and fantastically complicated electronic circuits that contain the core features of a computer. These are fabricated on a single slice of silicon, called an *integrated circuit* (IC). Some people, particularly those who are not engineers themselves, call these circuits *microchips,* or just *chips.*

What is less familiar to many people is the idea that instead of putting a microprocessor into a general-purpose computer, it can also be placed inside a product which has nothing to do with computing, like a washing machine, toaster, or camera. The microprocessor is then customized to control that product. The computer is there, inside the product; but it can't be seen, and the user probably doesn't even know it's there. Moreover, those add-ons which we normally associate with a computer, like a keyboard, screen, or mouse, are nowhere to be seen. We call such products *embedded systems,* because the microprocessor that controls them is embedded inside. Because such a microprocessor is developed to control the device, in many cases those used in embedded systems have different characteristics from the ones used in more general-purpose computing machines. We end up calling these embedded computers *microcontrollers.* Though much less visible than their microprocessor cousins, microcontrollers sell in far greater volume, and their impact has been enormous. To the electronic and system designer they offer huge opportunities.

Embedded systems come in many forms and guises. They are extremely common in the home, the motor vehicle, and the workplace. Most modern domestic appliances, like a washing machine, dishwasher, oven, central heating, or burglar alarm, are embedded systems. The motor car is full of them, in engine management, security (e.g., locking and anti-theft devices), crash sensing, air-conditioning, brakes, radio, and so on. They are

Fast and Effective Embedded Systems Design. https://doi.org/10.1016/B978-0-323-95197-5.00001-2

found across industry and commerce, in machine control, factory automation, robotics, electronic commerce, and office equipment. The list has almost no end, and it continues to grow.

Fig. 1.1 expresses the embedded system as a simple block diagram. A number of features appear; not every embedded system has all of them. There are one or more inputs from the external environment, for example, sensors providing voltages proportional to physical variables. The embedded computer runs a program dedicated to this application, permanently stored in its memory. Unlike the general-purpose desktop computer, which runs many programs, this is the only program it ever runs. Based on information supplied from the inputs, the microcontroller computes certain outputs, which may be connected to actuators and other devices within the system. Alternatively, that output may simply be data, transferred by cable or wireless to another subsystem elsewhere, or displayed to a user. The user may have further interaction—for example, via a keypad.

The actual electronic circuit, including the microcontroller and any electromechanical components, is often called the *hardware*; the program running on it is often called the *software*. There must be a supply of power to keep the circuits running. One other variable will affect all that we do in embedded systems, and this is time, represented as a dominating arrow which cuts across the figure. We will need to be able to measure time, make things happen at predetermined times, generate data streams or other signals with a strong time dependence, and respond to unexpected things in a timely fashion.

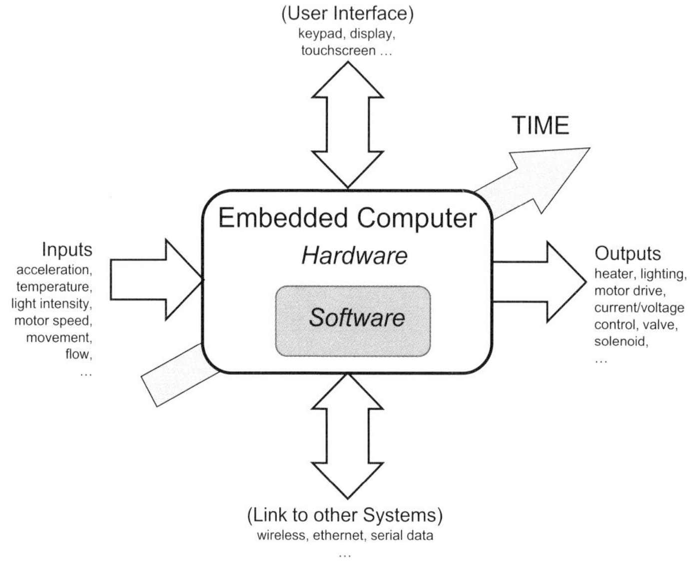

Figure 1.1
The embedded system.

A refrigerator controller provides us with an initial embedded system example. This has a temperature sensor in the main fridge compartment and one in the freezer. The user selects the required temperature or depends on default settings. Meanwhile, sensors are measuring the temperature in each fridge compartment. The embedded computer, a microcontroller, receives incoming signals from the sensors and control setting, compares these, and determines whether the compressor should be switched on or not. It will also display to the user the two temperatures measured, and may emit an alarm if a door is left open too long, or the temperature rises above a certain limit. In its traditional form the fridge does not have any external connection. However, a modern fridge may well be able to connect to the internet, so that the homeowner can check on its status while away—but more on this later.

1.1.2 Fast Forward to the Internet of Things

General-purpose computers, and now embedded systems, have become increasingly connected to the internet. The Internet of Things (IoT) is the name given to the resultant worldwide network of devices and artefacts—some everyday, some highly specialized—that are connected to the internet. These allow remote access to information that can be used to control and enhance diverse activities. The IoT generates huge quantities of real-time data that can potentially be accessed from anywhere in the world. IoT data now includes billions of sensors and databases that are made available for monitoring through the internet, from travel and transport data to environmental, medical, and industrial concerns large and small.

An integral part of the IoT is the concept of *cloud computing*, which uses servers and data memory storage locations that are only accessed through the internet. Companies such as Google, Microsoft, Apple, and Dropbox, among many others, provide their own cloud-based services, and it is anticipated that most, if not all, of our personal and professional data (and programs) will one day be stored in the cloud rather than on local computers and hard drives.

The IoT concept has expanded to include everyday objects attached to the internet. Examples include a washing machine that can alert the repair company to an impending fault, a vending machine that can tell the head office it is empty, a manufacturer who can download a new version of firmware to an installed burglar alarm, or a homeowner who can switch on the oven from the office or check that the garage door is closed.

1.1.3 The Electric Car—An Abundance of Embedded Systems

Technical developments in cars have been one of the driving forces behind the growth of embedded systems, with microcontrollers being applied in engine management, braking, diagnostics, and many other things. Perhaps nowhere has this outpouring of embedded control been more evident than in the modern electric vehicle (EV), with battery replacing gas tank, and DC (direct current) electric motors replacing the internal combustion engine (see Fig. 1.2). Such a vehicle represents an excellent opportunity to apply and exploit embedded systems, using those little microcontrollers to their best advantage.

The possible categories of electronic control in the EV are shown in Fig. 1.3. Most of these will overlap, and other categorizations could be made. Each category contains numerous embedded systems, some controlling one or few variables, others managing a sophisticated array.

Imagine Fig. 1.1 implemented dozens of times within the EV, and think of the sensors and actuators in operation. The battery management will need to sense battery voltage and current, sometimes current flowing from battery to motors, at other times from motors back to battery due to kinetic energy recovery, then from charge point to battery when charging is taking place. Safety features will detect closure of seat belts in occupied seats, tire pressure, nearness of vehicle to potential obstacles, and lane following on a motorway or freeway. Environment control will sense internal and external temperatures and the temperature demanded by the driver and manage seat heating. Within most of these are actuators moving, heating, or lighting things they control. There is also an enormous flow of data, within and between the subsystems.

Figure 1.2
The electric car, rich in embedded systems.

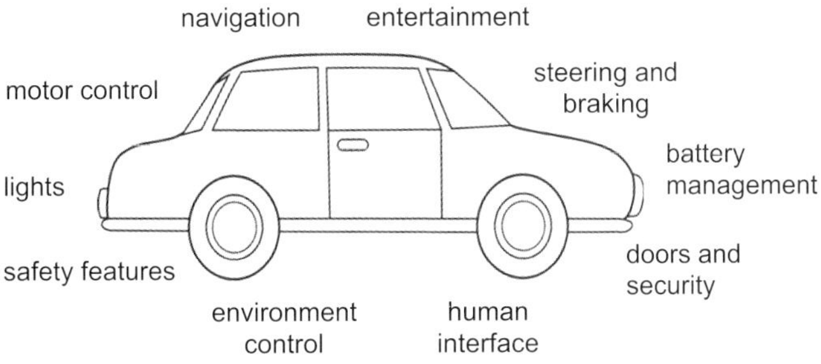

Figure 1.3
Electronic subsystems in the electric car.

Needless to say, data generated in the vehicle does not all stay within it. The owner is encouraged—indeed it's almost a necessity—to download a car-specific phone app, as seen in Fig. 1.4. This tells him or her many things, including the important features of whether the car is locked, the state of battery charge and available driving range, the

Figure 1.4
An electric car app.

vehicle state of health, and its location. From it the user can—if a charging cable is connected—start or stop charging. Also, on a cold day, the heating can be started and the windshield de-iced before a journey starts.

The actions just described place us firmly in the realm of the IoT. Data from the car has been transmitted wirelessly to the web, processed, and returned to the app, to be displayed in user-friendly form. Control data generated by the user with the app is then returned to the car for actuation.

This book takes us on a journey which—on completion—should allow you, the reader, to feel able to understand, apply, and even design the circuits and software found in embedded systems large and small. As each new chapter opens, we start at the beginning of that topic, initially with simple concepts.

This first chapter introduces or reviews many concepts relating to computers and embedded systems. It does this in overview form, to give a platform for further learning. We return to these concepts in later chapters, building on them and adding detail.

1.2 Microprocessors and Microcontrollers

Let's look more closely at the microcontroller, which sits at the heart of any embedded system. As the microcontroller is in essence a type of computer, it will be useful to get a grasp of basic computer details.

1.2.1 Some Computer Essentials

Fig. 1.5 shows the essential elements of any computer system. As its very purpose for existence, a computer can perform arithmetic or logical calculations. It does this in a

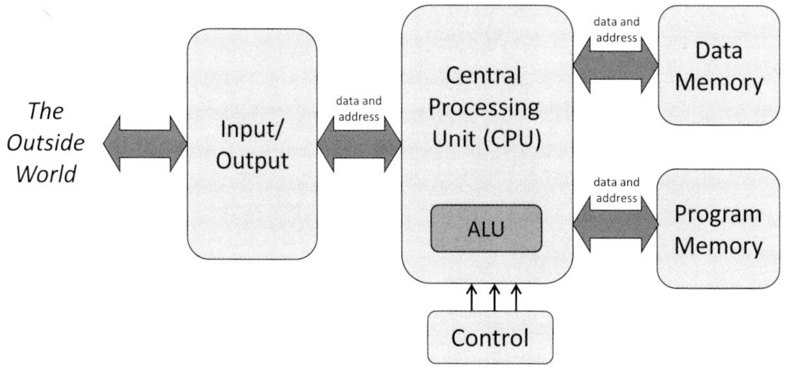

Figure 1.5
Essentials of a computer.

digital electronic circuit called the ALU, or *Arithmetic Logic Unit*. The ALU is placed within a larger circuit, called the *Central Processing Unit* (CPU), which provides some of the supporting features that it needs. The ALU can undertake a number of simple arithmetic and logic calculations. Which one it does depends on a digital code which is fed to it, called an instruction. If we can keep the ALU busy, by feeding it a sensible sequence of instructions, and also pass it the data it needs to work on, then we have the makings of a very useful computing machine.

The ability to keep feeding the ALU with instructions and data is provided by the control circuit which sits around it. It is worth noting that any one of these instructions performs a very simple function. However, because the typical computer runs so incredibly fast, the overall effect is one of very great computational power. The series of instructions is called a program, normally held in an area of memory called *program memory*. This memory needs to be permanent. If it is, then the program is retained indefinitely, whether power is applied or not, and it is ready to run as soon as power is applied. Memory like this, which keeps its contents when power is removed, is called *non-volatile memory*. The old-fashioned name for this is ROM—*Read Only Memory*. This terminology is still sometimes used, even though with new memory technology it is no longer accurate.

The control circuit of the CPU needs to keep accessing the program memory, to bring the next instruction and feed it to the ALU. The data that the ALU works on may be drawn from the data memory, with the result placed there after the calculation is complete. Usually this is temporary data. Therefore, this memory type need not be permanent, although there is no harm if it is. Memory which loses its contents when power is removed is called volatile memory. The old-fashioned name for this type of memory is RAM—*Random Access Memory*. This terminology is still used, though it conveys little useful information.

To be of any use, the computer must be able to communicate with the outside world, and it does this through its input/output. On a personal computer this implies human interaction, through things like a keyboard and mouse, a display screen, and a printer. In an embedded system, at least a simple one, the communication is likely to be primarily with the physical world around it, through sensors and actuators. Data coming in from the outside world might be quickly transferred to the ALU for processing, or it might be stored in data memory. Data being sent out to the outside world is likely to be the result of a recent calculation in the ALU.

Finally, there must be data paths between each of these main blocks, as shown by the block arrows in the diagram. These are collections of wires, which carry digital information in either direction. For example, one set of wires carries the data itself from program memory to the CPU; this is called the *data bus*. The other set of wires carries address information and is called the *address bus*. The address is a digital number which indicates which place in memory the data should be stored in or retrieved from. The wires in each of the data and address buses could literally be bundles of wires, but more likely are tracks on a printed circuit board, or interconnections within an IC.

One of the defining features of any computer is the size of its ALU. Older simpler processors are of 8-bit (binary digit) size, and some of that size still have useful roles to play. With their 8 bits, they can represent a number between 0 and 255. (Check Appendix A if you are unfamiliar with binary numbers.) More recent machines are 32- or 64-bit. This gives them far greater processing power, but of course adds to their complexity. Given an ALU size, it generally follows that many other features take the same size: memory locations, data bus, and so on.

As already suggested, the CPU has an *instruction set,* which is a set of binary codes that it can recognize and respond to. For example, certain instructions will require it to add or subtract two numbers, or store a number in memory. Many instructions must also be accompanied by data, or addresses of data, on which the instruction can operate. Fundamentally, the program that the computer holds in its program memory and to which it responds is a list of instructions taken from the instruction set, with any accompanying data or addresses that are needed.

1.2.2 The Microcontroller

A microcontroller takes the essential features of a computer as just described, and adds to these the features that are needed for it to perform its control functions. It is useful to think of it as being made up of three parts—core, memory, and peripherals—all contained in one IC, as shown in the block diagram of Fig. 1.6(a). The core is the CPU and its control circuitry. Alongside this go the program and data memories, already mentioned. The CPU gets a few more inputs we have not mentioned so far. There is a *reset*, a signal which can be used to force the CPU to start its program from the beginning again. Then there are *interrupts*, other inputs which do just what their name suggests: they can interrupt the action of the CPU while executing its main program, and force it to execute some other program section.

Finally, and importantly, there are the peripherals. These are the circuit blocks which distinguish a microcontroller from a microprocessor, for they are the elements which allow the wide-ranging interaction with the outside world that the microcontroller needs. Peripherals can include digital or analog input or output, ports to exchange serial data, counters, timers, and many other useful subsystems.

Using today's wonderful semiconductor technology, the whole complex circuit represented by Fig. 1.6(a) can be integrated onto one IC. An example is shown in Fig. 1.6(b), actually the LPC1768 device that is used on the Mbed development board of the same name.

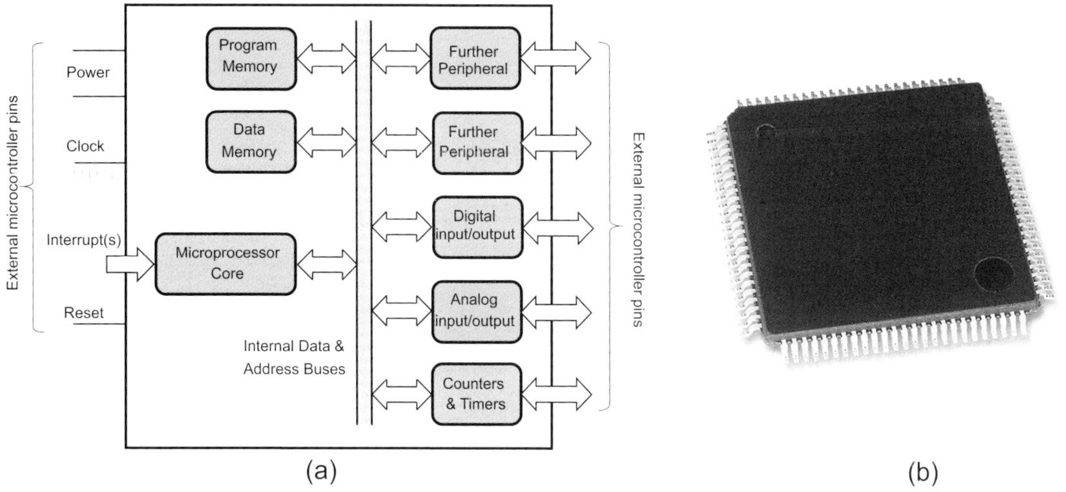

Figure 1.6
Example microcontroller. (a) Generic features: core + memory + peripherals; (b) An actual microcontroller, the LPC1768.

(Peeking forward, it can be seen in Fig. 2.1.) It looks like an unexciting black square, around 13 mm along each side, with a lot of little pins all around. Inside, however, is the whole complexity and cleverness of the microcontroller circuit. That huge number of pins, 100 in all, provides all the electrical connections that may be needed.

It almost goes without saying, but must not be forgotten, that the microcontroller needs power, in the form of a stable DC supply. A further requirement, as with any PC (personal computer), is to have a *clock* signal. This is the signal that endlessly steps the microcontroller circuit through the sequence of actions that its program dictates. The clock is often derived from a quartz oscillator, such as you may have in your wristwatch. This gives a stable and reliable clock frequency. It is not surprising to find that the clock often has a very important secondary function, of providing essential timing information for the activities of the microcontroller, for example, in timing the events it initiates, or in controlling the timing of a serial data message.

1.3 Code Development in Embedded Systems

1.3.1 Machine Code, Assembler, and High-level anguages

As programmers, it is our ultimate goal to produce a program which makes the microcontroller do what we want it to do; that program must be a listing of instructions, in binary, drawn from the microcontroller instruction set. We sometimes call the program in this raw binary form *machine code*. It is extremely tedious, to the point of being nearly impossible, for us as humans to work with the binary numbers which form the instruction set.

The problem of programming is represented in Fig. 1.7. We as humans express our ideas in complex and often loosely defined linguistic forms. A computer reads and "understands" binary, that is, a huge collection of 1s and 0s represented in electrical form. It responds in a precise way to precise instructions, drawn from its instruction set. It is ruthlessly logical, and does exactly what the instructions require.

Given this linguistic divide, how can a programmer write programs for a computer? Different ways of bridging the gap present themselves.

The idea of writing in the machine code itself having been discarded, a first step toward sanity in programming is called *Assembly Language*, or simply *Assembler*. In Assembler each instruction from the instruction set gets its own *mnemonic*, a little (usually 3-letter) word which a human can hope to remember and work with. For example, in the ARM Cortex M3 processor which we are about to use, the mnemonic **ADD** is used for the add instruction, **ADC** for add with carry, and **MUL** for multiply. That's not so difficult, but there can be many such mnemonics. Thus, the instruction set is represented by a set of mnemonics. The program is then written in strict format using these mnemonics, plus the items of data and memory addresses which need to go with them. A computer program called a *cross-assembler* converts this program version into the actual binary code that is loaded into program memory. Assembler has its uses; for one, it allows us to work very closely with the CPU capabilities; a well-written Assembler program can therefore be very fast and efficient as it executes, with the least possible demands on program memory. Yet it's easy to make mistakes as we program in Assembler, and hard to track them down. Thus, the programming process is time-consuming and error-prone.

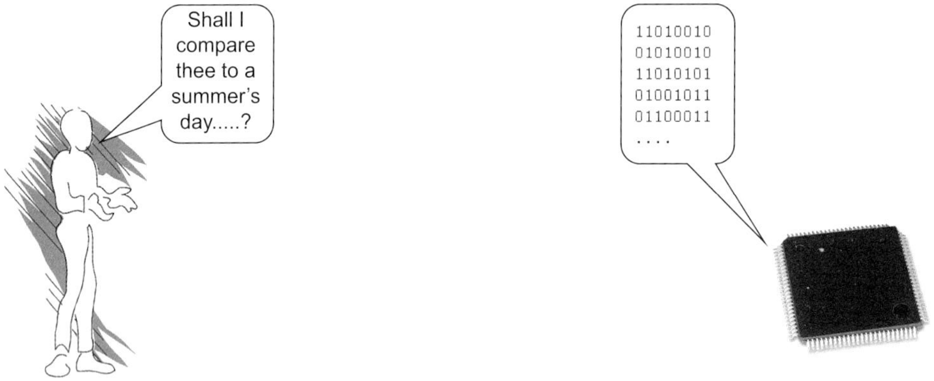

Figure 1.7
The problem of programming.

A further step to programming sanity is to use a *High-level Language* (HLL). In this case we use a language like C, Java, or Python to write the program, following the rules of that language. The HLL uses code words which are recognizably English, and structures are drawn from conventional mathematics. A computer program called a *compiler* then reads the program we have written, and—over several stages—converts (compiles) that into a listing of instructions from the instruction set, along with all necessary data and addresses. This assumes we have made no mistakes in writing the program, though the compiler can give useful error messages if we do. This list of instructions becomes the binary code that is then downloaded to the program memory.

Our needs in the embedded world, however, are not like those of other programmers. We want to be able to control the hardware we are working with and write programs which execute quickly in predictable times. Not all HLLs are equally good at meeting these requirements. People debate endlessly about what is the best language in the embedded environment. However, most agree that for many applications C has clear advantages. This is because it is simple and has features which allow us to get at the hardware when we need to. A step up from C is C++, which adds complexity but is widely used for more advanced embedded applications. As C is a direct subset of C++, this transition is a logical one, which can be managed without the need to start learning a completely new language.

It is important to say that a program can be written in an HLL, but can still have sections in Assembler. This proves to be a very useful compromise—for example, most of the program can be written in C/C++, but sections which are very time-conscious can be written in Assembler, thereby getting the best of both worlds. Such Assembler inserts may also be essential for very specific hardware-related actions, which the HLL simply does not have the capability to undertake.

1.3.2 Flow Diagrams and Simple Sequential Programming

Whatever programming strategy we select, we will be writing programs which read in data, make computations and decisions, and generate outputs. A simple example, which returns to an example in the opening section of this chapter, is shown in Fig. 1.8. Here a flow diagram is used to represent the program structure of the refrigerator controller. Flow diagrams help to picture the structure of a program or program section, and—if well constructed—can be turned fairly easily into code. They can take sophisticated forms, but here we use only two symbols: a box for any sort of action, and a diamond for decision. The diamond contains within it a yes/no choice, with two possible exit points and one entry point. As with many programs in the embedded world, which run on and on, this program is structured as a continuous loop.

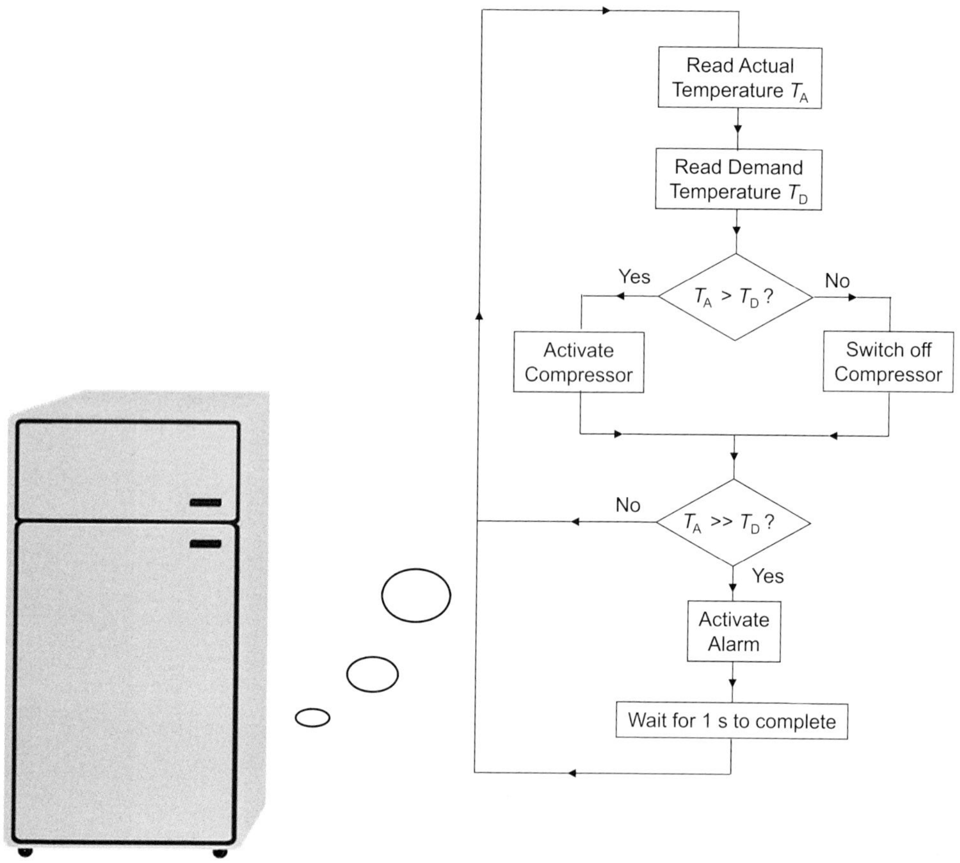

Figure 1.8
Flow diagram representation of a simple control program.

1.3.3 The Development Cycle

This book is about developing embedded systems quickly and reliably, so it is worth getting a first picture of what that development process is all about.

In the early days of embedded systems, the microcontrollers were very simple, without on-chip memory and with few peripherals. It took a lot of effort just to design and build the hardware. Moreover, memory was very limited so programs had to be short. The main development effort was often spent on hardware design, while programming consisted of writing rather simple programs in Assembler. However, over the years the microcontrollers became more and more sophisticated, and memory much more plentiful. Many suppliers started selling predesigned circuit boards containing the microcontroller and all associated circuitry. Where these were used, much less development effort was needed for the hardware. Now attention could be turned to writing complex and sophisticated programs, using all the memory which had become available. This tends to be the situation we find ourselves in today.

Despite these changes, the program development cycle is still based around the simple loop shown in Fig. 1.9. In this book, we will write source code using C or C++. The diagram of Fig. 1.9 shows the main stages we are likely to follow. Most of these stages are highly integrated, and in general are made available within an *Integrated Development Environment,* or IDE, a software package running on a computer—let's call it the *host computer.* Following consideration of overall program structure, for example with one or more flow diagrams, the source code is written in a text editor in the IDE. While the programmer will write original code, it is almost certain that he/she will make use of libraries of code already available, possibly in the form of APIs (Application Programming Interfaces). The source code will need to be converted into the binary machine code which is downloaded to the microcontroller itself, and placed within the circuit or system that it

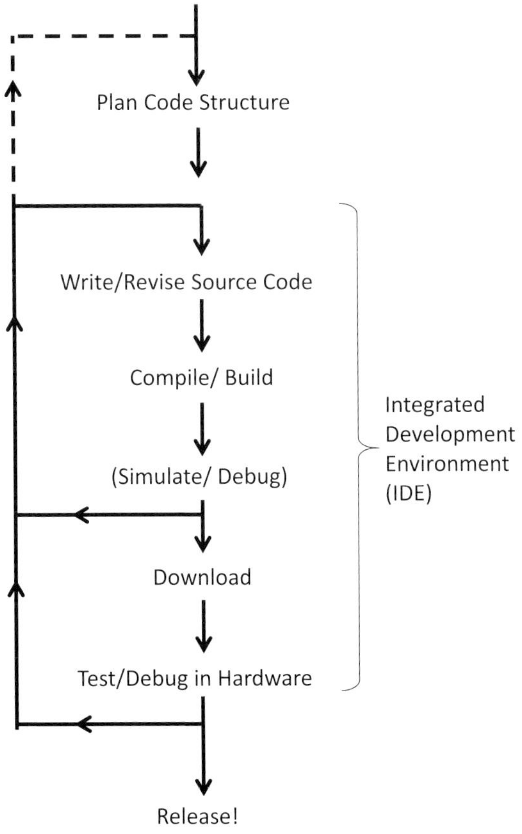

Figure 1.9
The embedded program development cycle.

will control (let's call this the *target system*). That conversion is done by the compiler, running on the host computer as part of the IDE.

Program download to the target system requires temporary connection to the host computer. There is considerable cleverness in how the program data is actually written into the program memory, but that need not concern us here. Before the program is downloaded, it is often possible to simulate it on the host computer; this allows a program to be developed to a good level before going to the trouble of downloading. We do not place great emphasis on simulation in this book, so we have put that stage in brackets in the diagram.

An important aspect of the development stage can be the debugger, generally integrated into the IDE. The program is now running in the hardware, but the debugger monitors and controls that program execution. Thus, for example, it can allow the program to run up to a certain point, and then display the values of internal registers or memory contents on the host computer screen. The final and true test of the program is to see it running correctly, independent of the host computer, in the target system. When program errors are found at these stages, as they inevitably are, the program can be revised, compiled, and downloaded once more.

1.4 The World of Arm

The development of computers and microprocessors has at different times been driven forward by giant corporations or by tiny start-ups; but it has always been driven forward by very talented individuals or teams. The development of the company Arm, whose designs form a core feature of this book, has seen a combination of all of these. The full history of Arm is a fascinating one, like so many other hi-tech start-ups in the past 40 or so years. A tiny summary is given below; you can read it in a fuller version on one of the several websites devoted to this topic. Then watch that history continue to unfold in the years to come!

1.4.1 A Little History

In 1981, the British Broadcasting Corporation launched a computer education project and asked for companies to bid to supply a computer which could be used for this. The winner was the Acorn computer. This became an extremely popular machine and was very widely used in schools and universities in the UK. The Acorn used a 6502 microprocessor, made by a company called MOS Technology. This was a little 8-bit processor, which we would not take very seriously these days, but was respected in its time. Responding to the growing interest in personal or desktop computers, IBM produced its very first PC in 1981, based on a more powerful Intel 16-bit microprocessor, the 8088. There were many companies producing similar computers at the time, including of course Apple. These early machines were pretty much incompatible with each other, and it was quite unclear whether one would finally dominate. Throughout the 1980s, however, the influence of the IBM PC grew, and its smaller competitors began to fade. Despite Acorn's UK success it did not export well, and its future no longer looked bright.

It was around this time that those clever designers at Acorn made three intellectual leaps. They wanted to launch a new computer. This would inevitably mean moving on from the 6502, but they just could not find a suitable processor for the sort of upgrade they needed. Their first leap was the realization that they had the capability to design the microprocessor itself and did not need to buy it elsewhere. Being a small team and experiencing intense commercial pressure, they designed a small processor, but one with real sophistication. The computer they built with this, the Archimedes, was a very advanced machine, but struggled against the commercial might of IBM. The company found themselves looking at computer sales which just were not sufficient, yet they held the design of an extremely clever microprocessor. They realized—their second leap—that their future might not lie in selling the completed computer itself. Therefore in 1990, Acorn Computers cofounded another Cambridge-based company, called Advanced RISC Machines Ltd., ARM for short. They also began to realize—their third leap—that you do not need to manufacture silicon to be a successful designer, but what mattered was the ideas inside the design. These can be sold as IP, intellectual property.

The ARM concept continued to prosper with a sequence of smart microprocessor designs being sold as IP to an increasing number of major manufacturers around the world. The company later dropped the original acronym meaning, but retained the name Arm. Now called Arm Holdings, they have enjoyed huge success. Those who buy the Arm designs incorporate them into their own products. For example, the Mbed LPC1768 target that we use in this book—among others—uses the Arm Cortex-M3 core. This is to be found in the LPC1768 microcontroller which sits in the Mbed. This microcontroller is *not* made by Arm, but by NXP Semiconductors. Arm sold NXP a license to include the Cortex core in their LPC1768 microcontroller, which Arm then bought back to put in their Mbed. Got that?! There is a similar pattern with the Nucleo F401RE board. Arm licensed the Cortex-M4 core to ST, who designed it into their 32F401RE microcontroller.

1.4.2 Some Technical Detail—What Does This Word RISC Mean?

Because Arm chose originally to place the RISC concept in its very name, and because it remains a central feature of Arm designs, it is worth checking out what RISC stands for. We have seen that any microcontroller executes a program which is drawn from its instruction set, and which is defined by the CPU hardware itself. In the earlier days of microprocessor development, designers were trying to make the instruction set as advanced and sophisticated as possible, with an instruction ready to meet almost every conceivable need. The price they were paying was that this was also making the computer hardware more complex, more expensive, and slower. Such a microprocessor is called a *Complex Instruction Set Computer* (CISC). Both the 6502 and 8088 mentioned above belong to the era when the CISC approach was dominant, and are CISC machines. One characteristic of the CISC approach is that instructions have different levels of complexity. Simple ones can be expressed in a short instruction code, say one byte of data, and execute quickly. Complex ones may need several bytes of code to define them, and take a long time to execute.

As compilers became better, and high-level computer languages developed, it became less useful to focus on the capabilities of the raw instruction set itself. After all, if you are programming in a high-level language, the compiler should solve most of your programming problems with little difficulty.

Another approach to CPU design is to insist on keeping things simple and to have a limited instruction set. This leads to the RISC approach—the *Reduced Instruction Set Computer*. The RISC approach looks like a "back-to-basics" move. A simple RISC CPU can execute code fast, but it may need to execute more instructions to complete a given task, compared with its CISC cousin. With memory becoming ever cheaper and of higher density, and more efficient compilers for program code generation, this disadvantage is diminishing. One characteristic of the RISC approach is that each instruction is contained within a single binary *word* (where "word" implies a binary number, of a size fixed for a particular computer). That word must hold all information necessary, including the instruction code itself, as well as any address or data information also needed.

A further characteristic, an outcome of the simplicity of the RISC approach, is that every instruction normally takes the same amount of time to execute. This allows other useful computer design features to be implemented. A good example is *pipelining*—as one instruction is being executed, the next is already being fetched from memory. It is easy to do this with a RISC architecture, where all (or most) instructions take the same amount of time to complete. This is illustrated in Fig. 1.10, for a 3-stage pipeline: the first stage fetches the instruction, the second decodes it, and the third executes. The diagram shows four cycles of operation. In every cycle one instruction from the program is being executed, while the next is being decoded, and the next after this is being fetched from memory. Pipelining can dramatically increase the speed of operation of a CPU because things can be made to happen simultaneously, rather than in turn.

An interesting subplot to the RISC concept is the fact that, due to its simplicity, RISC designs tend to lead to low-power consumption. This is hugely important for anything which is battery-powered, and helps to explain why Arm products find their way into so many mobile phones and tablets.

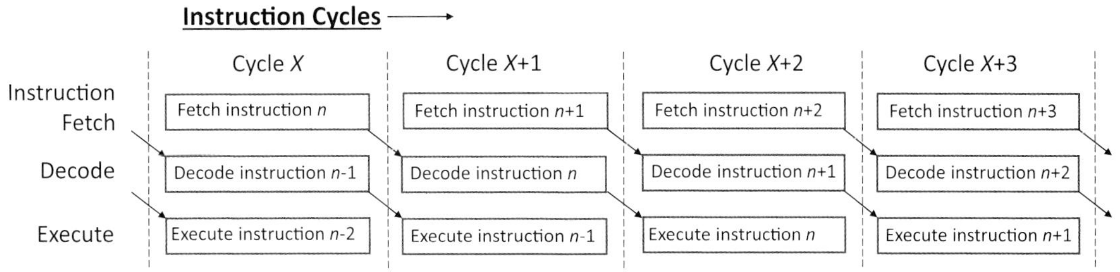

Figure 1.10
Example pipeline operation.

1.4.3 The Cortex Core

In the 1990s, Arm produced a series of evolutionary designs. These sustained and improved the fundamental Arm processor characteristics of efficient and low-power design, generally based on a RISC architecture. For these, a 32-bit instruction set was developed. A 16-bit subset of this, which allowed highly efficient coding, was called the Thumb instruction set. A further instruction set, called Thumb-2 (with the original Thumb set becoming Thumb-1), adds some 32-bit instructions to Thumb-1. Thumb-2 becomes close in capability to the original 32-bit Arm instruction set, with some greater coding efficiency possible.

The Cortex microprocessor family was introduced by Arm in 2004. Cortex-A and Cortex-R were introduced for higher-level applications, while Cortex-M was intended for smaller, microcontroller-oriented uses. Cortex-M is more than just a clever CPU design. Aside from the CPU, it contains an interrupt handling mechanism, memory protection, debug mechanisms, a bus interface, and power control. Significantly different versions of the Cortex-M have been produced, with development continuing. The main versions are summarized in Table 1.1. It is useful to recognize differences both in architecture and in instruction set size.

While a summary table like this can do very little justice to the huge complexity and sophistication of these devices, it is interesting to note some of the design trade-offs made. All except the M0+ apply a 3-stage pipeline; the M0+, the lowest power, has just a 2-stage pipeline. A disadvantage of pipelining occurs when a program branch occurs; an instruction may have been fetched which is not in fact needed. Therefore, the more powerful processors therefore incorporate branch prediction, a mechanism which aims to minimize this disadvantage. The inclusion of hardware multiply and divide, although demanding in hardware design, removes the need for time- and memory-consuming software routines to undertake these functions. The same is true of the floating point units (FPUs) in the M4 and M7, which greatly speed up the execution of more complex mathematical calculations.

Table 1.1: The ARM Cortex-M family.

Core	Main Characteristics	Instruction Set
Cortex-M0	With its minimum size and power consumption, the M0 is of special interest for low-cost, low-power, or small-size embedded systems. 3-stage pipeline. Up to 32 interrupts.	56 Thumb-1 and Thumb-2 instructions. Restricted multiply or divide instructions.
Cortex-M0+	Similar to Cortex-M0, but further reduced power consumption, partly due to 2-stage pipeline. Up to 32 interrupts.	56 Thumb-1 and Thumb-2 instructions. Restricted multiply or divide instructions.
Cortex-M1	Similar to Cortex-M0, but designed as a "soft core" which can be dropped into an FPGA (Field Programmable Gate Array) design. 3-stage pipeline. Up to 32 interrupts.	56 Thumb-1 and Thumb-2 instructions. Restricted multiply or divide instructions.
Cortex-M3	The first Cortex-M device to be launched, so often viewed as a point of reference for the others. Targeted at common embedded applications, including automotive and industrial. 3-stage pipeline with branch speculation. Up to 240 interrupts.	74 instructions including all Thumb-1 and Thumb-2. Multiply and divide instructions included.
Cortex-M4	All the features of Cortex-M3, with additional digital signal processing (DSP) instructions and optional hardware floating point unit (FPU). 3-stage pipeline with branch speculation. Up to 240 interrupts.	137 instructions, plus 32 for the FPU. Multiply, divide, and DSP instructions included.
Cortex-M7	A high-performance core with significant advances over the Cortex-M4. Includes a 6-stage pipeline with branch speculation and optional FPU. Up to 240 interrupts.	Multiply, divide, FPU, and DSP instructions available.
Cortex-M23 and upward	Ongoing developments and advances.	

Of particular interest to us in Table 1.1 are the Cortex-M3 and -M4, as these feature in the development boards we will be using.

A very simplified block diagram of the Cortex-M3 is shown in Fig. 1.11. This diagram allows us to see again some computer features we have already identified, and add some new ideas. Somewhere in the middle you can see the ALU, already described as the calculating heart of the computer. The instruction codes are fed into this through the *Instruction Fetch* mechanism, taking instructions in turn from the program memory. Pipelining—as seen in Fig. 1.10—is applied to instruction fetch, so that as one instruction is being executed, the next is being decoded, and the one after is being fetched from

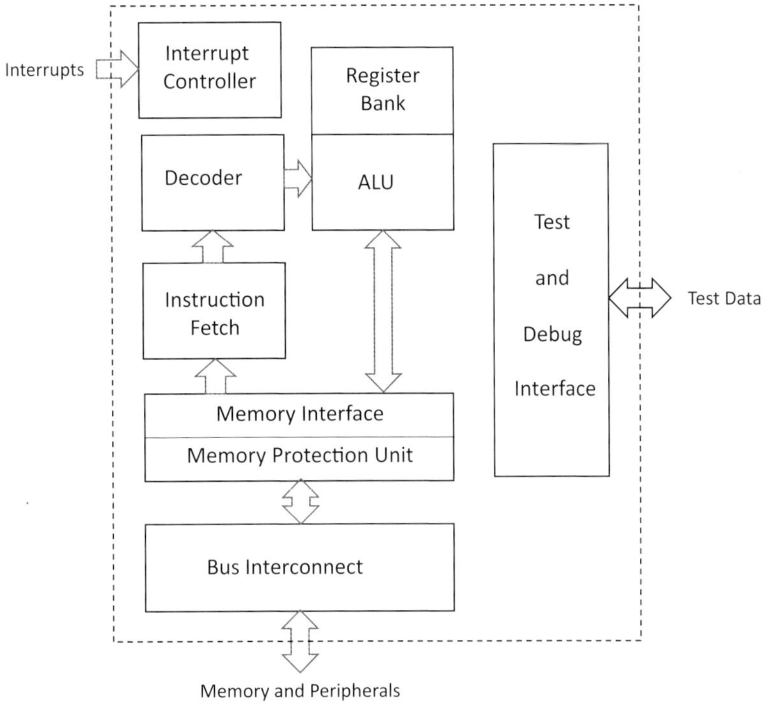

Figure 1.11
The Cortex-M3 core, simplified diagram.

memory. As it executes each instruction, the ALU simultaneously receives data from memory and/or transfers it back to memory. This happens through the interface blocks seen. The memory itself is not part of the Cortex core, although Arm does define the basic memory layout (known as the *memory map*) which must be implemented by users of the Cortex core. The ALU also has a block of registers associated with it. These act as a tiny chunk of local memory, which can be accessed quickly and used to hold temporary data as a calculation is undertaken.

The Cortex core also includes an *interrupt interface*. Interrupts are an important feature of any computer structure. They are external inputs which can be used to force the CPU to divert from the program section it is currently executing, and jump to some other section of code. The interrupt controller manages the various interrupt inputs. It should not be too difficult to imagine this microprocessor core being dropped into the microcontroller diagram of Fig. 1.6(a).

A detailed, yet readable, guide to the Cortex-M family is given in Reference 1.1. A guide to just the Cortex-M3 is given in Reference 1.2. It is not suggested that you read either of these books from cover to cover. However, if you do have access to them, you may like to refer to them from time to time at various stages of this book, for example, in Chapter 6 on interrupts or the hardware insights of Chapters 14 or 15.

Chapter Review

- An embedded system contains one or more tiny computers, which control it, and give it a sense of intelligence.
- The embedded computer usually takes the form of a microcontroller, which combines microprocessor core, memory, and peripherals, all on the same IC.
- Embedded system design combines hardware (electronic, electrical, and electromechanical) and software (program) design.
- The embedded microcontroller has an instruction set. The code developed by the programmer must, through one means or another, be converted to instructions from this instruction set.
- Most embedded system programming is done in a high-level language, with C or C++ often being preferred.
- A compiler is used to convert the program into binary code, drawn from the microcontroller instruction set.
- Arm has developed a range of effective microprocessor and microcontroller designs, widely applied in embedded systems.
- The Arm Cortex-M series microcontroller cores are intended specifically for microcontroller applications, with different versions targeted at different use scenarios.

Quiz

1. Explain the following acronyms: IC, ALU, CPU, FPU.
2. Describe an embedded system in less than 50 words.
3. What are the differences between a microprocessor and a microcontroller?
4. What range of numbers can be represented by a 16-bit ALU?
5. What is a "bus" in the context of embedded systems? Describe two types of buses that might be found in an embedded system.
6. Describe the term "instruction set" and explain how use of the instruction set differs for high- and low-level programming, from the programmer's point of view.
7. What are the main steps in the embedded program development cycle?
8. Explain the terms RISC and CISC. Give advantages and disadvantages for each.
9. What is pipelining? Explain why a branch instruction upsets the pipelining process.
10. What did the acronym and company name ARM stand for?
11. Name three instruction sets associated with Arm microprocessor cores. Briefly outline their characteristics.

12. Which Cortex-M core:
    a. has the lowest power consumption?
    b. is suitable for inclusion on an FPGA?
    c. is suitable for a DSP application?

References

1.1. Martin Trevor. *The Designers' Guide to the Cortex-M Processor Family*. 3rd ed. Oxford: Newnes; 2022.
1.2. Joseph Yiu. *The Definitive Guide to the ARM Cortex-M3*. 2nd ed. Oxford: Newnes; 2010.

CHAPTER 2

Introducing the World of Mbed

2.1 Introducing the Mbed Environment

2.1.1 Mbed, Mbed Enabled, and the Mbed "Ecosystem"

Chapter 1 reviewed some of the core features of computers, microprocessors, and microcontrollers. We now apply these findings to the Mbed environment. In the first edition of this book it was easy to introduce "the Mbed." It was a single device, based on the NXP LPC1768 microcontroller, with an on-line compiler and a software library. That early Mbed prospered, and a low-power version was added to it. The concept of *Mbed-enabled* then emerged. Partner companies were invited to produce their own Mbed-compatible devices according to criteria laid down by ARM. Many Mbed-enabled devices have now appeared. These can readily interact with each other and can all be developed using the same software development tools. An *ecosystem* emerged of compatible microcontrollers, platforms, and other system elements, backed up by the same development environment, an operating system, shared software libraries, and a community of developers interacting online.

The Mbed environment is used in the book as a means of developing the key concepts of embedded systems. By working through this book you are both learning those concepts and applying them directly to a certain microcontroller implementation. Whether you continue working with Mbed devices or move to an alternative, be sure that the knowledge gained can be adapted readily to other microcontroller types.

The Mbed ecosystem is now at the forefront of ARM's development of Internet of Things (IoT) systems. ARM themselves say, "Mbed gives you a free open source IoT operating system with connectivity, security, storage, device management, and machine learning. Build your next product with free development tools, thousands of code examples, and support for hundreds of microcontroller development boards" (Reference 2.1). This IoT orientation has major implications for the Mbed concept and its direction of travel, as design tools and processes are adapted to embrace the needs of IoT applications.

Links to extensive further information for the Mbed environment and its support tools can be found in Reference 2.1. While this book is intended to give you all the information that you need to start working in the Mbed environment, it will still he helpful to keep a close eye on this site.

Fast and Effective Embedded Systems Design. https://doi.org/10.1016/B978-0-323-95197-5.00002-4

2.1.2 Choosing Our Target Systems

In order to study microcontrollers and embedded systems in any detail, it is almost essential to adopt a microcontroller development board as a "target system" for study. In this third book edition, the choice of target system is far from simple. The original Mbed LPC1768 continues in use and still forms the basis for university courses worldwide. However, numerous other microcontrollers and Mbed-enabled development boards are now also used. Some of these aim for low cost and simplicity; others provide complete and sophisticated systems ready to be linked to the IoT.

For this book we choose three example development boards, summarized in Table 2.1. First of all, we keep the Mbed LPC1768 and use it to introduce many basic concepts. We parallel it with use of the popular STM32F401RE Nucleo-64 development board; this costs less than the LPC1768, though it does not have all of its capabilities. You can work through the first 10 chapters of this book using either of these devices or another Mbed-enabled target board. Finally, we choose the STM32 IoT Discovery Kit, particularly using it for more advanced IoT applications. It will be applied in some of the later chapters. We introduce each of these in turn below. Greatest detail is given to the LPC1768, as some of its concepts carry over to the other devices. This use of more than one board allows you to develop flexibility in moving between different hardware platforms. You will see that the same principles apply in all, and Mbed-related concepts can be transferred from one Mbed-enabled device to another.

Table 2.1: A summary comparison of three selected development boards.

	Mbed LPC1768	Nucleo F401RE	STM32 Discovery Kit IoT node
Microcontroller and manufacturer	LPC1768 (NXP)	STM32F401RE (ST)	STM32L475 (ST)
ARM core	Cortex-M3	Cortex-M4	Cortex-M4
Number of user LEDs	4	1	2
Number of user pushbuttons	None	1	1
External connection pins/ sockets	40	32 (Arduino headers) 76 (Morpho connectors)	32
Ethernet	Yes	No	No
Sensors	None	None	accelerometer, barometer, humidity, microphone, temperature, time-of-flight
Wireless	None	None	Bluetooth low energy, NFC (near field communication), Sub GHz RF, Wi-fi

2.2 Where Mbed Began—The Mbed LPC1768

2.2.1 Introducing the Mbed LPC1768

In very broad terms, the Mbed LPC1768 development board takes the LPC1768 microcontroller, such as we saw in simple form in Fig. 1.6, and surrounds it with some very useful support circuitry. It takes the form of a 2-inch by 1-inch (OK, 53 mm by 26 mm) printed circuit board (PCB), with 40 pins arranged in two rows of 20, with 0.1-inch spacing between the pins. This spacing is standard in many electronic components.

Fig. 2.1 shows different views of this Mbed. Looking at the main features, labeled in Fig. 2.1(b), we see that it is based on the LPC1768 microcontroller, hence the name. This is made by a company called NXP Semiconductors and contains an ARM Cortex-M3 core. Program download to the Mbed is achieved through a USB (universal serial bus) connector; this can also power the Mbed. Usefully, there are five LEDs on the board, one

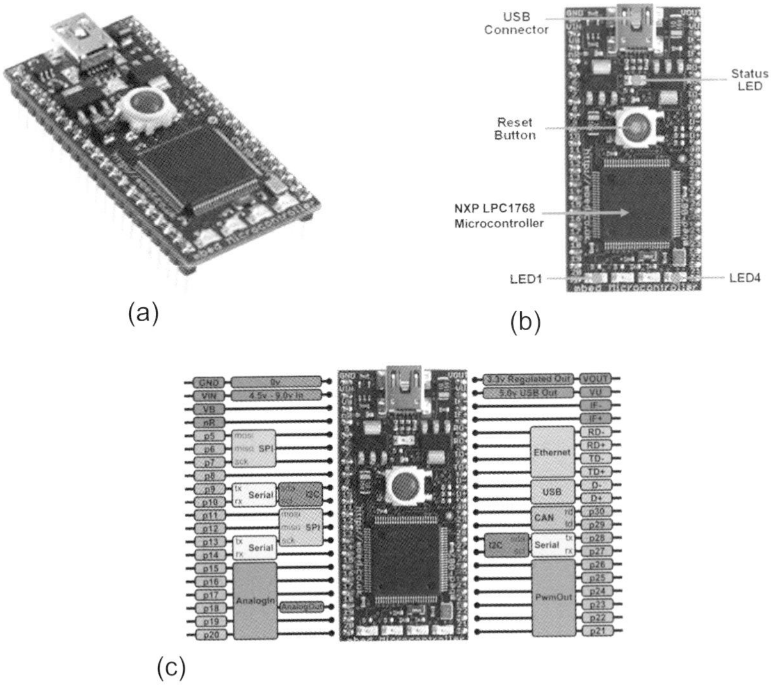

Key

CAN: Controller Area Network

PWM: Pulse Width Modulation

SPI: Serial Peripheral Interface

I^2C: Inter Integrated Circuit

USB: Universal Serial Bus

Note: Signal names in small text, for example mosi, miso or sck, are introduced in the chapter where they are considered

Figure 2.1

The ARM Mbed LPC1768. *Images courtesy of ARM.*

for status and four which are connected to four microcontroller digital outputs. These allow a minimum system to be tested with no external component connections needed. A reset switch is included to force restart of the current program.

The Mbed LPC1768 pins are clearly identified in Fig. 2.1(c), providing a summary of what each pin does. In many instances the pins are shared between several peripherals, to allow a number of design options. At top left we can see the ground and power supply pins. The actual internal circuit runs from 3.3 V; however, the board accepts any supply voltage within the range from 4.5 V to 9.0 V, while an onboard voltage regulator drops this to the required voltage. A regulated 3.3 V output voltage is available on the top right pin, with a 5 V output on the next pin down. The remainder of the pins connect to the Mbed peripherals. These are almost all the subject of later chapters; we will quickly overview them, though they may have limited meaning to you now. There are no less than five serial interface types: I^2C, SPI, CAN, USB, and Ethernet. Then there is a set of analog inputs essential for reading sensor values, and a set of PWM outputs useful for control of external power devices, for example, DC motors. While not immediately evident from the figure, pins 5 to 30 can also be configured for general digital input/output.

The Mbed LPC1768 is constructed to allow easy prototyping, which is of course its very purpose. While the printed circuit board itself is high density, interconnection is achieved through the very robust and traditional dual-in-line pin layout.

2.2.2 The Mbed LPC1768 Architecture

A block diagram representation of the Mbed LPC1768 architecture is shown in Fig. 2.2. It is possible, and useful, to relate the blocks shown here to the actual device. At the heart is the LPC1768 microcontroller, clearly seen in both Figs. 2.1 and 2.2. The signal pins of the Mbed, as seen in Fig. 2.1(c), connect directly to the microcontroller. Thus when, in the coming chapters, we use an Mbed digital input or output, or the analog input, or any other of the peripherals, we will be connecting directly to the microcontroller within the Mbed, and relying on its features. However, an interesting side to this is that the LPC1768 has 100 pins, but the Mbed has only 40. Therefore, when we get deeper into understanding this microcontroller, we will find that there are some microcontroller features that are simply inaccessible to us as Mbed users. However, this is unlikely to be a limiting factor.

There is a second microcontroller on this Mbed which interfaces with the USB. This is called the interface microcontroller in Fig. 2.2 and is the largest IC on the underside of the Mbed PCB. The cleverness of the Mbed hardware design is the way in which this device manages the USB link and acts as a USB terminal to the host computer. In most common uses it receives program code files through the USB and transfers those programs to a 16 Mbit memory (the 8-pin IC on the underside of the Mbed), which acts as the "USB Disk."

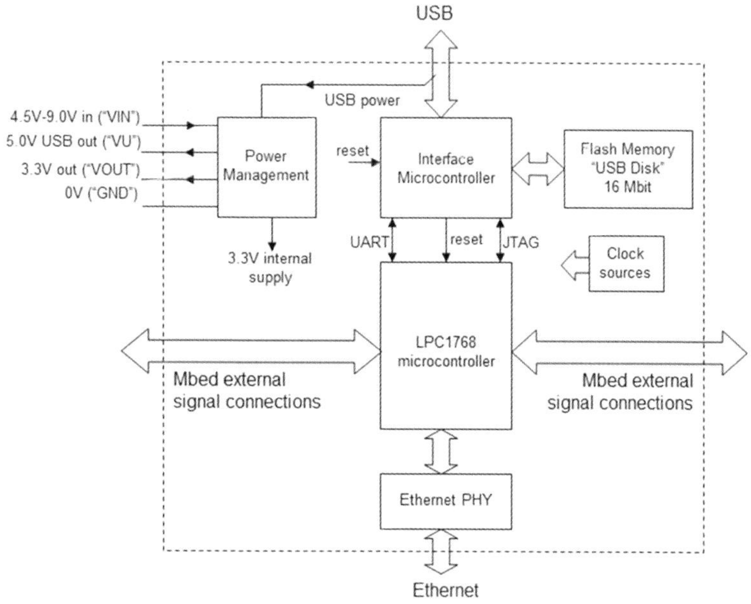

Key
PHY: Physical Interface
JTAG: Joint Test Action Group (Interface)
UART: Universal Asynchronous Receiver/Transmitter

Figure 2.2
Block diagram of the Mbed LPC1768 architecture.

When a program binary code is downloaded to the Mbed, it is placed in the USB disk. When the reset button is pressed, the program with the latest timestamp is transferred to the flash memory of the LPC1768, and program execution commences. Data transfer between the interface microcontroller and the LPC1768 goes as serial data through the UART (universal asynchronous receiver/transmitter—a serial data link, let's not get into the detail now) port of the LPC1768.

The "Power Management" unit is made up of two voltage regulators which lie on either side of the status LED. There is also a current limiting IC which sits at the top left of the Mbed. The Mbed *can* be powered from the USB; this is a common way to use it, particularly for simple applications. When you want or need to be independent from a USB link, or for more power-hungry applications or those which require a higher voltage, the Mbed can also be powered from an external 4.5 V to 9.0 V input, supplied to pin 2 (labeled VIN). Power can also be sourced from Mbed pins 39 and 40 (labeled VU and VOUT, respectively). The VU connection supplies 5 V, taken almost directly from the USB link; hence, it is only available if the USB is connected. The VOUT pin supplies a regulated 3.3 V which is derived either from the USB or from the VIN input.

The Mbed has several clock sources, symbolized by the block to the right of the diagram. The two microcontrollers share their main clock source; this is a crystal oscillator, the silvery rectangle just above LED4 in Fig, 2.1(b). The LPC1768 uses this to maintain an internal clock frequency of 96 MHz. The "Ethernet PHY," the IC toward the USB connector on the underside of the Mbed, has its own clock oscillator sitting between the memory IC and the interface microcontroller.

For those who are inclined—and we will refer to these from time to time—the Mbed LPC1768 circuit diagrams are available on the Mbed website (Reference 2.2).

2.2.3 The LPC1768 Microcontroller

A block diagram of the LPC1768 microcontroller is shown in Fig. 2.3. This looks complicated and we don't want to get into all of the details of what is a hugely sophisticated digital circuit. However, the figure is in a way the agenda for some of this book, as it contains all the capabilities of the Mbed LPC1768 and similar devices. Let's get a feel for the main features. If you want to get complete detail of this microcontroller, then consult one or more of References 1.2, 2.3, and 2.4. We do mention these references from time to time in the book; however, consulting them is *not* necessary for a complete reading of the book.

Remember again that a microcontroller is made up of a microprocessor core *plus* memory *plus* peripherals, as we saw in Fig. 1.6. Let's look for these. Top center in Fig. 2.3, contained within the dashed line, we see the core of this microcontroller, the ARM Cortex-M3. This is a compressed version of Fig. 1.11, the M3 outline which we considered in Chapter 1. To the left of the core are the memories: the program memory, made with flash technology and used for program storage; to the left of that is the static RAM used for holding temporary data. That leaves most of the rest of the diagram to show the peripherals, which give the microcontroller its embedded capability.

The peripherals lie in the center and lower half of the diagram and reflect almost exactly what the Mbed LPC1768 can do. It is interesting to compare the peripherals seen here with the Mbed inputs and outputs seen in Fig. 2.1(c). Finally, all these things need to be connected together, a task done by the address and data buses. Clever though they are, we have almost no interest in this side of the microcontroller design, at least not in this book. We can just note that the peripherals connect through something called the advanced peripheral bus. This in turns connects back through a bus interconnect called the advanced high-performance bus matrix, and from there to the CPU. This interconnection is not completely shown in this diagram, and we have no need to think about it further.

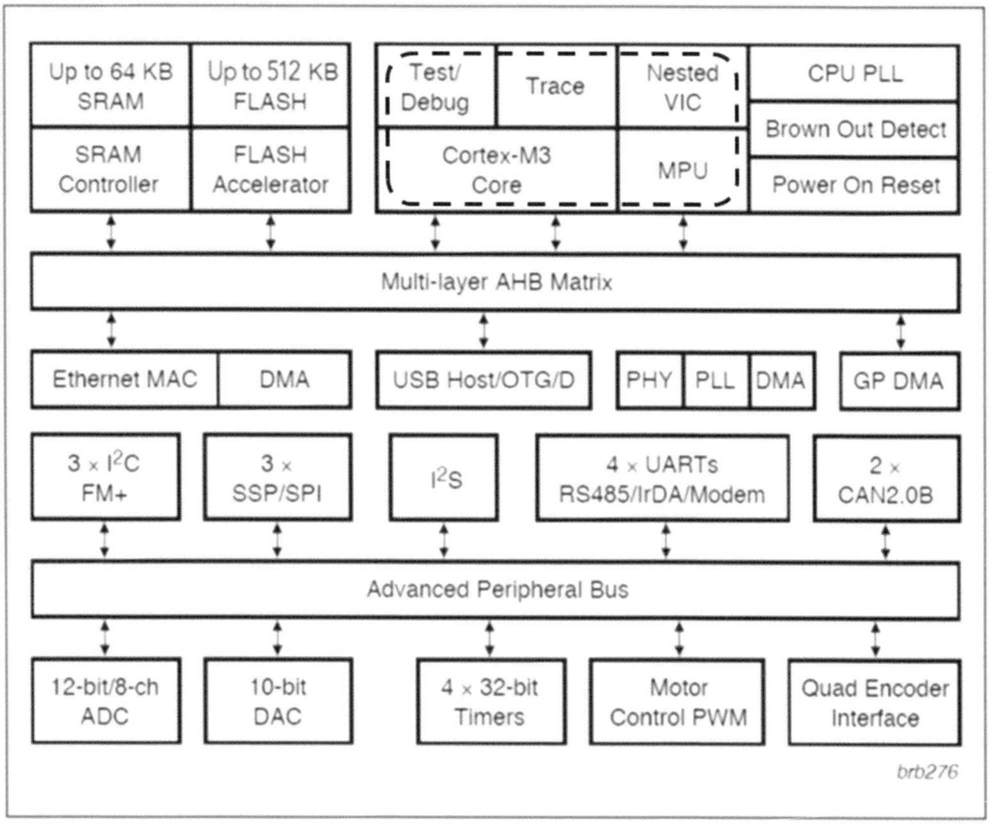

Key (see also key to Fig. 2.1)

ADC: Analog-to-Digital Converter
AHB: Advanced High-Performance Bus
CPU: Central Processing Unit
D: Device (USB)
DAC: Digital-to-Analog Converter
DMA: Direct Memory Access
FM+: Fast-mode Plus (I2C)
GP: General Purpose (DMA)
I^2S: Inter Integrated Circuit Sound
IrDA: Infrared Data Association
MAC: Media Access Control
MPU: Memory Protection Unit
PHY: Physical Layer
PLL: Phase Locked Loop
OTG: On the Go (USB)
SRAM: Static RAM
SSP: Synchronous Serial Port
UART: Universal Asynchronous Receiver/Transmitter
VIC: Vectored Interrupt Controller

Notes: The LPC1768 has 64 KB of static RAM and 512 KB of flash.
The Cortex core is made up of those items enclosed within the dashed line.

Figure 2.3
The LPC1768 block diagram. *Image courtesy of NXP. NXP B.V. Document no. 9397 750 16802, 2009.*

2.3 Nucleo and the F401RE Development Board

We now turn to the Nucleo F401RE development board, made by ST Microelectronics. It was produced much later than the original Mbed LPC1768 and has less on-board capability than that unit, but is considerably cheaper. It represents one of the many Mbed-enabled boards that have been produced by semiconductor manufacturers. The F401RE is one of a family of similar boards, each one based on a different microcontroller in the STM32 family, with each board reflecting those microcontroller features. Reference 2.5 provides a user manual for the full family of Nucleo boards. With a little care it is possible to pick out the detail that applies to our board, the F401RE.

At the heart of this Nucleo board is the 64-pin version of the STM 32F401RE 32-bit microcontroller (its full designation is STM32F401RET6U). Unlike the LPC1768, which has an ARM Cortex-M3 core, this has a Cortex-M4 core. The full power and versatility of the microcontroller can be glimpsed by viewing its data sheet and user manual, References 2.6 and 2.7, respectively. These are major documents and are absolutely not essential to make progress in the early chapters of this book. However, we do refer to them a little in later chapters, for example, Chapter 15.

The Nucleo F401RE is pictured in Fig. 2.4, with the main features identified in Fig. 2.5. An immediate point of interest is that the printed circuit board (PCB) is perforated, with

Figure 2.4
The Nucleo F401RE board. *Image courtesy of ST.*

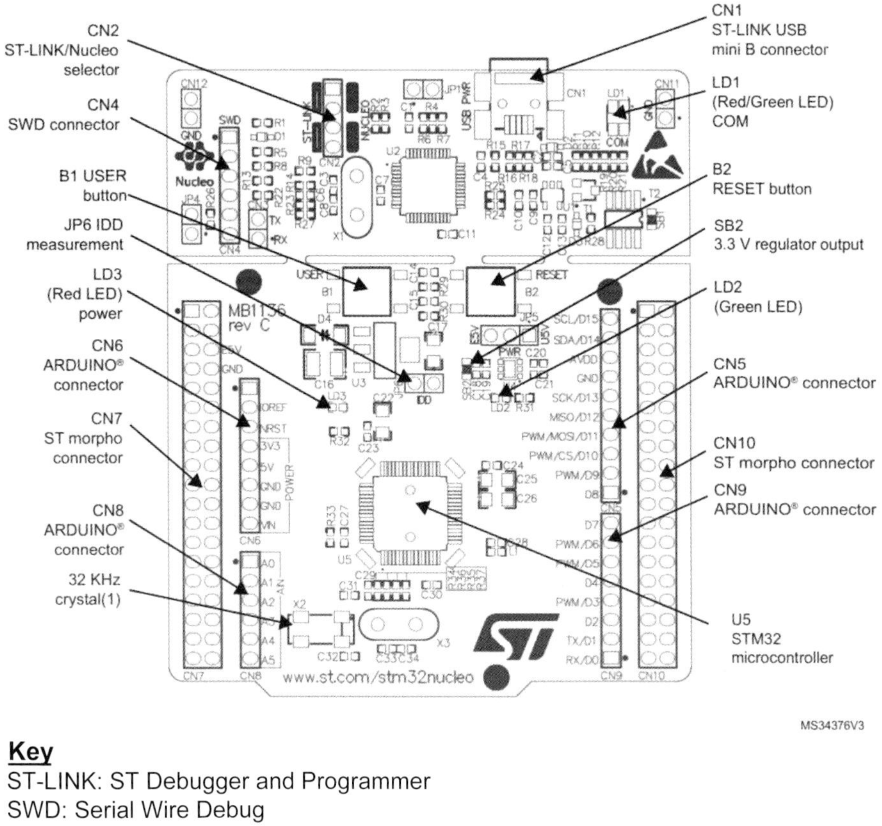

Figure 2.5
The Nucleo F401RE board, main features. *Image courtesy of ST.*

interface circuitry on an upper sub-board—which can be broken off if no longer needed—and the main circuit on a lower, larger one. Communication with the board is made through the USB connector at its top; this is used to download programs and as one means of providing power.

The user interface includes the bicolor (red/green) LED (LD1) at the top right of the Nucleo board, the power LED (LD3), the black reset button (B2), the green LED (LD2) below this, and the blue user button (B1). LD2 is the only LED on the board which can be controlled by the user. There are a number of jumper links on the board. For our purposes these should not be changed or moved.

The board has two main sets of connectors. One set (CN5, CN6, CN8, CN9) allow direct connection to Arduino expansion boards, while the other (CN7 and CN10) provide more complete interface capability. We will use the Arduino-compatible connectors; these are detailed in Fig. 2.6.

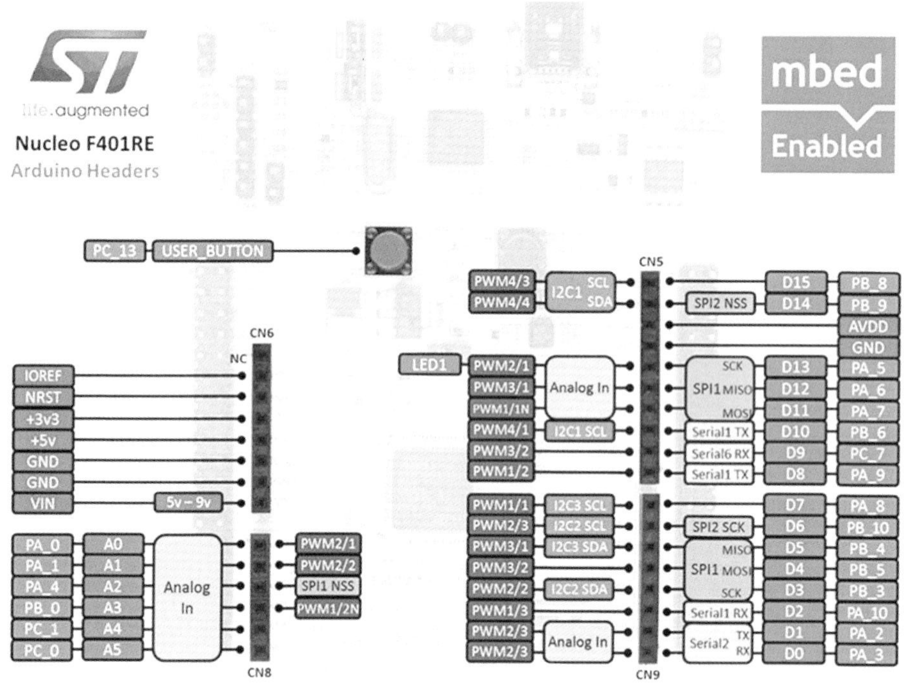

Note: Serial2 (nominally D0 and D1) is used for serial communication to the host computer.
It is connected to the external pins if solder bridges are made.
D0 and D1 are unavailable unless these bridges are made.

Figure 2.6

The Nucleo F401RE Arduino-compatible pin descriptions. *Image courtesy of ST.*

2.4 The STM32 Discovery Kit IoT Node

While the Mbed LPC1768 is the grandfather of all Mbed-enabled devices, this discovery kit represents what is more recently available in terms of sensors and sophisticated connectedness. The version used here has ST part number B-L475E-IOT01A. Full technical details can be found in Reference 2.8.

Fig. 2.7 shows the main features of the kit. At its heart is an STM32L475 microcontroller, based around an ARM Cortex-4 core. Sensors include accelerometer, barometer, humidity, microphone, temperature, and time-of-flight. Wireless connectivity includes Bluetooth low energy, Wi-Fi, and sub GHz RF. There are 7 single-color on-board LEDs and 1 bicolor. There is a single-user button.

A set of Arduino-compatible connectors, seen in Fig. 2.8, allow external connection.

We will explore the details of this kit later in the book and use it for all IoT experimentation.

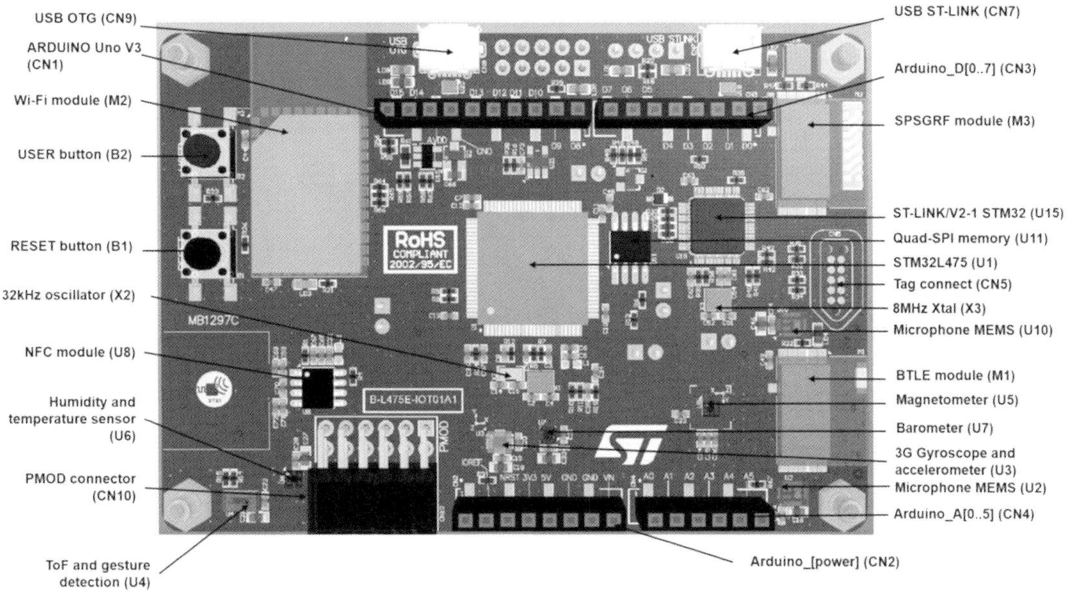

Key (see also Keys to earlier figures)

BTLE: Bluetooth Low Energy
NFC: Near Field Communication
SPSGRF: Sub-GHz RF transceiver module
PMOD: Peripheral Module Interface (Digilent open standard)
ToF: Time-of-Flight

Figure 2.7
The STM32L4 discovery kit overview. *Image courtesy of ST.*

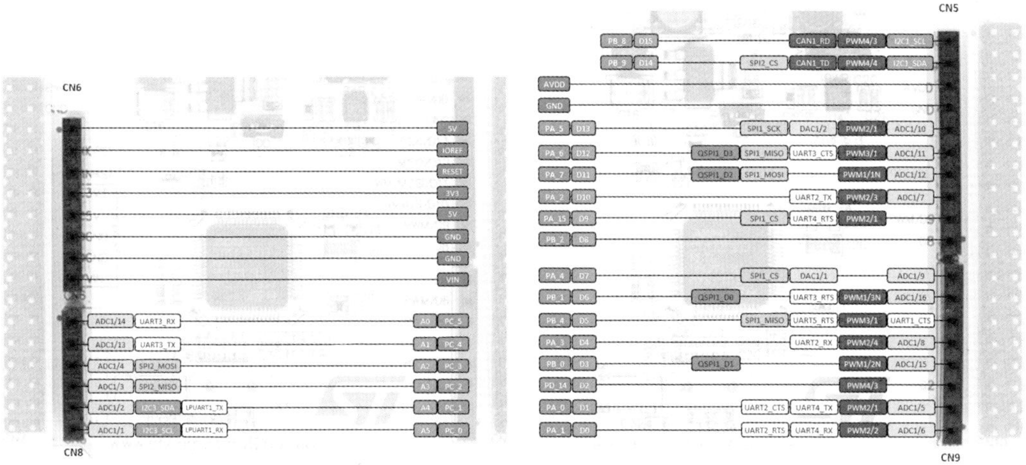

Figure 2.8
The STM32L4 discovery kit connection details. *Image courtesy of ST.*

2.5 The Framework for Mbed Code Development

2.5.1 Using C/C++

The programming language C has been the language of choice for embedded programs of low or medium complexity, but has increasingly been overtaken by the more complex and sophisticated C++. Yet C remains a well-defined subset of the larger C++ language. In the first two editions of this book, C proved to be more or less adequate for the programs being developed, even though the Mbed API is written in C++. Life has become a little more complex since then, however, and the C++ features of the API have become more prominent. In general, we will aim to use mainly C in the programs we develop, at least in the earlier chapters of the book. However, when a C++ feature is essential, we will adopt that without complaint. In a way, we could say that our programming is primarily in C, with some extra C++ features. However, because ultimately we are working in a C++ development environment, code files will carry the **.cpp** (C plus plus) extension.

This book does not assume that you have prior knowledge of C or C++, although you have an advantage if you do. We aim briefly to introduce new features as they come up, at least in the first half of the book, flagging this by using the symbol alongside. If you see that symbol and you are an experienced coder, then it means you can probably skim through that paragraph. If you are not an expert, you will need to read the section with care, and maybe refer across to Appendix B, which summarizes most features used. Even if you are a C/C++ expert, you may not have used it in an embedded context. As you work through the book, you will see a number of techniques that are used to optimize the language for this particular environment.

2.5.2 The Mbed API

A central feature of the Mbed ecosystem is its API. In brief, an API is a set of programming building blocks, appearing as C++ utilities and contained in software libraries, which allow programs to be devised quickly and reliably. These exist for the main features of the Mbed-enabled microcontrollers and target boards themselves, as well as for external devices which may be connected. The main elements of the Mbed API suite are summarized in Table 2.2. Not every detail of this table may be understandable at this stage, but a glance through it gives a glimpse of the range of features that are available. It is worth noticing how many of these relate to the microcontroller peripherals; for example, seen in Fig. 2.3; in other words, the API provides the software capability to control those peripherals. Full API details are to be found in Reference 2.9. Throughout the book we will be drawing heavily on these API features, so in a way Table 2.2 provides the framework for much of our programming work.

Table 2.2: Summary of the Mbed API.

API Category and Group	Summary
Scheduling	Contains the main features of the RTOS.
RTOS	The main RTOS features of threads, mutex, semaphores, and queues.
Event handling	Configures events and related timing.
Drivers	Driver APIs for main inputs and outputs of the target platform.
Serial (UART) drivers	Creates and operates universal asynchronous receive/transmit nodes.
SPI drivers	Creates and operates master and slave serial peripheral interface nodes.
Input/output drivers	Creates and operates analog and digital inputs and outputs, interrupt inputs, and pulse width modulation (PWM) outputs.
USB drivers	Creates and operates universal serial bus utilities.
Other drivers	Creates and operates CAN (controller area network), I2C (inter integrated circuit) serial interfaces, and watchdog timer.
Platform	Contains microcontroller management features, which may need to be adapted to the particular target.
Time	Creates and operates a range of timing facilities, including timer, ticker, timeout, and RTC (real-time clock).
Power	Provides power management and facilitates low-power consumption.
Memory	Manages microcontroller memory.
Other platform APIs	A range of platform-specific APIs, including debug and error handling.
Data Storage	APIs to manage data file systems
File system APIs	Includes management of directory, file, FAT (file allocation table) files, block devices, SD cards.
Connectivity	APIs to provide connectivity through different network protocols, examples appear below, others are available.
Network Interface	APIs to support internet connectivity, including for Ethernet and Wi-Fi.
Bluetooth	APIs to support Bluetooth connectivity.
Security	APIs to provide security, e.g., for IoT systems.

2.5.3 The Mbed Operating System

In the earlier days of Mbed, programming was undertaken using the APIs, with programs structured in a conventional sequential manner. With the advent of the IoT and the wide uptake that the Mbed concept has enjoyed, the APIs have been subsumed into an operating system, more specifically a real-time operating system (RTOS). At this stage we note the presence of the RTOS, but wait until Chapter 9 before delving into its details.

The Mbed OS that has emerged has many components, and contributors. It is represented in simplified form in Fig. 2.9. It is useful to get an overview impression of how it is structured. At the bottom of the diagram is the microcontroller hardware itself, centered around the ARM Cortex-M core. At the top is the code which you, the user, are writing. There is much going on between these two layers! Directly above the hardware is the HAL, the hardware application layer. This is the code which interfaces directly with target hardware. Often this code is contributed not by ARM, but by the manufacturers of the hardware. This includes the *bootloader*, that bit of code we don't usually see, which puts the microcontroller in working order as soon as power is applied and before any user code can start.

Above the HAL is the common microcontroller software interface standard (CMSIS), contributed by ARM. This acts as a standard interface between Cortex cores and the upper levels of an operating system. CMSIS merges with RTX, the RTOS created by the software company Keil (now owned by ARM); RTX forms an integral part of the Mbed RTOS. The APIs which we the user apply, represented in Table 2.2, then appear above this. The whole thing appears complex, and indeed it is. However, the beauty is that just as we don't need to understand the engine of a car in order to drive well, we can write good code without knowing every detail of the underlying OS.

We will make use of the features of the Mbed OS in all of the chapters which follow.

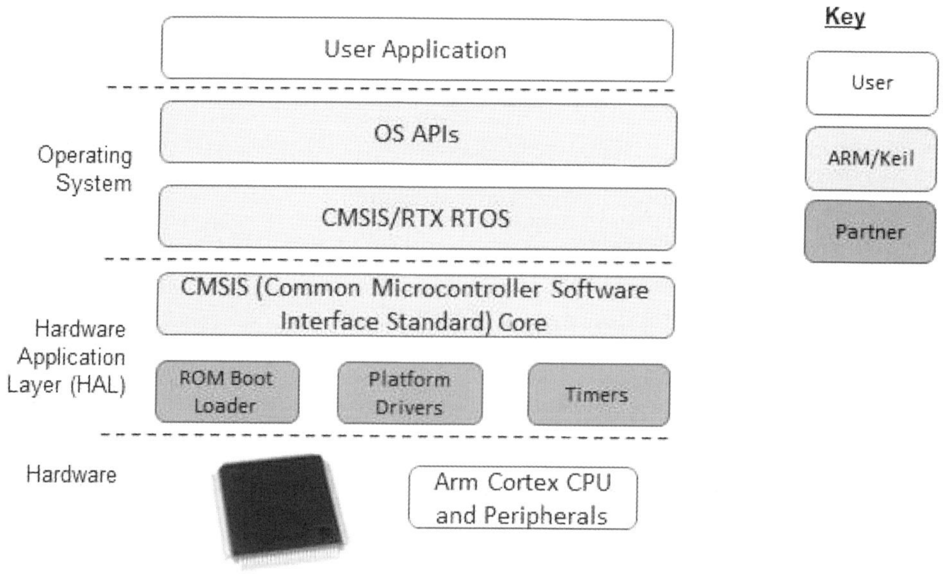

Figure 2.9
The Mbed OS conceptual architecture, simplified.

The whole of the code of the Mbed OS is publicly available and lodged on Github, Reference 2.10. A novice user should not expect to need to access this. As your expertise grows, however, there are times when it is useful to be able to check this source code, maybe to gain deeper understanding or to access code blocks which may be useful in a project you are developing.

2.5.4 The "Bare Metal" Profile

Applying the full Mbed OS for simple applications brings a level of sophistication which is not needed (the old English saying "a sledgehammer to crack a nut" springs to mind). Moreover, using the full OS takes up more program memory, takes longer to compile, and may then take longer to execute. To do without the full OS one can revert to the so-called "bare metal" profile. This is a subset of the Mbed OS and results in more efficient and compact code. It is not an RTOS and does not include the RTX element seen in Fig. 2.9.

Note that we used the full Mbed OS to compile all the C/C++ programs appearing in this book. In many cases this was not necessary, and we could have used the Bare Metal approach. You may wish to experiment with its use as we begin programming.

2.6 The Studio Development Environments, and a First Program

The online compiler was the first development environment offered for the original Mbed. Being online, it could be accessed wherever the internet was available. It was hugely successful, with developers writing programs on it around the world. However, to some extent it became a victim of Mbed's success. With many hardware platforms becoming available, it became increasingly difficult to update. The compiler was phased out at the end of 2022, although there remain numerous references to it on the internet.

The immediate successor development environment to the online compiler was Mbed Studio. This must be downloaded to the user's computer, with all development work subsequently undertaken on that machine. Software libraries and other programs can of course still be imported. Mbed Studio makes considerable memory demands on the host computer, and it is possible that a user may not have an adequate machine to run it.

A version of Mbed Studio, called Keil Studio Cloud, was then made available. This is more or less an online version of Mbed Studio, although there are some minor differences. Being online, Studio Cloud is accessible from any machine as long as reasonable internet access is available. Memory demands on the home computer are much less, as programs and OS are stored in the cloud. Studio Cloud has become the natural successor to the old online compiler and is the natural choice of IDE for most developers. For those for whom confidentiality is essential, or who wish or need to work off-line, Mbed Studio is a very good alternative.

2.6.1 Using Mbed Studio/Keil Studio Cloud

This section outlines the principles of getting an Mbed program running on an IDE, using Keil Studio Cloud as an example. Due to the similarities of the two environments, it will also be possible to follow these steps on Mbed Studio (Version 1.4.4 used here). A few features, however, are unique to Mbed Studio and are, for clarity, shaded in the rest of this section.

We focus here on initial familiarization with the IDE and will look into the details of the example program itself in the next chapter. With inevitable ongoing changes and upgrades in both IDE and host computer software, it is important to accept that implementation details will change; the underlying processes should, however, remain reasonably constant. There is a good and frequently updated user guide, available at Reference 2.12. This is written for Keil Studio Cloud, but has many insights useful for Mbed Studio.

A final caution before starting: because it is an older device, the Mbed LPC1768 does not have (at the time of writing) full connectivity with some of the features of the Studio IDEs. Don't worry too much, as you will get as much functionality as you are likely to need, certainly as much as with the old online compiler. The main impact for initial programming is that some features may not be available in all options, or be fully automated. We will aim to flag these. The User Guide, Reference 2.12, gives the latest information on this.

You will need:

- an Mbed development board (in our case the LPC1768, Nucleo F401RE or STM32 Discovery Kit) with its USB lead,
- a computer running Windows, Mac OSX, or GNU/Linux,
- and a web browser, for example, Internet Explorer, Safari, Chrome, or Firefox.

Keil Studio Cloud can be accessed directly from the Mbed website, Reference 2.1. You will need to create an account. Mbed Studio can be downloaded through Reference 2.11.

A typical Studio Cloud screen setup, with a program already active, is shown in Fig. 2.10. Notice the display of Active Project, Build Target, and Connected Device toward the top left of the screen. The important control buttons of Build (hammer), Run (arrowhead), Debug (the bug!), and Serial Monitor lie immediately below this. These are illuminated blue when there is an active program and a recognized target system has been connected and selected. Available files and folders are listed below left. Program source code is

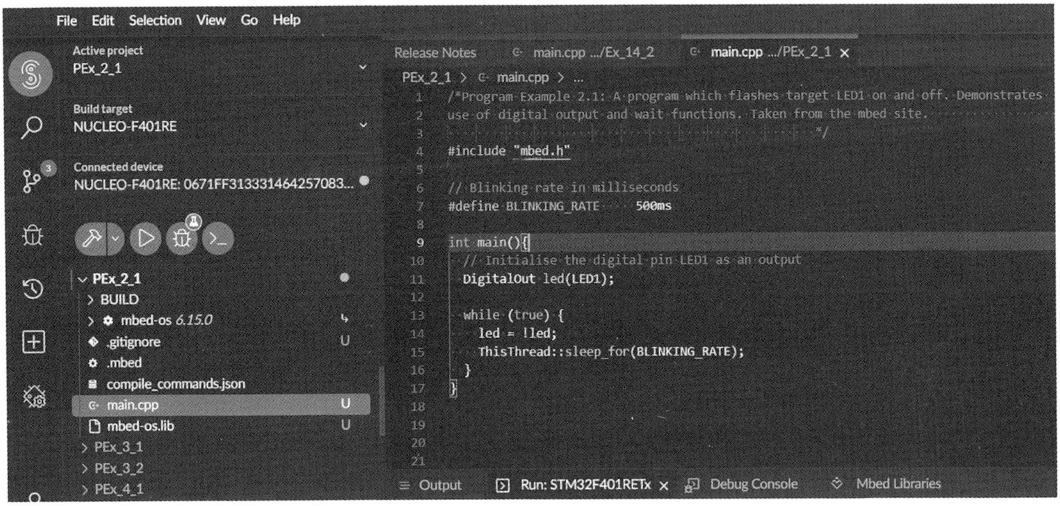

Figure 2.10
General screen in Studio Cloud. *Images courtesy of ARM.*

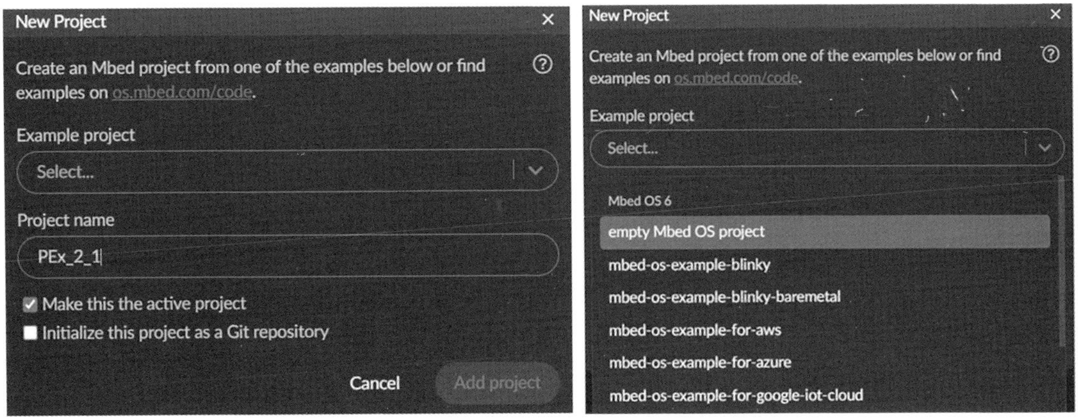

(a) (b)

Figure 2.11
Creating a new project in Studio Cloud: (a) opening dialog box; (b) template selection. *Images courtesy of ARM.*

shown opposite. Across the top of the screen are the familiar tabs of File, Edit, View, Help, and so on.

To develop a program, start with **File > New > Mbed Project**. You will be invited to select a program template, as seen in Fig. 2.11(a). Choices are shown in Fig. 2.11(b); for your first program select **Mbed-os-example-blinky**. Enter also your own program name (we have entered PEx_2_1, anticipating all the forthcoming program examples through the book!). You can follow this selection process for just about all the C/C++ programs in the book, unless you want to explore a "bare metal" alternative.

For Mbed Studio the program needs its own local copy of the OS. This can be stored automatically in the program folder using a dialog box option available as the new program is created. You will need to allow this for your very first program. This causes a complete copy of the Mbed OS to be downloaded to your computer, resulting in creation of a folder of slightly over 1 Gbyte size (a significant memory demand). To avoid the massive memory requirement of repeated copies of the OS, in subsequent programs you are advised to link to an existing copy of the OS on your computer (e.g., your first program). A dialog box option appears which allows you to do this. None of this is necessary for Studio Cloud, as everything is in the cloud!

If you plug in your target hardware through its USB cable, it should be detected by the computer as a USB device and displayed accordingly. The IDE should also detect it and ask if you want this to be your chosen target. This will then display under Build Target on the home screen. For Mbed LPC1768 you may need to select the target manually, and the hammer button may be the only one illuminated; you can, however, progress with this.

To show the source code, if it is not already there, click on **main.cpp** on the file listing for your current project, as highlighted toward the left in Fig. 2.10. You should now have arrived at a screen similar to the one seen in Fig. 2.10. The program in your selected template is repeated as Program Example 2.1.

```
#include "mbed.h"

// Blinking rate in milliseconds
#define BLINKING_RATE     500ms

int main(){
  // Initialize the digital pin LED1 as an output
  DigitalOut led(LED1);

  while (true) {
    led = !led;
    ThisThread::sleep_for(BLINKING_RATE);
  }
}
```

Program Example 2.1: An Mbed "Blinky" program

Three Build profiles are offered in Mbed Studio, Debug, Develop, or Release, seen at the top of Fig. 2.12. This selection is hidden in Studio Cloud. We recommend the Debug profile for introductory programs, though the differences are not significant in such applications. Further details can be found in Reference 2.13.

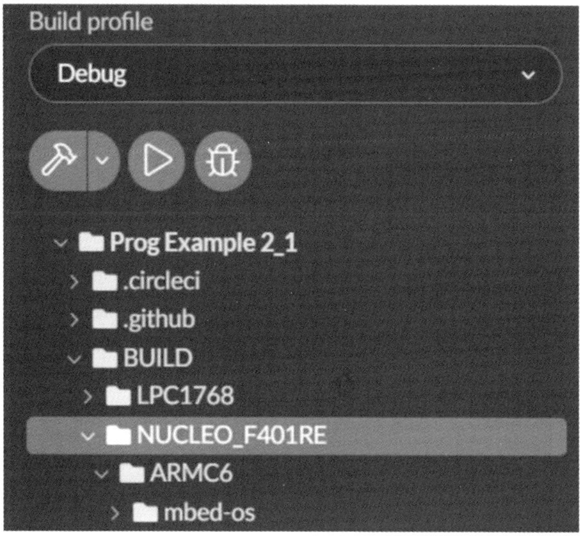

(some files omitted here in listing)

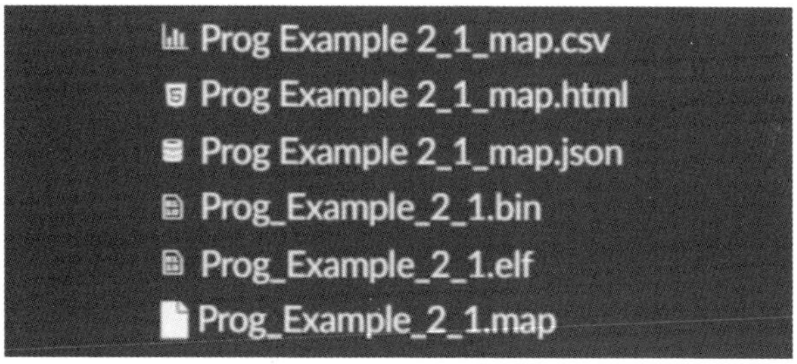

Figure 2.12
Mbed Studio screen, example files listing. *Images courtesy of ARM.*

Click the Build (hammer) button to compile the program (this may be the only option for the LPC1768) *or* the Run button to compile, download, and launch the program. Even for a tiny program such as this, compilation—at least for the first iteration of any program—is an extended process. A rolling screen, bottom right, indicates progress.

At the end of the Build process, if it has been successful, a binary file holding the program machine code has been created. This appears in the files listing to the left of the Mbed Studio screen, as shown in Fig. 2.12. It can also be found on your computer, for example in Windows Explorer, by following **C:\Users*User name*\Mbed Programs*program name*\BUILD*target name*\ARMC6**, where italiciszed entries are program- or target-dependent. For Studio Cloud, these files are held in the cloud. However, the binary file appears as a download to your computer.

If you have clicked Run, the binary program code is normally automatically downloaded to the flash memory in the target board, indicated by a flashing LED—look out for this! For many target boards, the program immediately starts running.

If automatic download does not occur (e.g., for the LPC1768), be ready to drag and drop the binary file across to the target board. For Studio Cloud, this is the file download already mentioned. For Mbed Studio, the file is in the folder identified above. For the LPC1768, you will need to press the reset button to start the program running.

Once the program is downloaded and started, the LED designated as LED1 on your chosen board should flash once per second, as prescribed in the program.

2.6.2 Adding Software Libraries

When you get to more advanced programming, you will import external software libraries into your program. You do this in the File drop-down menu, by selecting the **Add Mbed Library to Active Project** (Studio Cloud) or **Add Library to Active Program** (Mbed Studio). This is seen in Fig. 2.13(a). This then calls the dialog box of Fig. 2.13(b). The required URL can be found on the web page where the code is stored, often given as a reference in the later chapters of this book. If on the Mbed site it will appear in the format seen in Fig. 2.13(c)). The required URL is then found by clicking **Import into Keil Studio.**

2.7 The Mbed Application Board

App Board

In building circuits based on the Mbed LPC1768 or Nucleo F401RE, the early chapters of this book will use two approaches. One will require the Mbed to be used with a prototyping "breadboard," as seen in Fig. 3.5 and many figures thereafter. This

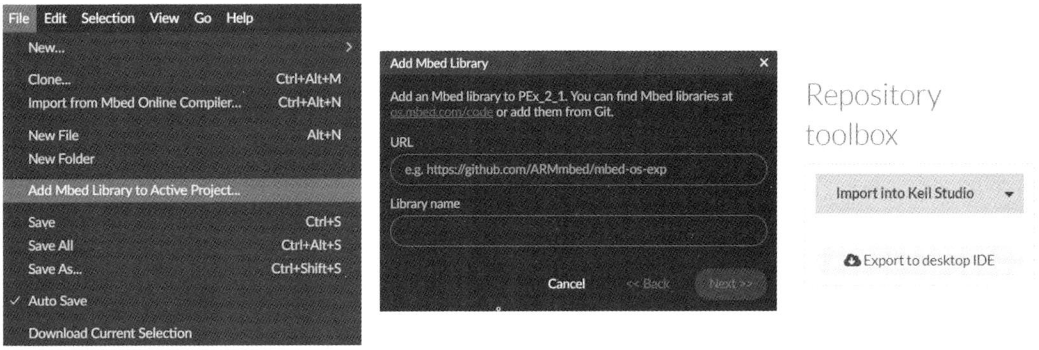

Figure 2.13
Adding an external library to a program. *Images courtesy of ARM.*

allows considerable flexibility, as almost any circuit can then be built around the Mbed. However, it can be time-consuming to build these circuits and easy to make mistakes. When built, the circuit can be unreliable, as wiring can easily be disturbed. An alternative is to use the Mbed *application board,* or *app board,* as seen in Fig. 2.14. This is not a new product, but it remains completely relevant to today's technology. It is not at all essential to have the app board to complete the practical work in this book, but it represents a convenient option.

The application board has a number of the most interesting peripheral devices that we might want to connect to an Mbed target, like display, LEDs, joystick, potentiometers, speaker, and certain sensors and connectors. An Mbed LPC1768 plugs into the board, and a physical connection with all the peripheral devices is then immediately made. Of course, the Mbed has to be programmed correctly before it can usefully drive these external devices. It is also possible to connect other development boards to the Mbed app board; this exploits the convenience of that collection of peripheral devices, though we are then back to making wired connections. One example of the Nucleo F401RE connected to the app board is seen in Fig. 3.9(b).

The application board gets its own page on the Mbed site, Reference 2.14. The full circuit diagram is also available there. This is useful to study if your interests include the hardware aspects, but it is not needed to make full use of the board. Fig. 2.15 and

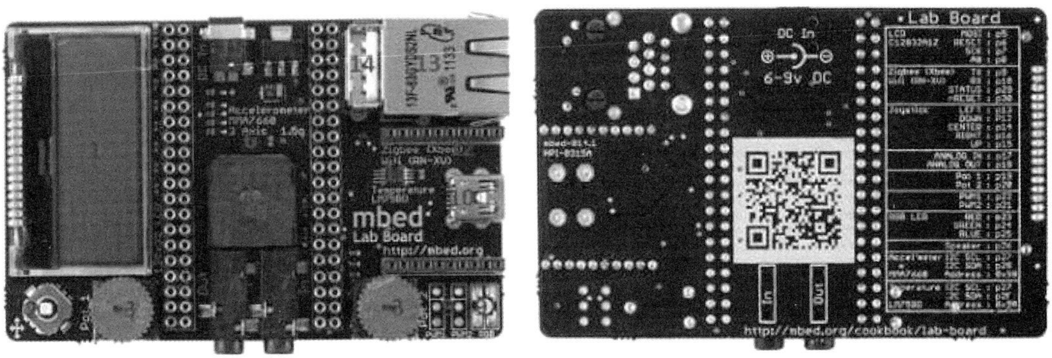

1. Graphics LCD (128x32)
2. 5-way Joystick
3. 2 x Potentiometers
4. 3.5mm Audio Jack (Analog Out)
5. Speaker, PWM Connected
6. 3 Axis +/1 1.5g Accelerometer
7. 3.5mm Audio Jack (Analog In)
8. 2 x Servo Motor Headers
9. RGB LED, PWM connected
10. USB-mini-B Connector
11. Temperature Sensor
12. Socket for Xbee (Zigbee) or RN-XV (Wifi)
13. RJ45 Ethernet Connector
14. USB-A Connector
15. 1.3mm DC Jack input

Figure 2.14
The Mbed application board. *Images courtesy of ARM.*

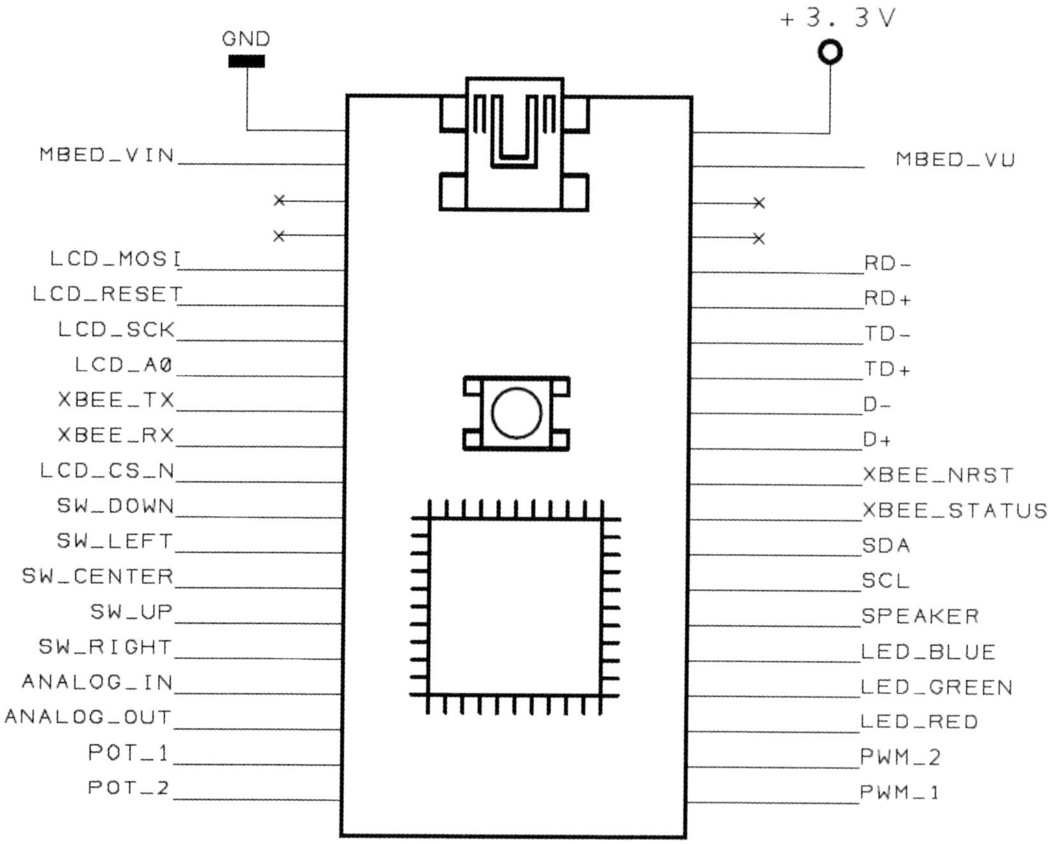

Figure 2.15
Application board connections to the Mbed LPC1768. *Image courtesy of ARM.*

Table 2.3 show how the Mbed LPC1768 connects to the devices on the application board. The information in Table 2.3 actually appears on the back of the application board PCB itself, but it is very small. It can be seen that many Mbed signals connect directly to the on-board sensors. Others, like the analog in and out, USB, and two PWM outputs, are available for external connection, through dedicated connectors.

It is useful to understand how power is distributed on the overall system when the Mbed LPC1768 is plugged into the app board. The app board has a DC input jack socket (item 15 on Fig. 2.14) and *two* USB connectors, a "USB—A" (item 14) and a "USB mini-B" (item 10) connector. The Mbed LPC1768 has its own USB mini-B connector. Power distribution is shown in Fig. 2.16, which shows the app board power circuit to the left, and the Mbed (in the form of half of Fig. 2.2, repeated) to the right. The interconnecting lines

Table 2.3: Mbed LPC1768 connections to the application board.

Device	Signal Name	Mbed pin	Device	Signal Name	Mbed pin
LCD	MOSI	5		Analog In	17
	Reset	6		Analog Out	18
	SCK	7	Potentiometer 1		19
	A0	8	Potentiometer 2		20
	nCS	11		PWM 1	21
Zigbee (Xbee)	TX	9		PWM 2	22
	RX	10	RGB LED	red	23
	Status	29		green	24
	nReset	30		blue	25
Joystick	Down	12	Speaker		26
	Left	13	Accelerometer (address 0x98)	I2C SCL	27
	Centre	14		I2C SDA	28
	Up	15	Temperature Sensor (address 0x90)	I2C SCL	27
	Right	16		I2C SDA	28

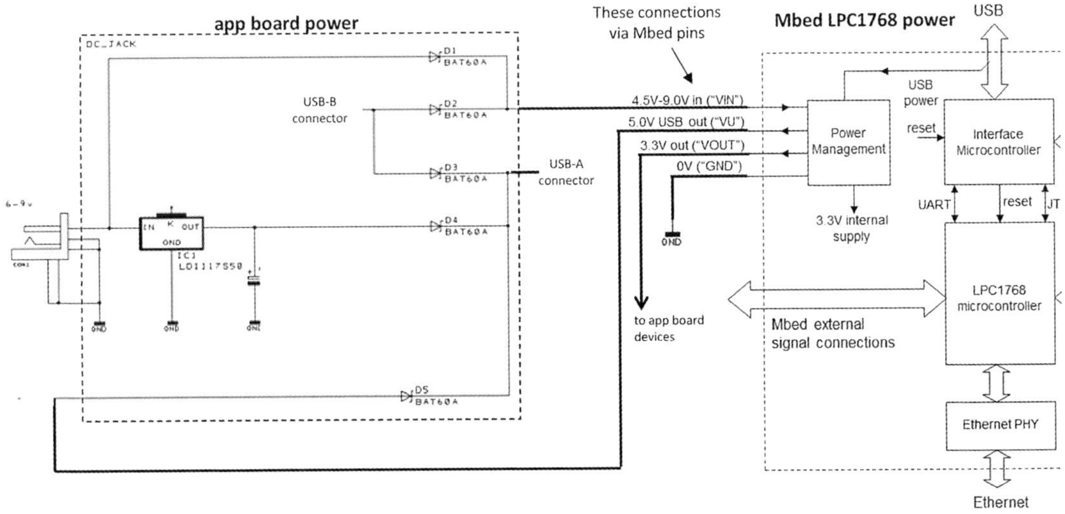

Figure 2.16
Application board/Mbed LPC1768 power distribution.

show the connections made once the Mbed is plugged in. The combination can be powered *three* possible ways:

- through the Mbed USB connector,
- through the app board DC jack—the Mbed is then powered through the upper diode in the figure and the Mbed "VIN" connection,
- and through the app board USB-B connector—the Mbed is then powered through the second diode, and then again through the Mbed "VIN" connection.

In each case the app board devices are powered from the Mbed 3.3 V output. The LD1117S50 on the app board, seen to the left of the figure, is a 5 V regulator. It is only used to power the USB-A connector, when the DC jack input is used. It is important to remember that although there are three ways to power the system, the *only* way of downloading program code to the Mbed LPC1768 is through its own USB connector.

Chapter Review

The Mbed ecosystem provides a complete environment for rapid development of embedded products.

- The Mbed LPC1768 and the Nucleo F401RE are compact, general-purpose Mbed-enabled development platforms, each capable of working in the Mbed ecosystem.
- The STM32 Discovery Kit IoT node is a sophisticated Mbed-enabled development platform, targeted toward the IoT environment.
- Communication to these development boards from a host computer is by USB cable; power can also be supplied through this link.
- Each development board above contains some on-board features which allow user interaction; each also provides connectors giving direct connection to many of the on-board microcontroller pins.
- The development boards above are each based around a microcontroller containing an ARM Cortex core, either M3 or M4. Many of the characteristics of any board derive directly from its microcontroller.
- The Mbed ecosystem includes an OS which is built from the API and other elements.
- Mbed programs invoke the APIs, and may (but don't have to) use the OS.
- Mbed Studio is one of several IDEs that can be used to develop programs in the Mbed ecosystem. Keil Studio Cloud is a convenient online equivalent of this, and can be viewed as a direct successor to the now-discontinued online compiler.
- The Mbed application board contains a set of popular peripheral devices; it is useful for a range of experiments, with both Mbed LPC1768 and Nucleo F401RE, but is constrained to the devices fitted and the interconnect available.

Quiz

1. What are the main elements of the Mbed "ecosystem"?
2. What do UART, CAN, I^2C, and SPI stand for, and what do these features have in common?
3. How many digital inputs are available on the Mbed LPC1768?
4. Which pins can be used for analog input and output for the Mbed LPC1768 and the Nucleo F401RE?
5. How many microcontrollers are on the Mbed LPC1768 PCB and what is their role?
6. Looking at a target development board, identify these elements:
   a. for the Mbed LPC1768, point out the LPC1768 microcontroller, interface microcontroller, voltage regulators, main clock oscillator, USB disk flash memory, and Ethernet PHY;
   b. for the Nucleo F401RE, point out the reset button, user push button, bicolor LED, 32 kHz crystal, and STM 32F401RE microcontroller;
   c. for the STM32 Discovery kit IoT node, point out the STM32L475 microcontroller, reset button, user push button, Wi-fi module, BTLE module, and humidity/temperature sensor.
7. What do API, HAL, CMSIS, and RTOS stand for? Briefly describe each.
8. An Mbed LPC1768 is part of a circuit which is to be powered from a 9 V battery. After programming, the Mbed is disconnected from the USB. One part of the circuit external to the Mbed needs to be supplied from 9 V, and another part from 3.3 V. No other battery or power supply is to be used. Draw a diagram which shows how these power connections should be made.
9. An Mbed target board is connected to a system and needs to connect with three analog inputs, one SPI port, and two PWM outputs. Draw a sketch showing how these connections can be made, for either LPC1768 or Nucleo F401RE. Indicate the relevant pin number in either case.
10. By not connecting all the LPC1768 microcontroller pins to the Mbed external pins, a number of microcontroller peripherals are "lost" for use. Identify which ones these are, for ADC, UART, CAN, I^2C, SPI, and DAC.

References

2.1. The ARM Mbed Site. https://os.Mbed.com/.

2.2. Mbed LPC1768 Circuit Diagrams. https://developer.Mbed.org/media/uploads/chris/Mbed-005.1.pdf.

2.3. LPC1768/66/65/64 32-bit ARM Cortex-M3 microcontroller, Objective Data Sheet, NXP B.V. Rev. 9.10, September 2020.

2.4. LPC176x/5x User Manual, NXP B.V. Rev. 3.1, April 2014, Doc. no. UM10360.

2.5. User Manual, STM32 Nucleo-64 boards (MB1136), Doc. no. UM1724, ST Microelectronics, August 2020.

2.6. Data Sheet, STM32F401xD STM32F401xE, 2015, STMicroelectronics, Doc. ID 025644.
2.7. Reference Manual STM32F401xB/C/D/E, Doc. no. RM0368 Rev. 5, ST Microelectronics, December 2018.
2.8. User Manual, Discovery Kit for IoT Node, Multi-channel Communication with STM32L4, Rev. 5, Doc. no. UM2153, ST Microelectronics, October 2019. (Also contains schematic diagrams.)
2.9. Full Mbed OS API List. https://os.Mbed.com/docs/Mbed-os/v6.15/apis/index.html.
2.10. ARM Mbed Repository. https://github.com/ARMMbed.
2.11. Mbed Studio download. https://os.Mbed.com/studio/.
2.12. ARM Keil Studio Cloud. User Guide, Issue 16. ARM Limited; 2022. https://developer.arm.com/documentation/102497/1-6/.
2.13. Build Profiles (Mbed OS). https://os.Mbed.com/docs/Mbed-os/v6.15/program-setup/build-profiles-and-rules.html.
2.14. Mbed application board. https://os.Mbed.com/components/Mbed-Application-Board/.

CHAPTER 3

Digital Input and Output

3.1 Starting to Program

In this chapter we will consider the most basic of microcontroller activity, the input and output of digital signals. Moving beyond this, we will see how we can make simple decisions within a program based on the value of a digital input. Fig. 1.1 also predicted the importance of time in the embedded environment; therefore, it is no surprise that at this early stage some timing activity is required as well. A number of programs will be trialed on the Mbed LPC1768 and Nucleo F401RE, either on their own or with a breadboard, or with the LPC1768 application board.

You may at this moment be embarking on C/C++ programming for the first time. Remember that C is a subset of C++. We try to use mainly C, but—as complexity increases—will increasingly need to apply some features of C++. This chapter takes you through a few of the key C concepts. If you are new to C, we suggest you now read through Sections B.1 to B.5, inclusive, of Appendix B.

One thing that is not expected of you in reading these early chapters is a deep knowledge of digital electronics. However, some understanding of electronic theory will be useful. If you need support in this area, you might want to have a book such as Reference 3.1 available, to access when needed. There are many good websites in this area as well.

3.1.1 Thinking About the First Program

We now take a look at our first program, introduced as Program Example 2.1, for convenience shown again below with comments inserted. Compare this with the original appearance of the program, shown in Chapter 2. Remember to check Appendix B when you need further C background.

C/C++ code feature

Comments are text messages which you write to yourself within the program; they have no impact on the actual working of the program. It is good practice to introduce comments widely; they help your own thought process as you write, remind you what the program was meant to do as you look back over it, and help others read and understand the program. This last point is essential for any sort of teamwork, or if you are handing in work to be marked!

Fast and Effective Embedded Systems Design. https://doi.org/10.1016/B978-0-323-95197-5.00003-6

There are two ways of inserting comments into a program, and both are used in this example. One is to place the comment between the markers /* and */. This is useful for a block of text information running over several lines. Alternatively, when two forward slash symbols (//) are used, the compiler ignores any text which follows on that line only; this can then be used for comment.

The opening comment in the program gives the program name and a brief summary of what it does. We adopt this practice in all subsequent programs in the book. It is also good practice to add the name of the author and the date, though we omit that information in the book program versions. Notice also that we have introduced some blank lines to make the program a little more readable. We then put a number of in-line comments to explain what individual lines of code do.

```
/*Program Example 2.1: A program which flashes target LED1 on and off. Demonstrates
use of digital output and wait functions.
Works with LPC1768, F401RE or STM32 IoT Discovery Kit as target systems*/

#include "mbed.h"  //include the Mbed header file as part of this program

//Blinking rate in milliseconds
#define BLINKING_RATE     500ms

int main() {          //the main function starts here
   // Initialise the digital pin LED1 as an output
   DigitalOut led(LED1);

   while(true) {       //a continuous loop is created
     led = !led;       //toggle led state
     ThisThread::sleep_for(BLINKING_RATE);  //no action for blinking rate duration
   }                   //end of while loop
}                      //end of main function
```

Program Example 2.1 Repeated (and further commented) for convenience

C/C++ code feature

Let's now identify all the C programming features of Program Example 2.1. The program starts by including the all-important Mbed header file, which is the connecting pathway to all Mbed library features. This means that the header file is literally inserted into the program, before compiling starts in earnest. The **#include** compiler directive is used for this (see Section B.2.3 in Appendix B). A second compiler directive, **#define,** is used to provide a name for a numerical constant; in this case, wherever **BLINKING_RATE** appears in the program, it is replaced by 500 ms.

C/C++ code feature

The action of *any* C or C++ program is contained within its **main( )** function, so that is always a good place to start looking. By the way, writing **main( )** with those brackets after it reminds us that it is the name of a C function; we will write all function names in this way in the book. The *function definition*—what goes on inside the function—is contained within the opening curly bracket or brace, appearing immediately after **main( )**, and goes on until the final closing brace. There are further pairs of braces inside these outer ones. Using pairs of braces like this is one of the ways that C groups blocks of code together and creates program structure.

Within **main( )** the program first applies the Mbed API utility **DigitalOut** to define a digital output, and gives it the name **led**. Once declared, **led** can be used as a variable within the program. The name LED1 is reserved by the API for the output associated with a user LED on many Mbed-enabled boards.

C/C++ code feature

Many programs in embedded systems are made up of an endless loop, a program which just goes on and on repeating itself indefinitely. Here is our first example. We create the loop by using the **while** keyword; this controls the code within the pair of braces which follow. Section B.7 tells us that normally **while** is used to set up a loop, which repeats if a certain condition is satisfied. However, if we write **while (1)** or **while (true)**, then we "trick" the **while** mechanism to repeat indefinitely.

C/C++ code feature

The real action of the program is contained in the two lines within the **while** loop. These are made up of a call to the API function **ThisThread::sleep_for(),** described in the following section, and a statement in which the value of **led** is changed. In the statement **led = !led** we apply two C operators. The first is the *assign* operator, the conventional equals sign. Note carefully that in C this operator means that a variable is set to, or assigned, the value indicated. Conventional "equals" is represented by a double equals sign, **==.** It is used later in this chapter. The second operator is the exclamation mark **!**, which is used to represent logical NOT. In this case **led = !led** means that the value of the binary variable **led** is reversed. There are many C operators, and it is worth noticing and learning each one as you meet it for the first time. They are summarized in Appendix B, Section B.5.

Notice that we indent the code by two spaces within **main( ),** and then two more within the **while** code block. This has no effect on the action of the program, but does help to make it more readable, which cuts down on programming errors, a practice we apply for programs in this book. Check Section B.12 for other code layout practices we adopt. Companies working with C or C++ often apply "house styles" to ensure that programmers write code in a readable and consistent fashion.

3.1.2 A First Word on Timing

We have stated from the beginning of Chapter 1 that timing is of great importance in embedded systems; we return to this theme repeatedly. Even in this very simple program, we need to specify how long the LED is switched on, and how long it is off. The traditional

Table 3.1: Example Mbed library wait and sleep functions.

API Function	Action
`wait_ns(int ns)`	waits for the number of nanoseconds specified
`wait_us(int us)`	waits for the number of microseconds specified
`thread_sleep_for(uint32_t ms)`	sleeps for the number of milliseconds specified
`ThisThread::sleep_for(chrono duration)`	sleeps for the specified integer duration, using chrono units of min, s, ms, μs, etc

`ThisThread::sleep_for(chrono::microseconds(int us))`
`ThisThread::sleep_for(chrono::milliseconds(int ms))`
`ThisThread::sleep_for(chrono::seconds(int seconds))`
`ThisThread::sleep_for(chrono::minutes(int minutes))`
	sleeps for the number of time units specified; now deprecated, but widely seen

way of establishing these time durations was to set the CPU into repeated program loops, running down a counter variable until a predetermined amount of time had elapsed. Earlier editions of this book and much legacy code followed this approach, and used the Mbed **wait( )** function. However, this is now deprecated (i.e., discontinued). It is more efficient in terms of power consumption to let the processor enter a "sleep" state (more on this in later chapters). The **ThisThread::sleep_for()** function, from the API library, does just this. Further options are shown in Table 3.1. We see here **chrono**, a mechanism used in C++ to define time durations, which is outlined in Section B.13.4 of Appendix B.

Exercise 3.1

Get familiar with Program Example 2.1 and the general process of compiling and running a program, by trying the following variations. Observe the effects.

1. Add the comments or others of your own given in the example. See that the program compiles and runs without change.
2. Vary the 500 ms value parameter in the **#define** statement.
3. Replace **ThisThread::sleep_for(BLINKING_RATE)** with **wait_us(500000)**.
4. Emit alternate long and short blinks.
5. Repeatedly send the Morse code signal for "SOS," blinks of short–short–short–long–long–long–short–short–short, with a longer break (e.g., two seconds) between each.

■

3.1.3 Using the Mbed API

The Mbed API was introduced in Section 2.5.2. It has a pattern which we will see repeated many times, so it is important to get used to it. The API is accessed through the header file **Mbed.h,** which is included in almost every example program (but not quite every!) in this book. The API is made up of many components, which are itemized in the complete API listing (Reference 2.8), and summarized in Table 2.2. Further header files are used for more advanced Mbed applications.

The first utility that we look at is **DigitalOut,** which has already been put to use. The API summary for this is given in Table 3.2. In a pattern which will become familiar, the **DigitalOut** API component is a C++ *class*. This appears at the head of the table. The class then has a set of member functions, which are listed below. The first of these is a C++ *constructor*, which must have the same name as the class itself. This can be used to create C++ *objects*. In the first example, we create the object **myled**. We can then write to it and read from it, using the member functions **write( )** and **read( )**. These are members of the class, so their format is **myled.write( ),** and so on. In the **DigitalOut** class there are also user-defined operators, a feature of C++. These also appear in Table 3.2 for example, the assign operator = . For instance, when we write

```
myled = 1;
```

the variable value is then not only changed (normal C usage), but also written to the digital output. This replaces **myled.write(1);**. We will find similar user-defined operators offered for all peripherals in the Mbed API.

Table 3.2: The Mbed digital output API summary.

Functions	Usage
Constructor	
`DigitalOut (PinName pin)`	create a DigitalOut object of specified name, connected to the specified pin
Member Functions	
`void write (int value)`	set the output, specified as 0 or 1
`int read()`	return the output setting, returned as 0 or 1
`Mbed-defined operator: =`	a shorthand for **write()**
`Mbed-defined operator: int()`	a shorthand for **read()**

3.1.4 Exploring the while Loop

Program Example 3.1 is the first "original" program in this book, although it builds directly on the previous example. Create a new project in your development environment (e.g., Studio Cloud), and copy this example into it.

C/C++ code feature

Look first at the structure of the program, derived from the *three* uses of **while**; one of these is a **while (true)**, which we are used to, and two are conditional. The latter are based on the value of a variable **i**. In the first conditional use of **while,** the loop repeats as long as **i** is less than 10. You can see that **i** is incremented within the loop, to bring up this value. In the second conditional loop, the value is decremented, and the loop repeats as long as **i** is greater than zero.

C/C++ code feature

An important new C feature seen in this program is the way that the variable **i** is *declared* at the beginning of **main( ).** It is essential in C to declare the data type of any variable before it is used. The possible data types are given in Section B.3. In this example, **i** is declared as a character, effectively an 8-bit number. In the same line, its value is initialized to zero. There are also four new operators introduced, +, -, <, >. The use of these is the same as in conventional algebra, so it should be immediately familiar.

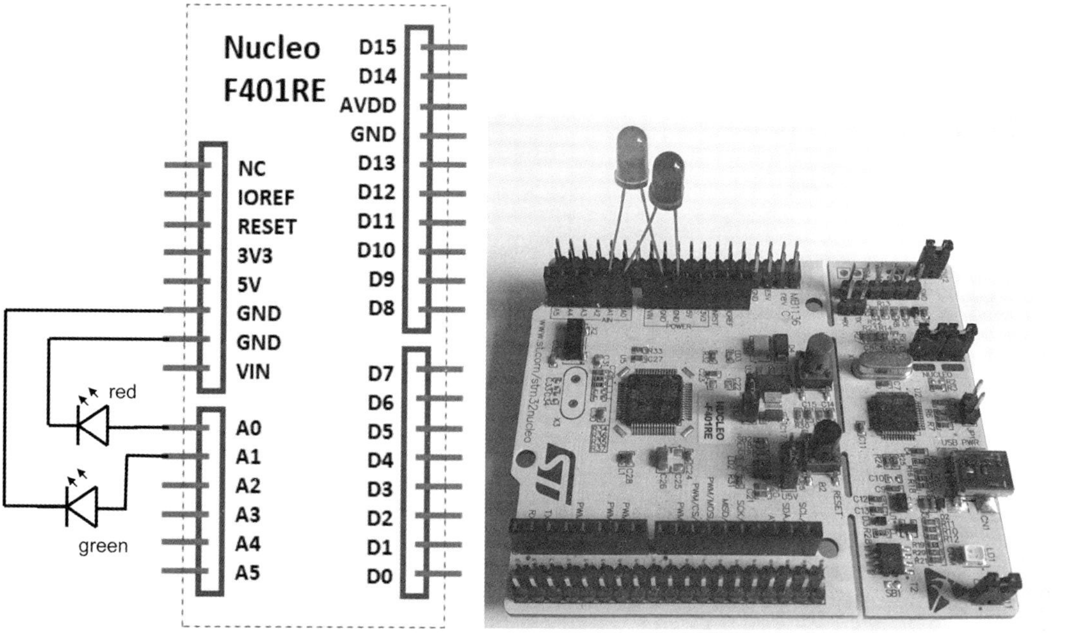

(a) (b)

Figure 3.1: Adding LEDs to the Nucleo F401RE board: (a) circuit; (b) circuit board.

If you run this program on the Mbed LPC1768, you do not need any external components. If you run it on the Nucleo F401RE, you can supplement its single user LED with external ones in a simple way. Fig. 3.1 shows two LEDs plugged directly into the board connectors, saving the need at this stage for an external prototyping breadboard. Attach the side of the LEDs with the longer leg (the anode) to the A0 and A1 pins, and the side with the shorter leg to the nearby GND pins. Together with the onboard user LED ("LED1"), three user LEDs are now readily available for experimentation. You only need one of these extra LEDs for Program Example 3.1, but we will use the second one later.

```
/*Program Example 3.1: Demonstrates use of while loops. Works with LPC1768 (no
external connection required) or F401RE (one external LED added, as in Fig. 3.1).
*/

#include "Mbed.h"
DigitalOut myled(LED1);
DigitalOut yourled(LED4); //replace LED4 with A0 for F401RE board

int main() {
  char i = 0;              //declare variable i, and set to 0
  while(true){             //start endless loop
    while(i < 10) {        //start first conditional while loop
      myled = 1;
      ThisThread::sleep_for(200ms);
      myled = 0;
      ThisThread::sleep_for(200ms);
      i = i + 1;          //increment i
    }                     //end of first conditional while loop
    while(i > 0) {        //start second conditional loop
      yourled = 1;
      ThisThread::sleep_for(200ms);
      yourled = 0;
      ThisThread::sleep_for(200ms);
      i = i - 1;
    }
  }                       //end infinite loop block
}                         //end of main
```

Program Example 3.1 Using *while*

Having compiled Program Example 3.1, download it to your target system and run it. You should see the two LEDs flashing 10 times in turn. Make sure you understand why this is so.

Exercise 3.2

C/C++ code feature

1. A slightly more elegant way of incrementing or decrementing a variable is by using the increment and decrement operators seen in Section B.5. Try replacing

   i = i+1; and i = i -1;

   with i++; and i - -; respectively,

   and run the program again.
2. Change the program so that the LEDs flash only 5 times each.
3. Try replacing the **myled = 1;** statement with **myled.write(1);**

3.1.5 Using Development Board External Pins

We now want to start systematically making connections to a development board and to identify the connections made in a program. This requires knowledge of which external pins are available, and how each should be named in a program. The Mbed OS allocates names for the pins of each Mbed-enabled development board. The most important ones for our two introductory development boards appear in Figs. 2.1(c) and 2.6. For example, you use p4 for pin 4 when programming for the Mbed LPC1768, or D4 for the F401RE, or LED1 for either; this is automatically recognized by the OS as an alias for the pin to which it is connected. However, LED2 *is* recognized by the LPC1768, but not for the F401RE, while BUTTON1 is recognized for the F401RE, but not for the LPC1768. The full listing for any board can be found in its **PinNames.h** file, which in turn can be found in the **Mbed-os** folder. Fig. 3.3 shows an example location for the F401RE when in Mbed Studio.

It is worth emphasizing that Figs. 2.1(c) and 2.6 give a connection overview, rather than every detail of what is available. In reality there are other uses for each pin not given in the diagrams, that would make them too complex. For example, we can see that the pin named "p18" on the LPC1768 Mbed can be an analog input or analog output; it can also be a digital input/output, though this is not directly stated here. Similarly, the LED connections suggested in Fig. 3.1 for the F401RE rely on using pins labeled only as analog input for digital output.

Finally, in Figs. 2.1(c) and 2.6, it is also possible to see 0 V or ground pins on both devices, as well as 3.3 V and 5 V outputs, and VIN inputs for external voltage supply. These will be important for us.

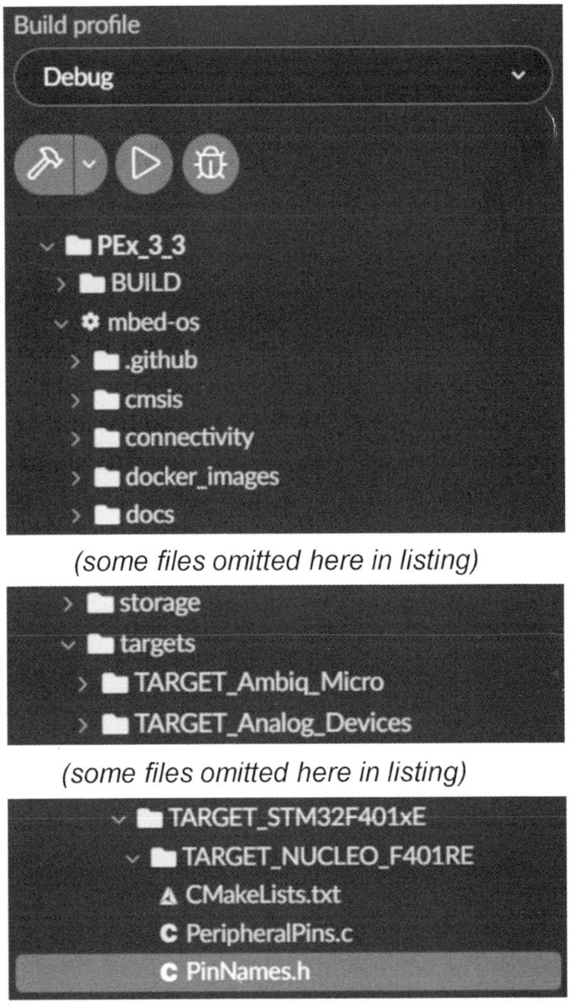

Figure 3.2
Locating the PinNames.h file in Mbed Studio. *Image courtesy of ARM.*

3.2 Voltages as Logic Values

We know that computers deal with binary numbers made up of lots of binary digits, or bits, and we know that each of these bits takes a logic value of 0 or 1. Now that we are starting to build hardware in earnest, it is worth thinking about how those logic values are actually represented inside the microcontroller electronic circuit, and at its connection pins.

In any digital circuit, logic values are represented as electrical voltages. Here is the *big* benefit of digital electronics: we do not need a precise voltage to represent a logic value.

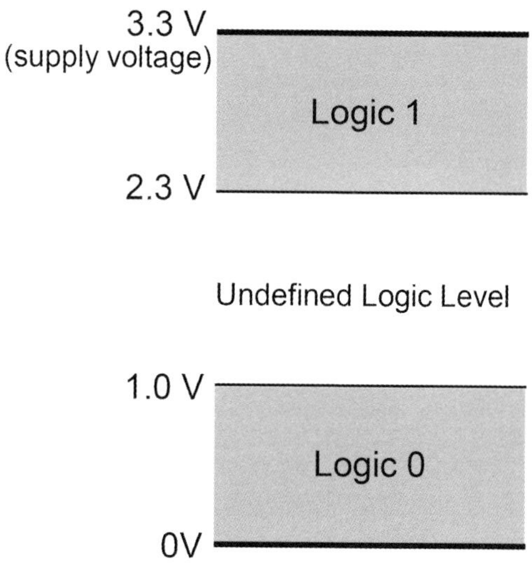

Figure 3.3: Input logic levels for the Mbed.

Instead, we accept a *range* of voltages as representing a logic value. This means that a voltage can pick up some noise or distortion and still be interpreted as the right logic value. For example, the LPC1768 microcontroller is powered from 3.3 V. We can find out which range of voltages it accepts as Logic 0, and which as Logic 1, by looking at its technical data. This is found in Reference 2.4, with important points summarized in Appendix C. (There will be moments when it will be useful to look at this Appendix, but don't rush to it now.) This data shows that, for most digital inputs, the LPC1768 interprets *any* input voltage below 1.0 V (specified as 0.3 x 3.3 V) as logic 0, and *any* input voltage above 2.3 V (specified as 0.7 x 3.3 V) as Logic 1. This idea is represented in the diagram of Fig. 3.3.

If we want to input a signal to the Mbed, hoping that it will be interpreted correctly as a logic value, then it will need to satisfy the requirements of Fig. 3.3. If we are outputting a signal from the Mbed, then we can expect it to comply with the same figure. The Mbed will normally output Logic 0 as 0 V, and Logic 1 as 3.3 V, as long as no electrical current is flowing. If current is flowing, for example into an LED, then we can expect some change in output voltage. We will return to this point. The neat thing is that when we output a logic value we are also getting a predictable voltage to make use of. We can use this to light LEDs, switch on motors, or many other things.

For many applications in this book we do not worry much about these voltages, as the circuits we build meet the requirements of Fig. 3.3. However, there are situations where it is necessary to take some care over logic voltage values. We mention them as they come up.

3.3 Digital Output

3.3.1 Using LEDs

We will now start connecting external devices to development boards more systematically. While you may not be an electronics specialist, it is important—whenever you connect anything to the board—that you have some understanding of what you are doing. The LED is a semiconductor diode, and behaves electrically as one. It will conduct current in one direction, sometimes called the *forward* direction, but not the other. What makes it so useful is that when it is connected so that it conducts, it emits photons from its semiconductor junction. The LED has the voltage/current characteristic shown in Fig. 3.4(a). A small forward voltage will cause very little current to flow. As the voltage increases there comes a point where the current suddenly starts flowing rather rapidly. For most LEDs this voltage is in the range shown, typically around 1.8 V.

Figs. 3.4(b) and 3.4(c) show circuits used to make direct connections of LEDs to the output of logic gates; for example, a microcontroller pin configured as an output. The gate here is shown as a logic buffer (the triangular symbol). If the connection of Fig. 3.4(b) is made, the LED lights when the gate is at logic high. Current then flows out of the gate into the LED. Alternatively, the circuit of Fig. 3.4(c) can be made; the LED now lights when the logic gate is at logic zero, with current flowing *into* the gate. Usually a current-limiting resistor needs to be connected in series with the LED to control how much current flows. The exception to this is if the combination of output voltage and gate internal resistance is such that the current is limited to a suitable value anyway. Some LEDs, such as the ones recommended for the next program, have a series resistance built into them. Therefore, they do not need any external resistor connected.

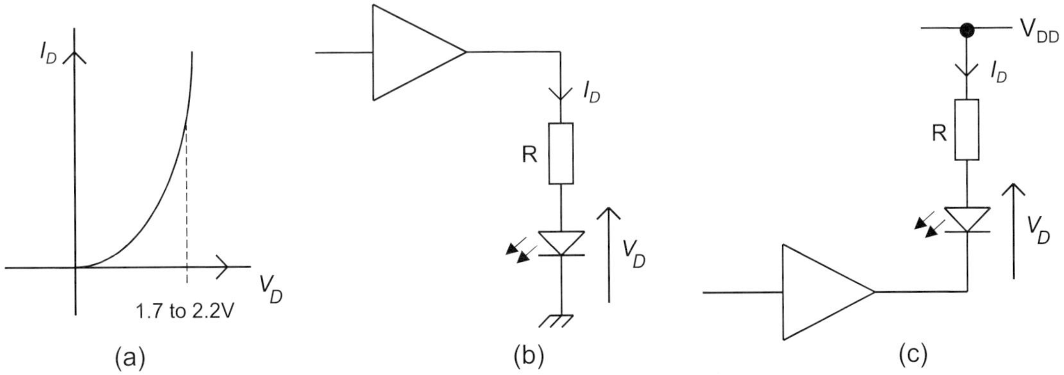

Key

I_D:	Current through LED	V_D:	Voltage across LED, for current I_D
V_S:	Supply voltage		

Figure 3.4

Driving LEDs from logic gates: (a) LED *V—I* characteristic; (b) Gate output sourcing current to LED load; (c) Gate output sinking current from LED.

LEDs now appear in all manner of shapes and sizes, the most familiar perhaps being the single LED, as seen in Fig. 3.1. Aside from having single LEDs like this, it is possible to place more than one diode within the same housing. If these are of different colors, then different light combinations can be obtained by lighting more than one of the diodes, and by varying their relative strengths. This is what happens in the bicolor LED on the F401RE, or the tricolor LED on the Mbed application board.

3.3.2 Using a Breadboard to Connect External Components

We have already connected two external LEDs to the F401RE board, as seen in Fig. 3.1. Here we start using an external prototyping breadboard to make external connections.

We will be applying the circuit of Fig. 3.5(a) for the Mbed LPC1768. Appendix D gives example part numbers for all parts used; equivalent devices can of course be implemented. Within this circuit we are applying the connection of Fig. 3.4(c). The particular LED recommended in Appendix D, however, has an internal series resistor of value around 240 Ω, so an external one is not required.

Place the Mbed LPC1768 in a breadboard, as seen in Fig. 3.5(c), and connect the circuit shown. Connect pin 1, the ground, across to one of the outer rows of connections of the breadboard, as seen in Fig. 3.5(b). You should adopt this as a habit for all similar circuits that you build. Remember to attach the anode of the LEDs (the side with the longer leg) to the Mbed pins. The negative side (cathode) should be connected to ground. For this and many circuits we will take power from the USB bus.

For the F401RE board, the LEDs shown in Fig. 3.1 can be retained and used. Alternatively, external breadboard connections can be made, as seen in Fig. 3.5(c).

Create a new project in your chosen Studio IDE, and copy across Program Example 3.2.

```
/*Program Example 3.2: Flashes red and green LEDs in simple time-based pattern
Works with LPC1768 or F401RE as target systems */
#include "Mbed.h"
//define and name digital outputs
DigitalOut redled(p5);      //Replace p5 with A0 for F401RE
DigitalOut greenled(p6);    //Replace p6 with A1 for F401RE

int main() {
   while(true) {
      redled = 1;
      greenled = 0;
      ThisThread::sleep_for(200ms);
      redled = 0;
      greenled = 1;
      ThisThread::sleep_for(200ms);
   }
}
```

Program Example 3.2 Flashing external LEDs

Compile, download, and run the code on your Mbed target. The code extends ideas already applied in Program Examples 2.1 and 3.1, so it should be easy to understand that the green and red LEDs are programmed to flash alternately. You should see this happen when the code runs.

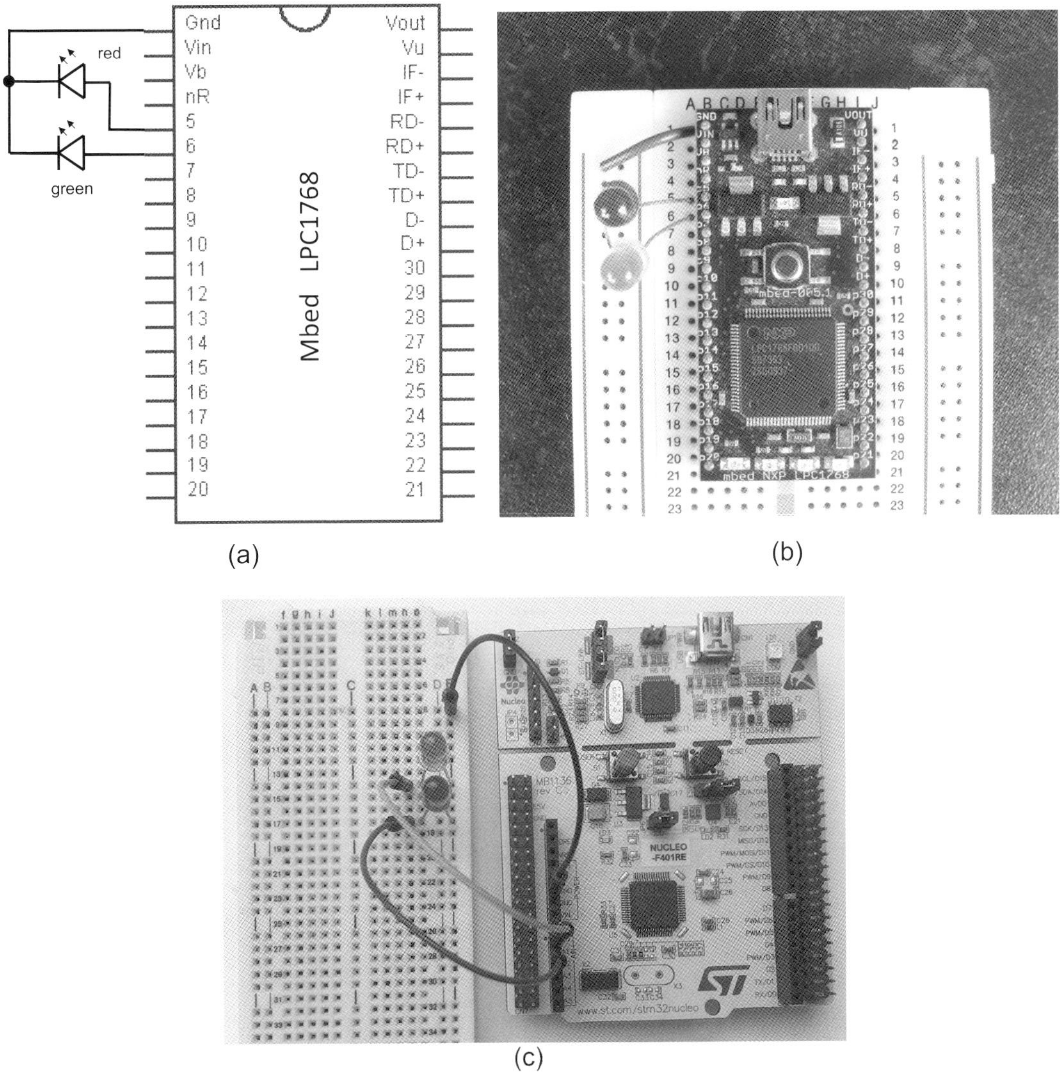

Figure 3.5

Simple connections to external components: (a) the connection diagram; (b) the breadboard build; (c) the Nucleo F401RE.

Exercise 3.3

Using any digital output pin, write a program which outputs a square wave, by switching the output repeatedly between Logic 1 and 0. Try different wait and sleep functions available in Table 3.1 to give a frequency of 100 Hz (i.e., a period of 10 ms). View the output on an oscilloscope. Measure the voltage values for Logic 0 and 1. How do they relate to Fig. 3.3? Does the square wave frequency agree with your programmed value?

3.4 Using Digital Inputs

3.4.1 Connecting Switches to a Digital System

We can use ordinary electromechanical switches to create logic levels, which will satisfy the logic level requirements seen in Fig. 3.3. Three commonly used ways are shown in Fig. 3.6. The simplest, Fig. 3.6(a), uses a SPDT (single pole, double throw) switch. This is what we will use in the next example. A resistor in series with the supply voltage is sometimes included in this circuit, as a precautionary measure. This is an effective and useful circuit. However, SPST (single pole, single throw) switches can be lower cost and smaller, and so are very widely used. They are connected to the supply voltage with a pull-up or pull-down resistor, as shown in Figs. 3.6(b) and 3.6(c). When the switch is open, the logic level is defined by the connection through the resistor. When it is closed, the switch asserts the other logic state. A wasted electrical current then flows through the resistor. This is kept small by selecting a high-value resistor. In many microcontrollers, pull-up and pull-down resistors are internally available, avoiding the need to make the external connection. Finally, as a practical example, the on-board user button on the

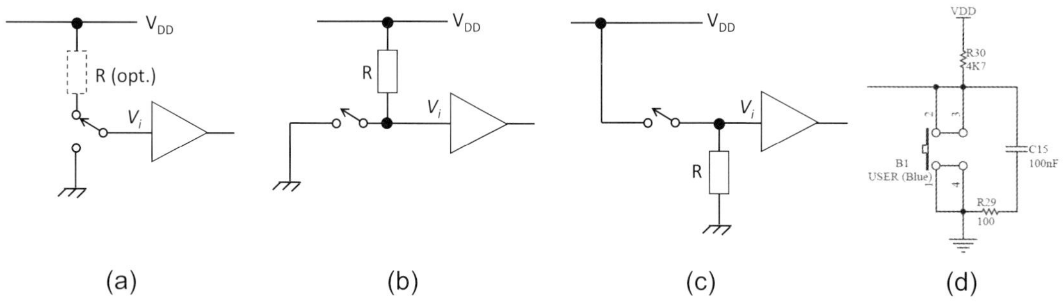

Figure 3.6
Connecting switches to logic inputs: (a) single-pole double-throw (SPDT) connection; (b) single-pole single-throw (SPST) connection with pull-up resistor; (c) SPST with pull-down resistor; (d) the user button on the Nucleo F401RE board. *Image courtesy of ST.*

F401RE board is shown in Fig. 3.6(d). This shows a pull-up resistor, equivalent to the circuit in 3.6(b). The extra capacitor and resistor are in place to minimize the effect of switch bounce.

3.4.2 The DigitalIn API

The Mbed API has the digital input functions listed in Table 3.3, with format identical to that of **DigitalOut,** which we have already seen. This section of the API creates a class called **DigitalIn,** with the member functions shown. The **DigitalIn** constructor can be used to create digital inputs, and the **read( )** function used to read the logical value of the input. In practice, the shorthand offered allows us to read input values through use of the digital object name. We will see this in the program example which follows. As with digital outputs, the same target board pins can be configured as digital inputs. Input voltages will be interpreted according to Fig. 3.3. Note that use of **DigitalIn** enables by default the internal pull-down resistor; that is, the input circuit is configured as Fig. 3.6(c). This can be disabled, or an internal pull-up enabled, using the **mode( )** function.

3.4.3 Using if to Respond to a Switch Input

We will now connect to our circuit a digital input, a switch, and use it to control LED states. In so doing, we make a significant step forward. For the first time, we will have a program which makes a decision based on an external variable, the switch position. This is the essence of many embedded systems. Program Example 3.3 is applied to achieve this.

C/C++ code feature

To make the decision within the program, we use the statement

```
if(switchinput==1).
```

Table 3.3: The Mbed digital input API summary.

Functions	Usage
Constructor	
`DigitalIn (PinName pin, PinMode mode)`	create a DigitalIn object of specified name, connected to the specified pin; input mode may also be specified
Member functions	
`int read()`	read the input, returned as 0 or 1
`void mode(PinMode)`	set the input pin mode, with parameter chosen from PullUp, PullDown, PullNone, OpenDrain
Mbed-defined `operator: int()`	shorthand for read ()

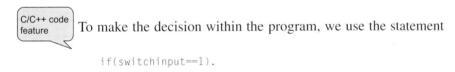

This line sees use of the C equal operator, == , for the first time. Read Section B.6.1 to review use of the **if** and **else** keywords. Using **if** in this way causes the line or block of code which follows it to execute, if the specified condition is met. In this case, the condition is that the variable **switchinput** is equal to 1. If the condition is not satisfied, then the code which follows **else** is executed. Looking now at the program, we can see that if the switch gives a value of logic 1, the green LED is switched off and the red LED is programmed to flash. If the switch input is 0, the **else** code block is invoked, and the roles of the LEDs reversed. Again, the **while (true)** statement is used to create an overall infinite loop, so the LEDs flash continuously.

```
/*Program Example 3.3: Flashes one of two LEDs, depending on the state of switch.
Works with either LPC1768 or F401RE using pins shown.
*/

#include "mbed.h"
DigitalOut redled(p5);       //Replace p5 with A0 for F401RE
DigitalOut greenled(p6);     //Replace p6 with A1 for F401RE
DigitalIn switchinput(p7);    //Replace p7 with BUTTON1 for F401RE

int main() {
  while(true) {
    if (switchinput == 1) {       //test value of switchinput
      //execute following block if switchinput is 1
      greenled = 0;            //green led is off
      redled = 1;              // flash red led
      ThisThread::sleep_for(200ms);
      redled = 0;
      ThisThread::sleep_for(200ms);
    }                          //end of if
    else {                     //here if switchinput is 0
      redled = 0;              //red led is off
      greenled = 1;            // flash green led
      ThisThread::sleep_for(200ms);
      greenled = 0;
      ThisThread::sleep_for(200ms);
    }                           //end of else
  }                             //end of while(true)
}                               //end of main
```

Program Example 3.3 Using *if* and *else* to respond to external switch

For the Mbed LPC1768, adjust the circuit of Fig. 3.5 to that of Fig. 3.7(a) by adding a SPDT switch as shown. The input it connects to is configured in the program as a digital input. The photograph of Fig. 3.7(b) shows the addition of the switch. If using the Nucleo F401RE, keep the connections of Fig. 3.1, and use the blue on-board push button switch—it is that simple! But remember that the switch on the F401RE board is SPST, so the two circuits are not identical. Create a new program, and copy across Program Example 3.3. Compile, download, and run the code on your target board.

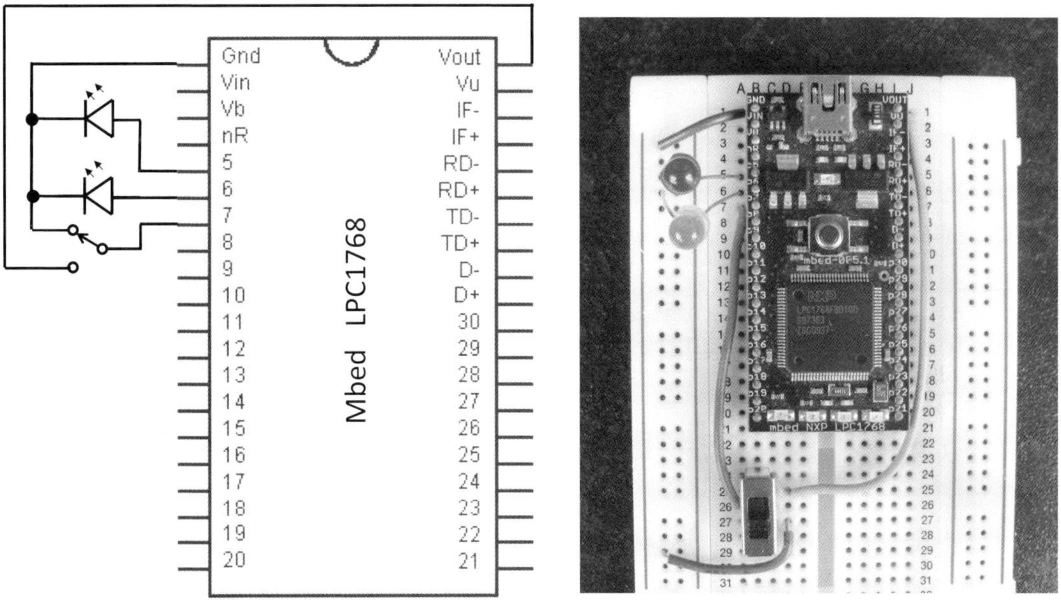

(a) (b)

Figure 3.7

Controlling the LED with a switch: (a) the LPC1768 connection diagram; (b) the breadboard build.

Exercise 3.4

Write a program which creates a square wave output, as in the previous exercise. Include now two possible frequencies, 100 Hz and 200 Hz, depending on an input switch position. Use the same switch connection you have just used. Observe the output on an oscilloscope.

3.5 Digital Input and Output with the Application Board

App Board

If you have an application board, you can now try applying the same program types to it. However, the application board has neither single switches available as inputs, nor single LEDs. Instead it has a joystick, which is a set of switches, and a tricolor LED. These are both symbolized in Fig. 3.8, with Mbed LPC1768 pin connections as on the app board. The joystick acts as a five-position multi-way switch, with the wiper connected to the supply voltage. The tricolor LED is of type CLV1A-FKB. It connects to external current-limiting resistors, which then connect to Mbed pins. The LED is a "common anode" type, which means the three internal LED anodes are connected

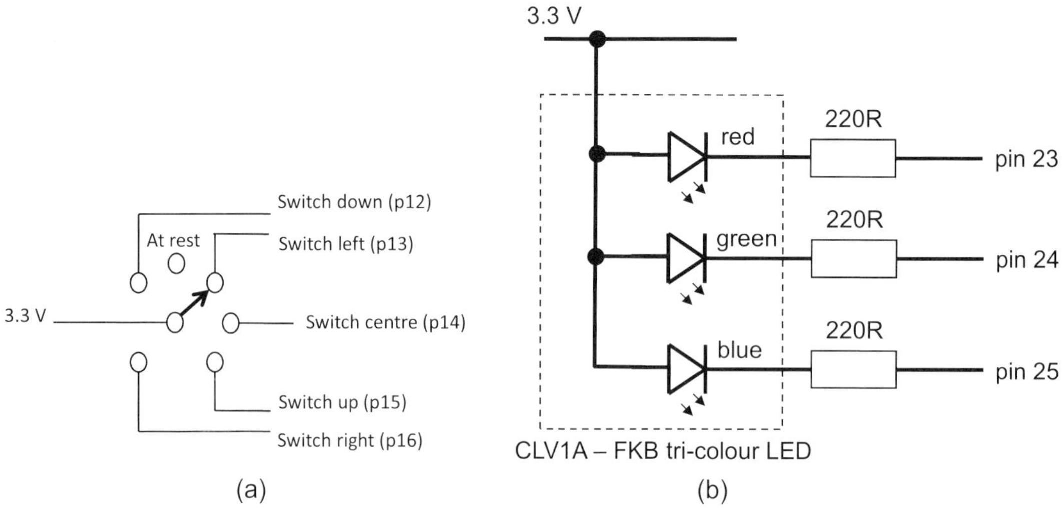

Figure 3.8
Digital input and output devices on the application board: (a) joystick; (b) tricolor LED.

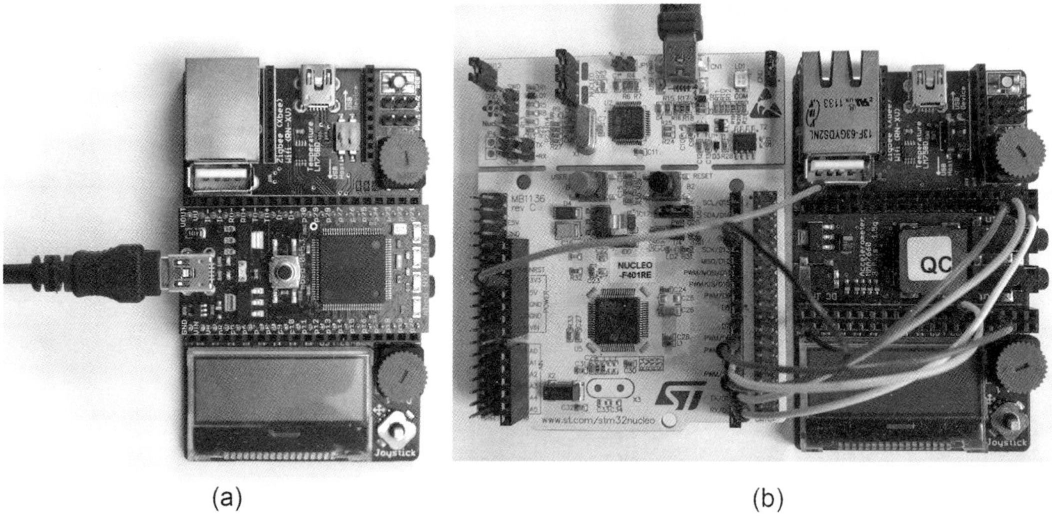

Figure 3.9
Use of the application board: (a) Mbed LPC1768 plugged into the application board; (b) F401RE connected to application board, for Program Example 3.4.

("commoned") together, and available externally as a single connection. It therefore requires a logic 0 on the Mbed pin to light the LED. Fig. 3.4(c) applies.

Plug an Mbed LPC1768 into the application board, as shown in Fig. 3.9(a) and link to the computer in the usual way. Make sure you connect the USB connector directly to the

Mbed, *not* to the application board USB connector. Connecting to the application board will power it and the Mbed, but will not allow the all-important connection to the Mbed program memory.

The application board can also be connected to the F401RE development board, as illustrated in Fig. 3.9(b). Connections are given in the program listing. The 3.3 V pins must be linked, that is, pin 40 of the empty LPC1768 socket to the 3.3 V pin of the F401RE, as well as the GND pins; the F401RE is now powering the app board.

Create a new project around Program Example 3.4, and download to your chosen Mbed target. Running the program should allow "left" and "right" movements of the joystick to light the tri-color LED red or green.

```
/*Program Example 3.4.
Uses Joystick values to switch tricolor led
Works with LPC1768 or F401RE connected to application board*/

#include "mbed.h"

//set up outputs to leds in tricolor led
DigitalOut redled(p23);   //Replace p23 with D2 for F401RE
DigitalOut greenled(p24); //Replace p24 with D3 for F401RE
DigitalOut blueled(p25);  //Replace p25 with D4 for F401RE
//set up joystick connections, left and right
DigitalIn joyleft(p13);   //Replace p13 with D5 for F401RE
DigitalIn joyright(p16);  //Replace p16 with D6 for F401RE
/*For F401RE board, connect the GND pins, and 3.3V output to app board pin 40.
Connect also the pins specified for each component, e.g. connect D2 to p23 of the
LPC1768 socket, D3 to p24 etc. */

int main() {
    greenled = redled = blueled = 1; //switch all leds off (logic 1 for off)
    redled = 0;          //switch red led on, diagnostic (logic 0 for on)
    ThisThread::sleep_for(1s);       //wait one second
    redled = 1;          //switch red led off
    while(true) {
      if (joyleft == 1) {     //test if the joystick is pushed left
        greenled = 0;         //switch green led on
        ThisThread::sleep_for(1s);       //wait one second
        greenled = 1;         //switch green led off
        }
      if (joyright == 1) {    //test if the joystick is pushed right
        redled = 0;           //switch red led on
        ThisThread::sleep_for(1s);
        redled = 1;           //switch red led off
        }
    }
}
```

Program Example 3.4 Controlling application board LED with the joystick

Exercise 3.5 (for app board)

Change Program Example 3.4 so that all directions of the joystick are used. Fig. 3.8(a) shows the five possible outputs from the joystick. As there are only three LEDs, you can make combinations of colors for two of the inputs.

Now that we're beginning to use more digital inputs and outputs, it is useful to be able to group them, so that we can read or switch whole groups at the same time. The Mbed API allows this, with two useful utilities, **BusIn** and **BusOut**. The first allows you to group a set of digital inputs into one bus, so that you can read a digital word direct from it. The equivalent output is **BusOut**. Applying either simply requires you to specify a name, and then list in brackets the pins which will be members of that bus, least significant bit first. This is illustrated in Program Example 3.5, taken from the Mbed website, where **BusIn** is now used for the left, right, up, and down switches of the joystick, and **BusOut** is used to group the four Mbed on-board LEDs. The center joystick switch gets its own digital input, labeled "fire."

Try running this program. If "fire" is pressed, then the hexadecimal number 0xf is transferred to the LEDs. Remember from Appendix A that "0x" is the identifier which specifies a hexadecimal number, which in this case is 15 in decimal, or 1111 in binary. Otherwise the value of the "joy" **BusIn** is simply transferred to the "leds" **BusOut.** As this program requires four LEDs, our little workaround of Fig. 3.1 will not be sufficient for the F401RE. However, it is easy to place LEDs on a breadboard, and to convert the program for F401RE use. Connections can be chosen for both digital input and output, following the pattern of the previous Program Example.

```
/*Program Example 3.5: Transfers the value of the joystick to target LEDs
Works with LPC1768 in application board, or F401RE with appropriate connections*/

#include "mbed.h"

BusIn joystick(p15,p12,p13,p16);   //Select own pins for F401RE (any digital I/O)
DigitalIn fire(p14);
BusOut leds(LED1,LED2,LED3,LED4); //Select own pins for F401RE (any digital I/O)

int main(){
  while(true){
    if (fire) {
      leds = 0xf;
    }
```

```
    else {
      leds = joystick;
      }
    ThisThread::sleep_for(100ms);
  }
}
```

Program Example 3.5 Controlling target LEDs from the app board joystick

3.6 Interfacing Simple Opto Devices

Given the ability to input and output bits of data, a wide range of possibilities are opened up. Many simple sensors can interface directly with digital inputs. Others have their own designed-in interfaces, which produce a digital output. In this section we look at some simple and traditional sensors and displays which can be interfaced directly to an Mbed development board. In later chapters we take this further, connecting to some very new and hi-tech devices.

3.6.1 Opto-reflective and Transmissive Sensors

Opto-sensors, such as we see in Figs. 3.10(a) and 3.10(b), are simple examples of sensors with outputs which can satisfy the logic level requirements of Fig. 3.3. When a light falls on the base of an opto-transistor, it conducts; when there is no light, it does not. In the reflective sensor, Fig. 3.10(a), an infrared LED is mounted in the same package as an opto-transistor. When a reflective surface is placed in front, the light bounces back, and the transistor can conduct. In the transmissive sensor, Fig. 3.10(b), the LED is mounted opposite the transistor. With nothing in the way, light from the LED falls directly on the

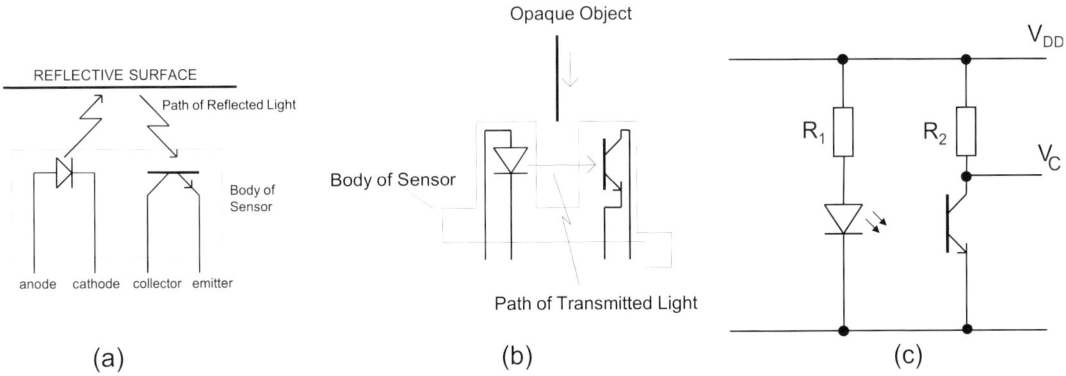

Fig 3.10: Simple opto-sensors: (a) the reflective opto-sensor; (b) the transmissive opto-sensor; (c) simple drive circuit for opto-sensor.

transistor, and so it can conduct. When something comes in the way, the light is blocked and the transistor stops conducting. This sensor is also sometimes called a slotted opto-sensor or photo-interrupter. Each sensor can be used to detect certain types of objects.

Either of the sensors shown can be connected in the circuit of Fig. 3.10(c). Here the resistor R_1 is calculated to control the current flowing in the LED, taking suitable values from the sensor data sheet. The resistor R_2 is chosen to allow a suitable output voltage swing, invoking the voltage thresholds indicated in Fig. 3.3. When light falls on the transistor base, current can flow through the transistor. The value of R_2 is chosen so that with that current flowing, the transistor collector voltage V_C falls almost to 0 V. If no current flows, then V_C rises to V_{DD}. In general, the sensor is made more sensitive by either decreasing R_1, or increasing R_2.

3.6.2 Connecting an Opto-sensor to an Mbed Development Board

Fig. 3.11 shows how a transmissive opto-sensor can be connected to an Mbed board. The device used is a KTIR0621DS, made by Kingbright; similar devices can be used. The particular sensor used has pins which can plug directly into the breadboard. Take care when connecting these four pins. Connections are indicated on the housing; alternatively, you can look these up in the data which Kingbright provides. This can be obtained from the Kingbright website, Reference 3.2, or from the supplier you use to buy the sensor.

Program Example 3.6 controls this circuit. The output of the sensor is configured in the program as a digital input. When there is no object in the sensor, light falls on the photo-transistor. It conducts, and the sensor output is at Logic 0. When the beam is

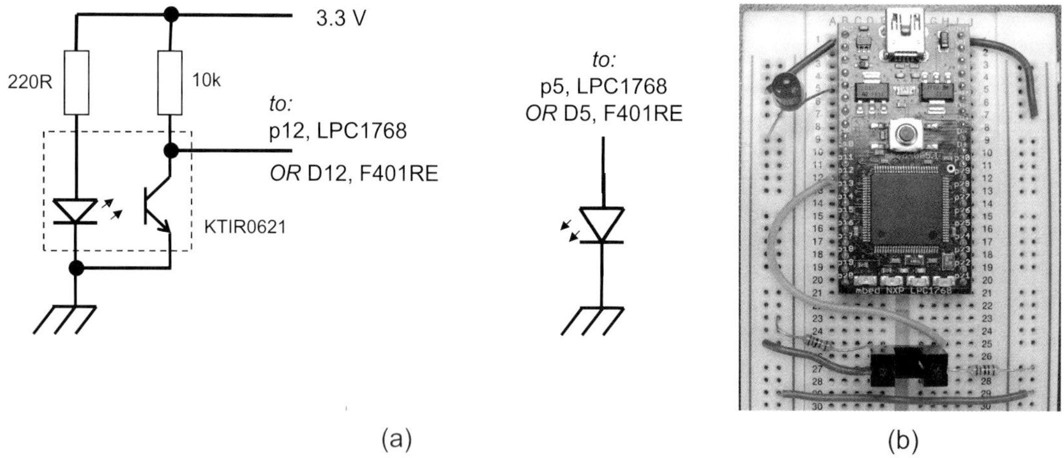

Fig. 3.11: Mbed connected to a transmissive opto-sensor: (a) the connection diagram; (b) breadboard build for Mbed LPC1768.

interrupted, then the output is at Logic 1. Therefore, the program switches the LED on when the beam is interrupted, that is, an object has been sensed. To make the selection, we use the **if** and **else** keywords, as in the previous example. Now, there is just one line of code to be executed for either state. It is not necessary to use braces to contain these single lines.

```
/*Program Example 3.6: Simple program to test KTIR slotted opto-sensor. Switches an
LED according to state of sensor
Works with LPC1768 or F401RE, using connections indicated*/

#include "mbed.h"
DigitalOut redled(p5);          //Replace p5 with D5 for F401RE
DigitalIn opto_switch(p12);     //Replace p12 with D12 for F401RE

int main() {
  while(true) {
    if (opto_switch == 1)       //input = 1 if beam interrupted
     redled = 1;                //switch led on if beam interrupted

    else
     redled = 0;                //led off if no interruption
  }                             //end of while
}
```

Program Example 3.6 Applying the transmissive opto-sensor

3.6.3 Seven-Segment Displays

We have now used single LEDs in several programs, and the application board tri-color LED. We often also find LEDs packaged together to form patterns, digits, or other types of display. There are a number of standard configurations which are very widely used; these include bar graph, seven-segment display, dot matrix, and "star-burst."

The seven-segment display is a particularly versatile configuration. An example single digit, such as is made by Kingbright (Reference 3.2), is shown in Fig. 3.12. By lighting different combinations of the seven segments, all numerical digits can be displayed, as well as a surprising number of alphabetic characters. A decimal point is usually included, as shown. This means that there are eight LEDs in the display, needing 16 connections. To simplify matters, either all LED anodes are connected together, or all LED cathodes. This is seen in Fig. 3.12(b); the two possible connection patterns are called *common cathode* and *common anode*. Now instead of 16 connections being needed, there are only nine, one

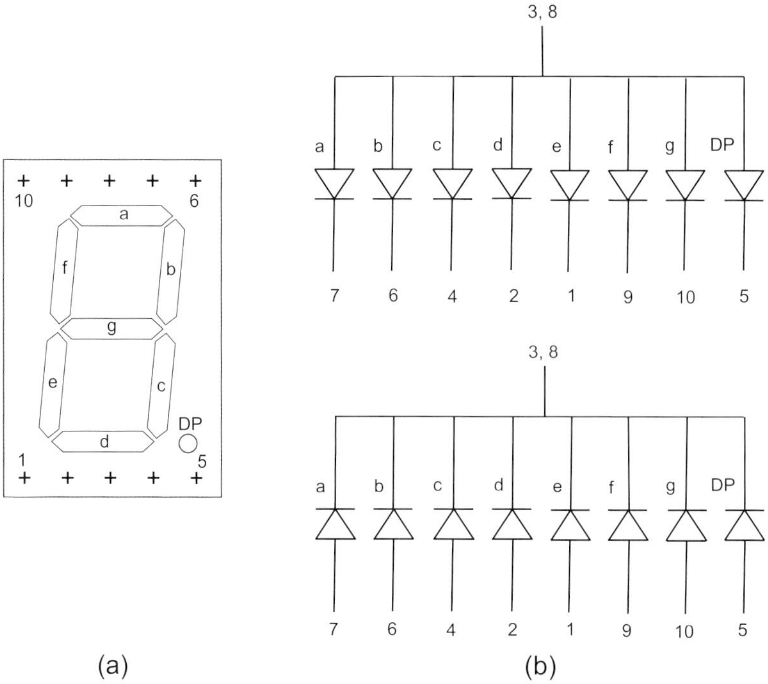

Figure 3.12
The seven-segment display: (a) a seven-segment digit (Kingbright, 12.7 mm); (b) electrical connection (upper, common anode; lower, common cathode). *Image courtesy of Kingbright Elec. Co.*

for each LED and one for the common connection. The actual pin connections in the example shown lie in two rows, at the top and bottom of the digit. There are 10 pins in all, with the common anode or cathode taking two pins.

A small seven-segment display as seen in Fig. 3.12 can be driven directly from a microcontroller. In the case of common cathode, the cathode is connected to ground, and each segment is connected to a port pin. If the segments are connected in this sequence to form a byte,

(MSB) DP g f e d c b a (LSB)

then the values shown in Table 3.4 apply. For example, if 0 is to be displayed, then all outer segments, that is, abcdef, must be lit, with the corresponding bits from the microcontroller set to 1. If 1 is to be displayed, then only segments b and c need to be lit. Note that larger displays have several LEDs connected in series for each segment. In this case a higher voltage is needed to drive each series combination and, depending on the supply voltage of the microcontroller, it may not be possible to drive the display directly.

Table 3.4: Example seven-segment display control values.

Display Value	0	1	2	3	4	5	6	7	8	9
Segment Drive (B) (MSB) (LSB)	0011 1111	0000 0110	0101 1011	0100 1111	0110 0110	0110 1101	0111 1101	0000 0111	0111 1111	0110 1111
Segment Drive (B) (hex)	0x3F	0x06	0x5B	0x4F	0x66	0x6D	0x7D	0x07	0x7F	0x6F
Actual Display										

3.6.4 Connecting a Seven-Segment Display to the Mbed Target

As we know, microcontrollers on either Mbed board we are using run from a 3.3 V supply. Looking at the data sheet of our display (accessed from Reference 3.2), we find that each LED requires around 1.8 V across it to light. This is within the capability of both LPC1768 and F401RE boards. However, if we had two LEDs in series in each segment, the Mbeds would barely be able to switch them into conduction. But can we just connect the Mbed output directly to the segment, or do we need a current-limiting resistor, as seen in Fig. 3.4(b)? Looking at the LPC1768 data summarized in Appendix C, we see that the output voltage of a port pin drops around 0.4 V when 4 mA is flowing. This implies an output resistance of 100 Ω. (Let's not get into the electronics of all of this; suffice it to say, this value is valid but approximate, and applies only in this region of operation.) Applying Ohm's Law, the current flow in an LED connected directly to a port pin of this type is given by:

$$I_D \cong \frac{3.3 - 1.8}{100} = 15 \text{ mA} \tag{3.1}$$

This current, 15 mA, will light the segments very brightly, but is acceptable. It can be reduced, for example for power-conscious applications, by inserting a resistor in series with each segment.

Connect a seven-segment display to an Mbed target, using one of the circuits of Fig. 3.13. In this simple application, the common cathode is connected directly to ground, and each segment is connected to one Mbed output.

The circuit can be driven by Program Example 3.7, which applies the **BusOut** Mbed API class.

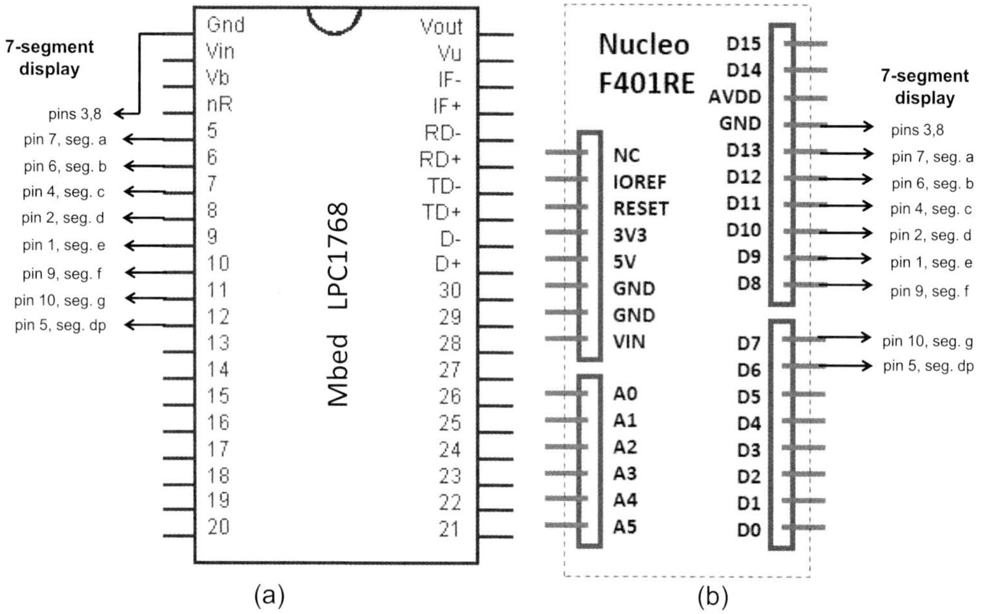

Figure 3.13

Mbed connections to a common cathode seven-segment display: (a) Mbed LPC1768; (b) Nucleo F401RE.

C/C++ code feature

Within this program we see for the first time a **for** loop. This is an alternative to **while**, for creating a conditional loop. Review its format in Section B.7.2. In this example, the variable **i** is initially set to 0, and on each iteration of the loop it is incremented by 1. The new value of **i** is applied within the loop. When **i** reaches 4, the loop terminates. However, as the **for** loop is the only code within the endless **while** loop, it simply starts again.

C/C++ code feature

The program goes on to apply the **switch, case,** and **break** keywords. Used together, these words provide a mechanism to allow one item to be chosen from a list, as described in Section B.6.2. In this case, as the variable **i** is incremented, its value is used to select the word that must be sent to the display in order for the required digit to illuminate. This method of choosing one value from a list is one way of achieving a *look-up table,* which is an important programming technique.

There are now quite a few nested blocks of code in this example. The **switch** block lies within the **for** block, which lies inside the **while** block, which finally lies within the **main( )** block! The closing brace for each is commented in the program listing. In writing more complex C programs, it becomes very important to ensure that each code block is terminated with a closing brace, at the right place.

```
/*Program Example 3.7: Simple demonstration of 7-segment display. Displays digits
0, 1, 2, 3 in turn.
Works for LPC1768 and F401RE, using connections indicated
                                                                              */
#include "mbed.h"
//Configure segments a,b,c,d,e,f,g,dp
//Pin listing for BusOut goes from lsb to msb, so lsb is pin 5
BusOut display(p5,p6,p7,p8,p9,p10,p11,p12);
                                  //D13,D12,D11,D10,D9,D8,D7,D6 for F401RE

int main() {
  while(true) {
    for(int i = 0; i < 4; i++) {
      switch (i){
        case 0: display = 0x3F; break;         //display 0
        case 1: display = 0x06; break;         //display 1
        case 2: display = 0x5B; break;
        case 3: display = 0x4F; break;
      }                                        //end of switch
      ThisThread::sleep_for(500ms);
    }                                          //end of for
  }                                            //end of while
}                                              //end of main
```

Program Example 3.7 Demonstrating a seven-segment display

Compile, download, and run the program. The display should appear similar to Fig. 3.14. Notice carefully, however, by looking at where pin 8 is connected, that this is a common anode connection. It does not exactly replicate Fig. 3.13(a). Pin 3 has been left unconnected.

Exercise 3.6

Adapt Program Example 3.7 to flash the letters H E L P in turn on a seven-segment display.

3.7 Switching Larger DC Loads

3.7.1 Applying Transistor Switching

A microcontroller can normally drive low-power, low-voltage DC loads directly with its digital I/O pins, as we saw with the LEDs. For example, the Mbed LPC1768 summary data (Appendix C) tells us that a port pin can source up to around 40 mA. However, this is a short-circuit current, so we are unlikely to be able to benefit from it with an actual electrical load connected to the port.

Figure 3.14
Mbed LPC1768 connected to a common anode seven-segment display.

If it is necessary to drive a load—say a motor—which needs more current than a microcontroller port pin can supply, or one which needs to run from a higher voltage, then an interface circuit will be needed. Three possibilities, which allow DC loads to be switched, are shown in Fig. 3.15. Each has an input labeled V_L, which is the logic voltage supplied from the port pin. The first two circuits show how a resistive load, like a heater, can be switched, using a bipolar transistor and MOSFET (metal oxide semiconductor field effect transistor—a mouthful, but a hugely important device in today's electronics), respectively. In the case of the bipolar transistor, the simple formula shown can be applied to calculate R_B, starting with knowledge of the required load current, and the value of the current gain (β) of the transistor. In the case of the MOSFET, there is a threshold gate-to-source voltage, above which the transistor switches on. This is a particularly useful configuration, as the MOSFET gate can so readily be driven by a microcontroller port bit output.

In Fig. 3.15C we see an inductive load, like a solenoid or DC motor, being switched. An important addition here is the *freewheeling diode*. This is needed because any inductance with current flowing in it stores energy in the magnetic field which surrounds it. When that current is interrupted, in this case by the transistor being switched off, the energy has to be returned to the circuit. This happens through the diode, which allows decaying

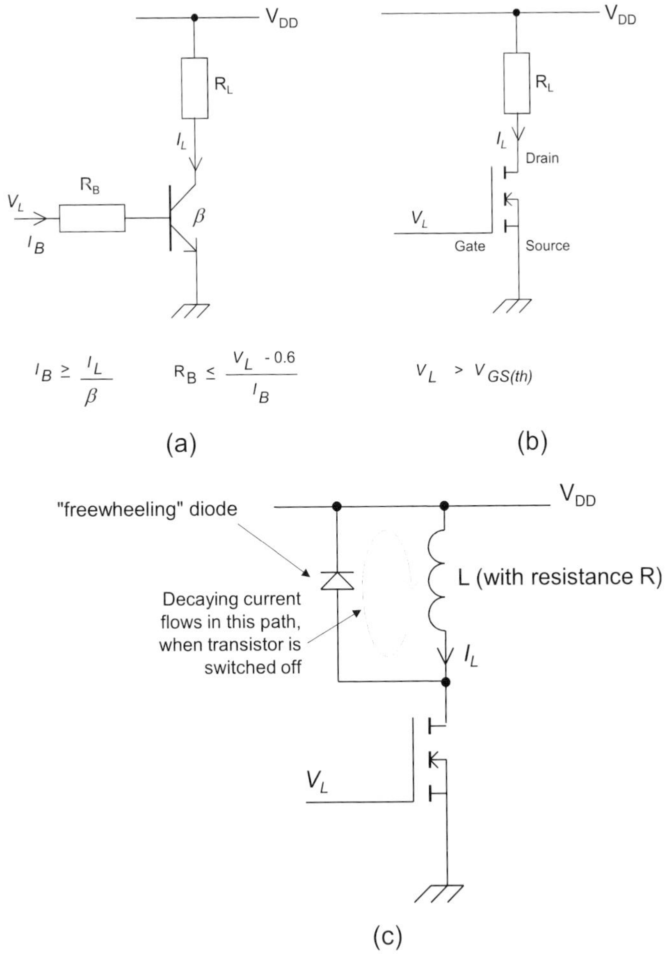

Figure 3.15

Transistor switching of DC loads: (a) resistive load, npn transistor; (b) resistive load, n-channel MOSFET; (c) inductive load, n-channel MOSFET.

current to continue to circulate. If the diode is not included, then a high voltage transient occurs, which can/will destroy the FET.

3.7.2 Switching a Motor with the Mbed

A good switching transistor for small DC loads is the ZVN4206A, whose main characteristics are listed in Table 3.5. An important value for Mbed interfacing use is the maximum V_{GS} threshold value, shown as 3 V. This means that the MOSFET will respond just to the 3.3 V Logic 1 output level of the Mbed LPC1768 or F401RE.

Table 3.5: Characteristics of the ZVN4206A n-channel MOSFET.

Characteristic	ZVN4206A
Maximum Drain-source Voltage V_{DS}	60 V
Maximum Gate-source Threshold $V_{GS(th)}$	3 V
Maximum Drain-source Resistance when 'On'. $R_{DS(on)}$	1.5 Ω
Maximum Continuous Drain current I_D	600 mA
Maximum Power Dissipation	0.7 W
Input Capacitance	100 pF

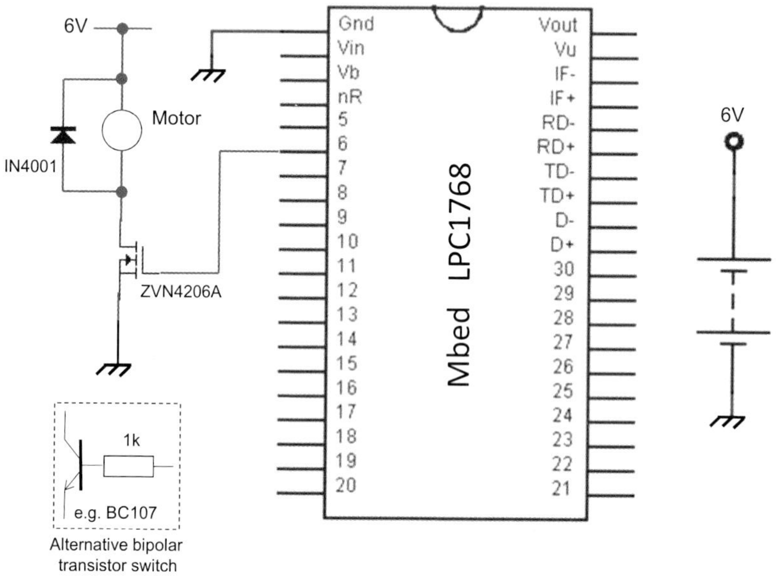

Figure 3.16
Switching a small DC motor. *Note:* For F401RE replace pin 6 of the LPC1768 with any digital output.

Exercise 3.7

Connect the circuit of Fig. 3.16, using a 6 V (or thereabouts) DC motor. For the F401RE, replace pin 6 with any available digital output. The 6 V for the motor can be supplied from an external battery pack or bench power supply. Its exact voltage is noncritical and depends on the motor you are using. Write a program so that the motor switches on and off continuously, say 1 s on, 1 s off. Increase the frequency of the switching until you can no longer detect that it is switching on and off. How does the motor speed compare with when the motor was just left switched on?

■

3.7.3 Switching Multiple Seven-segment Displays

We saw earlier in the chapter how a single seven-segment display could be connected to an Mbed development board. Each display required eight connections, and if we wanted many displays, then we would quickly run out of I/O pins. There is a very useful technique for getting around this problem, shown in Fig. 3.17. Each segment type on each display is wired together, as shown, and connected back to a microcontroller pin

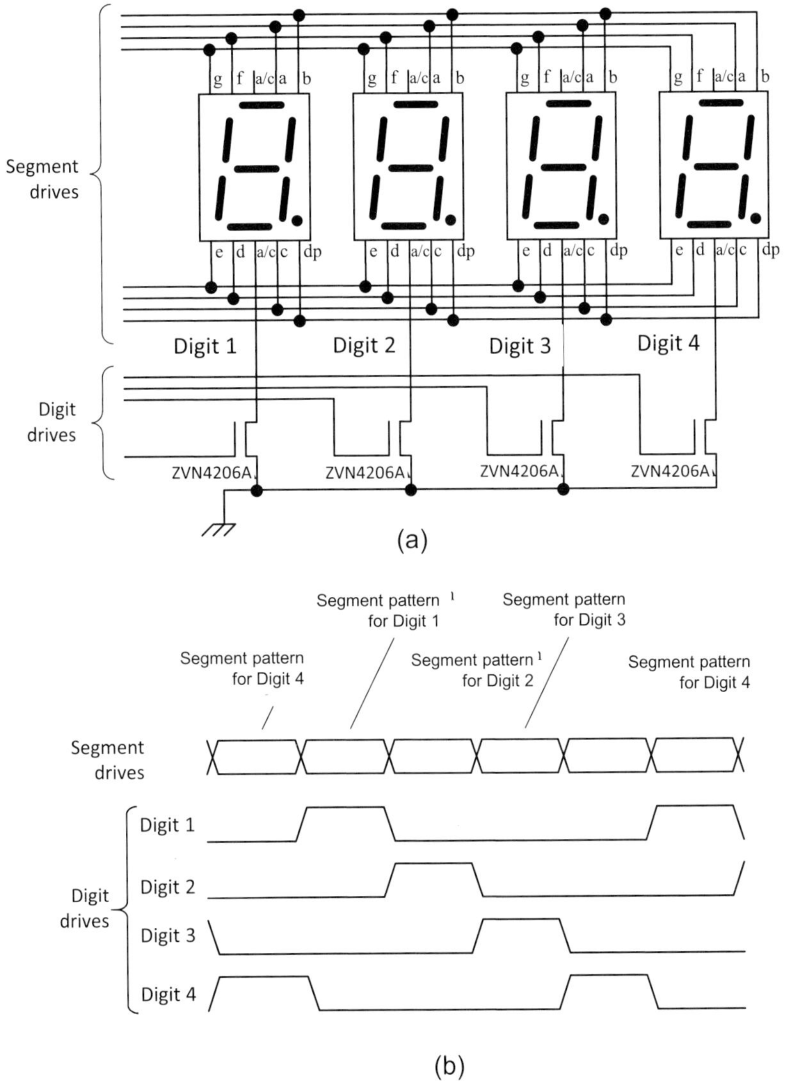

Figure 3.17
Multiplexing seven-segment displays: (a) display connections; (b) timing diagram.

configured as a digital output. The common cathode of each digit is then connected to its own drive MOSFET. The timing diagram shown then applies. The segment drives are configured for Digit 1, and that digit's drive transistor is activated, illuminating the digit. A moment later the segment drives are configured for Digit 2, and that digit's drive transistor is activated. This continues endlessly with each digit in turn. If it is done fast enough, then the human eye perceives all digits as being continuously illuminated; a useful rate is for each digit to be illuminated in turn is for around 5 ms.

3.8 Mini-project: Letter Counter

Use the slotted opto-sensor, push-button switch, and one seven-segment LED display to create a simple letter counter. Increment the number on the display by one every time a letter passes the sensor. Clear the display when the push button is pressed. Use an LED to create an extra "half" digit, so you can count from zero to 19.

Applying the circuit of Fig. 3.17(a), extend the letter counter above to have a three-digit display, counting up to 999.

Chapter Review

- Logic signals, expressed mathematically as 0 or 1, are represented in digital electronic circuits as voltages. One range of voltages represents 0, another represents 1.
- Mbed-enabled development boards have many digital I/O pins, which can be configured either as input or output.
- LEDs can be driven directly from the microcontroller digital outputs. They are a useful means of displaying a logic value and of contributing to a simple human interface.
- Electromechanical switches can be connected to provide logic values to digital inputs.
- Multi-colored LEDs can be made by enclosing several individual LEDs of different colors in the same housing. The Mbed application board contains one of these.
- A range of simple opto-sensors can be configured to connect directly to development board pins.
- Where the microcontroller pin cannot provide enough power to drive an electrical load directly, interface circuits must be used. For simple on/off switching, a transistor is often all that is needed.

Quiz

1. Complete Table 3.6, converting between the different number types. The first row of numbers is an example.

Table 3.6

Binary	Hexadecimal	Decimal
0101 1110	5E	94
1101		
	77	
		129
	6F2	
1101 1100 1001		
		4096

2. Is it possible to display unambiguously all of the capitalized alphabet characters A, B, C, D, E, and F on the seven-segment display shown in Fig. 3.12? For those that can usefully be displayed, determine the segment drive values. Use the connections of Table 3.4, and give your answers in both binary and hexadecimal formats.
3. A loop in an Mbed program is untidily coded as follows:

```
while(true){
  redled = 0;
  ThisThread::sleep_for(12ms);
  greenled = 1;
  ThisThread::sleep_for(2s);
  greenled = 0;
  wait_us(24000);
}
```

   What is the total period of the loop, expressed in seconds, milliseconds, and microseconds?
4. A friend enters the code shown below into Mbed Studio, but when it is compiled a number of errors are flagged. Find and correct the faults.

```
#include " Mbed"
Digital Out myled(LED1);
int main {
    while(1) {
      myled = 1;
      wait_us(200000)
      myled = 0;
      wait(0.5);
}
```

5. The circuit of Fig. 3.6(b) is used eight times over to connect eight switches to eight development board digital inputs. The pull-up resistors have a value of 10 kΩ and are connected to the 3.3 V supply. What current is consumed due to this circuit

configuration when all switches are closed simultaneously? If this current drain must be limited to 0.5 mA, to what value must the pull-up resistors be increased? Reflect on the possible impact of pull-up resistors in low-power circuits.

6. Estimate the current taken by the display connected in Fig. 3.13(a) or 3.13(b), when the digit 3 is showing.
7. If in Fig. 3.13(a) a segment current of approximately 4 mA was required, what value of resistor would need to be introduced in series with each segment?
8. A student builds an Mbed-based system. To one port he connects the circuit of Fig. 3.18(a), using LEDs of the type used in Fig. 3.5, but is then disappointed that the LEDs do not appear to light when expected. Explain why this is so, and suggest a way of changing the circuit to give the required behaviour.

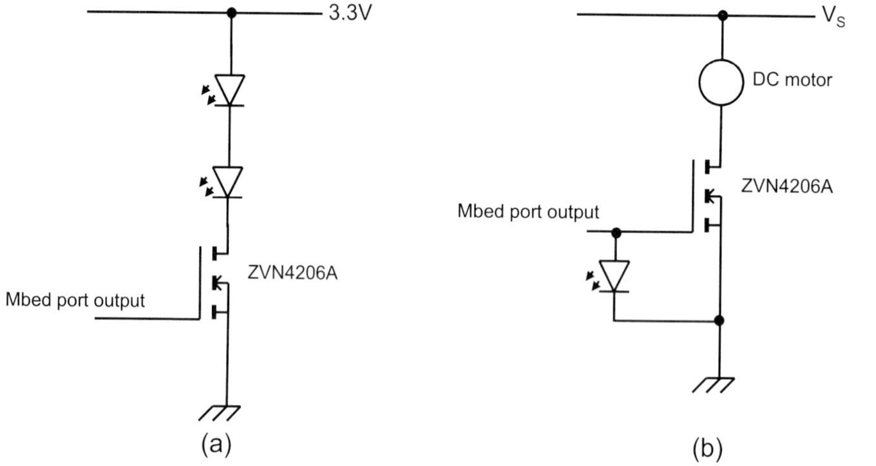

Figure 3.18
Circuits you should not build!

9. Another student wants to control a DC motor from the Mbed, and therefore builds the circuit of Fig. 3.18(b), where V_S is a DC supply of appropriate value. As a further indication that the motor is running, she connects a standard LED, as seen in Fig. 3.4, directly to the port bit. She then complains that the motor does not switch. Explain why this is, and any changes that should be made to the circuit.
10. a. Look at the Mbed LPC1768 circuit diagram, Reference 2.2, and find the four on-board LEDs. Estimate the current each one takes when "on," assuming a forward voltage of 1.8 V.
    b. What current flows in one of the LEDs in the tri-color application board, as seen in the circuit of Fig. 3.8(b), when lit?

References

3.1. Horowitz P, Hill W. *The Art of Electronics*. 3rd ed. Cambridge University Press; 2016.
3.2. The Kingbright website. http://www.kingbright.com/.

CHAPTER 4

Analog Output and Pulse Width Modulation

4.1 Introducing Data Conversion

Microcontrollers, as we know, are digital devices; but they spend most of their time dealing with an analog world. For example, to make sense of incoming analog signals from a microphone or temperature sensor, they must be able to convert them into digital form. After processing the data, they may then need to convert digital data back to analog form, (e.g., to drive a loudspeaker or DC motor). We give these processes the global heading "data conversion." Techniques applied in data conversion form a huge and fascinating branch of electronics. They are outside the subject matter of this book; however, if you wish to learn more about them, take a look at Chapter 13 of *The Art of Electronics,* Reference 3.1 (for example).

While conversion in both directions between digital and analog is necessary, it is conversion from analog to digital form that is more challenging; therefore, in this earlier chapter, we consider the easier option—conversion from digital to analog.

4.1.1 The Digital-to-Analog Converter

A digital-to-analog converter (DAC) is a circuit which converts a binary input number into an analog output. The actual circuitry inside a DAC is complex, and need not concern us. We can, however, represent the DAC as a block diagram, as in Fig. 4.1. This has a digital input, represented by D, and an analog output, represented by V_o. The yardstick by which the DAC calculates its output voltage is a *voltage reference* (a precise, stable, and known voltage).

Most DACs have a simple relationship between their digital input and analog output, with many (including the one inside the LPC1768) applying Eq. 4.1. Here, V_r is the value of the voltage reference, D is the decimal value of the binary input word, n is the number of bits in that word, and V_o is the output voltage. Fig. 4.2 shows this equation represented graphically. For each input digital value, there is a corresponding analog output. It's as if we are creating a voltage staircase with the digital inputs. The number of possible output values is given by 2^n, and the step size by $V_r / 2^n$; this is called the *resolution.* The maximum possible output value occurs when D reaches its maximum value (2^n - 1). So an

Fast and Effective Embedded Systems Design. https://doi.org/10.1016/B978-0-323-95197-5.00004-8

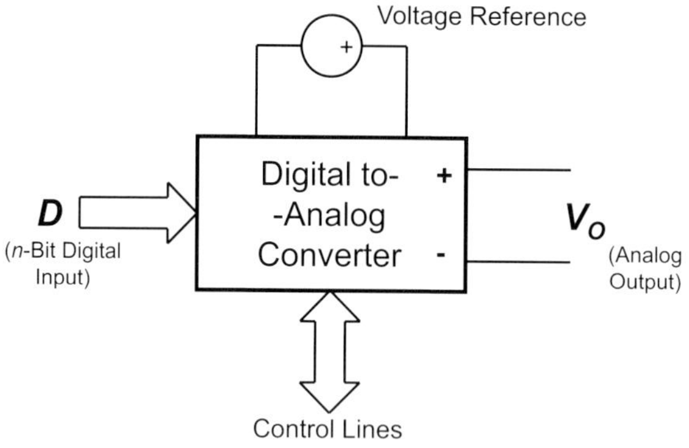

Figure 4.1
A digital-to-analog converter.

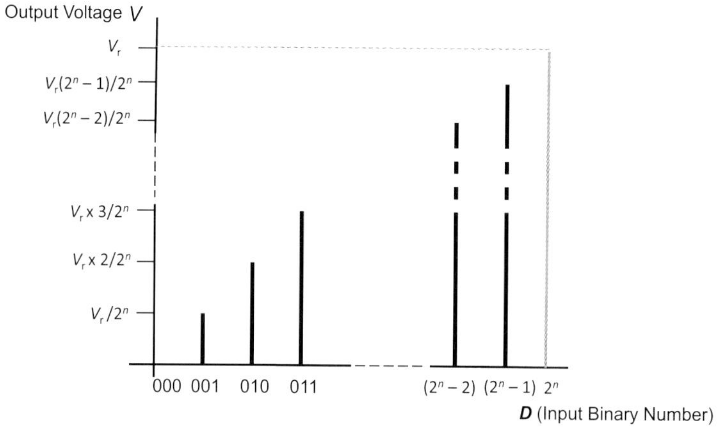

Figure 4.2
The DAC input/output characteristic.

output value of V_r is never quite reached. For interest, the figure also shows the theoretical case of D equaling 2^n, which would lead to an output value of V_r. The *range* of the DAC is the difference between its maximum and minimum output values. For example, a 6-bit DAC will have 64 possible output values; if it has a 3.2 V reference, it will have a resolution (step size) of 50 mV.

$$V_0 = \frac{D}{2^n} V_r \tag{4.1}$$

In Fig. 2.3, the LPC1768 block diagram, we see that the microcontroller has a 10-bit DAC; there will therefore be 2^{10} steps in its output characteristic, that is 1024. Reference 2.4 further tells us that it normally uses its own power supply voltage, 3.3 V, as voltage reference. The step size, or resolution, will therefore be 3.3/1024, that is, 3.22 mV. We note here that the F401RE does not have a DAC, so the sections which follow can only relate to the Mbed LPC1768, of our two introductory development board examples.

4.2 Analog Outputs on the Mbed LPC1768

The Mbed pin connection diagram, Fig. 2.1(c), has already shown us that this board is rich in analog input/output capabilities. We see that pins 15–20 can be analog input, while pin 18 is the only analog output.

The Application Program Interface (API) summary for analog output is shown in Table 4.1. It follows a pattern similar to the digital input and output utilities we have already seen. Here, with **AnalogOut,** we can initialize and name an output. Using **write( )** or **write_u16( )**, we can set the output voltage either with a floating point number, or with a hexadecimal number. Finally, we can simply use the = sign as a shorthand for write, which is what we will mostly be doing.

C/C++ code feature

Notice in Table 4.1 the way that data types *float*, and *unsigned short* value are invoked. In C/C++ all data elements have to be declared before use; the same is true for the return type of a function, and the parameters it takes. A review of the concept of *floating point* number representation can be found in Appendix A. Table B.4 in Appendix B summarizes the different data types that are available. The DAC itself requires an unsigned binary number as input, so the **write_u16( )** function represents the more direct approach of writing to it.

We try now a few simple programs which apply the LPC1768 DAC, creating first fixed voltages, and then waveforms.

Table 4.1: API Summary for Mbed Analog Output.

Functions	Usage
Constructor	
AnalogOut (PinName *pin*)	Create an AnalogOut object connected to the specified pin.
Member Functions	
void **write** (float *value*)	Set the output voltage, specified as a percentage (float).
void **write_u16** (unsigned short *value*)	Set the output voltage, represented as an unsigned short in the range 0x0 to 0xFFFF, producing respectively 0.0 V and 3.3 V.
float **read** ()	Return the current output voltage setting, measured as a percentage (float).
Mbed-defined operator: =	An operator shorthand for write().

4.2.1 Constant Output Voltages

In your chosen development environment, create a new program using Program Example 4.1. In this we specify an analog output labeled **Aout** by using the **AnalogOut** constructor. It's then possible to set the analog output simply by setting **Aout** to any permissible value; we do this three times in the program. By default **Aout** takes a floating point number between 0.0 and 1.0 and outputs this to pin 18. The actual output voltage on pin 18 is between 0 V and 3.3 V, so the chosen floating point number is scaled to this.

```
/*Program Example 4.1: Three values of DAC are output in turn on Pin 18. Read the
output on a DVM.
Works only on Mbed LPC1768. */

#include "mbed.h"
AnalogOut Aout(p18);    //create an analog output on pin 18
int main() {
  while(true) {
    Aout = 0.25;          // 0.25*3.3V = 0.825V
    ThisThread::sleep_for(2s); //wait 2 seconds
    Aout = 0.5;           // 0.5*3.3V = 1.65V
    ThisThread::sleep_for(2s);
    Aout = 0.75;          // 0.75*3.3V = 2.475V
    ThisThread::sleep_for(2s);
  }
}
```

Program Example 4.1 Trial DAC Output

Compile the program in the usual way, and let it run. Connect a digital voltmeter (DVM) between pins 1 and 18 of the Mbed LPC1768. You should see the three output voltages named in the comments of Program Example 4.1 being output in turn.

■ Exercise 4.1

Adjust Program Example 4.1 so that the **write_u16( )** function is used to set the analog output, giving the same output voltages.

■

4.2.2 Sawtooth Waveforms

Let's now make a sawtooth wave and view it on an oscilloscope. Create a new program and enter the code of Program Example 4.2. As in many cases previously, the program is made up of an endless **while(true)** loop. Within this we see a **for** loop. In this example, the variable **i** is initially set to 0, and on each iteration of the loop it is incremented by 0.1. The new value of **i** is applied within the loop. When **i** reaches 1, the loop terminates. As the **for** loop is the only code within the endless **while** loop, it then restarts, repeating this action continuously.

```
/*Program Example 4.2: Saw tooth waveform on DAC output. View on oscilloscope.
Only runs on LPC1768, as F401RE does not have a DAC.
                                                                              */
#include "mbed.h"
AnalogOut Aout(p18);
float i;

int main() {
  while(true){
    for (i = 0;i < 1;i = i+0.1){        // i is incremented in steps of 0.1
      Aout = i;
      ThisThread::sleep_for(1ms);       // wait 1 millisecond
    }
  }
}
```

Program Example 4.2 Sawtooth Waveform

Connect an oscilloscope probe to pin 18 of the Mbed, with its earth connection to pin 1. Check that you get a sawtooth waveform similar to that shown in Fig. 4.3. Ensure that the duration of each step is the 1 ms defined in the program, and try varying this. The waveform should start from 0 V and go up to a maximum of (0.9 x 3.3), or 2.97 V. Check for these values.

If you don't have an oscilloscope, you can set the sleep parameter to be much longer (say 100 ms) and use the DVM; you should then see the voltage step up from 0 V to 2.97 V and then reset back to 0 V again.

Exercise 4.2

Improve the resolution of the sawtooth by having more but smaller increments, that is, reduce the value by which **i** increments. The result should be as seen in Fig. 4.4. ■

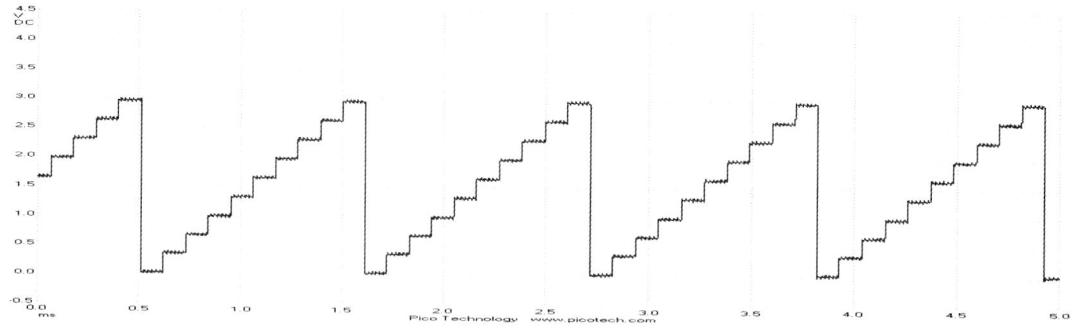

Figure 4.3
A stepped sawtooth waveform.

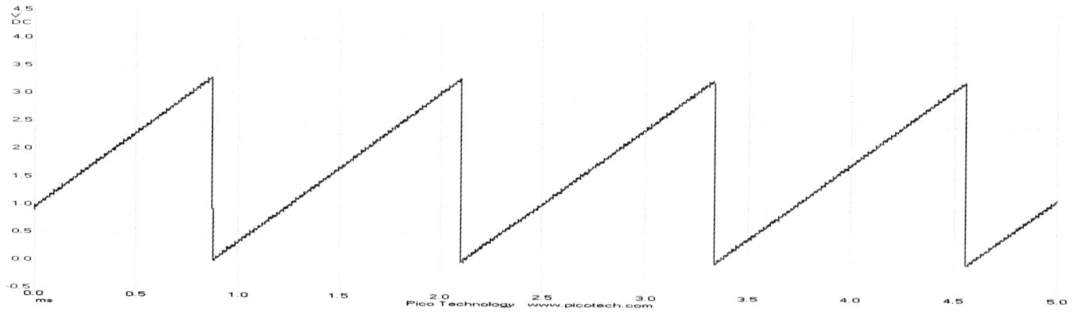

Figure 4.4
A smooth sawtooth waveform.

Exercise 4.3

Create a new project and devise a program which outputs a triangular waveform (i.e., one that counts down as well as up). The oscilloscope output should look like Fig. 4.5.

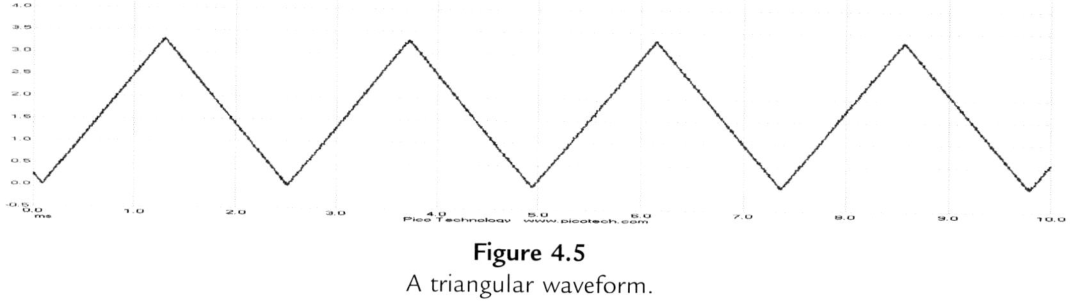

Figure 4.5
A triangular waveform.

4.2.3 Testing the DAC Resolution

Now let's return to the question of the DAC resolution, which we touched on in Section 4.1.1. Try changing the **for** loop in your version of Program Example 4.2 to this:

```
for (i = 0;i < 1;i = i+0.0001){
 Aout = i;
 ThisThread::sleep_for(1s);
  led1 = !led1;
}
```

We've also included a little LED indication here, so add this line before **main( )** to set it up:

```
DigitalOut led1(LED1);
```

This adjusted program will produce an extremely slow sawtooth waveform, which will take 10,000 steps to reach the maximum value, each one taking a second (hence the period of the waveform is 10,000s, or two and three-quarter hours!). Our purpose is not, however, to view the waveform, but to explore carefully the DAC characteristic of Fig. 4.2.

Switch your DVM to its finest voltage range, so that you have millivolt resolution on the scale; the 200 mV scale is useful here. Connect the DVM between pins 1 and 18 and run the program. The LED changes state every time a new value is output to the DAC. However, you will notice an interesting thing. Your DVM reading does not change with every LED change. Instead it changes state in distinct steps, with each change being around 3 mV. You will notice also that it takes around 5 LED "blinks," that is, 10 updates of the DAC value, before each DAC change. All this is what we anticipated in Section 4.1.1, where we predicted a step size of 3.22 mV; each step size is equal to the DAC resolution. The float value is rounded to the nearest digital input to the DAC, and it takes around 10 increments of the float value for the DAC digital input to be incremented by one.

4.2.4 Generating a Sine Wave

C/C++ code feature

It is an easy step from here to generate a sine wave. We will apply the **sin( )** function, which is part of the C standard library (see Section B9, Appendix B). Take a look at Program Example 4.3. To produce one cycle of the sine wave, we want to take sine values of a number which increases from 0 to 2π radians. In this program we use a **for** loop to increment variable **i** from 0 to 2 in small steps, and then multiply that number by π when the sine value is calculated. There are of course other ways of getting this result. The final challenge is that the DAC cannot output negative values. Therefore, we add a fixed value (an *offset*) of 0.5 to the number being sent to the DAC; this ensures that all output values lie within its available range. Notice that we are using the multiply operator, *, for the first time.

```
/*Program Example 4.3: Sine wave on DAC output. View on oscilloscope
Only runs on LPC1768, as F401RE does not have a DAC.
                                                                  */
#include "mbed.h"
AnalogOut Aout(p18);
float i;
int main() {
  while(true) {
    for (i = 0;i < 2;i = i + 0.05) {
      // Compute the sine value, + halve the range
      Aout = 0.5 + (0.5 * sin(i * 3.14159));
      ThisThread::sleep_for(1ms);       // Controls the sine wave period
    }
  }
}
```

Program Example 4.3 Generating a Sinusoidal Waveform

Exercise 4.4

Observe on the oscilloscope the sine wave that Program Example 4.3 produces. Estimate its frequency from information in the program, and then measure it. Do the two values agree? Try varying the frequency by varying the sleep parameter. What is the maximum frequency you can achieve with this program?

■

4.3 Another Way of Replicating Analog Output: Pulse Width Modulation

The DAC is a fine system element, and we have many uses for it. Yet it adds complexity to a microcontroller and sometimes moves us to the analog domain before we're ready to go. *Pulse width modulation* (PWM), an alternative, represents a neat and remarkably simple way of getting a rectangular digital waveform to control an analog variable, usually voltage or current. PWM control is used in a variety of applications, ranging from

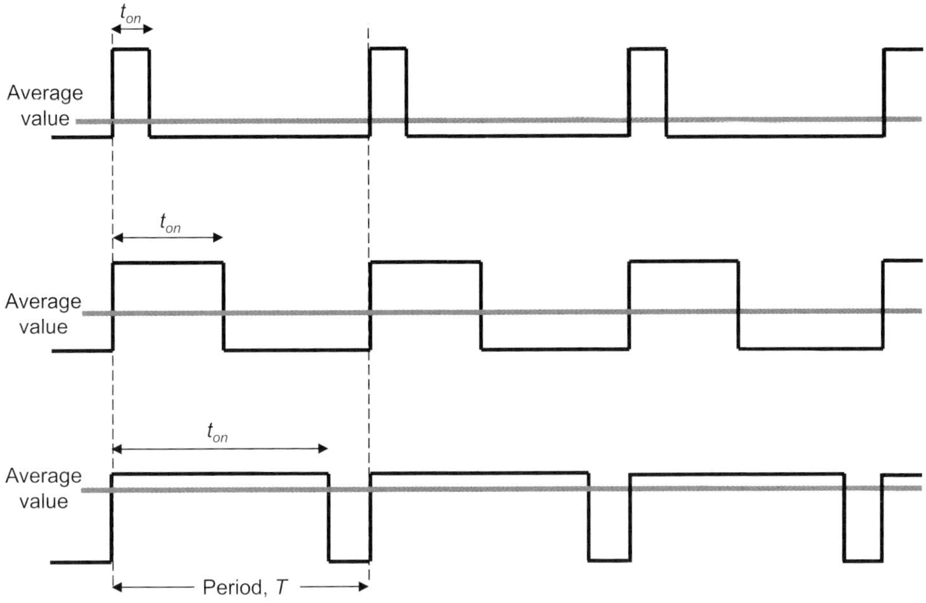

Figure 4.6
PWM Waveforms.

telecommunications to robotic control. Its importance is reflected in the fact that the Mbed LPC1768 has *six* PWM outputs (Fig. 2.1), compared with its one analog output. The F401RE claims no fewer than 19.

Three example PWM signals are shown in Fig. 4.6. In keeping with a typical PWM source, each has the same period, but a different pulse width, or "on" time; the pulse *width* is being modulated. The *duty cycle* is the proportion of time that the pulse is "on" or "high," and is expressed as a percentage; that is applying symbols from Fig. 4.6,

$$\text{duty cycle} = \frac{\text{pulse on time}}{\text{pulse period}} * 100\% = \frac{t_{\text{on}}}{T} * 100\% \quad 4.2$$

Therefore, a 100% duty cycle means "continuously on" and a 0% duty cycle means "continuously off." PWM streams are easily generated by digital counters and comparators, which can readily be designed into a microcontroller. They can also be produced simply by program loops and a standard digital output, with no dedicated hardware at all. We see this later in the chapter.

Whatever duty cycle a PWM stream has, there is an average value, as indicated in the figure. If the on time is short, the average value is low; if the on time is long, the average

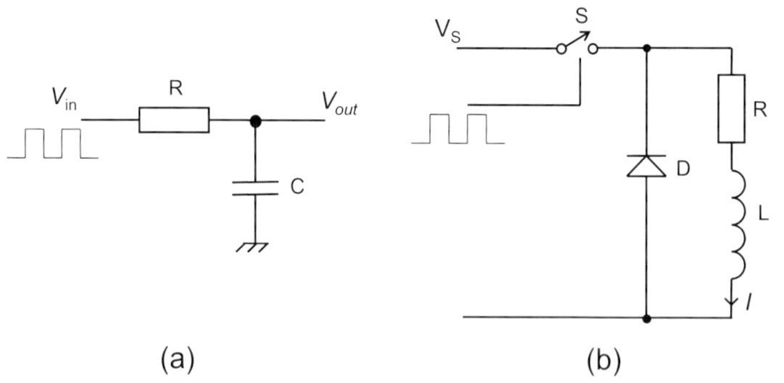

Figure 4.7
Simple averaging circuits: (a) A resistor—capacitor low-pass filter; (b) An inductive load.

value is high. By controlling the duty cycle, we control this average value. When using PWM, it is this average that we are usually interested in. It can be extracted from the PWM stream in a number of ways. Electrically, we can use a low-pass filter, such as the resistor—capacitor combination of Fig. 4.7(a). In this case, and as long as PWM frequency and values of R and C are appropriately chosen, V_{out} becomes an analog output, with a bit of ripple, and the combination of PWM and filter acts like a simple DAC. Alternatively, if we switch the current flow in an inductive load, as seen in Fig. 4.7(b), then the inductance has an averaging effect on the current flowing through it. This is very important, as the windings of any motor are inductive, so we can use this technique for motor control. The switch in Fig. 4.7(b) is controlled by the PWM stream and can be a transistor. We need, incidentally, to introduce the freewheeling diode, just as we did in Fig. 3.15(c), to provide a current path when the switch is open.

In practice, this electrical filtering is not always required. Many physical systems have internal inertias which, in reality, act like low-pass filters. We can, for example, dim a conventional filament light bulb with PWM. In this case, varying the pulse width varies the average temperature of the bulb filament, and the dimming effect is achieved.

As an example, the control of a DC motor is a very common task in robotics; the speed of a DC motor is proportional to the applied DC voltage. We *could* use a conventional DAC output, drive it through an expensive and bulky power amplifier, and use the amplifier output to drive the motor. Alternatively, a PWM signal can be used to drive a power transistor directly, which replaces the switch of Fig. 4.7(b); the motor is the inductor/ resistor combination in the same circuit. This technique is taken much further in the field of power electronics, with the PWM concept being taken far beyond these simple but useful applications.

4.4 Pulse Width Modulation in the Mbed Environment

4.4.1 Using the Mbed PWM Sources

As mentioned, it is easy to generate PWM pulse streams using simple digital building blocks. We don't explore how that is done in this book, but you can find the information elsewhere if you wish, for example in Reference 3.1. Fig. 2.1(c) shows that the Mbed LPC1768 has six PWM outputs available, from pin 21 to 26 inclusive. Note that on the LPC1768 microcontroller, and hence the Mbed target, the PWM sources all share the same period/frequency; if the period is changed for one, then it is changed for all. The Nucleo F401RE, Fig. 2.6, appears to have 19 PWM outputs, though beware of pins D0 and D1, for reasons stated in the figure. What is interesting about this multiplicity of outputs is that they come from *four* sources. Hence—unlike the LPC1768—the F401RE can generate PWM signals at up to four different frequencies. Thus, the PWM sources on the F401RE are labeled, for example, PWM2/1, indicating output 1 of PWM source 2.

As with all peripherals, Mbed PWM ports are supported by library utilities and functions, as shown in Table 4.2. This is rather more complex than the API tables we have seen so far, so we can expect a little more complexity in its use. Notably, instead of having just one variable to control (for example, a digital or analog output), we now have two, period and pulse width, or duty cycle derived from this combination. Similar to previous examples, a PWM output can be established, named, and allocated to a pin using **PwmOut.** Subsequently, it can be varied by setting its period, duty cycle, or pulse width. As shorthand, the **write( )** function can simply be replaced by =.

4.4.2 Some Trial PWM Outputs

As a first program using an Mbed PWM source, let's create a signal which we can see on an oscilloscope. Make a new project, and enter the code of Program Example 4.4. This will generate a 100 Hz pulse with 50% duty cycle, that is, a perfect square wave.

```
/*Sets PWM source to fixed frequency and duty cycle. Observe output on
oscilloscope.
Runs on LPC1768 or F401RE, using pins shown.                              */

#include "mbed.h"

//create a PWM output called PWM1
PwmOut PWM1(p21);              //Use D2 for F401RE
int main() {
   PWM1.period(0.010);         // set PWM period to 10 ms
   PWM1 = 0.5;                // set duty cycle to 50%
}
```

Program Example 4.4 Trial PWM Output

In this program example we first set the PWM period. There is no shorthand for this, so it is necessary to use the full **PWM1.period(0.010);** statement. The duty cycle is then defined as a decimal number, between the values of 0 and 1. We could also set the duty cycle as a pulse time with the following:

```
PWM1.pulsewidth_ms(5);                  // set PWM pulsewidth to 5 ms
```

When you run the program, you should be able to see the square wave on the oscilloscope and verify the output frequency.

Table 4.2: API Summary for PWM Output.

Functions	Usage
Constructor	
PwmOut(PinName *pin*)	Create a PwmOut object connected to the specified pin.
Member Functions	
void **write**(float *value*)	Set the output duty-cycle, in range 0.0 (0% on) to 1.0 (100% on).
float **read**()	Return the current output dutycycle setting, measured as a normalized float (0.0 – 1.0)
void **period**(float *seconds*)	Set the PWM period, specified in seconds (float), keeping the duty cycle the same.
void **period_ms**(int *ms*)	Set the PWM period, specified in milliseconds (int), keeping the duty cycle the same.
void **period_us**(int *us*)	Set the PWM period, specified in microseconds (int), keeping the duty cycle the same.
void **pulsewidth**(float *seconds*)	Set the PWM pulse width, specified in seconds (float), keeping the period the same.
void **pulsewidth_ms**(int *ms*)	Set the PWM pulse width, specified in milliseconds (int), keeping the period the same.
void **pulsewidth_us**(int *us*)	Set the PWM pulse width, specified microseconds (int), keeping the period the same.
Mbed-defined operator: =	An operator shorthand for write().

Exercise 4.5

(i) Change the duty cycle of Program Example 4.4 to some different values, say 0.2 (20%) and 0.8 (80%), and check that the correct display is seen on the oscilloscope, for example as shown in Fig. 4.8.

(ii) Change the program to give the same output waveforms, but using **period_ms( )** and **pulsewidth_ms( )**.

(iii) Add another PWM output to Program Example 4.4, setting it at a different frequency than the first. If using the F401RE, make these from two different PWM sources, for example PWM1 and PWM2. Note that on the LPC1768, both outputs will run at the frequency of the one which appears second in the program. With the F401RE, you will see that different frequencies are possible.

4.4.3 Speed Control of a Small Motor

Let us now apply the Mbed PWM source to control motor speed. Use the simple motor circuit already applied in the previous chapter (Fig. 3.16), but move the motor drive to a target board PWM output. Create a project using Program Example 4.5. This ramps the PWM up from a duty cycle of 0% to 100%, using programming features that are already familiar.

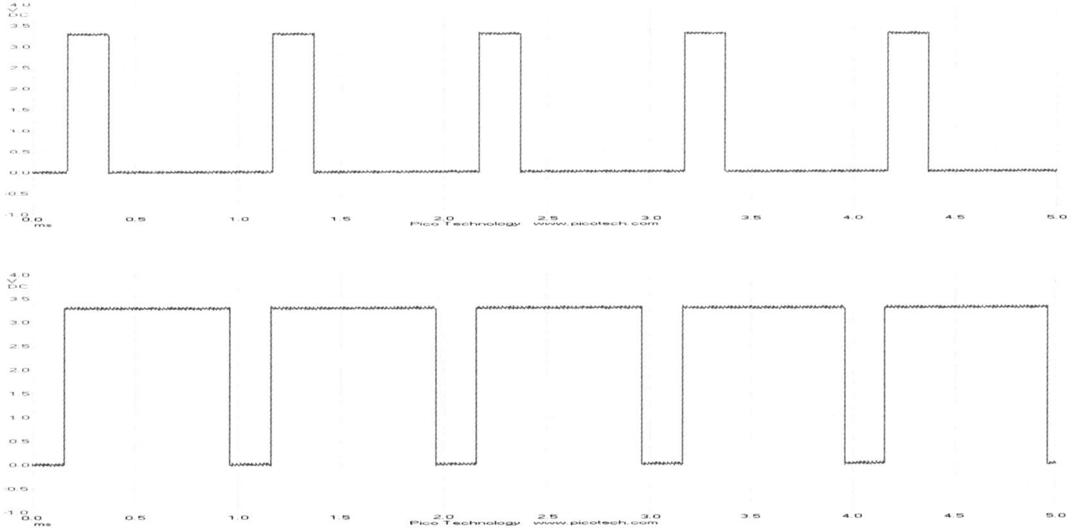

Figure 4.8
PWM observed from an Mbed target output.

```
/*Program Example 4.5: PWM control to DC motor is repeatedly ramped
Runs on LPC1768 or F401RE, using pins shown.
*/

#include "mbed.h"
PwmOut PWM1(p21);      //Use D2 for F401RE
float i;
int main() {
  PWM1.period(0.010);                //set PWM period to 10 ms
  while(true) {
    for (i = 0;i < 1;i = i + 0.01) {
      PWM1 = i;                      // update PWM duty cycle
      ThisThread::sleep_for(1ms);
      }
  }
}
```

Program Example 4.5 Controlling Motor Speed with the Mbed PWM Source

Compile, download, and run the program. See how the motor performs, and observe the waveform on the oscilloscope. You are seeing one of the classic applications of PWM, controlling motor speed through a simple digital pulse stream.

4.4.4 Generating PWM in Software

Although we have just used PWM sources that are available on the development board, it is useful to realize that these aren't essential for creating PWM; we can actually do it just with a digital output and some timing. In Exercise 3.7 you were asked to write a program which switched a small DC motor on and off continuously, 1 second on, 1 second off. If you speed this switching up, to say 1 ms on, 1 ms off, you will immediately have a PWM source.

Create a new project with Program Example 4.6 as source code. Notice carefully what the program does. There are two **for** loops in the main **while** loop. The motor is initially switched off for 5 s. The first **for** loop then switches the motor on for 400 μs, and off for 600 μs; it does this 5000 times. This results in a PWM signal of period 1 ms, and duty cycle 40%. The second **for** loop switches the motor on for 800 μs, and off for 200 μs; the period is still 1 ms, but the duty cycle is 80%. The motor is then switched full on for 5 s. This sequence continues indefinitely.

```
/*Program Example 4.6: Software generated PWM. 2 PWM values generated in turn, with
full on and off included for comparison.
Runs on LPC1768 or F401RE, using pins shown.
*/
```

```
#include "Mbed.h"

DigitalOut motor(p6);      //use any digital output for F401RE
int i;
int main() {
  while(true) {
    motor = 0;                    //motor switched off for 5 secs
    ThisThread::sleep_for(5s);
    for (i = 0; i < 5000;i = i + 1) {    //5000 PWM cycles, low duty cycle
      motor = 1;
      wait_us(400);                      //output high for 400us
      motor = 0;
      wait_us(600);                      //output low for 600us
    }
    for (i = 0; i < 5000;i = i + 1){     //5000 PWM cycles, high duty cycle
      motor = 1;
      wait_us(800);                      //output high for 800us
      motor = 0;
      wait_us(200);                      //output low for 200us
    }
    motor = 1;                           //motor switched fully on for 5 secs
    ThisThread::sleep_for(5s);
  }                       //end of while loop
}
```

Program Example 4.6 Generating PWM in Software

Compile and download this program, and apply the circuit of Fig. 3.16. You should find that the motor runs with the speed profile indicated. PWM period and duty cycle can be verified readily on the oscilloscope. You may wonder for a moment if it's necessary to have dedicated PWM ports, when it seems quite easy to generate PWM with software. Remember, however, that in this example the CPU becomes totally committed to this task, and can do nothing else. With the hardware ports, we can set the PWM running, and the CPU can then get on with some completely different activity.

■ Exercise 4.6

Vary the motor speeds in Program Example 4.6 by changing the duty cycle of the PWM, initially keeping the frequency constant. Depending on the motor you use, you will probably find that for small values of duty cycle the motor will not run at all, due to its own friction. This is particularly true of geared motors. Observe the PWM output on the oscilloscope, and confirm that on and off times are as indicated in the program. Try also at much higher and lower frequencies.

■

4.4.5 Servo Control

A servo is a small, lightweight, rotary position control device, often used in radio-controlled cars and aircraft to control angular positions of variables such as steering, elevators, and rudders. The Hitec HS-422 servo is shown in Fig. 4.9(a). Servos are now popular in a range of robotic applications. The servo shaft can be positioned to specific angular positions by sending the servo a PWM signal. As long as the modulated signal exists on the servo input, it will maintain the angular position of the shaft. As the duty cycle changes, the angular position of the shaft changes. This is illustrated in Fig. 4.9(b). Many servos use a PWM signal with a 20 ms period, as is shown here. In this example, the pulse width is modulated from 1.25 ms to 1.75 ms to give the full 180° range of the servo. Servos are usually supplied with a three-way connector, as seen in Figs. 4.9(a) and 4.10. These connect to 0 V, 5 V, and the PWM signal.

Connect a servo to an Mbed development board, using either a breadboard or the LPC1768 application board. The servo requires a higher current than the USB standard can provide, so it is essential that you power it using an external supply. A 4xAA (6 V) battery pack meets the supply requirement of most small servos.

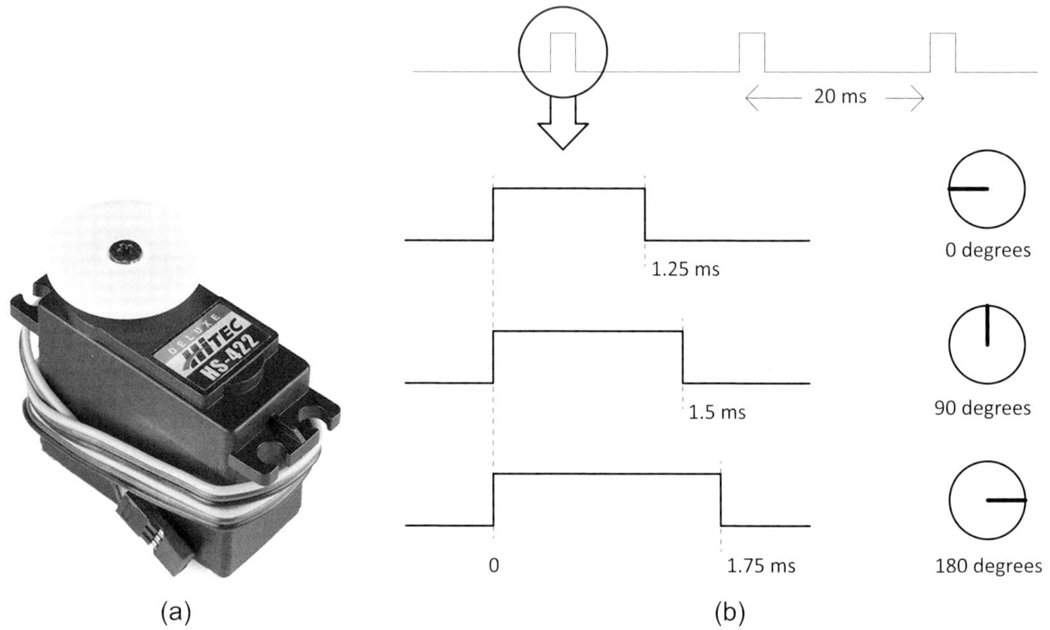

Figure 4.9
Servo essentials: (a) The Hitec HS-422 servo; (b) Example servo drive PWM waveforms.

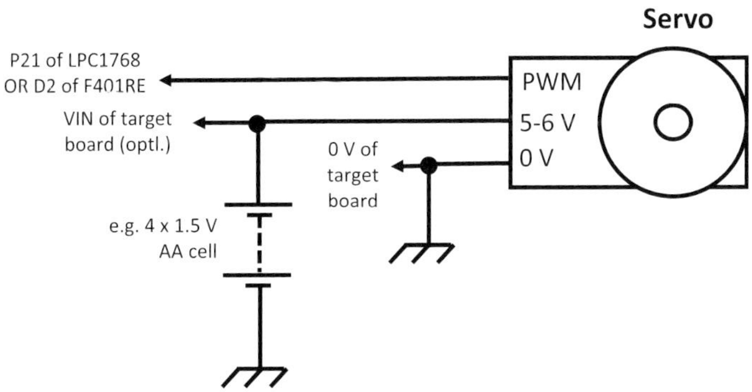

Figure 4.10
Using PWM to drive a servo: connection diagram.

The connections shown in Fig. 4.10 should be applied if using the breadboard; the development board itself can still be supplied from the USB. Alternatively, it can also be supplied from the battery pack, through VIN. Either Mbed LPC1768 or F401RE then regulates the incoming 6 V to the 3.3 V that it requires. In this case, the USB can be left disconnected once the program has been downloaded.

App Board

Conveniently, the LPC1768 app board has two PWM connectors which match the standard connector used on most servos. These are seen as Item 8 in Fig. 2.14; they are labeled PWM1 and PWM2, and link to Mbed LPC1768 pins 21 and 22, respectively. To align with the circuit diagram of Fig. 4.10, connect to PWM1, the one next to the potentiometer. The external supply connects onto the app board through a jack plug immediately below the Mbed USB connector. A battery pack connected to this jack can power the whole system, so again the USB connector can be disconnected following program download.

■ Exercise 4.7

(for app board or breadboard)

Create a new project and write a program which sets a PWM output on the pin indicated in Fig. 4.10. Set the PWM period to 20 ms. Try a number of different duty durations, taking values from Fig. 4.9, and observe the servo's position. Then write a program which continually moves the servo shaft from one limit to the other.

■

4.4.6 Producing Audio Output

We can use the PWM source simply as a variable-frequency signal generator. In this example we use it to sound a piezo transducer or speaker and play the start of an old London folk song called "Oranges and Lemons." This song imagines that the bells of each church in London call a particular message. If you are a reader of music you may recognize the tune in Fig. 4.11; if you are not, don't worry! You just need to know that any note which is a minim ("half note" in the United States) lasts twice as long as a crotchet ("quarter note" in the United States), which in turn lasts twice as long as a quaver ("eighth note" in the United States). Put another way, a crotchet lasts one beat, a minim two, and a quaver one-half. The pattern for the music is as shown in Table 4.3. Note that here we are simply using the PWM as a variable-frequency signal source, and not actually modulating the pulse width as a proportion of frequency at all.

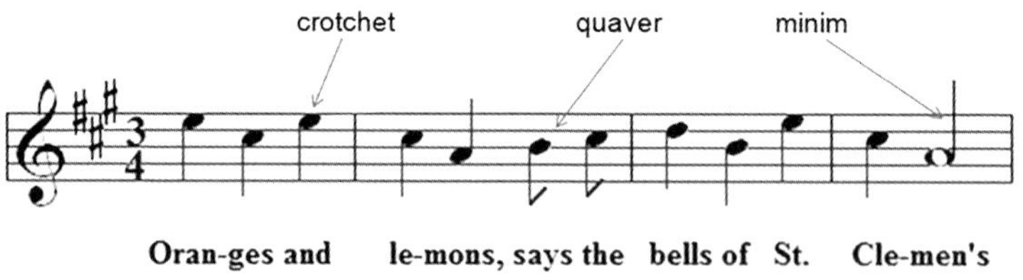

Figure 4.11
The "Oranges and Lemons" tune.

Table 4.3: Frequencies of Notes Used in Tune.

Word/Syllable	Musical Note	Frequency (Hz)	Beats
Oran-	E	659	1
ges	C#	554	1
and	E	659	1
le-	C#	554	1
mons,	A	440	1
says	B	494	½
the	C#	554	½
bells	D	587	1
of	B	494	1
St	E	659	1
Clem-	C#	554	1
en's	A	440	2

C/C++ code feature

Create a new program and enter Program Example 4.7. This introduces an important new C feature, the *array*. If you are unfamiliar with this, then read the review in Section B8.1. The program uses two arrays, one defined for frequency data, the other for beat length. There are 12 values in each, one for each of the 12 notes in the tune. The program is structured around a **for** loop, with variable **i** as counter. As **i** increments, each array element is selected in turn. Notice that **i** is set just to reach the value 11; this is because the value 0 addresses the first element in each array, and the value 11 hence addresses the 12th. From the frequency array the PWM period is calculated and set, always with a 50% duty ratio. The beat array determines how long each note is held, using the **wait_us( )** function.

```
/*Program Example 4.7: Plays the tune "Oranges and Lemons" on a piezo buzzer, using
PWM.
Runs on LPC1768 or F401RE, using pins shown.                        */

#include "Mbed.h"
PwmOut buzzer(p26); //Use D2, or PWM output of choice, for F401RE

                                                            //frequency array
float frequency[]={659,554,659,554,440,494,554,587,494,659,554,440};
float beat[]={1,1,1,1,1,0.5,0.5,1,1,1,1,2};                 //beat array
int main() {
  while (true) {
    for (int i = 0;i <= 11;i++) {
      buzzer.period(1/(2 * frequency[i]));              // set PWM period
      buzzer = 0.5;                                     // set duty cycle
      wait_us(400000 * beat[i]);                            // hold for beat period
    }
  }
}
```

Program Example 4.7 "Oranges and Lemons" Program

App Board

Compile the program and download. If you are using the app board, then the on-board speaker is hard wired to pin 26 of the Mbed (item 5 of Fig. 2.7), and the sequence should play on reset. Otherwise, connect a piezo transducer, as seen in Fig. 4.12, between pins 1 and 26. Let the program run. The transducer seems very quiet when held in air. You can increase the volume significantly by fixing or holding it to a flat surface like a table top.

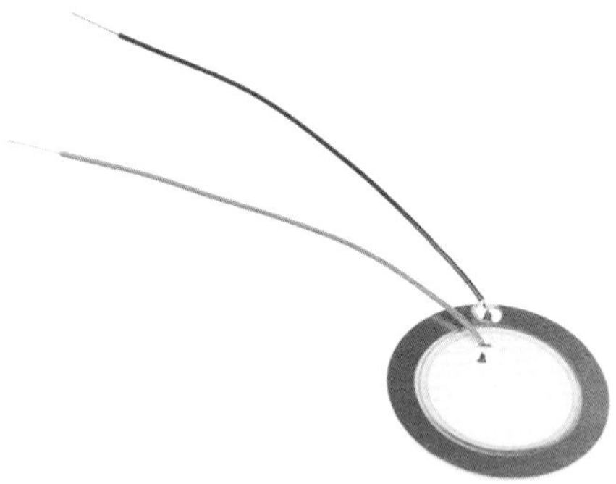

Figure 4.12
A piezo transducer.

Exercise 4.8

Try the following:

i) Make the "Oranges and Lemons" sequence play an octave higher by doubling the frequency of each note.
ii) Change the tempo by modifying the multiplier in the wait command.
iii) (For the more musically inclined) Change the tune, so that the Mbed plays the first line of "Twinkle Twinkle Little Star." This uses the same notes, except that it also needs F#, of frequency 699 Hz. The tune starts on A. Because there are repeated notes, consider putting small pauses between notes.
iv) (For the musically competent) Change the tune to any that you wish. It's easy to look up the frequency of any note you want on the web.

■

Chapter Review

A digital-to-analog converter (DAC) converts an input binary number to an output analog voltage, which is proportional to that input number.

- DACs are widely used to create continuous varying voltages; for example, to generate analog waveforms.
- The Mbed LPC1768 has a single DAC and an associated set of library functions. The Nucleo F401RE does not have a DAC.
- Pulse width modulation (PWM) provides a way of controlling certain analog quantities by varying the pulse width of a fixed-frequency rectangular waveform.
- PWM is widely used for controlling the flow of electrical power; for example, LED brightness or motor control.

- The Mbed LPC1768 has six possible PWM outputs. They can all be individually controlled, but must all share the same frequency. The Nucleo F401RE shows 19 possible PWM outputs; these are from four different sources, so four different frequencies are possible.

Quiz

1. A 7-bit DAC obeys Eq. 4.1, and has a voltage reference of 2.56 V.
   (a) What is its resolution?
   (b) What is its output if the input is 100 0101?
   (c) What is its output if the input is 0x2A?
   (d) What is its digital input in decimal and binary if its output reads 0.48 V?
2. What is the Mbed LPC1768's DAC resolution and what is the smallest analog voltage step increase or decrease which it can output?
3. What is the output of the LPC1768 DAC, if its input digital word is:
   (a) 00 0000 1000
   (b) 0x080
   (c) 10 1000 1000 ?
4. What output voltages will be read on a DVM while this program loop runs on the Mbed LPC1768?

```
while(true){
   for (i = 0; i < 1; i = i + 0.2){
     Aout = i;
     ThisThread::sleep_for(200ms);
   }
}
```

5. The waveform in Question 4 gives a crude saw tooth. What is its period?
6. What are the advantages of using PWM for control of analog actuators?
7. A PWM data stream has a frequency of 4 kHz, and a duty cycle of 25%. What is its pulse width?
8. A PWM data stream has a period of 20 ms, and an on time of 1 ms. What is its duty cycle?
9. The PWM on an Mbed development board is set up with these statements. What is the on time of the waveform?

```
PWM1.period(0.004);          // set PWM period
PWM1 = 0.75;                 // set duty cycle
```

10. How long does Program Example 4.7 take to play through the tune once? Calculate this by checking the program, and then check while the tune plays.

CHAPTER 5

Analog Input

5.1 Analog-to-Digital Conversion

The world around the embedded system is largely an analog one, and sensors—of temperature, sound, acceleration, and so on—mostly have analog outputs. Yet it is essential for the microcontroller to have these signals available in digital form. This is where the *analog-to-digital converter* (ADC) comes in. We can convert analog signals into digital representations, with an accuracy determined by the characteristics of the ADC in use. Having performed this analog-to-digital conversion, we can then use the microcontroller to process or analyze this information, based on the value of the analog input.

5.1.1 The Analog-to-Digital Converter

An ADC is an electronic circuit whose digital output is proportional to its analog input. It effectively "measures" the input voltage and gives a binary output number proportional to its size. The list of possible analog input signals is endless, including such diverse sources as audio and video, medical or climatic variables, and a host of industrially generated signals. Of these, some change very slowly (e.g., temperature). Others are periodic, with a frequency range up to tens of kiloHertz (e.g., sound and other vibrations). Still others have a very high-frequency content (e.g., video or radar). These examples display very different signal characteristics, and it is not surprising to discover that many types of ADCs have been developed, with characteristics optimized for these differing applications.

The ADC almost always operates within a larger environment, often called a *data acquisition system*. Some features of a general-purpose data acquisition system are shown in Fig. 5.1. To the right of the diagram is the ADC itself. This has an analog input and a digital output. It is under computer control; the computer can start a conversion. The conversion takes finite time, maybe some microseconds or more, so the ADC needs to signal when it has finished. The output data can then be read. The ADC works with a voltage reference—an accurate and stable voltage source. Think of this as a ruler or tape measure. In one way or another the ADC compares the input voltage with the voltage reference, and comes up with the output number, based on this comparison. As with so many digital or digital/analog subsystems, there is also a clock input—a continuously

Fast and Effective Embedded Systems Design. https://doi.org/10.1016/B978-0-323-95197-5.00005-X

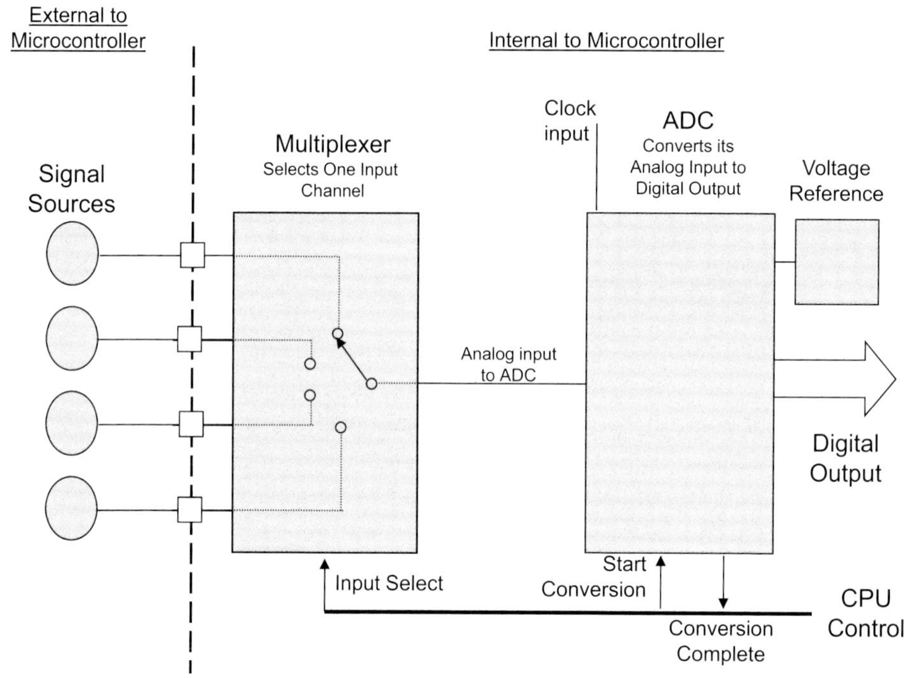

Figure 5.1
An example data acquisition system.

running square wave, which sequences the internal operation of the ADC. The clock frequency, subject to its own set of constraints, determines how fast the ADC operates.

Once we appreciate the usefulness of the ADC, we usually find that we want to work with more than just one signal. We *could* just use more ADCs, but this is costly, and takes up semiconductor space. Instead, common practice is to put an *analog multiplexer* in front of the ADC. This acts as a selector switch. The user can then select any one of several inputs to the ADC. If this is done quickly, it is as if all inputs are being converted at the same time. Many microcontrollers, including the LPC1768 and the F401RE, include an ADC and multiplexer on the chip. The inputs to the multiplexer are connected to microcontroller pins, and multiple inputs can be used. This is shown in Fig. 5.1, for a 4-bit multiplexer. We return to some of the detail of Fig. 5.1 in Chapter 14.

5.1.2 Range, Resolution, and Quantization

Many ADCs obey Eq. 5.1, where V_i is the input voltage, V_{ref} the reference voltage, n the number of bits in the converter output, and D the n-bit digital output value. The output number D is an integer, and for an n-bit number can take any value from 0 to (2^n-1). The internal ADC process effectively rounds or truncates the calculation in Eq. 5.1 to produce the integer output:

$$D = \frac{V_i}{V_{ref}} \times 2^n \tag{5.1}$$

Clearly, the ADC cannot convert just any input voltage, but has maximum and minimum permissible input values. The difference between this maximum and minimum is called the *range*. Often, the minimum value is 0 V, so the range is then equal to the maximum possible input value. Analog inputs which exceed the maximum or minimum permissible input values are likely to be digitized as the maximum and minimum values, respectively; that is, a limiting (or "clipping") action takes place. The input range of the ADC is directly linked to the value of the voltage reference. It can be calculated by rearranging Eq. 5.1 and setting D to its maximum value, 2^n-1. This is seen in Eq. 5.2, where V_{rge} is the range. For any practical value of n, typically 8 or higher, V_{rge} is nearly equal to V_{ref}:

$$V_{rge} = \frac{(2^n-1)}{2^n} \times V_{ref} \tag{5.2}$$

Eq. 5.1 is represented in graphical form in Fig. 5.2 for a 3-bit converter. If the input voltage is gradually increased from 0 V, and the converter is running continuously, then the ADC's output is initially 000. If the input slowly increases, there comes a point where the output will take the value 001. As the input increases further, the output changes to 010, and so on. At some point it reaches 111, i.e. 7 in decimal, or (2^3-1). This is the

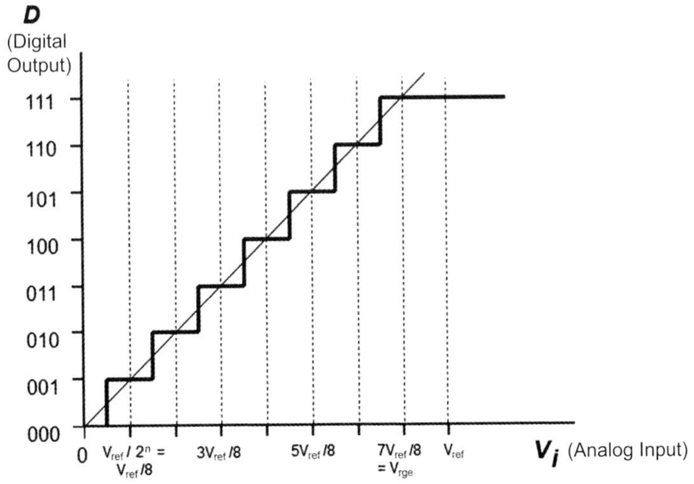

Figure 5.2
A 3-bit ADC characteristic.

maximum possible output value, whose input is V_{rge}. The input may be increased further, but it cannot force any increase in output value.

As Fig. 5.2 demonstrates, by converting an analog signal to digital, we run the risk of approximating it. This is because any one digital output value has to represent a small range of analog input voltages, that is, the width of any of the steps on the "staircase" of Fig. 5.2. For example, the digital output 001 in the figure should ideally represent $V_{ref}/8$, as the figure shows. In reality it must represent *any* analog voltage along its step. The width of this step is called the *resolution* of the ADC; resolution is a measure of how precisely an ADC can convert and represent a given input voltage. Clearly, the more steps we have representing the range, the narrower the steps will be, and hence, the resolution is reduced (and improved). We get more steps by increasing the number of bits in the ADC process. Hence, a common shorthand is to say that a certain ADC has, for example, a 12-bit resolution. Increasing the number of bits inevitably increases the complexity and cost of the ADC, and usually the time it takes to complete a conversion. Eq. 5.3 shows how resolution relates to reference voltage and number of bits, for an ideal ADC.

$$\text{resolution} = \frac{V_{ref}}{2^n} \tag{5.3}$$

To take this discussion a little further, as an example, let us return to the step represented by 001 in Fig. 5.2. If the output value of 001 is precisely correct for the input voltage at the middle of the step, then as we move away from the center, a measurement error occurs. The greatest error appears at either end of the step. This is called *quantization error*. Following this line of reasoning, the greatest quantization error is one-half of the step width which is one-half of the resolution, or half of one least significant bit (LSB) equivalent of the voltage scale.

As a further example, if we want to convert an analog signal that has a range of 0 to 3.3 V to an 8-bit digital signal, then there are 256 (i.e., 2^8) distinct output values. Each step has a width of 3.3/256 = 12.89 mV, which is the ADC resolution. The worst case quantization error is half of this, or 6.45 mV. As far as the Mbed LPC1768 is concerned, Fig. 2.3 shows that the LPC1768 ADC is 12-bit. This leads to a resolution of $3.3/2^{12}$, or 0.8 mV, with a worst-case quantization error of 0.4 mV.

For many applications an 8-, 10-, or 12-bit ADC allows sufficient resolution; it all depends on the required accuracy. For example, in certain audio applications, listening tests have demonstrated that 16-bit resolution is adequate; however, improved quality can be noticed using 24-bit conversion.

All of this assumes that all other aspects of the ADC are perfect, which they are not. The reference voltage can be inaccurate, or can drift with temperature, and the staircase pattern can have nonlinearities. Furthermore, different methods of analog-to-digital conversion each bring their unique inaccuracies to the equation. Reference 3.1 discusses in detail a number of ADC designs, including *successive approximation*, *flash*, *dual slope*, and *delta–sigma* methods, which all have their own relative advantages. However, discussing the specific design of the ADC system is not necessary for us to understand the concepts of data conversion and implementing ADCs in the embedded environment.

5.1.3 Sampling Frequency

When a changing analog signal is converted to digital form, in most cases a sample—that is, an ADC conversion—is taken repeatedly, as illustrated in Fig. 5.3. Generally, sampling occurs at a fixed rate, called the *sampling frequency*. Once a sample is taken, it holds the value of the signal until the next one is taken. The more samples that are taken, the more accurate the digital representation is likely to be.

The *minimum* possible sampling frequency depends on the maximum frequency of the signal being digitized. If the sampling frequency is too low, then rapid changes in the analog signal will not be represented in the resulting digital data. The *Nyquist sampling criterion* states that the sampling frequency must be at least double that of the highest signal frequency. For example, the human auditory system is known to extend up to approximately 20 kHz, so standard audio CDs are sampled and played back at 44.1 kHz in order to adhere to the Nyquist sampling criterion. If the sampling criterion is not satisfied, then a phenomenon called *aliasing* occurs—a new, lower frequency is generated, called the *alias*. This is illustrated in Fig. 5.4, and demonstrated later in the chapter. Aliasing is very damaging to a signal, and must always be avoided. A common approach is to use an *anti-aliasing filter*, which limits all signal components to those which satisfy the sampling criterion.

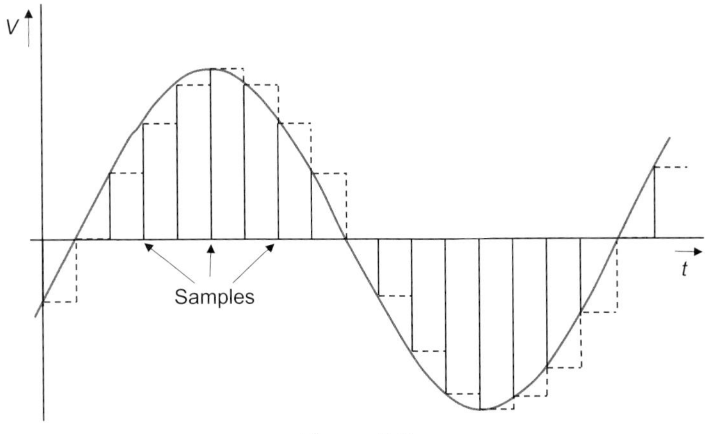

Figure 5.3
Digitizing a sine wave.

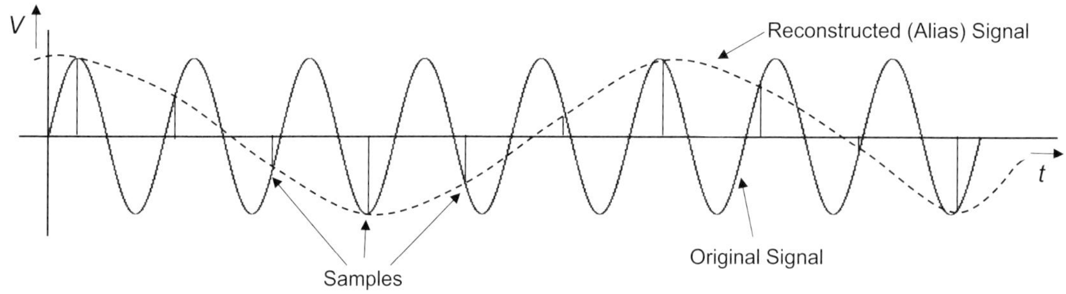

Figure 5.4
The effect of aliasing.

5.1.4 Analog Input with the Mbed

Fig. 2.1 shows that the Mbed LPC1768 has up to six analog inputs, on pins 15 to 20; one of these, pin 18, can also be the analog output, as described in Chapter 4. Fig. 2.6 shows a surprising 11 possible analog inputs for the F401RE; however, two of these, on pins D0 and D1, are not connected by default.

Fig. 2.15 (or Table 2.3) shows how these LPC1768 analog inputs are used on the application board. Pins 19 and 20 are usefully connected to two on-board potentiometers, while pins 17 and 18 are available externally through two audio jack connectors (items 4 and 7 in Fig. 2.14); these sit between the potentiometers on the board. Pins 15 and 16 are hard-wired to the joystick, so they are not available for analog input.

The analog input API summary, following a pattern which is quite familiar, is shown in Table 5.1. It is useful to note that the ADC output is available to the program, either as an unsigned integer (as it would be at the ADC output) or as a floating point number.

5.2 Combining Analog Input and Output

The ADC is an input device, which transfers data *into* the microcontroller. If it is used on its own, we will have no idea of what output values it has created. Therefore, we go on to do two things in order to make that data visible. First, we use the ADC output values to control an output variable immediately (e.g., via digital-to-analog converter [DAC] or PWM). We will later transfer its output values to the PC screen and explore some measurement applications.

5.2.1 Controlling LED Brightness by Variable Voltage

Let us start with a simple program which reads the analog input and uses it to control the brightness of an LED by varying the voltage drive to the LED. Here we will use a

Table 5.1: API Summary for Analog Input.

Functions	Usage
Constructor	
AnalogIn (PinName *pin,* float *vref*)	Create an AnalogIn object, connected to the specified pin; a reference voltage value can also be defined.
Member Functions	
float **read**()	Read the input voltage, represented as a float in the range from 0.0 to 1.0.
unsigned short **read_u16**()	Read the input voltage, represented as an unsigned short in the range from 0x0 to 0xFFFF, normalized to a 16-bit value.
float **read_voltage**()	Read the input voltage in volts; output depends on the target board's ADC reference voltage (typically equal to the supply voltage).
void **set_reference_voltage**(float *vref*)	Set the AnalogIn reference voltage.

potentiometer to generate the analog input voltage, and then pass the value read straight to the analog output.

For the Mbed LPC1768, connect up the simple circuit of Fig. 5.5. As the F401RE does not have a DAC, this program cannot be applied to it. The circuit shown uses pin 20 as the analog input, connecting the potentiometer across 0–3.3 V. The potentiometer type shown here conveniently plugs directly into the breadboard. The LED is connected to Pin 18, the analog output. Start a new program and copy into it the very simple code of Program Example 5.1. This just sets up the analog input and output, and then continuously transfers the input to the output.

```
/*Program Example 5.1: Uses analog input to control LED brightness, through DAC
output.
This program is not available for F401RE, as it has no DAC.
                                                                              */
#include "mbed.h"
AnalogOut Aout(p18);          //defines analog output on Pin 18
AnalogIn Ain(p20);            //defines analog input on Pin 20

int main() {
  while(true) {
    Aout = Ain;      //transfer analog in value to analog out, both are type float
  }
}
```

Program Example 5.1 Controlling LED brightness by variable voltage

Compile the program and download to the Mbed. With the program running, the potentiometer should control the brightness of the LED. However, you will probably find that there is a range of potentiometer rotation where the LED is off. The LED will be following the curve of Fig. 3.4(a), and there will be negligible illumination when the drive voltage is low.

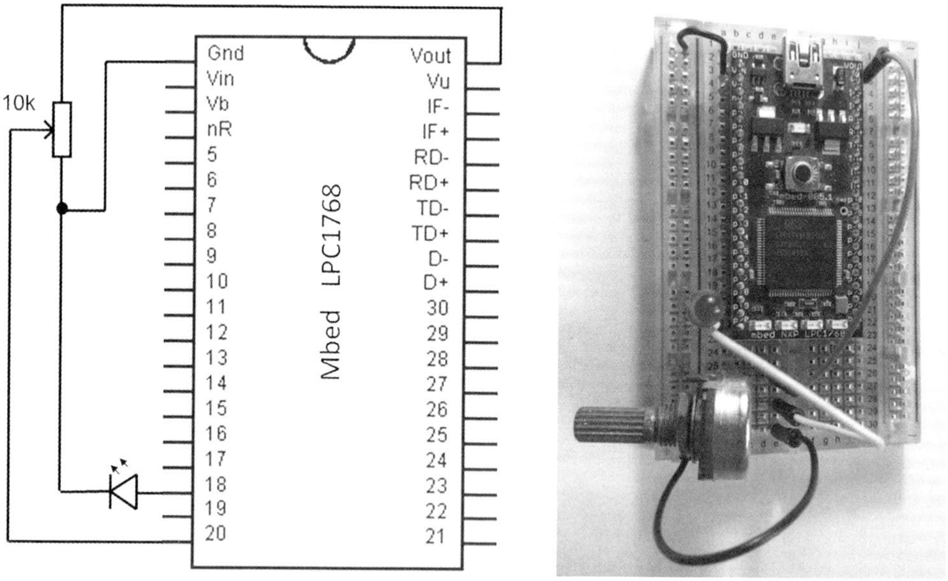

(a) (b)

Figure 5.5
A potentiometer controlling LED brightness: (a) circuit diagram; (b) construction detail.

Exercise 5.1

Measure the DAC output voltage at pin 18 as you adjust the potentiometer. You will find that when this exceeds around 1.8 V, the LED will be lit, with varying levels of brightness. When it is below 1.8 V, the LED conducts very little, and there is negligible illumination.

5.2.2 Controlling LED Brightness by PWM

App Board The potentiometer can be used instead to alter the PWM duty cycle, using the same approach as for the analog output. Program Example 5.2 can run on the F401RE or the Mbed LPC1768, with the latter on a breadboard or app board. If the app board is used, the pin chosen in the program connects to the red onboard LED. For breadboard builds, adapt the circuit of Fig. 5.5a, connecting the LED to the PWM output used. Create a new program, and enter the code of Program Example 5.2. Here we see the analog input value being transferred to the PWM duty cycle.

```
/*Program Example 5.2: Uses analog input to control PWM duty cycle, fixed period
Runs on both LPC1768 and F401RE, using pins shown                    */

#include "mbed.h"

PwmOut PWM1(p23);          //use D2 for F401RE
AnalogIn Ain(p20);         //use A0 for F401RE

int main() {
  PWM1.period(0.010);      // set PWM period to 10 ms
  while(true){
    PWM1 = Ain;            //Analog in value becomes PWM duty, both are type float
    ThisThread::sleep_for(100ms);
  }
}
```

Program Example 5.2 Controlling PWM pulse width with potentiometer

The LED brightness should again be controlled by the potentiometer. While the outcome is very similar to that of the previous program, the means of doing it is quite different. In practice, we would normally not wish to commit a whole DAC to controlling the brightness of an LED, but would be more ready to make use of the simpler PWM source.

5.2.3 Controlling PWM Frequency

App Board Instead of using the potentiometer to control the PWM duty cycle, we can use it to control the PWM frequency. Use the same hardware as in the previous section, either app board or breadboard. Create a new program and enter the code of Program Example 5.3.

Notice that the PWM period is calculated in the line:

```
PWM1.period(Ain/10 + 0.001);        // set PWM period
```

C/C++ code feature

It is first worth noting that a calculation is placed where we might have expected a simple parameter to be placed. This is not a problem for C. The program will first evaluate the expression inside the brackets, and then call the **period( )** function. This calculation invokes the divide operator, /, for the first time. It also raises the thorny little question, which all children face when they learn arithmetic, of what order operators should be evaluated in. In C this is very clearly defined, and can be seen by checking Table B.5 in Appendix B. This shows a precedence for each operator, with / having precedence 3, and + having precedence 4. Therefore, the division will be done before the addition, and there will be no uncertainty in evaluating the expression. The values used mean that the minimum period, when **Ain** is zero, is 0.001 s (1000 Hz). The maximum is when **Ain** is 1, leading to a period of 0.101 s, around 10 Hz.

Having said all of this about operators, there is a view that you should not depend too much on your knowledge of the precedence of each operator—it is too easy to make a mistake. For a more secure approach, where more than one operator is invoked, you can put brackets a round each subexpression. In this case, that line of code becomes:

```
PWM1.period((Ain/10) + 0.001);         // set PWM period
```

Now the bracketing enforces the precedence that is required in a visible way.

```
/*Program Example 5.3: Uses analog input to control PWM period.
Runs on both LPC1768 and F401RE, using pins shown           */

#include "mbed.h"
PwmOut PWM1(p23);        //use D2 for F401RE
AnalogIn Ain(p20);       //use A0 for F401RE

int main() {
  while(true){
    PWM1.period(Ain/10 + 0.001);           // set PWM period
    PWM1 = 0.5;                              // set duty cycle
    ThisThread::sleep_for(500ms);
  }
}
```

Program Example 5.3 Controlling PWM frequency with potentiometer

When running the program, you should be able to see the frequency change as the potentiometer is adjusted.

Exercise 5.2

Observe the PWM waveform on an oscilloscope, setting the time base initially to 5 ms/div.

1. Adjust the values in the PWM period calculation to give different ranges of frequency output.
2. At what frequency does the LED appear not to flash, but seem to be continuously on? Take note of this, and see if the perceived frequency varies between different people. This is an important question, as knowing its value allows us to estimate at what frequency we can "trick" the eye into thinking that a flashing image is continuous (e.g., in multiplexed LED displays).

The 0.5 s delay in Program Example 5.3 is added to the loop so that each new PWM period is implemented before the next update. If it were omitted, then the PWM would potentially be updated repeatedly within each cycle, which would lead to a large amount of instability or *jitter* in the output. Try removing the delay in order to see the effect. Notice there can be discontinuities in the PWM output as the frequency values are updated. With care (and especially if you are using a storage oscilloscope) you can see this as the PWM is updated. This is one reason that it is good to fix the frequency for a PWM signal.

Exercise 5.3

Connect a servo to your Mbed development board as indicated in Fig. 4.10, with the potentiometer connected as in Section 5.2.1. Write a program which allows the potentiometer to control the servo position. Scale values so that the full range of potentiometer adjustment leads to the full range of servo position change.

5.3 Processing Data from Analog Inputs

5.3.1 Displaying Data on the Computer Screen—printf( ) and the Serial Terminal

We turn now to the second way of making use of the ADC output, promised at the beginning of Section 5.2. It is possible to read analog input data through the ADC, and then print the value to the PC screen. This is an important step forward, as it gives the possibility of displaying on the computer screen any data we are working with in the Mbed environment. This is made possible because Mbed development boards are generally equipped with a virtual serial port, communicating through the USB connection, which the

PC can recognize and link to, through one of its "COM" ports. We return to this in Section 7.8.3. To enable the link, both the Mbed development board and the host computer need to be configured correctly, and we need the host computer to be able to display that data.

Very conveniently, the Studio Integrated Development Environments (IDEs) have built-in serial monitors, as seen in Fig. 5.6a. Serial data coming via the USB link will be displayed here. There is also the possibility of downloading and installing a standalone *terminal emulator*. Possibilities are Tera Term (as seen in Fig. 5.6b) or PuTTY for Microsoft Windows users, and CoolTerm for Apple OS X developers; Appendix E gives a little further background.

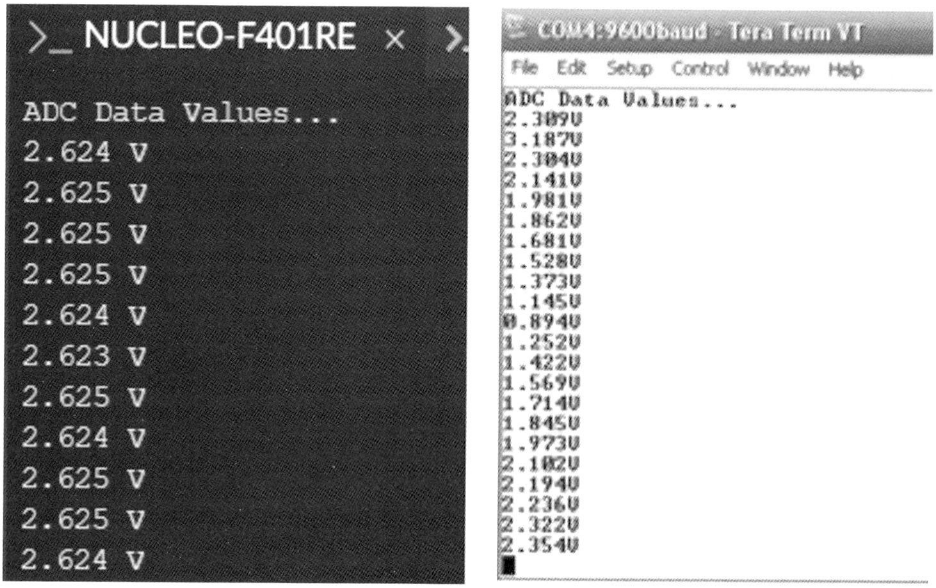

(a) (b)

Figure 5.6
Logged data on serial terminals: (a) on Mbed Studio; (b) on Tera Term.

C/C++ code feature

Start a new Mbed project and enter the code of Program Example 5.4a. Writing to the computer and the terminal emulator is achieved using the **printf( )** function. The Mbed OS directs any **printf( )** output to the USB serial port, and back to the host PC. (In previous OS versions, the link had to be set up within the program; you still see this in legacy code.) We see here **printf( )** for the first time, along with some of its far-from-friendly format specifiers. Check Section B9 for some background on this.

```
/*Program Example 5.4: Reads input voltage through the ADC, and transfers to PC
terminal.
Works for both LPC1768 (App Board and Breadboard) and F401RE, with pin allocations
shown.                                                                        */

#include "mbed.h"

AnalogIn Ain(p20);    //Replace p20 with A5 for F401RE

float ADCdata;

int main() {
  printf("ADC Data Values...\n\r");    //send an opening text message
  while(true){
    ADCdata = Ain;
    printf("%1.3f \n\r",ADCdata);     //send the data to the terminal
    ThisThread::sleep_for(500ms);
  }
}
```

(A)

a) the main.cpp file

There is a final complication in all of this. Because the use of **printf( )** can be very hungry in both program memory and execution time, the Mbed OS applies the "minimal" **printf( )** implementation (Reference 5.1). Among other things, this does not allow printing of floating point variables. As Program Example 5.4 requires such variables, their use needs to be enabled. This is done through the **Mbed_app.json** file. Understand for now that this file can be implemented to override configuration settings which already exist in an **Mbed_lib.json** file, which is part of the OS. JSON stands for Javascript Object Notation. It is a data interchange format which is programming-language independent, but with a structure similar to any in the C family of languages. Read Reference 5.2 for more detail. You will need to create a file using **File -> New File** in Mbed Studio, name the file **Mbed_app.json,** and place it in the program folder. Enter into it the small code block in Program Example 5.4B.

```
{
  "target_overrides":{
    "*": {
       "target.printf_lib": "std"
       }
    }
}
```

(B)

Program Example 5.4 Logging data into the PC: (A) the main.cpp file; (B) the Mbed_app.json file (for printing of floating point numbers)

Connect a potentiometer to your Mbed board, adapting the circuit of Fig. 5.5 (with LED not needed). Ensure that the analog input chosen aligns with that specified in the program, or adjust accordingly. You should be able to compile and run the code to give an output on your Studio IDE, Tera Term, or CoolTerm. If you have problems, check Appendix E or the Mbed site to ensure that you have set up the host terminal correctly.

5.3.2 Scaling ADC Outputs to Recognized Units

The data displayed through Program Example 5.4 is just a set of numbers proportional to the voltage input, in the range from 0 to 1. Yet they represent a range of voltages in the range of 0 to 3.3 V. The numbers can therefore be scaled readily to give a voltage reading, by multiplying by 3.3. Substitute the code lines below into the **while** loop of Program Example 5.4 to do this, and to place a unit after the voltage value.

```
ADCdata = Ain * 3.3;      //read and scale the data
printf("%1.3f",ADCdata); //send the data to the terminal
printf(" V\n\r");        //insert a unit
ThisThread::sleep_for(500ms);
```

Run the adjusted program; its output should appear similar to Fig. 5.6a or 5.6b. View the measured voltage on the PC screen, and read the actual input voltage on a digital voltmeter. How well do they compare?

■ Exercise 5.4

There is a simpler way of scaling the analog input to a voltage measurement, shown in Table 5.1. Insert the two lines:

```
set_reference_voltage(3.3);
Ain.read_voltage();
```

at appropriate places in Program Example 5.4, the first before the **while(true)**, the second within it. Remove any line no longer needed. Run the program again, and check that you are getting properly scaled voltage readings.

■

5.3.3 Applying Averaging to Reduce Noise

If you leave Program Example 5.4 running, with a fixed input and values displayed on a serial terminal, you may be surprised to see that the measured value is not always the same, but varies around some average value. You may already have noticed that the PWM value in Section 5.2.2 or 5.2.3 also appeared to fluctuate, even when the potentiometer was not being moved. Several effects may be at play here, but almost certainly you are seeing the effect of some interference, and all the problems it can bring. If you look with the oscilloscope at the ADC input (i.e., the "wiper" of the potentiometer), you are likely to see some high-frequency noise superimposed on this; exactly how much will depend on what equipment is running nearby, how long your interconnecting wires are, and a number of other things.

A very simple first step to improve this situation is to average the incoming signal. This should help to find the underlying average value and remove the high-frequency noise element. Try inserting the **for** loop shown below, replacing the `ADCdata=Ain;` line in Program Example 5.4. You will see that this code fragment sums 10 ADC values and takes their average. Try running the revised program, and see if a more stable output results. Note that while this sort of approach gives some benefit, the actual measurement now takes 10 times as long. This is a very simple example of digital signal processing.

```
for (int i = 0;i <= 9;i++) {
  ADCdata = ADCdata + Ain * 3.3;          //sum 10 samples
}
ADCdata = ADCdata / 10;                   //divide by 10
```

5.4 Some Simple Analog Sensors

Now that we are equipped with analog input, it is appropriate to explore some analog sensors. We begin with the simpler and more traditional ones, which have an analog output voltage that can be connected to the Mbed board ADC inputs. Later in the book other sensors will be introduced, which can communicate with an Mbed target board by a digital interface. Although the application board has a number of sensors, they all communicate digitally. Therefore, it is simplest if the following builds are done with a breadboard.

5.4.1 The Light-Dependent Resistor

The *light-dependent resistor* (LDR) is made from a piece of exposed semiconductor material. When light falls on it, its energy flips some electrons out of the crystalline structure; the brighter the light, the more electrons are released. These electrons are then available to conduct electricity, and the resistance of the material falls as a result. If the light is removed, the electrons pop back into their places, and the resistance goes up again.

The NORPS-12 LDR (Reference 5.3), made originally by Silonex, is readily available and low-cost. It is shown in Fig. 5.7, connected in a simple potential divider, giving a voltage output. Indicative data appears in Table 5.2. This shows that it has a resistance when completely dark of at least 1.0 MΩ, falling to a few hundred ohms when very brightly illuminated. The value of the series resistor, shown here as 10 kΩ, is chosen to give an output value of approximately mid-range for normal room light levels. It can be adjusted

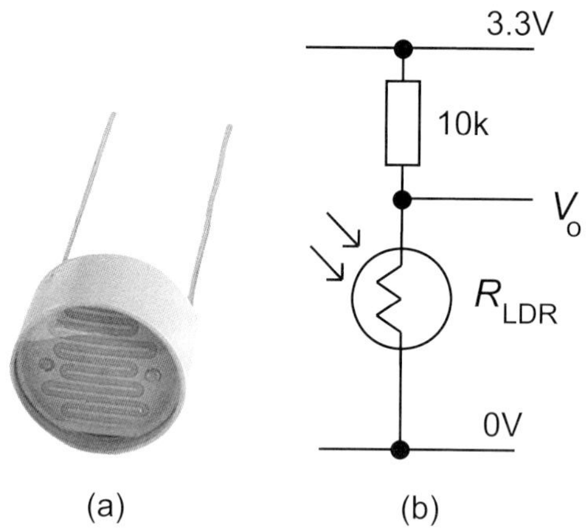

Figure 5.7
The NORPS-12 LDR: (a) the NORPS-12 LDR; (b) connected in a potential divider.

Table 5.2: NORPS-12 LDR—indicative resistance and output values.

Illumination (lux)	R_{LDR} (Ω)	V_o
Dark	≥1.0 M	≥3.27 V
10	9k	1.56 V
1,000	400	0.13 V

to modify the output voltage range. Putting the LDR at the bottom of the potential divider, as shown here, gives a low output voltage in bright illumination and a high output voltage in low illumination. This can be reversed by putting the LDR at the top of the divider.

The LDR is a simple, effective, and low-cost light sensor. Its output is not however linear, and each device tends to give a slightly different output from another. Hence, it is not used for precision measurements.

Exercise 5.5

Using the circuit of Fig. 5.7, connect a NORPS-12 LDR to an Mbed target using any analog input. Write a program to display light readings on the serial terminal. You will not be able to scale these into any useful unit. Try reversing resistor and LDR locations, and note the effect.

5.4.2 Integrated Circuit Temperature Sensors

Semiconductor action is highly dependent on temperature, so it is not surprising that semiconductor temperature sensors are made. A very useful form of sensor is one which is contained in an integrated circuit (IC), such as the LM35, Fig. 5.8. This device has an output of 10 mV/°C, with operating temperature up to 110°C (for the LM35C version). Thus, it is immediately useful for a range of temperature sensing applications. The simplest connection for the LM35, which we can use with the Mbed, is shown in Fig. 5.8. A range of more advanced connections (e.g., to get an output for temperatures below 0°C) are shown in the data sheet, Reference 5.4.

■ Exercise 5.6

Design, build, and program a simple temperature measurement system using an LM35 sensor, which displays temperature on the computer screen. The V_S pin of the sensor is the power supply (4 V to 20 V), which can be connected to pin 39 of the Mbed LPC1768, or the "+5 V" pin of the F401RE. When the sensor is connected, it is possible to plug it directly into a suitable location in a breadboard build. However, each terminal can also be soldered to a wire, so that remote sensing can be undertaken. If these wires are insulated appropriately, for example by coating connections with silicone rubber at the sensor end, then the sensor can be used to measure liquid temperatures.

Noting its maximum operating temperature, how well does the LM35C exploit the input range of your Mbed ADC?

■

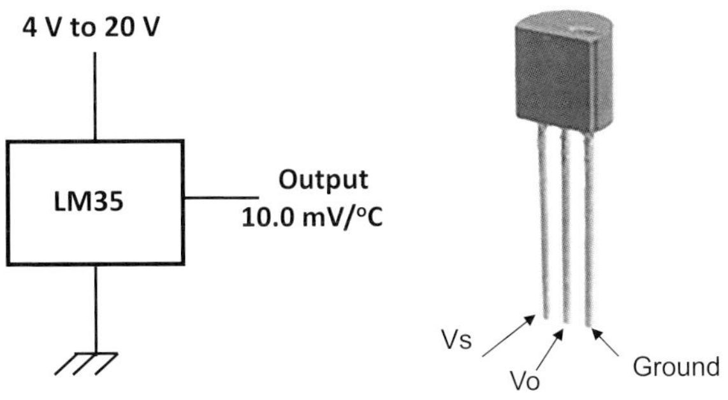

Figure 5.8
The LM35 IC temperature sensor.

5.5 Exploring Data Conversion Timing

Nyquist's sampling theorem suggests that a slow ADC will only be able to convert low-frequency signals. In designing a carefully specified system it is therefore very important to know how long each data conversion takes. For this reason, it is interesting to make a measurement of Mbed LPC1768 ADC and DAC conversion times and then put Nyquist to the test.

5.5.1 Estimating Conversion Time and Applying Nyquist

Program Example 5.5 provides a very simple mechanism for measuring conversion times, and then viewing Nyquist's sampling theorem in action. It adapts Program Example 5.1, but pulses a digital output between each two stages. Enter this as a new program, compile, and run.

```
/*Program Example 5.5: Inputs signal through ADC, and outputs to DAC. View DAC
output on oscilloscope. To demonstrate Nyquist, connect variable frequency signal
generator to ADC input. Allows measurement of conversion times, and explores
Nyquist limit.
Works for LPC1768. Not available for F401RE, as it has no DAC.
*/

#include "mbed.h"
AnalogOut Aout(p18);        //defines analog output on Pin 18
AnalogIn Ain(p20);          //defines analog input on Pin 20
DigitalOut test(p5);
float ADCdata;

int main() {
  while(true) {
    ADCdata = Ain; //starts A-D conversion, and assigns analog value to ADCdata
    test = 1;      //switch test output, as time marker
    test = 0;
    Aout = ADCdata;   //transfers stored value to DAC, and forces D-A conversion
    test = 1;         //a double pulse, to mark the end of conversion
    test = 0;
    test = 1;
    test = 0;
    // wait_us(1000);   //optional wait state,
                        //to explore different cycle times
  }
}
```

Program Example 5.5 Estimating data conversion times

This program allows a number of measurements to be made which are of great importance, and which require careful use of the oscilloscope. The measurements are presented in the two exercises which follow.

Exercise 5.7

Note that for this test you do not need anything connected to the ADC input. Running Program Example 5.5, observe carefully the waveform displayed by the "test" output (i.e., pin 5) on an oscilloscope—a digital storage oscilloscope will give best results. This may require some patience—they are very narrow pulses. You can widen the pulses if needed by inserting a tiny wait while the output is high.

You will be able to detect the single pulse at the end of the analog-to-digital conversion, and the double one at the end of the loop. Measure the time duration of the analog-to-digital conversion and the digital-to-analog conversion. What comment can you make on these? Note that the conversion times you measure are not the actual conversion times of the ADC and DAC themselves; they include all associated programming overheads. Keep a note of these values as we aim to account for them later in Exercises 5.8 and 14.8.

Exercise 5.8

Armed with the knowledge of the conversion times, connect a signal generator as input to the ADC. Choose sine wave and set the signal amplitude so that it is just under 3.3 V peak to peak; apply a DC offset so that the voltage value never goes below 0 V. This facility is available on most signal generators. Insert the **wait_us(1000);** line at the end of the loop ("commented out" in Program Example 5.5). This will give a sampling frequency of a little below 1 kHz. Nyquist's sampling theorem predicts that the maximum signal frequency that we can digitize will be 500 Hz for this sampling frequency. Let's test it.

Start initially with an input signal of around 200 Hz. Observe the input signal and DAC output on the two beams of the oscilloscope. You should see the input signal and a reconstructed version of it, something like Fig. 5.3, with a new conversion approximately every millisecond. Now gradually increase the signal frequency toward 500 Hz. As you approach Nyquist's limit, the output becomes a square wave. When input frequency equals sampling frequency, a straight line on the oscilloscope should occur, though in practice it may be difficult to find this condition exactly. As the input frequency increases further, an *alias* signal (as illustrated in Fig. 5.4) appears at the output.

Decrease the duration of the wait state, and predict and observe the new Nyquist frequency. Finally, remove the wait state altogether. The data conversion should now be taking place at the highest possible rate, with conversion time corresponding to your earlier measurement. The Nyquist limit that you now find is the limit for this particular hardware/software configuration.

■

5.6 Introducing the Debugger

At this point we have introduced most of the simpler I/O interfaces available on a microcontroller, and made some progress in programming. It is a good moment to create a program which combines these features, and then use it to introduce an important feature of any good IDE, the debugger. The debugger is also introduced in Reference 2.12, the IDE user guide. We use a motor control program to introduce the debugger. At the time of writing, the debugger was not fully available with the Mbed LPC1768, although its rollout was anticipated (check the most recent version of Reference 2.12 for updates on this). Therefore, it may not be possible to undertake this section fully with that device. It is still useful to build a circuit for Program Example 5.6, and to trial it. You should not find it difficult to choose connections for the circuit.

5.6.1 A Motor Control Program

The motor control program appears as Program Example 5.6. It uses digital I/O, PWM, and analog input. The motor has start and stop buttons, a safety guard, and a temperature sensor. The motor must not start if the safety guard is open, or if the motor is too hot. It must be stopped if the stop button is pressed, if the guard is opened, or if the operating temperature exceeds a certain value. The motor is driven by PWM, and its speed must be ramped up from rest when starting, and ramped down when stopping. A diagnostic word gives status information. The lower five bits count the number of times the motor has been started; the upper three bits indicate current position in the program.

To simulate the hardware (i.e., without a real motor, temperature sensor, or guard), a circuit can be built with four LEDs (Ready, Running, Fault, and motor PWM) and three switches (Start, Stop, and Guard). Switches can all be pushbuttons, though the guard switch can also be two-way. A potentiometer is needed to represent the temperature. A possible circuit diagram, for the Nucleo F401RE board, is given in Fig. 5.9.

The program moves through a series of **while** loops, first waiting to start, then running; it should be comparatively easy to follow from the in-line comments. As it enters the

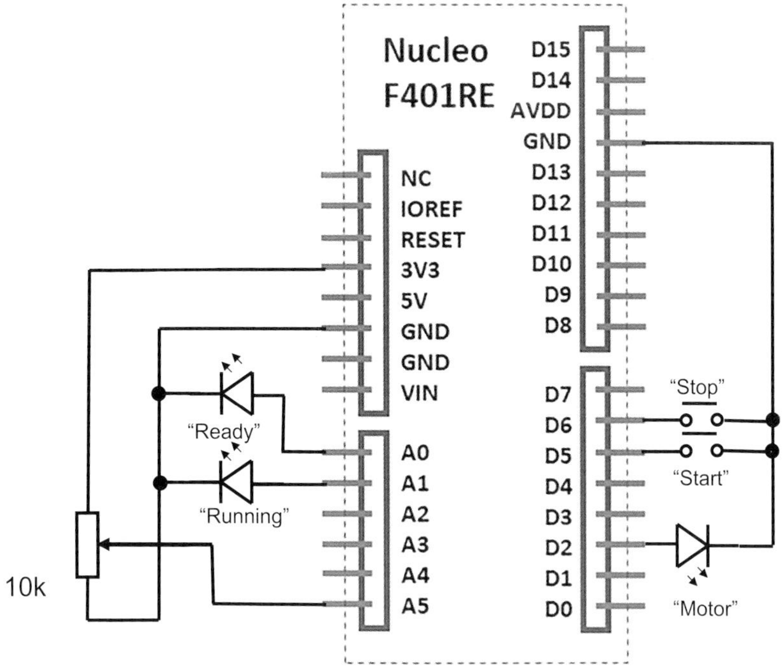

Figure 5.9
Motor control build for Nucleo REF401RE. *Note:* Onboard LED is used for "Fault." Onboard pushbutton is used for "Guard."

running mode, a function is called to ramp up the motor speed; on exiting this mode, the motor speed is ramped down by a similar function.

```
/* Program Example 5.6
Controls DC motor with PWM. Motor has guard and temp sensor.
User control is via start and stop buttons.
A diagnostic word is updated based on program execution; the lower five bits
count number of times motor has been started, upper 3 give status.
Runs on F401RE and LPC1768, but debug capability may not be available for the
latter.
*/

#include "mbed.h"

// Define the Input pins
DigitalIn START(D5,PullDown);    //Start control, start = 1
DigitalIn STOP(D6,PullDown);     //Stop control, stop = 1
DigitalIn GUARD(BUTTON1);        //Machine guard sensor; guard closed = 1
AnalogIn TEMP(A5);               //Temperature sensor; temp must be <= 0.75

//Define the Output pins
DigitalOut READY(A0);        //Motor is ready to run, i.e. guard closed, temp OK.
```

```
DigitalOut RUNNING(A1);      //Motor is running
DigitalOut FAULT(LED1);      //Guard is opened while motor running
PwmOut Mot_Drive(D2);        //Motor drive, 1kHz.
char diagnostic = 0;

void motor_rundown(void);    //Function prototypes
void motor_runup(void);

void motor_runup(void){     //runs motor up to speed, still checking for Stop button
  for (int j = 2;j <= 8;j++){
    Mot_Drive.pulsewidth_us(j * 100);
    ThisThread::sleep_for(100ms);

    if (STOP == 1) break;
  }
}
void motor_rundown(void){  //runs motor down from speed
  for (int j = 7;j >= 0;j-){
    Mot_Drive.pulsewidth_us(j * 100);
    ThisThread::sleep_for(100ms);
  }
}

int main(){
  Mot_Drive.period_ms(1);      //configure PWM
  Mot_Drive.pulsewidth_ms(0);
    while(true) {
      RUNNING = FAULT = 0;     //clear all indicators
      diagnostic=diagnostic&0b00011111; //clear status info, keep cycle counter
      //A wait to start state
      while(RUNNING == 0){
        while ((GUARD==0)||(TEMP>0.75)){     //wait in loop until machine is ready
          diagnostic = diagnostic|0b00100000;
          RUNNING = FAULT = 0; //clear all displays
          READY = !READY;       //flash READY, to indicate not ready
          ThisThread::sleep_for(200ms);
        }
        READY = 1;   //motor is ready to run, READY LED holds steady!
        while((START == 0) && (GUARD == 1)) //wait for START to be pressed
        diagnostic = diagnostic|0b01000000;
        //Either Guard has opened, or correct Start
        if(START == 1){
          RUNNING = 1;      //Start the motor up, and indicate
          motor_runup();
          diagnostic = diagnostic|0b01100000;
          if((diagnostic & 0b00011111) <( 0b00011111))
            diagnostic++;           //increment count, unless already full
      }
    }     //Return to start of loop if RUNNING still 0, ie Guard reopened

    //Here if running stay in this loop while running
```

```
      while((GUARD == 1) && (TEMP < 0.75) && (STOP == 0)){
        READY = 0;
      }
      //here if exiting Running
      RUNNING = 0;
      motor_rundown();
      int tempval = TEMP;
      if ((tempval > 0.75)||(GUARD == 0)){      //check for a fault condition
        diagnostic = diagnostic|0b10000000;
        FAULT = 1;    //light fault LED
        ThisThread::sleep_for(500ms);
        FAULT=0;
      }
    }   //end of while(true)
  }     //end of main()
```

Program Example 5.6 A simple motor control program

5.6.2 Simple Use of the Debugger

Create a new program around Example 5.6. To compile the program and then enter debug, press the debug symbol . This compiles and downloads the program, and then opens the debugger screen—the debug session has started. The top half of this is seen in Fig. 5.10. An olive bar across the screen indicates where program execution is placed. On start-up, it waits at the beginning of **main( )**, as seen.

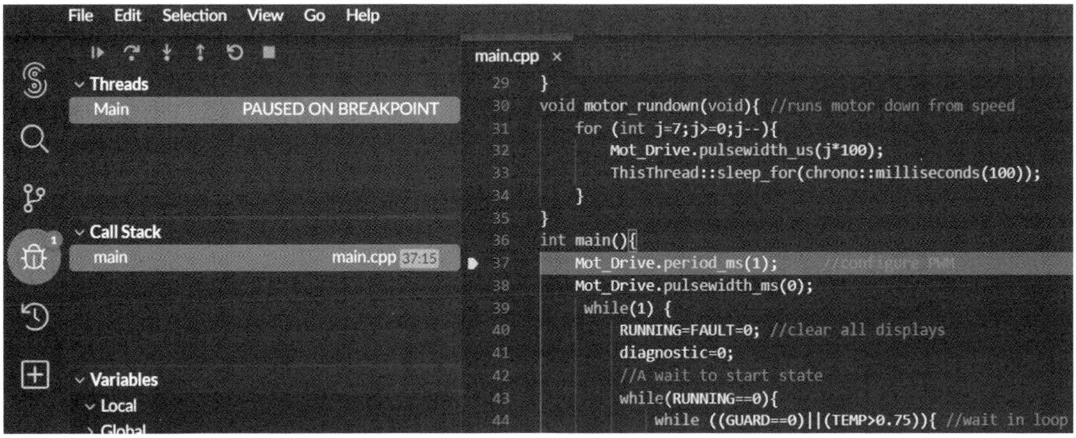

Figure 5.10
Debugger screen. *Image courtesy of ARM.*

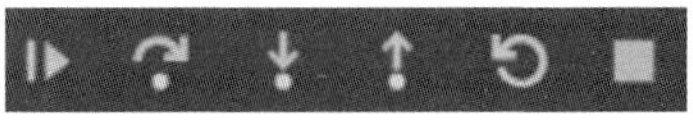

Figure 5.11
Debugger controls. *Image courtesy of ARM.*

The control buttons shown in Fig. 5.11 appear top left of the debugger screen. Program execution can now be controlled by using the buttons. **Continue**, **Restart**, and **Stop** are self-explanatory. **Step Over** advances program execution to the next line of source code. **Step Into** advances the debugger into a function (which may include a "hidden" API function), while **Step Out** takes it out of a function. The latter two are useful, as you are unlikely to want to step through every function in the program (especially not all the API ones); once in a function, you may just want to leave it to return to the bigger program picture.

With your circuit hardware in place, set the Guard open (press the onboard pushbutton) and try single-stepping through the program, using **Step Over.** You should find yourself looping within the first loop, that is, the second use of **while**. If you release the Guard, ensuring that temperature is on the low side, you should then be able to enter the next waiting state, waiting for Start to be pressed. You can explore moving through the program in this way. If you advance to the **motor_runup( )** function, you will see that **Step Over** does just that—you do not enter the function, but instead it executes in real time, as you step over it. If instead you approach the same function, but then press **Step Into** when that line is highlighted, the debugger enters the function. In order to follow your own lines of code, you should then return to using **Step Over.** Notice that on any line of source code, if you do press **Step Into**, then you are likely to enter execution of the underlying API code. This can be an informative and/or a scary process! Either way, a mass of coding detail is suddenly presented. Use **Step Out** to get out of a function whose detail you do not wish to see.

Of course, it is useful to be able to step through a program in this way, but in anything but the smallest of programs you do not want to labor through every code line. **Breakpoints** can be inserted into the code, and you can then run the program to the next breakpoint, examine operating conditions once there, or run on to the next breakpoint. A first breakpoint has already been inserted at the start of **main( )**; we see this in Fig. 5.10. Breakpoints are easily inserted elsewhere in the code, simply by left-clicking against a line of code, to the left of the line numbers. Right-click on the breakpoint to remove or disable it; see Fig. 5.12.

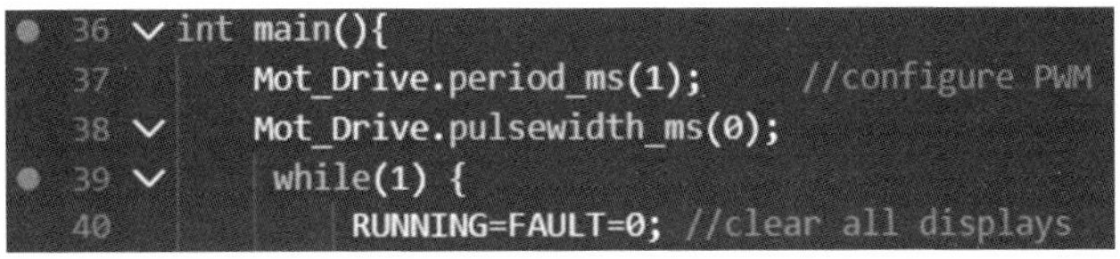

36 int main(){
37 Mot_Drive.period_ms(1); //configure PWM
38 Mot_Drive.pulsewidth_ms(0);
39 while(1) {
40 RUNNING=FAULT=0; //clear all displays

Figure 5.12
Inserting breakpoints. *Image courtesy of ARM.*

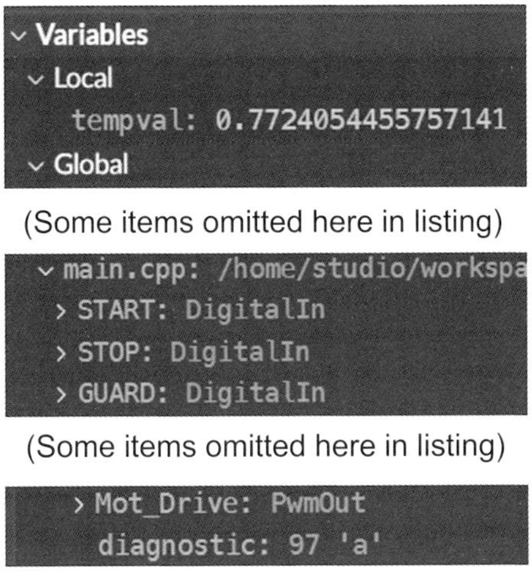

Figure 5.13
View of selected variables. *Image courtesy of ARM.*

When you run to a breakpoint, you can then inspect variable values to the left of the screen, as seen in Fig. 5.13. This is taken from a breakpoint inserted where the motor is exiting running. You can see that the motor temperature has just exceeded 0.75 (with a meaningless string of decimal places), which has forced the exit from run. Notice that **tempval** is a local variable, being declared as it is used. Global variables can be sought out in the Global listing, as shown. The ones relating to the program are grouped under **main.cpp**.

This completes a debugger introduction. Develop your confidence with it by using it for further program builds. Refer to Reference 2.12 for details on more advanced features, like RAM debugging, or Memory view.

5.7 Mini-Projects

5.7.1 Two-Dimensional Light Tracking

Light-tracking devices are very important for the capture of solar energy. Often, they operate in three dimensions, and tilt a solar panel so that it is facing the sun as accurately as possible. To start rather more simply, create a two-dimensional light tracker by fitting two LDRs, angled away from each other by around 90°, to a servo. Connect the LDRs to two ADC inputs using the circuit of Fig. 5.7b. Write a program which reads the light values sensed by the two LDRs and rotates the servo so that each is receiving equal light. The servo can only rotate 180°. However, this is not unreasonable, as a sun-tracking system will be located to track the sun from sunrise to sunset, that is, not more than 180°. Can you think of a way of meeting this need using only one ADC input?

5.7.2 Temperature Alarm

Using an LM35 and a piezo transducer (Fig. 4.12), make a temperature alarm. Define two threshold temperatures, which should be above room temperature, but not hazardous. When the lower threshold is passed, the transducer should beep at a low rate (e.g., once per second). When the higher one is passed, it should beep at a high rate (e.g., 10 times a second). Example thresholds could be 26°C and 32°C. Then, assuming a room temperature of around 20°C, it should be possible to hand-warm the sensor to these temperatures. Those working in hotter environments may wish to adjust these temperatures. Different heat sources and temperatures can also be explored, ensuring that safe operating conditions are maintained at all times.

Chapter Review

- It is important to understand ADC characteristics, in terms of input range, resolution, and conversion time.
- An ADC is available in both the Mbed LPC1768 and the Nucleo F401RE; they can be used to digitize analog input signals.
- Nyquist's sampling theorem must be understood and applied with care when sampling AC signals. The sampling frequency must be at least twice that of the highest-frequency component in the sampled analog signal.
- Aliasing occurs when the Nyquist criterion is not met; this can introduce false frequencies into the data. Aliasing can be avoided by applying an anti-aliasing filter to the analog signal before it is sampled.
- Data gathered by the ADC can be further processed and displayed or stored.
- There are numerous sensors available which have an analog output; in many cases this output can be connected directly to the Mbed ADC input.
- A debugger allows the user to run the program to a breakpoint, single step, or step into or out of a function. At each point, internal registers, program variables, or memory locations may be inspected. More advanced features are also available.

Quiz

1. Give three types of analog signals which might be sampled through an ADC.
2. An ideal 8-bit ADC has an input range of 5.12 V. What are its resolution and greatest quantization error?
3. Explain how a single ADC can be used to sample four different analog signals.
4. An ideal 10-bit ADC has a reference voltage of 2.048 V, and behaves according to Eq. 5.1. For a particular input its output reads 10 1110 0001. What is the input voltage?
5. What will be the result if an Mbed LPC1768 or F401RE is required to sample an analog input value of 4.2 V?
6. An ultrasound signal of 40 kHz is to be digitized. Recommend the minimum sampling frequency.
7. The conversion time of an ADC is found to be 7.5 μs. The ADC is set to convert repeatedly, with no other programming requirements. What is the maximum-frequency signal it can digitize?
8. The ADC in Question 7 is now used with a multiplexer, so that four inputs are repeatedly digitized in turn. A further time of 2,500 ns per sample is needed to save the data and switch the input. What is the maximum frequency of the signal that can now be digitized?
9. An LM35 temperature sensor is connected to an Mbed LPC1768 ADC input and senses a temperature of 30°C. What is the binary output of the ADC?
10. What will be the value of integer **x** for input voltages of 1.5 V and 2.5 V sampled by an Mbed board using the following program code?

```
#include "mbed.h"
AnalogIn Ain(p20);       //Use A5 for F401RE
int main(){
  int x = Ain.read_u16();
}
```

References

5.1. printf and Reducing Memory (in Mbed OS). https://os.Mbed.com/docs/Mbed-os/v6.6/apis/printf-and-reducing-memory.html.

5.2. The (Mbed OS) Configuration System. https://os.Mbed.com/docs/Mbed-os/v6.15/program-setup/advanced-configuration.html.

5.3. The NORPS-12 data sheet. http://www.farnell.com/datasheets/409710.pdf.

5.4. LM35 Precision Centigrade Temperature Sensors. Texas Instruments; 2017. http://www.ti.com/lit/ds/symlink/lm35.pdf.

CHAPTER 6

Interrupts and Timers

6.1 Thinking about Time

The very first diagram in this book, Fig. 1.1, shows the key features of an embedded system. Among these is *time.* Embedded systems have to respond in a timely manner to events as they happen, *or* they need to be able to initiate events at the right time. Usually, this means they have to be able to do the following:

- Respond with appropriate speed to external events, which occur at unpredictable times;
- Measure time durations;
- Generate time-based activities, which may be single or repetitive.

In doing these things, the system may find that it has a conflict of interest, with two actions needing attention at the same time. For example, an external event may demand attention just when a periodic event needs to take place. Therefore, the system may need to distinguish between events which have a high level of urgency, and those which do not, and take action accordingly.

It follows that we need a set of tools and techniques to allow effective time-based activity to occur. Key features of this toolkit are interrupts and timers, the subjects of this chapter. In brief, a timer is just what its name implies, a digital circuit which enables time measurement, and can make things happen when a certain time has elapsed. An interrupt is a mechanism whereby a running program can be interrupted by an external input, with the CPU then being required to jump to some other activity. The two often work hand-in-hand, and greatly improve our capability to develop effective embedded systems.

We are going to see some simple activities triggered by external events, and some triggered by time. In keeping with well-used terminology, if a little informally to begin with, we will call these activities *tasks*, and refer to *event-triggered* and *time-triggered* tasks. Concepts which emerge here are developed much further in Chapter 9, on real-time operating systems (RTOSs).

Fast and Effective Embedded Systems Design. https://doi.org/10.1016/B978-0-323-95197-5.00006-1

6.2 Responding to External Events

6.2.1 Polling

A simple example of an event-triggered task is when a user presses a pushbutton. This can happen at any time, without warning, but when it does the user expects a response. One way of programming for this is to test the external input continuously. This is illustrated in Fig. 6.1, where a program is structured as a continuous loop. Within this, it tests the state of two input buttons and responds to them if activated. This way of checking external events is called *polling*; the program ensures that it periodically checks input states, and responds if needed. This is the sort of approach we have used so far in this book. It works well for simple systems; but it is not adequate for more complex programs.

Suppose the program of Fig. 6.1 is extended, so that a microcontroller has 20 input signals to test in each loop. On most loop iterations, the input data may not even change, so we are running the polling for no apparent benefit. Worse, the program might spend time checking the value of less important inputs, while not responding quickly when a major fault condition has arisen.

There are two main problems with polling:

1. The processor cannot perform any other operations during a polling routine.
2. All inputs are treated as equal; the urgent input has to wait its turn before it is recognized.

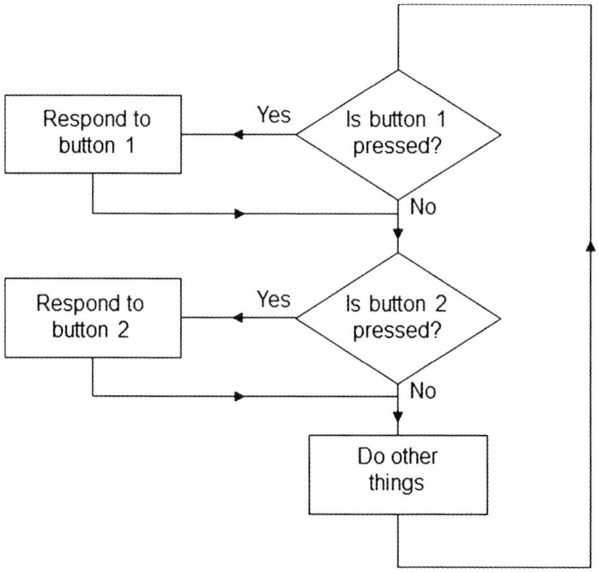

Figure 6.1
A simple program using polling.

A better solution is for input changes to announce themselves; time is not wasted finding out that there is no change. The difficulty lies in knowing when the input value has changed. This is the purpose of the interrupt system.

6.2.2 Introducing Interrupts

The interrupt represents a radical alternative to the polling approach just described. With an interrupt, the hardware is designed so that an external input to the CPU can stop it in its tracks and demand attention. Suppose you lived in a house, and were worried that a thief might come in during the night. You *could* arrange an alarm clock to wake you up every half hour to check there was no thief, but you would not get much sleep. In this case you would be *polling* the possible thief "event". Alternatively, you could install a burglar alarm. You would then sleep peacefully, *unless* the alarm went off and interrupted your sleep, and you would jump up and chase the burglar. In very simple terms, this is the basis of the computer interrupt.

Interrupts have become a very important part of the structure of any microprocessor or microcontroller, allowing external events and devices to force a change in CPU activity. In early processors, interrupts were mainly used to respond to major external events; designs allowed just one, or a small number of interrupt sources. However, the interrupt concept was found to be so useful that more and more possible interrupt sources were introduced. Many of these are now generated by the microcontroller peripherals, signaling, for example, that an ADC conversion is complete, or a new serial message has arrived.

In responding to interrupts, most microprocessors follow the pattern of the flow diagram shown in Fig. 6.2. The interrupt appears as an external signal, recognized by the CPU. On receiving an interrupt, the CPU completes the instruction it is currently executing. As it is about to go off and find a completely different piece of code to execute, it must save key information about what it has just been doing; this is called the *context.* The context includes at least the value of the *program counter* (this tells the CPU where it should come back to when the interrupt has completed) and generally a set of key registers, for example, those holding current data values. All this is saved on a small block of memory local to the CPU, called the *stack*. The CPU then runs a section of code called an *interrupt service routine* (ISR); this has been specifically written to respond to the interrupt which has occurred. The address of the ISR is found through a memory location called the *interrupt vector.* On completing the ISR, the CPU returns to the point in the main code immediately after the interrupt occurred, finding this by retrieving the program counter from the stack where it left it. It then continues program execution as if nothing happened.

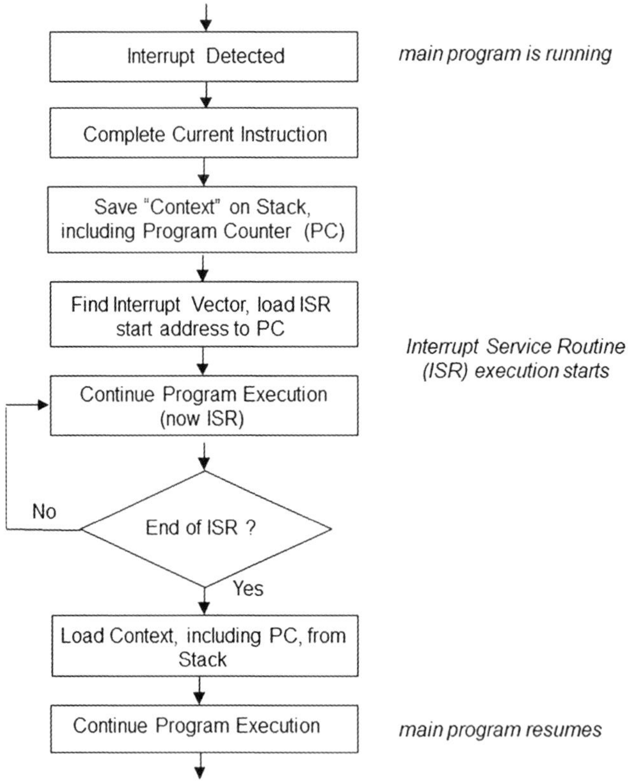

Figure 6.2
A typical microprocessor interrupt response.

6.3 Interrupts with the Mbed Operating System

The interrupt capability of most microcontrollers, like the LPC1768 or F401RE, is extremely sophisticated and complex. The Mbed API exploits only a small subset of this capability, focussing on the external interrupts. Some of the API functions are shown in Table 6.1. Using these we can create an interrupt input, write the corresponding ISR, and link the ISR with the interrupt input. Most digital I/O can be used for interrupts. For example, on the LPC1768, any of pins 5 to 30 can be used as interrupt inputs, except only pins 19 and 20.

Program Example 6.1 is a simple interrupt program, adapted from the Mbed website. It is made up of a continuous loop, which switches an LED labeled **flash** on and off. The interrupt input is labeled **button**. A tiny ISR is written for it, called **ISR1**, structured exactly as a function. The address of this function is attached to the rising edge of the interrupt input, in the line.

Table 6.1: Mbed Interrupt API Summary.

Function	Usage
Constructor	
InterruptIn (PinName *pin*)	Create an InterruptIn connected to the specified pin, of given name.
Member Functions	
void **rise** (callback<void()> *func*)	Attach a function to call when a rising edge occurs on the input.
void **fall** (callback<void()> *func*)	Attach a function to call when a falling edge occurs on the input.
void **enable_irq**()	Enables the interrupt.
void **disable_irq**()	Disables the interrupt.

```
button.rise(&ISR1);
```

A fuller callback mechanism may also be invoked, as seen in the Table. When the interrupt is activated by this rising edge, the ISR executes and the LED toggles. This can occur at any time in program execution. The program has effectively one time-triggered task, the continuous LED switching, and one event-triggered task, the response to the interrupt.

```
/* Program Example 6.1: Simple interrupt example. External input causes interrupt,
while led flashes.
Runs on either LPC1768 or F401RE with pin connections shown below
                                                        */

#include "mbed.h"
//define and name the interrupt input
InterruptIn button(p5);     //Replace p5 with BUTTON1 for F401RE
DigitalOut led(LED1);
DigitalOut flash(LED4);     // Replace LED4 with A0 for F401RE

void ISR1() {               //this is the response to interrupt, i.e. the ISR
  led = !led;
}
int main(){

button.rise(&ISR1);     // attach the address of the ISR function to the
                            // interrupt rising edge
       while(true) {        // continuous loop, ready to be interrupted
       flash = !flash;
       ThisThread::sleep_for(100ms);
   }
}
```

Program Example 6.1 Introductory use of an interrupt.

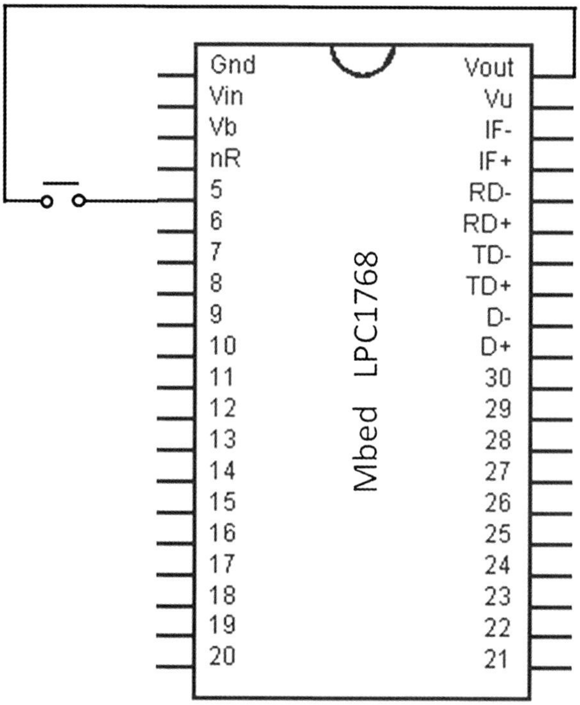

Figure 6.3
Mbed LPC1768 circuit for Program Example 6.1. Use on-board pushbutton for Nucleo F401RE, with extra LED on pin A0.

Compile and run Program Example 6.1. For the Mbed LPC1768, apply the very simple build of Fig. 6.3; for the F401RE, apply Fig. 3.1. The LPC1768 build depends on the internal pull-down resistor, which is enabled by default during initialization. Notice that when you push the button the interrupt is taken high, and LED1 changes state; LED4 meanwhile continues its flashing, almost unperturbed. On the Nucleo board, which applies Fig. 3.5(d) for its push button, the situation is reversed. When you push the button the interrupt is taken low, and there is no change to LED1; it changes state when the button is released. *If* you experience erratic behavior with this program, you may be experiencing switch bounce. In this case, fast forward to Section 6.10 for an introduction to this important topic.

■ Exercise 6.1

Change Program Example 6.1 so that:

i) The interrupt is triggered by a falling edge on the input.
ii) There are two ISRs, from the same pushbutton input. One toggles LED1 on a rising interrupt edge, and the other toggles a further LED on a falling edge.

■

6.4 Getting Deeper into Interrupts

We can take our first generalized understanding of interrupts further and try to understand a little more of what goes on inside the microcontroller. This is not essential as far as using the Mbed API is concerned. In fact, although both the LPC1768 and the F401RE microcontrollers have very sophisticated interrupt structures, the Mbed API exploits only a small part of this. Therefore, go immediately to the next section if you do not want to get into any deeper interrupt detail.

Okay, so you are still here! Let's extend our ideas of interrupts. While we said earlier that an interrupt was possibly like a thief coming in the night, imagine now a different scenario. Suppose you are a teacher of a big class made up of enthusiastic but poorly behaved kids; you have given them a task to do but they need your help. Tom calls you over, but while you are helping him Jane starts clamoring for attention.

Do you:

> Tell Jane to be quiet and wait until you have finished with Tom?
> or
> Tell Tom you will come back to him, and go over to sort Jane out?

To make matters worse, your school principal has asked you to let the members of the school band out of class half an hour early, but you really want them to finish their work before they go. This influences the above decision. Suppose Jane is in the band, but Tom is not. In this situation, you decide you must leave Tom in order to help Jane.

This school classroom situation is reflected in almost any embedded system. There could be a number of interrupt sources, all possibly needing attention. Some will be of great importance, others much less. Therefore, most processors contain four important mechanisms:

1. Interrupts can be *prioritized*; in other words, some are recognized as more important than others. If two occur at the same time, then the higher-priority one executes first.
2. Interrupts can be *masked*, in other words switched off, if they are not needed, or are likely to get in the way of more important activity. This masking could be just for a short period, for example, while a critical program section completes.
3. Interrupts can be *nested*. This means that a higher-priority interrupt can interrupt one of lower priority, just like the teacher leaving Tom to help Jane. Working with nested interrupts increases the demands on the programmer, and is strictly for advanced players

only. Not all processors permit nested interrupts, and some allow you to switch nesting on or off.

4. The location of the ISR in memory can be selected to suit the memory map and programmer wishes.

Let's take on just a couple more important interrupt concepts, these from the point of view of the interrupt source. Go to the moment in the above scenario when Jane suddenly realizes she needs help and puts her hand up. Some short time later the teacher comes over. The delay between her putting up her hand and the teacher actually arriving is called the interrupt *latency*. Latency may be due to a number of things. In this case, the teacher has to notice Jane's hand in the air, may need to finish with another pupil, and then actually has to walk over. Once the teacher arrives, Jane puts her hand down. While Jane is waiting with her hand in the air, patiently we hope, her interrupt is said to be *pending*.

These concepts and capabilities hint at some of the deep magic that can be achieved with advanced interrupt structures.

We can put all of this in more technical terms, refining our understanding of interrupt action. This was first illustrated in the flow diagram of Fig. 6.2. Further detail on part of this figure is now shown in Fig. 6.4. The interrupt being asserted is like Jane putting up

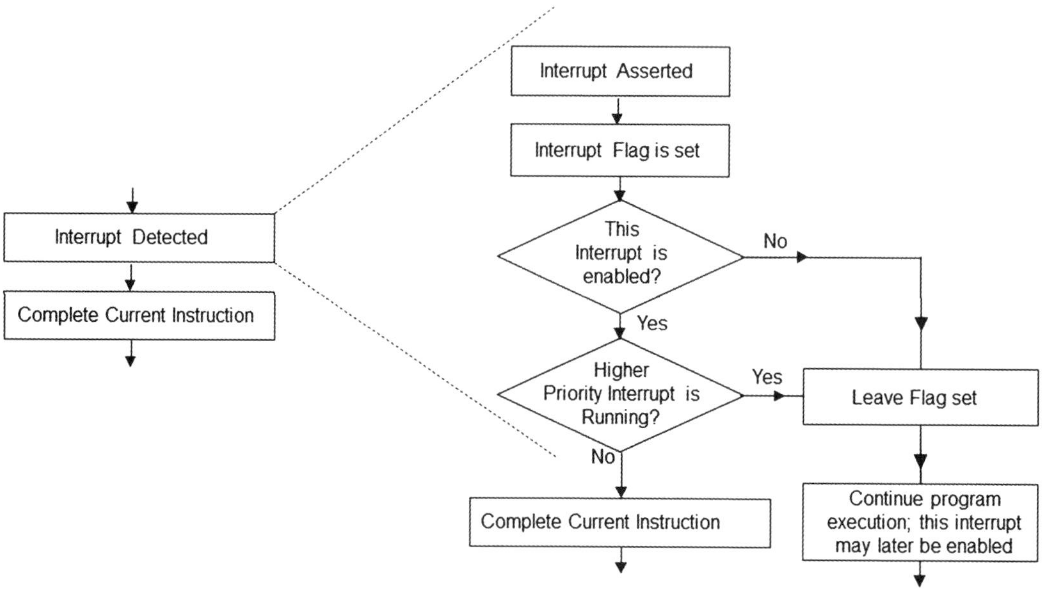

Figure 6.4
A typical microprocessor interrupt response—some greater detail.

her hand. In a microprocessor, the interrupt input will be a logic signal; depending on its input configuration, it may be active high or low, or triggered by a rising or falling edge. This input will cause an internal *flag* to be set. This is normally just a single bit in a register, linked to that interrupt source, which records the fact that it has occurred. This doesn't necessarily mean that the interrupt automatically gets the attention it seeks. If it is not enabled (i.e., it is masked), then there will be no response. However, the flag is left high as the program might later enable that interrupt, or the program may just poll the interrupt flag. Back to the flow diagram—if another ISR is already running, then again the incoming interrupt may not get a response, at least not immediately. If it is a higher priority, and nested interrupts are allowed, then it will be allowed to run. If it is a lower priority, it will have to wait for the other ISR to complete. The subsequent actions in the flow diagram, as already seen in Fig. 6.2, then follow. Note that the figure is potentially misleading, as it implies that these actions happen in turn. To get low latency, they should happen as fast as possible; a good interrupt management system will allow some of the actions to take place in parallel. For example, the interrupt vector could be accessed while the current instruction is completing.

6.4.1 Interrupts on the LPC1768

Now we will get back to microprocessor hardware. Using the LPC1768 as an example, recall that it contains the ARM Cortex core. Back in Fig. 2.3, we saw the Cortex core set within the LPC1768 microprocessor. Management of all interrupts in the Cortex is undertaken by the formidable-sounding *nested vectored interrupt controller* (NVIC). You could think of this as managing the processes overviewed in Figs. 6.2 and 6.4. The NVIC is also a bit like an electronic control box with a lot of unconnected wires hanging out. When the Cortex core is embedded into a microcontroller, such as the LPC1768, the chip designer assigns and configures those features (through the "loose wires") of the NVIC which are needed for that application. For example, this Cortex core allows 240 possible interrupts, both external and from the peripherals, and theoretically 256 possible priority levels. The LPC1768, however, has "only" 33 interrupt sources, with 32 possible programmable priority levels. The F401RE microcontroller, also having a Cortex core, depends in a similar way on the NVIC to manage its interrupts.

6.4.2 Testing Interrupt Latency

Now that we have met the concept of interrupt latency, we will test it in the Mbed environment. Program Example 6.2 adapts Program Example 6.1, but the interrupt is now generated by an external square wave, instead of an external button push. This makes it easier to see on an oscilloscope. When an interrupt occurs, the digital output **flash** is pulsed high for a fixed duration.

```
/* Program Example 6.2: Tests interrupt latency. External input causes interrupt,
which pulses external LED while LED1 flashes continuously.
Runs on either LPC1768 or F401RE with pin connections shown below
                                                                            */
#include "mbed.h"
//Connect input square wave here
InterruptIn squarewave(p5);        //Replace p5 with D5, for Nucleo F401RE
DigitalOut led(p6);                //Replace p6 with A0, for Nucleo F401RE
DigitalOut flash(LED1);

void pulse() {                     //ISR sets external led high for fixed duration
  led = 1;
  wait_us(2000);
  led = 0;
}

int main() {
  squarewave.rise(&pulse); // attach the address of the pulse function to
                                                            // the rising edge
  while(true) {                 // interrupt will occur within this endless loop
    flash = !flash;
    ThisThread::sleep_for(250ms);
  }
}
```

Program Example 6.2 Testing interrupt latency

Create a new project from Program Example 6.2. Connect an external LED between the LED output and ground, ensuring correct polarity. Connect to the *square wave* input a signal generator set to logic-compatible square wave output (probably labeled "TTL compatible"), running initially at around 10 Hz. Once connected, with the program running, the external LED should flash at this rate, around 10 times a second.

Exercise 6.2

Connect two inputs of an oscilloscope to the interrupt input and the LED output triggering from the interrupt. Increase the input frequency from the signal generator to around 50 Hz. Set the oscilloscope time base to 5 μs per division. You should be able to see the rising edge of the interrupt input, and a few microseconds later the LED output rising. The time delay between the two is an indication of latency. The rise of the LED will be flickering a little, as the delay will depend on what the CPU is doing at the instant the interrupt occurs. It is important to note that the latency as measured here depends on both hardware and software factors.

■

6.4.3 Disabling Interrupts

Interrupts are an essential tool in embedded design. But because they can occur at any time, they can have unexpected or undesirable side effects. There may be code sections—sometimes called *critical regions*—where it is essential to disable (mask) the interrupt. This can include when you are undertaking a time-sensitive activity, or a complex calculation which must be completed in one go. In general, an incoming interrupt which is masked will leave its flag set, so a response can be made once the interrupt is enabled again. Of course, a delay in response has been introduced, and the latency much compromised.

The compiler allows interrupts to be disabled, as seen in Table 6.1, and below. This may be hardware-specific, so it needs checking. For the interrupt used in Program Example 6.2, these lines can be used:

```
squarewave.disable_irq();                //disable squarewave interrupt
... //insert here activity which must not be interrupted
squarewave.enable_irq();                //enable squarewave interrupt
...
```

■ Exercise 6.3

Using Program Example 6.2, disable the interrupt for the duration of the **This-Thread::sleep_for()** in the **while** loop. Connect the oscilloscope as in Exercise 6.2, and observe the output again. Comment on how the ISR responds in this new program version.

Experiment with different values for this sleep function. Then try splitting it into two, running immediately after each other, with one having interrupts disabled, and the other having them enabled.

■

6.4.4 Interrupts from Analog Inputs

Aside from digital inputs, it is useful to generate interrupts when analog signals change—for example, if an analog temperature sensor exceeds a certain threshold. One way to do this is by applying a *comparator*. A comparator is just that: it compares two input voltages. If one input is higher than the other, then the output switches to a high state; if it is lower, the output switches to a low state. A comparator can easily be configured from an operational amplifier (op amp), as shown in Fig. 6.5. Here, an input voltage, labeled V_{in}, is compared with a threshold voltage derived from a potential

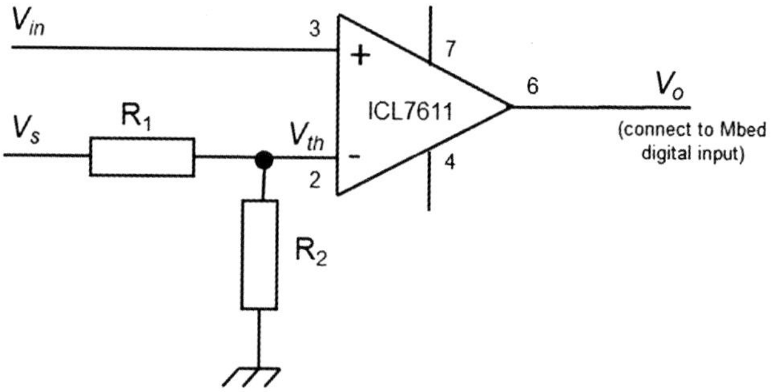

Figure 6.5
A simple comparator circuit.

divider, made from the two resistors R_1 and R_2. These are connected to the supply voltage, labeled V_S. For the right choice of op amp or comparator, and with suitable supply voltages, the output is a Logic 1 when the input is above the threshold value, and Logic 0 otherwise. The threshold voltage just mentioned, labeled V_{th} in the diagram, is calculated using Eq. 6.1.

$$V_{th} = V_s\, R_2/(R_1 + R_2) \tag{6.1}$$

As an example, we might want to generate an interrupt from a temperature input, using an LM35 temperature sensor (Fig. 5.8), with the interrupt triggered if the temperature exceeds 30°C. This sensor has an output of 10 mV/°C, which would lead to an output voltage of 300 mV at the trigger point proposed. To set V_{th} to 300 mV in Fig. 6.5, we apply Eq. 6.1, with a V_S of 3.3 V. This leads to $R_2 = 0.1\ R_1$. Values of $R_1 = 10k$ and $R_2 = 1k$ could therefore be chosen.

■ Exercise 6.4

Unlike many op amps, the ICL7611 can be run from very low supply voltages, even the 3.3 V available on the Mbed platforms. Using an LM35 IC temperature sensor and an ICL7611 op amp connected as a comparator, design and build a circuit which causes an interrupt when the temperature exceeds 30°C. Write a program which lights an LED when the interrupt occurs. The pin connections shown in Fig. 6.5 can be applied. Pin 7 is the positive supply, and can be connected to the Mbed platform 3.3 V; pin 4 is the negative supply, and is connected to 0 V. For this op amp, also connect pin 8 to the positive supply rail. (This pin controls the output drive capability; check the data sheet for more information.)

■

6.4.5 Conclusion on Interrupts

We have had a good introduction to interrupts and the main concepts associated with them. They are an essential part of the toolkit of any embedded designer. We have so far limited ourselves to single-interrupt examples. Where multiple interrupts are used, the design challenges become considerably greater—interrupts can have a very destructive effect if not used with care. We return to interrupts and their use in an RTOS context in Chapter 9.

6.5 An Introduction to Timers

As we write more demanding programs, we need the right tools to measure time and to trigger time-based activities. Ideally, there should be a way of letting timing activity go on in the background, while the program continues to do other useful things. We turn to the digital hardware for this.

6.5.1 The Digital Counter

It is an easy task in digital electronics to make electronic counters; you simply connect together a series of bistables or flip-flops—check Reference 3.1 for details. Each holds one bit of information, and all those bits together form a digital word. This is illustrated in very simple form in Fig. 6.6, with each little block representing one flip-flop, each holding one bit of the overall number. If the input of this arrangement is connected to a clock signal, then the counter will count, in binary, the number of clock pulses applied to it. It is easy to read the overall digital number held in the counter, and it is not difficult to arrange the necessary logic to preload it with a certain number, or to clear it to zero.

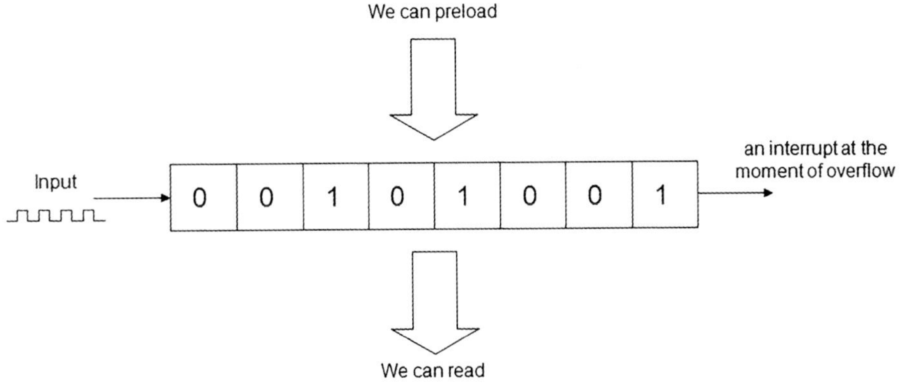

Figure 6.6
A simple 8-bit counter.

The number that a counter can count up to is determined by the number of bits in the counter. In general, an n-bit counter can count from 0 to (2^n-1). For example, an 8-bit counter can count from 0000 0000 to 1111 1111, or 0 to 255 in decimal. Similarly, a 16-bit counter can count from 0 to 65,535. If a counter reaches its maximum value, and the input clock pulses keep on coming, then it overflows back to zero and starts counting up all over again. All is not lost if this happens—in fact, we have to be ready to deal with it. Many microcontroller counters cause an interrupt as the counter overflows; this interrupt can be used to record the overflow, and the count can continue in a useful way.

6.5.2 Using the Counter as a Timer

The input signal to a counter can be a series of pulses coming from an external source, for example, counting people going through a door. Alternatively, it can be a fixed-frequency logic signal, such as the clock source within a microcontroller. Very importantly, if that clock source is at a known and stable frequency, then the counter becomes a timer. As an example, if the clock frequency is 1.000 MHz (with a period of 1 μs), then the count will update every microsecond. If the counter is cleared to zero, and then starts counting, the value held in the counter will give the elapsed time since the counting started, with a resolution of 1 μs. This can be used to measure time, or trigger an event when a certain time has elapsed. It can also be used to control time-based activity, for example, serial data communication or a PWM stream.

Alternatively, if the counter is just free-running with a continuous clock signal, then the "interrupt on overflow" occurs repeatedly. This becomes very useful where a periodic interrupt is needed. For example, if an 8-bit counter is clocked with a clock frequency of 1 MHz, it will reach its maximum value and overflow back to zero in 256 μs (it is the 256th pulse which causes the overflow from 255 to 0). If it is left running continuously, then this train of interrupt pulses can be used to synchronize timed activity; for example, it could define the baud rate of a serial communication link.

Timers based on these principles are an essential feature of any microcontroller. Indeed, most microcontrollers have many timers applied to a variety of different tasks. These include general purpose timing, as well as generating timing in PWM or serial links, or measuring the duration of external events.

6.5.3 Timers in the Mbed Environment

Let's see what hardware timers the LPC1768 microcontroller has, by turning back to Fig. 2.3 and the data and user manuals. We find that the microcontroller has four general-purpose timers, a *repetitive interrupt timer*, and a *system tick timer*. Very similar timing capabilities are found in other microcontrollers. The Mbed API makes use of these in three distinct applications, described in the sections which follow. These are Timer (used for simple timing applications), Timeout (which calls a function after a pre-determined delay), and Ticker (which repeatedly calls a function at a pre-determined rate). It also has a *real-time clock* (RTC) to keep track of time of day and date. As expected, we do not need knowledge of the actual hardware timer details at all; that is all handled by the operating system!

6.6 Using the Mbed Timer

6.6.1 The Timer API

The Mbed Timer allows stopwatch-like timing activities to take place. A Timer can be created, started, stopped, read, and reset. There is no limit to the number of Timers that can be set up. The API summary is shown in Table 6.2. The Mbed website indicates that Timers are based on 64-bit signed microsecond counters, giving a timing range of up to 250,000 years!

Program Example 6.3 gives a simple but interesting timing example, and is taken from the Mbed site. It measures the time taken to write a message to the PC screen. It displays that message on screen and displays the time taken to write it and the overall elapsed time.

C/C++ code feature

We mentioned timing delays in Section 3.1.2, without much explanation. Use of Mbed timing functions, starting with Program Example 6.3, draws us into this in greater detail. Section B13 of Appendix B gives a little more background on **chrono** and the **duration_cast( )** function.

Table 6.2: API Summary for Timer, with duration_cast( ) example formats.

Function	Usage
Constructor	
Timer *name*	Create a Timer, called *name*.
Member Functions	
void **start**()	Start the Timer
void **stop**()	Stop the Timer
void **reset**()	Reset the Timer to 0
duration_cast<milliseconds>(*timer*.elapsed_time()).count()	Read from *Timer*, if integer milliseconds are needed
duration_cast<microseconds>(*timer*.elapsed_time()).count()	Read from *Timer*, if integer microseconds are needed

```
/* Program Example 6.3: A simple Timer example, from mbed web site.
Works on LPC1768 and F401RE, without adjustment or extra components
*/
#include "mbed.h"

using namespace std::chrono;     //Invoke chrono in C++ standard library

Timer timer_1;     //Create two timers
Timer timer_2;
DigitalOut LED (LED1);

int main(){
  timer_1.reset();
  timer_1.start();
  while(true){
    timer_2.start();
    printf("Hello World!\n");
    timer_2.stop();
    printf("The time taken was %llu milliseconds\n", duration_cast<milliseconds>(timer_
2.elapsed_time()).count());   //!!place statement on single line in IDE
    printf("The total time elapsed is %llu milliseconds\n", duration_cast<milliseconds>
(timer_1.elapsed_time()).count());
    ThisThread::sleep_for(500ms);
    timer_2.reset();
    LED = !LED;
  }
}
```

Program Example 6.3 A simple Timer application

Compile and run the program. Observe the output either on the integral screen in Studio IDE, or on a standalone serial monitor such as Tera Term.

Exercise 6.5

Run Program Example 6.3, and note from the computer screen readout the time taken for the message to be written. Write some other messages, of differing lengths, and record in each case the number of characters and the time taken. Can you make any deductions?

6.6.2 Using Multiple Mbed Timers

We now apply the Timer in a different way, to run one function at one rate and another function at another. Two LEDs will be used to show this; you will quickly realize that the principle is powerful, and can be extended to more tasks and more activities. Program Example 6.4 shows the program listing. The program creates two Timers, named

timer_fast and **timer_slow**. The main program starts these running, and tests when each exceeds a certain number. When the time value is exceeded, a function is called, which flips the associated LED.

```
/*Program Example 6.4: Program which runs two time-based tasks
Works on both LPC1768 and F401RE, with connections as indicated below

                                                                          */
#include "mbed.h"
Timer timer_fast;          // define Timer with name "timer_fast"
Timer timer_slow;          // define Timer with name "timer_slow"
DigitalOut ledA(LED1);    //Use A0 for F401RE
DigitalOut ledB(LED4);    //Use A1 for F401RE

using namespace std::chrono;

void task_fast(void);               //function prototypes
void task_slow(void);

int main() {
  timer_fast.start();    //start the Timers
  timer_slow.start();
  while (true){
    //test fast Timer value
    if (duration_cast<milliseconds>(timer_fast.elapsed_time()).count()>110){
      task_fast();              //call the task if trigger time is reached
      timer_fast.reset();             //and reset the Timer
    }
    //test slow Timer value
    if (duration_cast<milliseconds>(timer_slow.elapsed_time()).count()>200){
      task_slow();
      timer_slow.reset();
    }
  }
}
void task_fast(void){
    ledA = !ledA;
}
void task_slow(void){
    ledB = !ledB;
}
```

Program Example 6.4 Running two timed tasks

Create a project around Program Example 6.4, and run it on the standalone Mbed LPC1768 or the F401RE (in which case, use the extra LEDs shown in Fig. 3.1). Check the timing with a stopwatch or oscilloscope.

Exercise 6.6

Experiment with different repetition rates in Program Example 6.4, including ones which are not multiples of each other. Add a third and then fourth timer to it, flashing other LEDs at different rates.

6.7 Using the Mbed Timeout

Program Example 6.4 showed the Mbed Timer being used to trigger time-based events in an effective way. However, we needed to poll the Timer value to know when the event should be triggered. The Timeout allows an event to be triggered by an interrupt, with no polling needed. Timeout sets up an interrupt to call a function after a specified delay. There is no limit on the number of Timeouts created. The API summary is shown in Table 6.3.

6.7.1 A Simple Timeout Application

A simple first example of Timeout is shown in Program Example 6.5. This causes an action to be triggered a fixed period after an external event. The program is made up of the **main( )** and **blink( )** functions. A **Timeout** object is created, named **Response**, along with some familiar digital input and output. In the **main( )** function an **if** statement tests if the button is pressed. If it is, the **blink( )** function is attached to the **Response** Timeout. We can expect that two seconds after this attachment is made, the **blink( )** function will be called. To aid in our diagnostics, the button also switches on LED3. As a continuous task, the state of LED1 is reversed every 0.2 s. Thus, this program is a microcosm of many embedded systems programs. A time-triggered task needs to keep going, while an event-triggered task needs to take place at unpredictable times.

Table 6.3: API Summary for Timeout.

Function	Usage
Constructor	
Timeout *name*	Create a Timeout, called *name*.
Member Functions	
void **attach**(*function address*, float *delay*)	Attach a function to be called by the Timeout, specifying the **std::chrono** delay to occur before function is called.
void **detach**()	Detach the function.

```
/*Program Example 6.5: Demonstrates Timeout, by triggering an event a fixed
duration after a button press.
Works on both LPC1768 and F401RE, with connections as indicated below
                                */

#include "mbed.h"
Timeout Response;          //create a Timeout, and name it "Response"
DigitalIn button (p5);     //Replace p5 with BUTTON1, for F401RE
DigitalOut led1(LED1);
DigitalOut led2(LED2);     //Replace LED2 with A0, for F401RE
DigitalOut led3(LED3);     //Replace LED3 with A1, for F401RE

void blink() {             //this function is called at the end of the Timeout
  led2 = 1;
  wait_us(400000);         //light LED for 400 ms (avoid sleep functions in ISRs)
  led2 = 0;
}

int main() {
  while(true) {
    if(button == 1){
      //attach blink function to Response Timeout, to occur after 2 seconds
      Response.attach(&blink,chrono::milliseconds(2000));
      led3 = 1;                    //show button state
    }
    else led3 = 0;
    led1 = !led1;     //indicate looping in progress
    ThisThread::sleep_for(200ms);
  }
}
```

Program Example 6.5 Simple Timeout application

Compile Program Example 6.5, and download to a target system, using the connections given in the program listing. The LPC1768 will require an external pushbutton, as in Fig. 6.3. The F401RE needs the extra LEDs, as seen in Fig. 3.1.

■ Exercise 6.7

With Program Example 6.5 running, answer the following questions:

1. Is the 2-s Timeout timed from when the button is pressed, or when it is released? Why is this? Note that the answer is different for LPC1768 and F401RE.
2. When the event-triggered task occurs (i.e., the delayed blinking of the LED), what impact does it have on the time-triggered task (i.e., the flashing LED)?
3. If you tap the button very quickly, you will see that it is possible for the program to miss it entirely (even though electrically we can prove that the button has been pressed). Why is this?

■

6.7.2 Further Use of Timeout

Program Example 6.5 has demonstrated use of the Timeout nicely, but the questions in Exercise 6.7 bring up some of the classic problems of task timing, notably that execution of the event-triggered task can interfere with the timing of the time-triggered task.

Program Example 6.6 does the same thing as the previous example, but in a better way. Glancing through it, we see two Timeouts created and an interrupt. The latter connects to the pushbutton. There are now three functions, in addition to **main( ).** This has become extremely short and simple, and is concerned primarily with keeping the time-triggered task going. Now, response to the pushbutton is by interrupt, and it is within the interrupt function that the first Timeout is set up. When it is spent, the **blink( )** function is called. This sets the LED output, but then enables the second Timeout, which will trigger the end of the LED blink.

```
/*Program Example 6.6: Demonstrates the use of Timeout and interrupts, to allow
response to an event-driven task while a time-driven task continues.
Works on both LPC1768 and F401RE, with connections indicated.
*/

#include "mbed.h"
void blink_end (void);
void blink (void);
void ISR1 (void);
InterruptIn button (p5);     //Replace p5 with BUTTON1, for Nucleo F401RE
DigitalOut led1(LED1);
DigitalOut led2(LED2);       //Replace LED2 with A0, for Nucleo F401RE
DigitalOut led3(LED3);       //Replace LED3 with A1, for Nucleo F401RE
Timeout Response;              //create a Timeout named Response
Timeout Response_duration;     //create a Timeout named Response_duration

void blink() {               //This function is called when Timeout is complete
  led2 = 1;
  //set the duration of the led blink, with another timeout, duration 1 s
  Response_duration.attach(&blink_end, chrono::milliseconds(1000));
}

void blink_end(){  //A function called at the end of Timeout Response_duration
  led2 = 0;
}

void ISR1(){
  led3 = 1;      //shows button is pressed; diagnostic and not central to program
  //attach blinkl function to Response Timeout, to occur after 2 seconds
  Response.attach(&blink, chrono::milliseconds(2000));
}

int main() {
  button.rise(&ISR1);    //attach the address of ISR1 function to the rising edge
```

```
    while(true) {
      led3 = 0;                //clear LED3
      led1 = !led1;
      ThisThread::sleep_for(200ms);
    }
  }
```

Program Example 6.6 Improved use of Timeout

Compile and run the code of Program Example 6.6, using the same build options as in the previous program.

Exercise 6.8

Repeat all the questions of Exercise 6.7 for the most recent program example, noting and explaining the differences you find.

6.7.3 Timeout Used to Test Reaction Time

C/C++ code feature

Program Example 6.7 shows an interesting recreational application of Timeout, in which the Timeout duration is itself a variable. It tests reaction time by blinking an LED and timing how long it takes for the player to hit a switch in response. To add challenge, a "random" delay is generated before the LED is lit. This uses the C library function **rand( ).** The program should be understandable from the comments it contains.

```
/*Program Example 6.7: Tests reaction time, and demos use of Timer and Timeout
functions
Works on both LPC1768 and F401RE, with connections as indicated below
                                                                              */
#include "mbed.h"
#include <stdio.h>
#include <stdlib.h>              //contains rand() function

DigitalOut led1(LED1);
DigitalOut led4(LED4);  //Replace LED4 with A1, for Nucleo F401RE

//the player hits the switch connected here to respond
DigitalIn responseinput(p5);  //Replace p5 with BUTTON1, for Nucleo F401RE
Timer t;                //Timer used to measure the response time
Timeout action;         //Timeout used to initiate the response speed test
bool test_run = 0;      //Indicates reaction test triggered to start

//Function prototypes
void trigger (void);
void measure (void);
```

```
int main (){
  printf("Reaction Time Test\n\r");
  printf("------------------\n\r");
  while (true) {
    int r_delay;        //this will be the "random" delay before the led is blinked
    printf("New Test\n\r");
    led4 = 1;                      //warn that test will start
    ThisThread::sleep_for(200ms);
    led4 = 0;
   r_delay = rand()%20+1;  // generates a pseudorandom number range 1-20
   //Print random number for diagnostic, remove for normal play
   printf("random number is %i\n\r", r_delay);
   // set up Timeout to call measure()after random delay
   action.attach(&trigger,chrono::milliseconds(r_delay*200));
   //Wait for signal that test run is starting
   while (test_run == 0) {    //this loop exits when test_run set to 1
     wait_us(10);             //put something in loop, or compiler may optimise out
   }
   measure(); //the test is located in this function
   test_run = 0;
   ThisThread::sleep_for(1000ms);
  }
}
//Short ISR, which simply indicates that a test run has started
void trigger(){      //called at end of "action" Timeout
  test_run = 1;
}

void measure(){      //called when the led blinks, and measures response time
  if (responseinput == 1){                     //detect cheating!
   printf("Don't hold button down!\n\r");
  }
  else{
   t.start();            //start the timer
   led1 = 1;               //blink the led for 50 ms
   ThisThread::sleep_for(50ms);
   led1 = 0;
   while (responseinput == 0) {   //wait here for response
   }
   t.stop();                      //stop the timer once response detected
   printf("Your reaction time was %llu milliseconds\n\r",
           duration_cast<std::chrono::milliseconds>(t.elapsed_time()).count());
   t.reset();
  }
}
```

Program Example 6.7 Reaction time test: applying Timer and Timeout

The circuit build for Program Example 6.7 is the same as for Program Example 6.1. Now the pushbutton is the switch the player must hit to show a reaction. The result is shown on the Studio IDE terminal or Tera Term or equivalent.

Exercise 6.9

Run Program Example 6.7 for a period of time, and note the sequence of "random" numbers. Run it again. Do you recognize a pattern in the sequence? In fact, it is difficult for a computer to generate true random numbers, though a number of tricks and algorithms are used to create *pseudorandom* sequences. Read up about the concept of "seeding" a pseudorandom sequence, and explore use of the **srand( )** function.

6.8 Using the Mbed Ticker

The Mbed Ticker class sets up a recurring interrupt, which can be used to call a function periodically, at a rate decided by the programmer. There is no limit on the number of Tickers that can be created. The API summary is shown in Table 6.4.

We can demonstrate Ticker by returning to our very first program example, number 2.1. This simply flashed an LED periodically. Creating a periodic event is one of the most natural and common requirements in an embedded system, so it is not surprising that it appeared as a first program. We created the period by using a **sleep_for( )** function. The Ticker alternative allows the processor to continue with other productive activity, while it runs in the background. Program Example 6.8 replaces the effect of a wait/sleep function with the Ticker.

Table 6.4: API Summary for Ticker.

Function	Usage
Constructor	
Ticker *name*	Create a Ticker, called *name*.
Member Functions	
void **attach**(*function address, interval*)	Attach a function to be called by the Ticker, specifying the **std::chrono** *interval*.
void **attach** (callback<void()> *func*, std::chrono::microseconds *interval*)	Attach a function to be called by the Ticker, specifying the **std::chrono** *interval*.
Void **detach**()	Detach the function.

```
/* Program Example 6.8: Simple demo of "Ticker".
Replicates behaviour of "blinky", the first led flashing program.
Works standalone on LPC1768 and F401RE, without adjustment                    */

#include "mbed.h"
void led_switch(void);
Ticker time_up;                   //define a Ticker, with name "time_up"
DigitalOut myled(LED1);

void led_switch(){                //the function that Ticker will call
    myled = !myled;
}

int main(){
    //initialises the ticker
    time_up.attach(&led_switch,(500ms));
    while(true){    //sit in a loop doing nothing, waiting for Ticker interrupt
    }
}
```

Program Example 6.8 Applying Ticker to the original "blinky" program

It should be easy to follow what is going on in this program. The major step forward is that the CPU is now freed to do anything that is needed, while the task of measuring the time between LED changes is handed over to the Timer hardware, running in the background.

We have already called functions periodically with the Timer feature, so at first Ticker does not seem to add anything really new. Remember, however, that we have to poll the Timer value in Program Example 6.4 to test its value, and instigate the related function. Ticker is an interrupt that calls the associated function when time is up; this is a more efficient use of programming resources.

6.8.1 Using Ticker for a Metronome

Program Example 6.9 creates a metronome, using the Ticker facility. If you have not met one before, a metronome is an aid to musicians, setting a steady beat, against which they can practice their music. The musician selects a beat rate, traditionally in a range between 40 and 208 beats per minute. Old metronomes were based on elegant clockwork mechanisms, with a swinging pendulum arm. These days, most are electronic, including apps for mobile phones. Normally, the indication given to the musician is a loud audible "tick" sometimes accompanied by an LED flash. Here we restrict ourselves to the LED.

The breadboard build for the metronome is simple, and shown in Fig. 6.7. In each case, the beat is shown by an on-board LED, although an external one can be substituted if

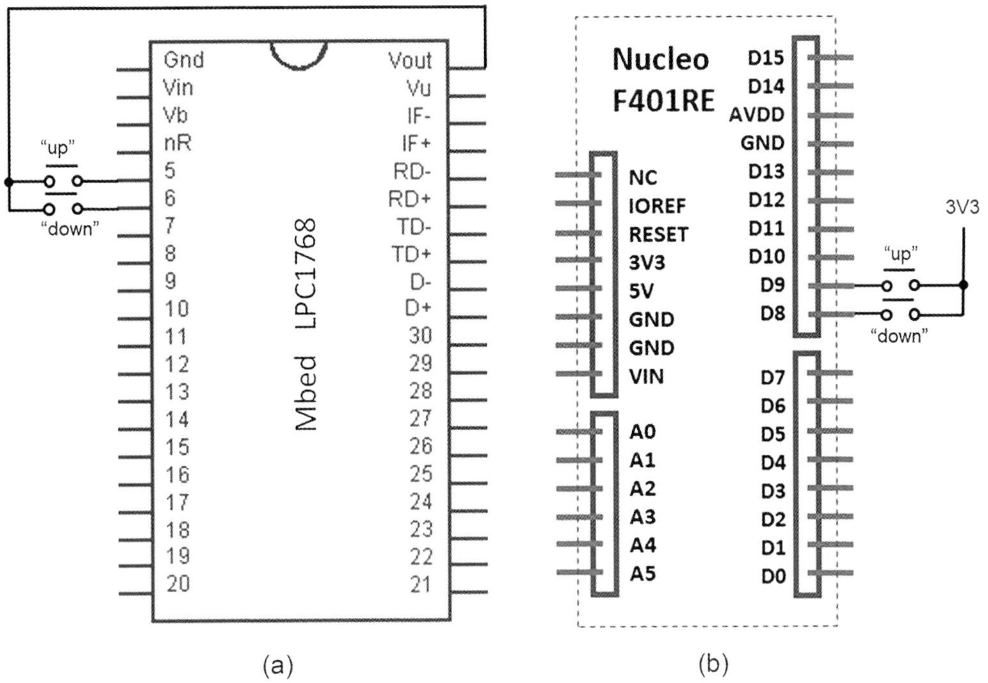

Figure 6.7
Metronome builds: (a) Mbed LPC1768; (b) Nucleo F401RE.

desired. Up and down pushbuttons are used to adjust the beat rate. If you are using the LPC1768 app board you can configure the joystick up and down switch positions and an on-board LED instead of the connections shown in the figure.

The program initializes a Ticker called **beat_rate**. It creates two related variables: **beat** (number of beats per minute) and **period** (time between beats). The main program **while** loop checks the up and down buttons, adjusts the beat rate accordingly, and displays the current rate to the host terminal screen. Meanwhile, the Ticker runs continuously, calling the function **beat( )**, and using **period** as the delay. The **beat( )** function updates the Ticker, possibly with a new value of **period**, and the LED is flashed. Program execution then returns to the main **while** loop until the next Ticker occurrence.

```
/*Program Example 6.9: Metronome. Uses Ticker to set beat rate
Works on LPC1768 or F401RE, using pin connections indicated.          */

#include "mbed.h"
#include <stdio.h>

DigitalIn up_button(D9,PullDown);         //for Nucleo F401RE
DigitalIn down_button(D8,PullDown);       //for Nucleo F401RE
/*OR use these for LPC1768
```

```
DigitalIn up_button(p5);            //for LPC1768
DigitalIn down_button(p6);          //for LPC1768
*/

DigitalOut beat_led(LED1);          //displays the metronome beat
Ticker beat_rate;                   //define a Ticker, with name "beat_rate"

void beat(void);
int period(500);                //metronome period in milliseconds
int rate (120);                 //metronome rate, initial value 120

int main() {
  printf("\r\n");
  printf("Mbed Metronome!\r\n");
  printf("_______________\r\n");
  //period = 1000;
  beat_led = 1;         //diagnostic
  ThisThread::sleep_for(100ms);
  beat_led = 0;
  beat_rate.attach(&beat, chrono::milliseconds(period));   //initialises the beat rate

  //main loop checks buttons, updates beat rate and display
  while(true){
    if (up_button == 1)   //increase rate by 4
      rate = rate + 4;
    if (down_button == 1) //decrease rate by 4
      rate = rate - 4;
    if (rate > 208)       //limit the maximum beat rate to 208
      rate = 208;
    if (rate < 40)        //limit the minimum beat rate to 40
      rate = 40;
    period = 60000/rate;  //calculate the beat period, result in ms
    printf("metronome rate is %i\r\n", rate);
    //printf("metronome period is %i\r\n", period);     //optional check
    ThisThread::sleep_for(500ms);
  }
}
//This is the metronome beat, called by ticker
void beat() {
  //update beat rate at this moment
  beat_rate.attach(&beat, chrono::milliseconds(period));
  beat_led = 1;        //show the beat
  wait_us(50000);      //50 ms flash
  beat_led = 0;
}
```

Program Example 6.9 Metronome, applying Ticker

Build the hardware, and compile and download the program. With a stopwatch or other timepiece, check that the beat rates are accurate.

Exercise 6.10

We have left a bit of polling in Program Example 6.9. Rewrite it so that response to the external pins is done with interrupts.

6.8.2 Reflecting on Multi-tasking in the Metronome Program

Over the last few examples, we have developed a useful program structure, where activities or tasks which are triggered by time can be linked to a Timer or Ticker function, and activities triggered by external events can be linked to one or more interrupts. Routine program activity then sits within an endless loop. In the metronome example, the program has to keep a regular beat going. While doing this, it must also respond to inputs from the user, calculate beat rates, and write to the display. Therefore, there is one time-triggered task (the beat) and at least one event-triggered task (the user input). This program structure is useful for simple or medium-complexity programs. In Chapter 9 we meet the RTOS, which provides a radical alternative to this simple program structure.

6.9 The RTC

The Real Time Clock (RTC) is an ultra-low-power peripheral implemented on many microcontrollers, including the LPC1768 and F401RE. The RTC is a timing/counting system which maintains a calendar and time-of-day clock, with registers for seconds, minutes, hours, day, month, year, day of month, and day of year. It can also generate an alarm for a specific date and time. It runs from its own 32-kHz crystal oscillator and can have its own independent battery power supply. Thus, it can be powered, and continue in operation, even if the rest of the microcontroller is powered down.

The Mbed API for the RTC implements functions from the standard C library, for example as shown in Table 6.5; it does not create any C++ objects. It applies the **tm** structure,

Table 6.5: Example C/C++ Functions for RTC.

Function	Usage
time_t **time**(time_t **t*)	Get the current time (number of seconds since January 1, 1970).
void **set_time**(time_t *t*)	Set the current time (number of seconds since January 1, 1970).
time_t **mktime**(struct tm **t*)	Converts a tm structure to a timestamp.
struct tm **localtime**(const time_t **t*)	Converts a timestamp to a tm structure.
char **ctime**(const time_t **t*)	Converts a timestamp to a human-readable string.
size_t **strftime**(char **str*, size_t *maxsize*, const char **format*, const struct tm **t*)	Converts a tm structure to a custom-format human-readable string, in destination pointed to by *str*, with specified maximum size and format.

which is a format for storing seconds, minutes, hours, dates, and so on. By convention, time is measured in seconds since "the epoch," 00:00:00 UTC (Coordinated Universal Time), January 1, 1970.

Program Example 6.10, slightly adapted from an official Mbed example, is a simple illustration of some RTC features. It uses many of the functions in Table 6.5. The program starts by loading the timer with a calculated value for an example time and date in October 2022. At 2 s intervals it then prints out the current time and date in several different formats.

```
/*Program Example 6.10. Demonstrates the Real Time Clock
Works on LPC1768 or Nucleo F401RE, with no external circuit
*/

#include "mbed.h"

int main(){
  set_time(1666697756);  // Set RTC time to Tues, 25 Oct 2022 11:35:56
  while (true) {
    time_t seconds = time(NULL);   //return elapsed seconds
    printf("Time as seconds since January 1, 1970 = %u\n", (unsigned int)seconds);
    printf("Time as a basic string = %s", ctime(&seconds));
    char buffer[32];
    strftime(buffer, 32, "%I:%M %p\n", localtime(&seconds));
    printf("Time as a custom formatted string = %s\n", buffer);
    ThisThread::sleep_for(2s);
  }
}
```

Program Example 6.10 A simple RTC application

An example printout for Program Example 6.10 is seen in Fig. 6.8. The while loop is iterated every 2 s, and three lines are then printed. The first line gives a raw seconds count, starting with the preloaded value, and counting up by two each time. The count is

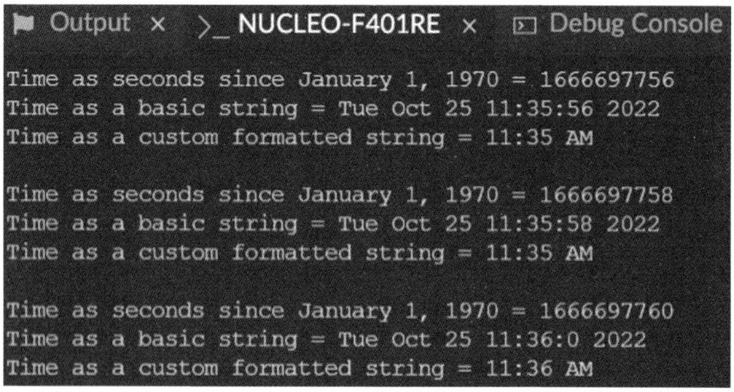

Figure 6.8
RTC output for Program Example 6.10.

then converted into two human-readable formats. The seconds count increments by 2 in every readout, while the minute count is seen to increment in the third.

Exercise 6.11

1. Calculate the number of seconds in a year and a leap year. Try adjusting the **set_time( )** entry in the program example to align with the exact date and time of the moment you read this. Run the program again, and check the print output aligns with your expected time and date.
2. By referring to Reference B.5 or equivalent, explore the formats and use of the functions in Table 6.5. For example, try changing the settings in the **strftime( )** call.

■

6.10 Switch Debouncing

With the introduction of interrupts, we now have some choices to make when writing a program to a particular specification. For example, Program Example 3.3 uses a digital input to determine which of two LEDs to flash. The digital input value is continuously polled within an infinite loop. However, we could equally have designed this program with an event-driven approach, to flip a control variable every time the digital input changes. Importantly, there are some inherent timing constraints within Program Example 3.3 which have not previously been discussed. One is that the frequency of polling is actually quite low, because once the switch input has been tested a 0.4 s flash sequence is activated. This means that the system has a response time of at worst 0.4 s, because it only tests the switch input once for every program loop. When the switch changes position, it could take up to 0.4 s for the LED to change, which is very slow in terms of embedded systems.

With interrupt-driven systems, we can have much quicker rates of response to switch presses, because response to the digital input can take place while other tasks are running. However, when a system can respond very rapidly to a switch change, we see a new issue which needs addressing, called *switch bounce*. This is because the mechanical contacts of an electromechanical switch literally bounce together as the switch closes. This can cause a digital input to swing wildly between Logic 0 and Logic 1 for a short time after a switch closes, as illustrated in Fig. 6.9. The solution to switch bounce is a set of techniques called *switch debouncing*.

We can identify the problem with switch bounce by running a simple event-triggered program. Program Example 6.11 attaches a function to an interrupt input, which simply toggles (flips) the state of the target's onboard LED1 for every rising edge.

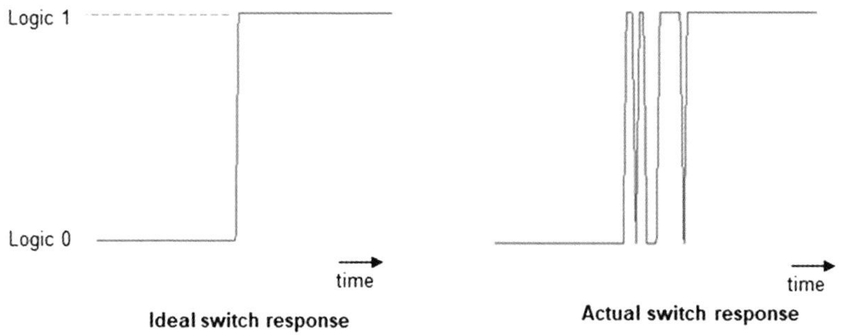

Figure 6.9
Demonstrating switch bounce.

```
/* Program Example 6.11: Demonstrates switch bounce.
Toggles LED1 every time switch input goes high.
Works on LPC1768 (build shown in Figure 6.3) or F401RE (no external circuit)
                                                                          */
#include "mbed.h"
InterruptIn button(p5);      // Replace p5 with BUTTON1, for F401RE
DigitalOut led1(LED1);
void toggle(void);           // function prototype

int main() {
  button.rise(&toggle);      // attach the address of the toggle
}                                       // function to the rising edge

//This is the ISR
void toggle() {
  led1 = !led1;
}
```

Program Example 6.11 Switch bounce demonstration

Implement Program Example 6.11. For the LPC1768, use the simple circuit of Fig. 6.3. For the F401RE, use the onboard pushbutton. Then, depending a little on the type of switch you use, you will see that the program does not work very well. A button press may produce no apparent response, though this may be due to the LED state being reversed twice or four times, or there may be a response on the edge when nothing is expected. This demonstrates the problem with switch bounce.

From Fig. 6.9, it is easy to see how a single button press or change of switch position can cause multiple interrupts and the LED can get out of synch with the button. There are a number of ways to implement switch debouncing. In hardware, there are simple configurations of logic gates which can be used (see Reference 3.1). In software, we can use timers or delay routines. The debounce feature needs to ensure that once the rising edge has been seen, no further rising edge interrupts should be implemented until a time period has elapsed. In reality, some switches have less contact bounce than others, so the exact timing required needs some tuning. To assist with this, switch manufacturers often provide data on switch bounce duration.

A simple solution to the switch bounce issue is to replace the ISR in Program Example 6.11 with this revision:

```
void toggle() {
  wait_us(4000);
  if(button == 1)
    led1 = !led1;
}
```

Program execution reaches the ISR on the first rising edge of the switch input, but a timing delay is then inserted. This delay time should exceed the switch bounce duration. The button is then tested again, and if it is still at Logic 1 the ISR is then completed. If the button at this point was found to be at Logic 0, then it is likely that a voltage spike had been detected at the input, for example, due to electromagnetic interference, rather than a switch transition. The downside of this simple debouncing method is that time is lost during the **wait_us( )** function.

Adjust Program Example 6.11 as indicated, and confirm that switch bounce is no longer experienced.

■ Exercise 6.12

Experiment with modifying the debounce time to shorter or longer values. There comes a point where the delay does not reliably correct the debouncing, and at the other end of the scale responsiveness is reduced, because short switch pushes are not detected. What is the minimum debounce time for the switch you are using? Try with a few different switches—can you detect a difference?

■

Chapter Review

- Signal inputs can be repeatedly tested in a loop, a process known as polling.
- An interrupt allows an external signal to interrupt the action of the CPU and start code execution from elsewhere in the program.
- Interrupts are a powerful addition to the structure of the microprocessor and allow almost immediate response to external events.
- It is easy to make a digital counter circuit, which counts the number of logic pulses presented at its input. Such a counter can easily be integrated into a microcontroller structure.

- Given a clock signal of known and reliable frequency, a counter can readily be used as a timer.
- Timers can be configured through hardware and software in different ways to address different needs; this chapter introduced the Timer, Timeout, Ticker, and RTC.
- Switch debounce is required in many cases to avoid multiple or unreliable responses being triggered by a single switch press.

Quiz

1. Explain the differences between using polling and interrupts to test the state of one of the digital input pins on a microcontroller.
2. List the most significant actions that a CPU takes when it responds to an enabled interrupt.
3. Explain the following terms with respect to interrupts:
   a. Priority
   b. Latency
   c. Nesting
4. A comparator circuit and LM35 temperature sensor are to be used to create an interrupt source, using the circuit of Fig. 6.5. The comparator is supplied from 5.0 V, and the temperature threshold is to be approximately 38°C. Suggest values for R_1 and R_2. Resistor values of 470, 680, 820, 1k, 1k2, 1k5, and 10k are available (where 1k2 = 1200 ohms and so on).
5. Describe in overview how a timer circuit can be implemented in hardware as part of a microprocessor's architecture.
6. What are the maximum values, in decimal, that a 12-bit and 24-bit counter can count up to?
7. A 4.0 MHz clock signal is connected to the inputs of a 12-bit and 16-bit counter. Each starts counting from zero. How long does it take before each reaches its maximum value?
8. A 10-bit counter, clocked with an input frequency of 512 kHz, runs continuously. Every time it overflows, it generates an interrupt. What is the frequency of that interrupt stream?
9. a. What is the purpose of a Real Time Clock? Give an example of when it might be used.
   b. A microcontroller RTC is loaded with the time value of 946,684,800. What date and time of day does this represent? Calculate your answer with care; then check by entering the value into Program Example 6.10.
10. Describe the issue of switch bounce and explain a simple software means of overcoming it. What is the disadvantage of this approach?

CHAPTER 7

Starting with Serial Communication

7.0 A Word on Inclusive Language

It is now widely recognized that the terminology used in some aspects of electronics and computing contains words that can be viewed as non-inclusive. This is particularly true in serial communication, with repeated use of master/slave terminology. This is deeply embedded in the literature and device nomenclature. The industry as a whole is attempting to address this, but at the time of writing, a consensus on new terms has not yet been reached, particularly across the different protocols which are in use. To adopt replacements which are not yet fully agreed upon, or to propose our own, would run the risk of creating unneeded confusion. Therefore, we have decided for this edition to retain conventional terminology. We will be pleased to adopt new terminology when a clear consensus has been reached and accepted.

7.1 Introducing Synchronous Serial Communication

There is an unending need in computer systems to move data around—lots of it. In Chapters 1 and 2 we came across the idea of data buses, on which data flies backward and forward between different parts of a computer. In these buses, data is transferred in *parallel*. There is one wire for each bit of data, and one or two more to provide synchronization and control; data is transferred a whole data word at a time. This works well, but it requires a lot of wires, and a lot of connections on each device that is being connected. It is bad enough for an 8-bit device; for 16 or 32 bits the situation is far worse. An alternative to parallel communication is *serial*. Here we effectively use a single wire for data transfer, with bits being sent in turn. A few extra connections are almost inevitably needed; for example, for earth return and synchronization and control.

Once we start applying the serial concept, a number of challenges arise. How does the receiver know when each bit begins and ends, and how does it know when each word begins and ends? There are several ways of responding to these questions. A straightforward approach is to send a clock signal alongside the data, with one clock pulse per data bit. The data is *synchronized* to the clock. This idea, called *synchronous serial*

Fast and Effective Embedded Systems Design. https://doi.org/10.1016/B978-0-323-95197-5.00007-3

communication, is represented in Fig. 7.1. When no data is being sent, there is no movement on the clock line. Every time the clock pulses, however, one bit is output by the transmitter, and should be read by the receiver. Generally the receiver synchronizes its reading of the data with one edge of the clock. In this example, it is the rising edge, highlighted by a dotted line.

A simple serial data link is shown in Fig. 7.2. Each device which connects to the data link is sometimes called a *node*. In this figure, Node 1 is designated *Master*; it controls what is going on as it controls the clock. The *Slave* is similar to the Master, but receives the clock signal from the master.

An essential feature of the serial link shown, and indeed of most serial links, is a *shift register*. This is made up of a string of digital bistables, each connected to the same clock

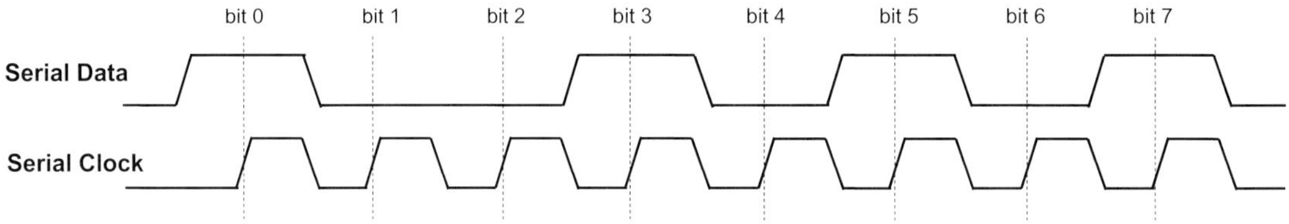

Figure 7.1
Synchronous serial data.

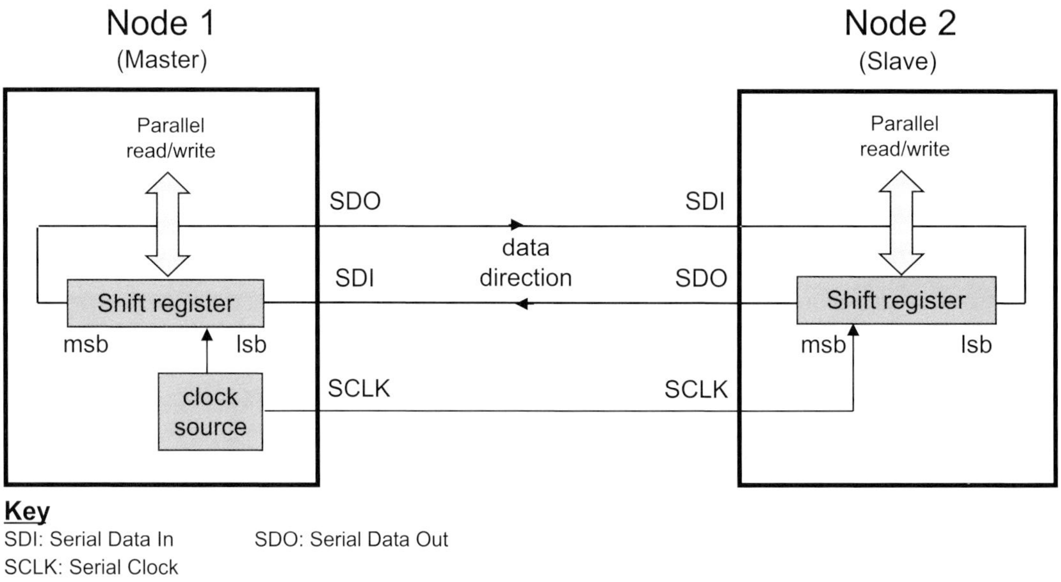

Figure 7.2
A simple serial link.

source, and arranged so that the output of one is connected to the input of the next. Each bistable holds one bit of information. Every time the shift register is pulsed by the clock signal, each bistable passes its bit on to its neighbor on one side, and receives a new bit from its other neighbor. The one at the input end clocks in data received from the outside world, and the one at the output end outputs its bit to the outside world. Therefore, as the clock pulses occur, the shift register can be feeding in external data and outputting data. The data held by all the bistables in the shift register can, moreover, be read all at the same time, as a parallel word, or a new value can be loaded in. In summary, the shift register is an exceptionally useful subsystem: it can convert serial data to parallel data, and vice versa, and it can act as a serial transmitter and/or a serial receiver.

As an example, suppose both shift registers in Fig. 7.2 are 8-bit, that is, each has eight bistables. Each register is loaded with a new word, and the Master then generates eight clock pulses. For each clock cycle, one new bit of data appears at the output end of each shift register, indicated by the SDO (serial data output) label. Each SDO output is, however, connected to the input (SDI—serial data input) of the other register. Therefore, as each bit is clocked out of one register, it is clocked into the other. After eight clock cycles, the word that was in the Master shift register is now in the Slave, and the word that was in the Slave shift register is now held in the Master.

Generally, the circuitry for the Master is placed within a microcontroller. The Slave might be another microcontroller, or some other peripheral device. The hardware circuitry which allows serial data to be sent and received, and which interfaces between the microcontroller CPU and the outside world, is usually called a serial port.

7.2 Serial Peripheral Interface

To ensure that serial data links are reliable and can be applied by different devices in different places, a number of standards or *protocols* have been defined. One that is very widely used currently is of course the USB (universal serial bus). A protocol defines details of timing and signals, and may go on to define other things, like type of connector used. SPI is a simple protocol which has had a large influence in the embedded world.

7.2.1 Introducing SPI

In the early days of microcontrollers, both National Semiconductors and Motorola started introducing simple serial communication, based on Fig. 7.2. Each formulated a set of rules which governed how their microcontrollers worked, and allowed others to develop devices which could interface correctly. These became *de facto* standards; in other words, they were never initially designed as standards, but were adopted by others to the point where they acted as formal standards. Motorola called its standard *Serial Peripheral Interface (SPI)*, and National Semiconductors called its *Microwire*. They are very similar to each other.

It was not long before both SPI and Microwire were adopted by manufacturers of other ICs, who wanted their devices to be able to work with the new generation of microcontrollers. SPI has become one of the most durable standards in the world of electronics, applied to short-distance communications, typically within a single piece of equipment. There is not a formal document defining SPI, but data sheets for the Motorola 68HC11 (now a very old microcontroller) effectively define it in full. Good related texts also do so, such as Reference 7.1.

In SPI communication one device, usually a microcontroller is designated the Master; it controls all activity on the serial interconnection. The Master communicates with one or more Slaves. A minimum SPI link uses just one Master and one Slave, and follows the pattern of Fig. 7.2. The Master generates and controls the clock and all data transfer. The SDO of one device is connected to the SDI of the other, and vice versa. In every clock cycle one bit is moved from Master to Slave, and one bit from Slave to Master; after eight clock cycles a whole byte has been transferred. Thus, data is actually transferred in both directions—in old terminology this is called *full duplex*. If data in only one direction is wanted, then the data transfer line which is not needed can be omitted.

If more than one Slave device is needed, then the approach of Fig. 7.3 can be used. Only one Slave is active at any time, determined by which Slave select ($\overline{SS}$) line the Master activates. Note that writing $\overline{SS}$ indicates that the line is active when low; if it were active

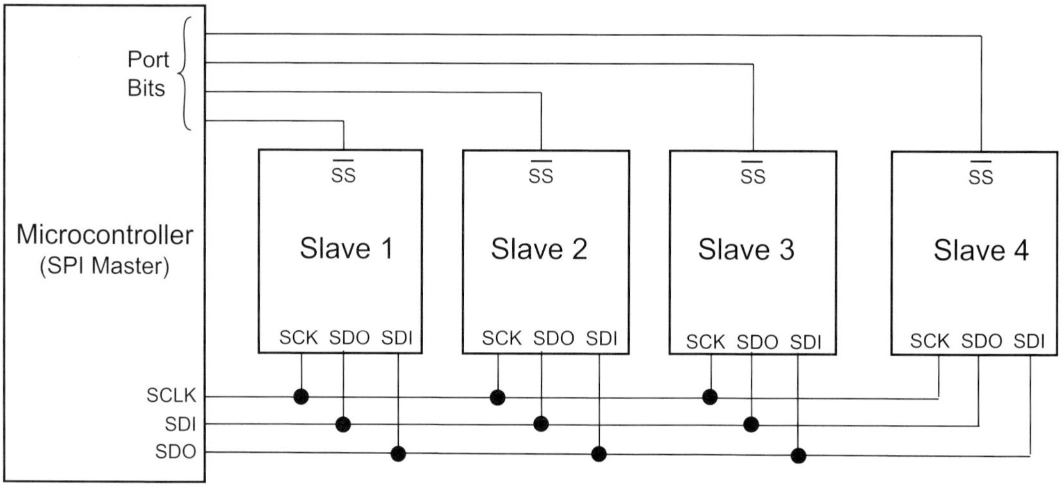

Figure 7.3
Example SPI interconnection for multiple Slave devices.

high, it would simply be SS. The terminology chip select ($\overline{\text{CS}}$) is also sometimes used for the same role, including later in this chapter. Only the Slave activated by its $\overline{\text{SS}}$ input responds to the clock signal. The Master then communicates with this one active Slave device just as in Fig. 7.2. Notice that for n Slaves, the microcontroller needs to commit $(3 + n)$ lines. One of the advantages of serial communication, the small number of interconnections, is beginning to disappear.

7.2.2 SPI in the Mbed Environment

As the diagram of Fig. 2.1 shows, the Mbed LPC1768 has two SPI ports, one appearing on pins 5, 6, and 7, and the other on pins 11, 12, and 13. On this Mbed, as with many SPI devices, the same pin is used for SDI if in Master mode, or SDO if Slave. Hence, this pin is often called MISO (or miso), that is, Master in, Slave out. Its partner pin is MOSI (Master out, Slave in). Fig. 2.6 similarly shows an SPI port for the Nucleo F401RE on pins D3, D4, and D5, which can also be mapped to D11, D12, and D13. A signal labeled NSS ("not Slave select") appears on pin A2. Fig. 2.6 does not give the full story of possible connections, as a reading of the microcontroller data sheet, Reference 2.6, reveals. There are other serial ports not shown in the figure, and other possible mappings of the serial connections that are shown.

Fig. 2.15 shows that the application board commits one of the two LPC1768 SPI ports for the all-important connection to the liquid crystal display (LCD). The pins of the second port are hard-wired to other things, so they are not available for SPI use.

The API summary available for SPI Master is shown in Table 7.1.

Table 7.1: Mbed SPI Master API Summary.

Functions	Usage
Constructor	
`SPI (PinName mosi, PinName miso, PinName sclk, PinName ssel)`	Create an SPI Master connected to the specified pins (which must be a valid SPI port for the target board in use).
Member Functions	
`void format(int bits,int mode)`	Configure the data length and transmission mode (see Table 7.2).
`void frequency(int Hz)`	Set the SPI bus clock frequency.
`virtual int write(int value)`	Write to the SPI Slave and return the response.

7.2.3 Setting up an Mbed SPI Master

Program Example 7.1 shows a very simple setup for an SPI Master. The program initializes the SPI port, choosing for it the name **ser_port**, with the pins of one of the possible ports being selected. The **format( )** function is used here for illustration only, as it asserts default values. It requires two variables: the first is the number of bits, and the second is the mode. The mode is a feature of SPI which is illustrated in Fig. 7.4, with associated codes in Table 7.2. It allows choice of which clock edge is used to clock data

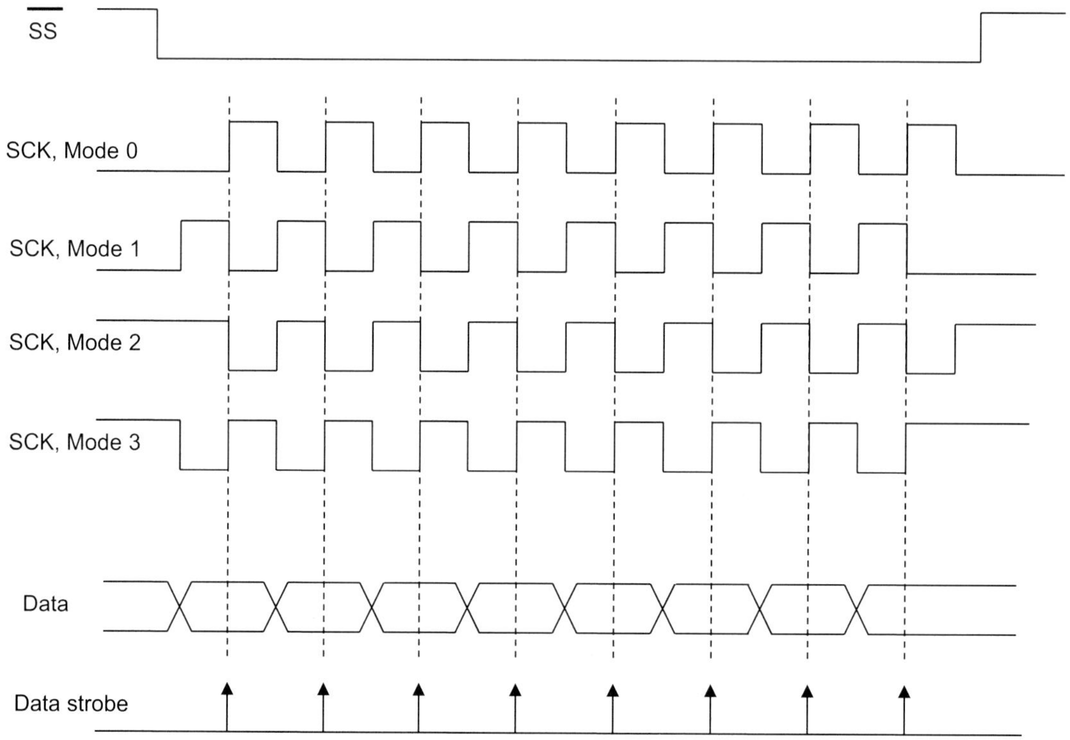

Figure 7.4
Polarity and phase.

Table 7.2: SPI modes.

Mode	Polarity	Phase
0	0	0
1	0	1
2	1	0
3	1	1

into the shift register (indicated as "Data strobe" in the diagram), and whether the clock idles high or low. For most applications, the default mode, Mode 0, is acceptable.

```
/* Program Example 7.1: Sets up target as SPI Master, and continuously sends
a single byte
Works for both LPC1768 and F401RE, with pin allocations shown
                                                                              */

#include "mbed.h"
//Apply line below for F401RE
SPI ser_port(D11, D12, D13);      //mosi, miso, sclk
//Apply line below for LPC1768
//SPI ser_port(p11, p12, p13);    //mosi, miso, sclk
char switch_word;                 //word we will send

int main() {
  ser_port.format(8,0);           // Setup the SPI for 8 bit data, Mode 0 operation
  ser_port.frequency(1000000);    // Clock frequency is 1 MHz
  while (true){
    switch_word = 0xA1;             //set up word to be transmitted
    ser_port.write(switch_word);    //send switch_word
    wait_us(50);
  }
}
```

Program Example 7.1 Minimal SPI Master Application

Compile, download, and run Program Example 7.1 on a single Mbed target, and observe the data (mosi) and clock (sck) lines simultaneously on an oscilloscope. See how clock and data are active at the same time, and verify the clock data frequency. Check that you can read the transmitted data byte, 0xA1. Is the most or least significant bit sent first?

■ Exercise 7.1

1. Try each of the different SPI modes in Program Example 7.1, observe both clock and data waveforms on the oscilloscope, and check how they compare with Fig. 7.4.
2. Set the SPI format of Program Example 7.1 to 12 and then 16 bits, sending the words 0x8A1 and 0x8AA1 respectively (or to your choice). Check each on an oscilloscope.

■

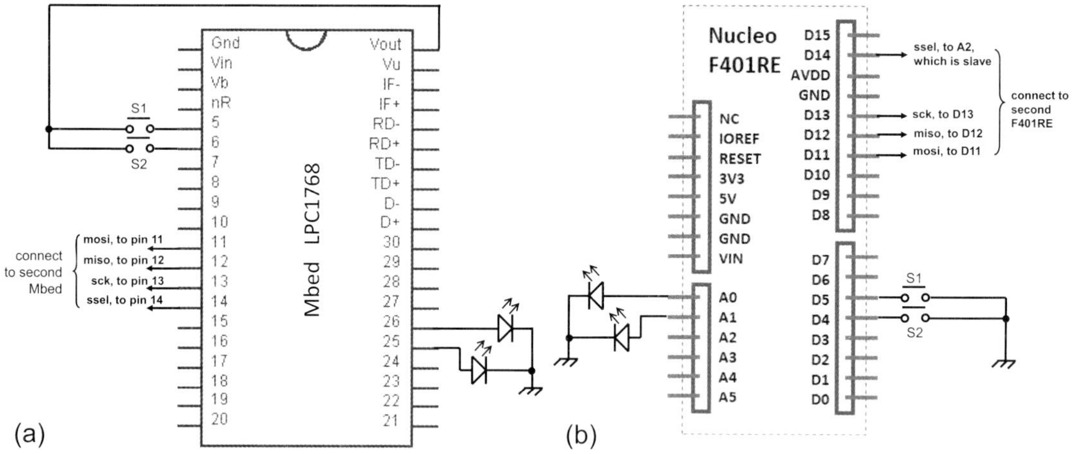

Figure 7.5
Using SPI to link two Mbed targets: (a) with LPC1768 Mbed; (b) with Nucleo F401RE.

7.2.4 Creating an SPI Data Link

We will now develop two programs, one Master and one Slave, and get two Mbeds to communicate. Each will have two switches and two LEDs; the aim will be to get the switches of the Master to control the LEDs on the Slave, and vice versa. When you apply the programs, be sure to select the correct initializations (toward the top of the programs) for the board you will be using, and delete the others.

The program for the Master is shown as Program Example 7.2. This is written for the circuit of Fig. 7.5. It sets up the SPI port as before and defines the switch inputs on pins 5 and 6. It declares a variable **switch_word,** the word that will be sent to the Slave, and a variable **recd_val,** which is the value received from the Slave. For this application, the default settings of the SPI port are chosen, so there is no further initialization in the program. Once in the main loop, the value of **switch_word** is established. To give a pattern to this that will be recognizable on the oscilloscope, the upper four bits are set to hexadecimal A. The two switch inputs are then tested in turn; if they are found to be high, the appropriate bit in **switch_word** is set, by ORing with 0x01 or 0x02. The **cs** line is set low, and the command to send **switch_word** is made. The return value of this function is the received word, which is read accordingly.

```
/*Program Example 7.2. Sets the Mbed up as Master, and exchanges data with a Slave,
sending its own switch positions, and displaying those of the Slave.
Works for both LPC1768 and F401RE, with pin allocations shown      */
```

```
#include "mbed.h"
//Use the following 6 lines for F401RE
SPI ser_port(D11, D12, D13);       // mosi, miso, sclk
DigitalOut red_led(A0);            //red led
DigitalOut green_led(A1);          //green led
DigitalOut cs(D14);                //used as "Slave select", under program control
DigitalIn switch_ip1(D4,PullUp);
DigitalIn switch_ip2(D5,PullUp);

/* Use the following 6 lines for LPC1768
ser_port(p11, p12, p13);    // mosi, miso, sclk
DigitalOut red_led(p25);    //red led
DigitalOut green_led(p26); //green led
DigitalOut cs(p14);         //used as "Slave select", under program control
DigitalIn switch_ip1(p5,PullUp);
DigitalIn switch_ip2(p6,PullUp);       *//end of LPC1768 block

char switch_word ;          //word we will send
char recd_val;              //value return from Slave

int main() {
  while (true){
    //Default settings for SPI Master chosen, no need for further configuration
    //Set up the word to be sent, by testing switch inputs
    switch_word = 0xa0;              //set up a recognisable output pattern
    if (switch_ip1 == 1)
      switch_word = switch_word | 0x01;       //OR in lsb
    if (switch_ip2 == 1)
      switch_word = switch_word | 0x02;       //OR in next lsb
    cs = 0;                                   //select Slave
    recd_val = ser_port.write(switch_word);  //send switch_word and receive data
    cs = 1;
    wait_us(10000);

    //set leds according to incoming word from Slave
    red_led = 0;                //preset both to 0
    green_led = 0;
    recd_val = recd_val&0x03; //AND out unwanted bits
    if (recd_val == 1)
      red_led = 1;
    if (recd_val == 2)
      green_led = 1;
    if (recd_val == 3){
      red_led = 1;
      green_led = 1;
    }
  }
}
```

Program Example 7.2 Mbed target set up as SPI master, for bidirectional data transfer

Table 7.3: Mbed SPI Slave API Summary.

Functions	Usage
Constructor	
SPISlave (PinName mosi, PinName miso, PinName sclk, PinName ssel)	Create an SPI Slave connected to the specified pins.
Member Functions	
void **format**(int *bits*, int *mode*)	Configure the data length and transmission mode (see Table 7.2).
int **receive**(void)	Poll the SPI to see if data has been received: 0=no data waiting 1=data waiting.
int **read**(void)	Retrieve data from receive buffer as Slave, return data in Slave buffer.
void **reply**(int *value*)	Fill the transmission buffer with the data *value* to be clocked out on the next received message from the Master.

The slave program draws upon the Mbed OS functions shown in Table 7.3, and appears as Program Example 7.3. It is almost the mirror image of the Master program, with small but key differences. This emphasizes the very close similarity between the Master and Slave roles in SPI. Let's check the differences. The serial port is initialized with **SPISlave.** Now four pins must be defined, the extra being the Slave Select input, here called **ssel.** As the slave will also be generating a word to be sent, and receiving one, it also declares variables **switch_word** and **recd_val.** Change these names if you would rather have something different. The slave program configures its **switch_word** just like the Master. Now comes a difference. While the Master initiates a transmission when it wishes, the Slave must wait. The Mbed API does this with the **receive( )** function. This returns 1 if data has been received, and 0 otherwise. Of course, if data has been received from the Master, then data has also been sent from the Slave to the Master. If there is data, then the Slave reads this and sets up the LEDs accordingly. It also sets up the next word to be sent to the Master by transferring its **switch_word** to the transmission buffer, using **reply( ).**

```
/*Program Example 7.3: Sets the Mbed up as slave, and exchanges data with a master,
sending its own switch positions, and displaying those of the Master. as SPI slave.
Works for both LPC1768 and F401RE, with pin allocations shown. F401RE interconnect
may give improved performance with slave select link not connected.     */

#include "mbed.h"

//The following block for Nucleo F401RE
SPISlave ser_port(D11,D12,D13,A2);//mosi, miso, sclk, ssel
DigitalOut red_led(A0);            //red led
DigitalOut green_led(A1);          // green led
DigitalIn switch_ip1(D4,PullUp);
```

```
DigitalIn switch_ip2(D5,PullUp);
/* The following block for LPC1768
SPISlave ser_port(p11,p12,p13,p14);   // mosi, miso, sclk, ssel
DigitalOut red_led(p25);              //red led
DigitalOut green_led(p26);            //green led
DigitalIn switch_ip1(p5);
DigitalIn switch_ip2(p6);             */ //end of LPC1768 block

char switch_word ;                    //word we will send
char recd_val;                        //value received from Master

int main() {
  //default formatting applied
  while(true) {
    //set up switch_word from switches that are pressed
    switch_word = 0xa0;             //set up a recognisable output pattern
    if (switch_ip1 == 1)
      switch_word = switch_word|0x01;
    if (switch_ip2 == 1)
      switch_word = switch_word|0x02;

    if(ser_port.receive()) {        //test if data transfer has occurred
      recd_val = ser_port.read();   // Read byte from Master
      ser_port.reply(switch_word);  // Make this the next reply
    }

    //now set leds according to received word
    ...
    (continues as in Program Example 7.2)
    ...
  }
}
```

Program Example 7.3 Mbed set up as SPI Slave, for bidirectional data transfer

Now connect two Mbed targets together, applying one of the circuits of Fig. 7.5. It is simplest if each is powered individually through its own USB cable. If you do not want to set up the full circuit immediately, then connect just the switches to the Master, and just the LEDs to the Slave, or vice versa.

Compile and download Program Example 7.2 into one Mbed (which will be the Master), and Program Example 7.3 into the other. Once you run the programs, you should find that pressing the switches of the Master controls the LEDs of the Slave, and vice versa. This is another big step forward; we are communicating data from one microcontroller to another, or from one system to another.

Exercise 7.2

Try the following, and be sure you understand the result. In each case, test for data transmission from Master to Slave, and from Slave to Master.

(i) Remove the data link from Master to Slave.
(ii) Remove the data link from Slave to Master.
(iii) Remove the clock link.
(iv) Remove the chip select line.

Notice that in some cases if you disconnect a wire, but leave it dangling in the air, you may get odd intermittent behavior which changes if you touch the wire. This is an example of the impact of electromagnetic interference, which is being interpreted by the SPI port as a clock or data signal.

7.3 Intelligent Instrumentation and an SPI Accelerometer

Despite (and indeed because of) its age, the SPI standard is wonderfully simple and widely used. It is embedded into all sorts of electronic devices, ICs, and gadgets. Given an understanding of how SPI works, we can now set up communication between an Mbed target and any SPI-compatible device.

With the very high level of integration found in modern ICs, it is common to find sensor, signal conditioning, ADC, and data interface all combined on a single chip. Such devices are part of the new generation of *intelligent instrumentation.* Instead of just having a standalone sensor, as we did with the light-dependent resistor in Chapter 5, we can now have a complete measurement subsystem integrated with the sensor, with a convenient serial data output. Such sensors have become surprisingly cheap and are becoming the option of choice in many microcontroller-based systems.

In this chapter, we meet a number of these sensors, including an accelerometer and a temperature sensor. The accelerometer is an example of a *microelectromechanical system* (MEMS); the accelerometer mechanics is actually fabricated within the IC structure. The accelerometer has an internal capacitor mounted in the plane of each axis. Acceleration causes the capacitor plates to move, hence changing the output voltage proportionally to the acceleration or force. The accelerometer output is analog in nature, and measures acceleration on three axes. The on-board ADC on the accelerometer converts the analog voltage fluctuations to digital, and can output these values over a SPI serial link.

7.3.1 Introducing the ADXL345 Accelerometer

The ADXL345 accelerometer, made by Analog Devices, is an example of an integrated intelligent sensor. Its data sheet appears as Reference 7.2. Control of the ADXL345 is done by writing to a set of registers through the serial link. This link is SPI, but can also be I^2C (inter-integrated circuit—our next topic). Example registers are shown in Table 7.4. It is clear that the device goes well beyond just making direct measurements. It is possible to calibrate it, change its range, and get it to recognize certain events, for example, when it is tapped or in free fall. Measurements are made in terms of g (where $1 \times g$ is the value of acceleration due to earth's gravity, 9.81 $m.s^{-2}$).

The ADXL345 IC is extremely small, and designed for surface mounting on a printed circuit board; therefore, we use it ready-mounted on a "breakout" board, as shown in Fig. 7.6. It would otherwise be difficult to handle. Notice how the acceleration axes are marked on the circuit board. Pin connections are shown here (for SPI only) and in Table 7.5.

Table 7.4: Selected ADXL345 Registers.

Address*	Name	Description
0x00	DEVID	Device ID
0x1D	THRESH_TAP	Tap threshold
0x1E/1F/20	OFSX, OFSY, OFSZ	X, Y, Z axis offsets
0x21	DUR	Tap duration
0x2D	POWER_CTL	Power-saving features control. Device powers up in standby mode; setting bit 3 causes it to enter measure mode.
0x31	DATA_FORMAT	Data format control Bits 7: force a self test by setting to 1 6: 1 = 3-wire SPI mode; 0 = 4-wire SPI mode 5: 0 sets interrupts active high, 1 sets them active low 4: always 0 3: 0 = output is 10-bit always; 1 = output depends on range setting 2: 1 = left justify result; 0 = right justify result 1-0: range - 00 = ± 2 *g*; 01 = ± 4 *g*; 10 = ± 8 *g*; 11 = ± 16 *g*;
0x33:0x32	DATAX1:DATAX0	X axis data, formatted according to DATA_FORMAT, in two's complement.
0x35:0x34	DATAY1:DATAY0	Y axis data, as above.
0x37:0x36	DATAZ1:DATAZ0	Z axis data, as above.

*In any data transfer the register address is sent first, and formed:
bit 7 = R/$\overline{W}$ (1 for read, 0 for write); bit 6: 1 for multiple byte, 0 for single;
bits 5-0: the lower six bits found in the address column.

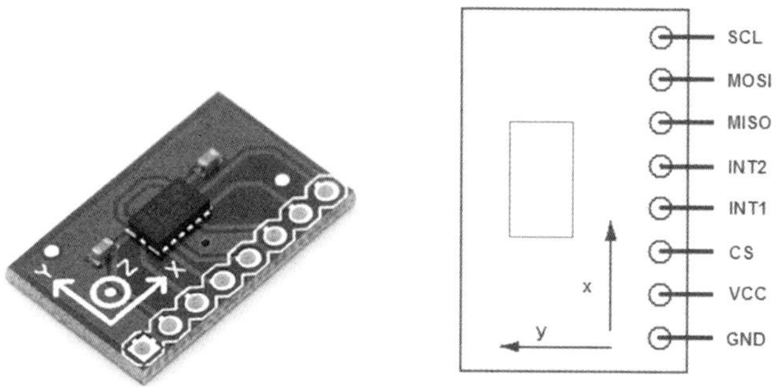

Figure 7.6
The ADXL345 accelerometer on breakout board. *Image courtesy of Sparkfun.*

Table 7.5: ADXL345 Pin Connections to Target Board.

ADXL345 signal name	Mbed LPC1768 pin	F401RE pin
VCC	Vout	3V3
GND	GND	GND
SCL/SCK	13	D13
MOSI/SDO	11	D11
MISO/SDI	12	D12
CS	14	D10

Note: There are differences in detail between naming of breakout board connections, and ADXL345 connections.

7.3.2 Developing a Simple ADXL345 Program

Program Example 7.4 applies the ADXL345, reading acceleration on three axes, and outputting the data to the host computer screen. We use a target board SPI port for connecting the accelerometer, applying the connections shown in Table 7.5. Because floating point numbers are to be printed, we need to create an **Mbed_app.json** file, as described in Section 5.3.

C/C++ code feature

The program initializes a Master SPI port, which we have chosen to call **acc.** It further declares two arrays; one is a buffer (called **buffer**), which will hold data read direct from the accelerometer's registers, two for each axis. The second array, **data**, applies the **int16_t** specifier. This is from the C standard library **stdint**. Use of **int16_t** tells the compiler that exactly 16-bit (signed) integer type data is being declared. This array will hold the full accelerometer axis values, each combined from two bytes received from the registers.

The main function initializes the SPI port in a manner with which we are familiar. It then loads two of the accelerometer registers, writing the address first, followed by the data byte. It should be possible to work out what is being written, by looking either at the data sheet itself, or at Table 7.4. A continuous **while** loop is then initiated. In this, a multi-byte read is set up, using an address word formed from information shown in Table 7.4. This fills the **buffer** array. The **data** array is then populated, concatenating (i.e., combining to form a single number) pairs of bytes from the **buffer** array. These values are then scaled to actual *g* values, using the conversion factor from the data sheet, of $0.004 \times g$ per unit. Results are then displayed on screen.

```
/*Program Example 7.4: Reads values from accelerometer through SPI, and outputs
continuously to terminal screen.
Works for both LPC1768 and F401RE, with pin allocations shown
*/

#include "mbed.h"

SPI acc(D11,D12,D13);              //Use pins p11,p12,p13 for LPC1768
DigitalOut cs(D10);                //p14 for LPC1768. For chip/slave select
char buffer[6];                    //raw data array type char
int16_t data[3];                   //16-bit twos-complement integer data
// floating point data, to be displayed on-screen
float x, y, z;

int main() {
  cs = 1;                            //cs idles high
  acc.format(8,3);                 //8 bit data, Mode 3
  acc.frequency(2000000);          //2MHz clock rate
  cs = 0;                            //start a data transfer
  acc.write(0x31);                 //data format register
  acc.write(0x0B);                 //format +/-16g, 0.004g/LSB
  cs = 1;                            //end of transfer
  cs = 0;                            //start a new transfer
  acc.write(0x2D);                 //power ctrl register
  acc.write(0x08);                 //measure mode
  cs = 1;                            //end of transfer
  while (true) {                   //infinite loop
    cs = 0;                          //start a transfer
    acc.write(0x80|0x40|0x32);       //RW bit high, MB bit high, plus address
    for (int i = 0;i <= 5;i++) {
      buffer[i] = acc.write(0x00);             //read back 6 data bytes
    }
    cs = 1;                          // end of transfer
    data[0] = buffer[1]<<8 | buffer[0];      //combine MSB and LSB
    data[1] = buffer[3]<<8 | buffer[2];
    data[2] = buffer[5]<<8 | buffer[4];
```

```
        //convert to g value, float
        x = 0.004*data[0]; y = 0.004*data[1]; z = 0.004*data[2];
        printf("x = %+1.2fg\t y = %+1.2fg\t z = %+1.2fg\n\r", x, y,z); // print
        ThisThread::sleep_for(500ms);
    }
}
```

(A)

```
{
    "target_overrides":{
        "*": {
            "target.printf_lib": "std"
        }
    }
}
```

(B)

Program Example 7.4 Accelerometer continuously outputs three-axis data to terminal screen: (A) the main.cpp file; (B) the Mbed_app.json file (for printing of floating point numbers)

Carefully make the connections of Table 7.5 between the accelerometer and your target board, and compile, download, and run the code on the target. Accelerometer readings should be displayed to a terminal emulator, or the IDE serial monitor. You will see that when the accelerometer is flat on a table the *z* axis should read approximately 1 g, with the *x* and *y* axes reading approximately 0 g. As you rotate and move the device, the *g* readings will change. If the accelerometer is shaken or displaced at a high rate, *g* values in excess of $1 \times g$ can be observed. Note that there are some inaccuracies in the accelerometer data, which can be reduced in a real application by developing configuration/calibration routines and data averaging functions.

7.4 Evaluating SPI

The SPI standard is extremely effective. The electronic hardware is simple and cheap, and data can be transferred rapidly. However, it does have its disadvantages. There is no acknowledgment from the receiver, so in a simple system the Master cannot be sure that data has been received. Also, there is no addressing. In a system where there are multiple Slaves, a separate select line must be run to each one, as we saw in Fig. 7.3. Therefore, we begin to lose the advantage that serial communications should give us, that is, a limited number of interconnect lines. Finally, there is no error checking. Suppose some electromagnetic interference was experienced in a long data link: data or clock would be corrupted, but the system would have no way of detecting this, or correcting for it. You may have experienced this in a small way in Exercise 7.2. Overall we could grade SPI as simple, convenient, and low-cost, but not appropriate for complex or high-reliability systems.

7.5 The Inter-integrated Circuit (I^2C) Bus

7.5.1 Introducing the I^2C Bus

This standard was developed by Philips, to resolve some of the perceived weaknesses of SPI and its equivalents. As its name suggests, it was intended for interconnection over short distances, and generally within a piece of equipment. It uses only two interconnecting wires, no matter how many devices are connected to the bus. These lines are called SCL—serial clock, and SDA—serial data. All devices on the bus are connected to these two lines, as shown in Fig. 7.7. The SDA line is bidirectional, so data can travel in either direction, but only one direction at any one time. In the jargon, this is called *half duplex*. Like SPI, it is a synchronous serial standard.

One of the interesting features of I^2C, and one which makes it versatile, is that any node connected to it can only pull down the SCL or SDA line to Logic 0; it cannot force the line up to Logic 1. This role is played by a single pull-up resistor connected to each line. When a node pulls a line to Logic 0, and then releases it, it is returned to Logic 1 by the action of the pull-up resistor. There is, however, capacitance associated with the line. Although this is labeled "stray" capacitance in the figure, it is in reality mainly unavoidable capacitance which exists in the semiconductor structures connected to the

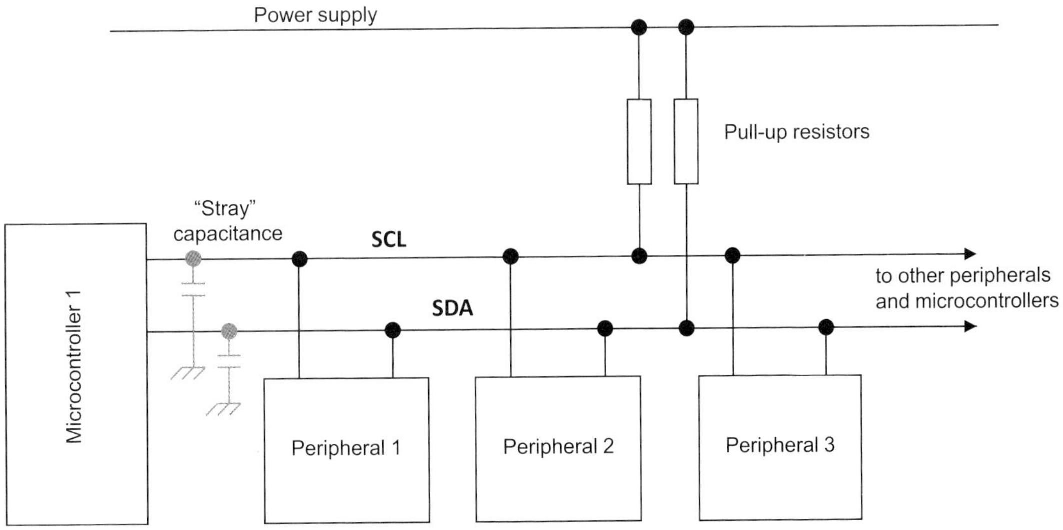

Figure 7.7
An I^2C-based system.

line. Thus, this capacitance is higher if there are many nodes connected and/or if the interconnecting wires are long, and lower otherwise. The higher the capacitance and/or pull-up resistance, the longer the rise time of the Logic transition from 0 to 1. The I^2C standard requires that the rise time of a signal on SCL or SDA must be less than 1000 ns. Given a known bus setup, it is possible to do reasonably precise calculations of pull-up resistors required, particularly if you need to minimize power consumption. For simple applications, default pull-up resistor values, in the range from 2.2 to 10 kΩ, are usually acceptable.

The I^2C protocol has been through several revisions, which have dramatically increased the possible speeds, and reflect technological changes, for example, in reduced minimum operating voltages. The original standard mode version allowed data rates up to 100 kbit/s. In 1992, Version 1.0 increased the maximum data rate to 400 kbit/s. This latter is very well established, and still probably accounts for most I^2C implementation. In 1998, Version 2.0 increased the possible bit rate to 3.4 Mbit/s. The I^2C protocol is defined in a surprisingly readable manner in Reference 7.3, and forms the basis of the description which follows.

Nodes on an I^2C bus can act as Master or Slave. The Master initiates and terminates transfers, and generates the clock. The Slave is any device addressed by the Master. A system may have more than one Master, although only one may be active at any time. Therefore, more than one microcontroller could be connected to the bus, and they can claim the Master role at different times, when needed. An arbitration process is defined if more than one Master attempts to control the bus.

A data transfer is made up of the Master signaling a *start condition*, followed by one or two bytes containing address and control information. The start condition, Fig. 7.8a, is defined by a high-to-low transition of SDA when SCL is high. All subsequent data transmission follows the pattern of Fig. 7.8b. One clock pulse is generated for each data bit, and data may only change when the clock is low.

The byte following the start condition is made up of seven address bits and one data direction bit, as shown in Fig. 7.8c. Each Slave has a predefined device address; the Slaves are responsible for monitoring the bus and responding only to commands associated with their own addresses. A Slave device which recognizes its address will then be readied either to receive data, or to transmit it onto the bus.

It is important to be clear that some manufacturers and users quote the Slave address in its 7-bit format. This needs to be located in the upper 7 bits of the I^2C address byte, often done by shifting the address left by one bit, leaving bit 0 available for the $R/\overline{W}$ bit. Others, however, quote the address in 8-bit format. In this case, the address must be an even number, again leaving the LSB clear for the $R/\overline{W}$ bit. A 10-bit addressing mode is also available.

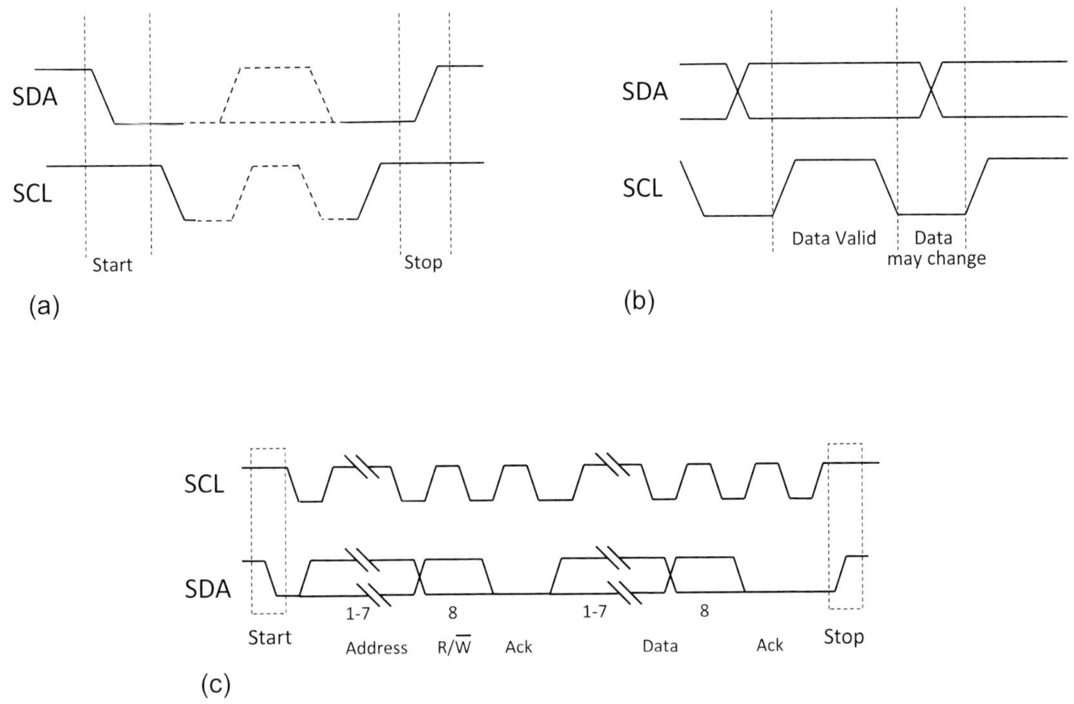

Figure 7.8
I²C data transfer: (a) start and stop conditions; (b) clock and data timing; (c) a complete transfer of one byte.

All data transferred is in units of one byte, with no limit on the number of bytes transferred in one message. Each byte must be followed by a 1-bit acknowledge from the receiver. During this time, the transmitter relinquishes SDA control and the addressed Slave acknowledges by taking the line low. A low-to-high transition of SDA while SCL is high defines a *stop* condition. Fig. 7.8c illustrates the complete transfer of a single data byte, preceded by the address byte.

7.5.2 I^2C with the Mbed OS

Fig. 2.1 shows that the Mbed LPC1768 offers two I^2C ports, on pins 9 and 10, or 27 and 28. Fig. 2.6 shows that the Nucleo F401RE offers three such ports. Their use follows the pattern of other Mbed peripherals, with available functions summarized in Tables 7.6 and 7.7.

7.5.3 Setting Up an I^2C Data Link

We will now replicate the action of Section 7.2.4, but using I^2C as the communication link, rather than SPI. On the Mbed LPC1768 we will use the I^2C port on pins 9 and 10, or

Table 7.6: Mbed OS I^2C Master API Summary.

Functions	Usage
Constructor	
I2C (PinName sda, PinName scl)	Create an I^2C Master interface, connected to the specified pins.
Member Functions	
void **frequency** (int *Hz*)	Set the frequency of the I^2C interface.
int **read** (int *address*, char **data*, int *length*, bool *repeated*)	Full read from I^2C Slave. The address lsb is forced to 1 to indicate a read; *data* is pointer to byte array to be sent, *length* is number of bytes, *repeated* = 1 for repeated start. Returns 0 for success.
int **read**(int *ack*)	Reads single byte from !2C Slave; ack is 1 for acknowledge.
int **write** (int *data*)	Write single byte to I^2C Slave; returns 1 for ack received, 0 for no ack, 2 for Timeout.
int **write** (int *address*, const char **data*, int *length*, bool *repeated*)	Write a complete message to I^2C Slave; *data* is pointer to byte array to be sent, *length* is number of bytes, *repeated* = 1 for repeated start; returns 0 for success.
void **start** (void)	Create a start condition on the I^2C bus.
void **stop** (void)	Create a stop condition on the I^2C bus.

Table 7.7: Mbed OS I^2C Slave API Summary.

Function	Usage
Constructor	
I2CSlave (PinName sda, PinName scl)	Create an I^2C Slave interface, connected to the specified pins.
Member Functions	
void **address** (int *address*)	Sets the I^2C Slave address.
void **frequency** (int *Hz*) int **read**(void)	Set the frequency of the I^2C interface. Read single byte from I^2C Master.
int **read** (char **data*, int *length*)	Read specified number of bytes from I^2C Master; *data* is pointer to buffer to read data into, *length* is number of bytes.
int **receive** (void)	Check if this I^2C Slave has been addressed. Return value: NoData = 0 ReadAddressed = 1. WriteGeneral = 2 WriteAddressed = 3.
int **write** (int *data*)	Write a single byte to I^2C Master.
int **write**(const char **data*,int *length*)	Write specified number of bytes to I^2C Master; *data* is pointer to buffer holding data to be sent, *length* is number of bytes; returns 0 for success.
void **stop**(void)	Reset the I^2C Slave into the known ready receiving state.

D5 and D7 on the F401RE. Program Example 7.5, the Master, undertakes the same actions as Program Example 7.2, except that SPI-related sections are replaced by those which relate to I^2C. It also rewrites some of the program features in a more elegant and compact way; for example, using **BusOut** for the two LEDs.

Early in the program, an I^2C serial port is configured using the Mbed constructor **I2C**, choosing the name **i2c_port**. An arbitrary Slave address is chosen, 0x52. This is in the 8-bit address format already mentioned, so it must be an even number. Following determination of the variable **switch_word,** the I^2C transmission can be seen. This is created from the separate components of a single-byte I^2C transmission; that is, start—send address—send data—stop, as allowed by the Mbed functions. Table 7.6 shows that these can also be grouped together in a single line of code. The program also tests for acknowledgement from the Slave, using LED1 to indicate this.

Further down the program we see a request for a byte of data from the Slave, with similar message structure. Now the Slave address is ORed with 0x01, which sets the $R/\overline{W}$ bit in the address word to indicate Read. The received word is then interpreted in order to set the LEDs, just as we did in the SPI program earlier.

```
/*Program Example 7.5: I2C Master, transfers switch state to second Mbed acting as
Slave, and displays state of Slave's switches on its leds.
Works with LPC1768 and F401RE, with pin allocations shown. Action for F401RE may be
incomplete, as this style of I2C implementation, with separate start, write, stop
statements, does not map ideally to F401RE I2C architecture.
                                                                                */
#include "mbed.h"

//Configure a serial port, naming sda, scl
I2C i2c_port(p9, p10);       //Use D5, D7 for F401RE
BusOut leds(p25, p26);       //Use A0, A1 for F401RE
DigitalOut error_light(LED1);
DigitalIn switch_ip1(p5,PullUp);   //Replace p5 with D3 for F401RE
DigitalIn switch_ip2(p6,PullUp);   //Replace p6 with D4 for F401RE

char switch_word ;           //word we will send
char recd_val;               //value received from Slave
const int addr = 0x52;       //the I2C Slave address, an arbitrary even number

int main() {
    char ack;
  while(true) {
    switch_word = 0xa0;              //set up a recognisable output pattern
    if (switch_ip1 == 1)
      switch_word = switch_word|0x01; //OR in lsb
    if (switch_ip2 == 1)
      switch_word = switch_word|0x02; //OR in next lsb
```

```
    //send a single byte of data, in correct I2C package
    i2c_port.start();                 //force a start condition
    i2c_port.write(addr);             //send the address
    ack = i2c_port.write(switch_word);  //send byte of data, and save ack value
    error_light = !ack;               //LED lights if no acknowledge
    i2c_port.stop();                  //force a stop condition
    wait_us(200);
    //request and receive a single byte of data, in correct I2C package
    i2c_port.start();
    i2c_port.write(addr|0x01);        //send address, with R/W bit set to Read
    recd_val = i2c_port.read(1);      //read and save the received byte
    i2c_port.stop();                  //force a stop condition
    leds = recd_val & 0x03;   //set leds according to incoming word from slave
  }
```

Program Example 7.5 I^{2}C data link Master

The Slave program is shown in Program Example 7.6. It is similar in effect to Program Example 7.3, with SPI features replaced by I^{2}C, and a more compact coding style. As in SPI, the I^{2}C Slave just responds to calls from the Master. The Slave port is defined with the Mbed OS constructor **I2CSlave,** with **Slave** chosen as the port name. Just within the **main()** function the Slave address is defined, importantly the same 0x52 as we saw in the Master program. A variable **how_addressed** is also declared; its value will indicate the action the Master is requiring of the Slave. As before, the **switch_word** value is set up from the state of the switches. The **receive( )** function is used to test if an I^{2}C transmission has been received. The program then responds to any read or write request.

```
/*Program Example 7.6: I2C Slave, when called transfers switch state to master, and
displays state of master's switches on its leds.
Applies to LPC1768 and F401RE, with pin allocations shown. See also note in master
program header.     */
#include "mbed.h"

I2CSlave slave(p9,p10);               //Use D5, D7 for F401RE
BusOut leds(p25,p26);                 //Use A0, A1 for F401RE
DigitalIn switch_ip1(p5,PullUp);      //Replace p5 with D3 for F401RE
DigitalIn switch_ip2(p6,PullUp);      //Replace p6 with D4 for F401RE
char switch_word;                     //word we will send
char recd_val;                        //value received from Master
const int addr = 0x52;

int main() {
  int how_addressed;
  slave.address(addr);
  while (true) {
    //set up switch_word from switches that are pressed
    switch_word = 0xa0;
    if (switch_ip1) switch_word = switch_word|0x01;
```

```
      if (switch_ip2) switch_word = switch_word|0x02;
      how_addressed = slave.receive();       //check if addressed, and if so, how.
      if (how_addressed == I2CSlave::ReadAddressed) {
        slave.write(switch_word);
        slave.stop();
      }
      if (how_addressed == I2CSlave::WriteAddressed) {
        recd_val = slave.read();
        slave.stop();
        leds = recd_val & 0x03;
      }
    }
  }
```

Program Example 7.6 I^{2}C data link Slave

Connect two Mbed targets together with an I^{2}C link, applying one of the circuit diagrams of Fig. 7.9. These are of course similar to Fig. 7.5, with SPI connections removed and replaced by the I^{2}C connection. It is essential to include the pull-up resistors; values of 4.7 kΩ are shown, but in this non-critical application they can be anywhere in the range from 2.2 to 10 kΩ. Note that each Mbed target should have two switches and two LEDs connected, but there should just be one pair of pull-up resistors between them. Compile and download Program Example 7.5 to either target, and Example 7.6 to the other. With both programs running, you should find that the switches of one Mbed target control the LEDs of the other, and vice versa.

The action of the Master LED1 is used diagnostically to check correct circuit operation. It will remain unlit if the Slave acknowledges correctly, but will light if no acknowledgment is received.

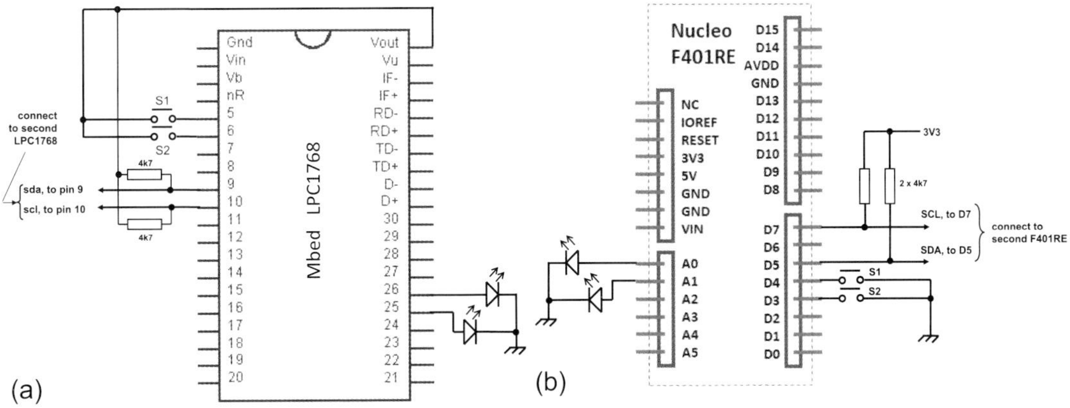

Figure 7.9
Linking two Mbed targets with I^{2}C: (a) with Mbed LPC1768; (b) with Nucleo F401RE.

Exercise 7.4

Monitor SCL and SDA lines on an oscilloscope. This will require careful oscilloscope triggering. With appropriate setting of the oscilloscope time base, you should be able to see the two messages being sent between the Mbeds. Identify as many features of I^2C as you can, including the idle high condition, the start and stop conditions, and the address byte with the $R/\overline{W}$ bit embedded.

7.6 Communicating with I^2C-Enabled Sensors

7.6.1 The TMP102 Sensor

Just as we did with the SPI port and the accelerometer, we can use the I^2C port to communicate with a very wide range of peripheral devices, including many intelligent sensors. The Texas Instruments TMP102 temperature sensor (Reference 7.4) has an I^2C data link. This is similar to the accelerometer that we have just met, in that an analog sensing device is integrated with an ADC and a serial port, producing an ideal and easy-to-use system element. Note from the data sheet that the TMP102 actually makes use of the SMBus (system management bus). This was defined by Intel in 1995, and is based on I^2C. The differences are very usefully described in Reference 7.3.

The TMP102 itself is a tiny device, just as we would want of a temperature sensor. Like the accelerometer before, we use it mounted on a small breakout board, seen in Fig. 7.10. It has six possible connections, shown in Table 7.8. The address pin, ADD0, is used to select the address of the device, as seen in the table. This allows four different address options; hence

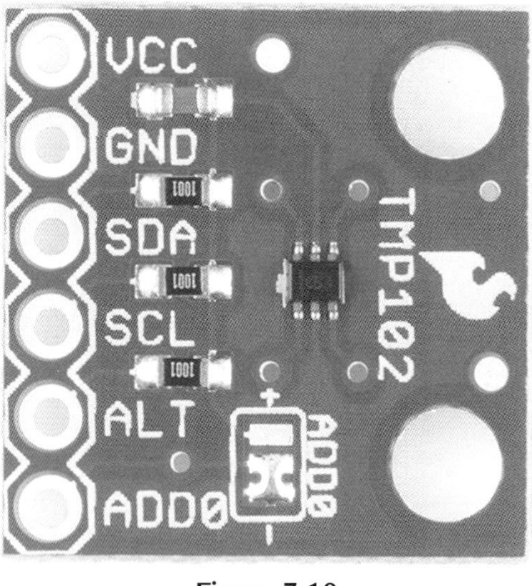

Figure 7.10
The TMP102 temperature sensor on breakout board. *Image courtesy of Cool Components.*

Table 7.8: Connecting the TMP102 Sensor to the Mbed.

Signal	Mbed LPC1768 Pin	F401RE Pin	Notes
VCC (3.3V)	40	+3V3	
SDA	9	D5	Note presence of on-board pull-up.
SCL	10	D7	Note presence of on-board pull-up.
GND (0V)	1	GND	
ALT (Alert)	1	GND	
ADD0	1	GND	Connect to 8-bit Slave address 0V 0x90 Vcc 0x92 SDA 0x94 SCL 0x96

four of the same sensor can be used on the same I^2C bus. The current version of the board has built-in pull-up resistors, so there is no need to connect these externally.

Program Example 7.7 can be applied to link an Mbed target to the sensor. It defines an I^2C port and names it **tempsensor**. As sensor pin ADD0 is tied to ground, Table 7.8 shows that the sensor address will be 0x90; this is defined in the program, with name **addr**. Two small arrays are also defined, one to hold the sensor configuration data, and the other to hold the raw data read from the sensor. A further variable, **temp,** will hold the scaled decimal equivalent of the reading.

The configuration options can be found from the TMP102 data sheet. To set the configuration register we first need to send a data byte of 0x01 to specify that the pointer register is set to "Configuration Register." This is followed by two configuration bytes, 0x60 and 0xA0. These select a simple configuration setting, initializing the sensor to normal mode operation. These values are sent at the start of the **main( )** function. Note the format of the write command used here, which is able to send multi-byte messages. This is different from the approach used in Program Example 7.5, where only a single data byte was sent in any one message. Using the I^2C **write( )** function, we need to specify the device address, the data array, followed by the number of bytes to send.

The sensor will now operate and acquire temperature data, so we simply need to read the data register. To do this, first we need to set the pointer register value to 0x00. In this command we have only sent one data byte to set the pointer register. The program now starts an infinite loop to read the 2-byte temperature data continuously. This data is then converted from a 16-bit reading to an actual temperature value. The conversion required (as specified by the data sheet) is to shift the data right by 4 bits (it is actually only 12-bit data held in two 8-bit registers) and to multiply by the specified conversion factor, 0.0625°C per LSB. The value is then displayed on the PC screen.

```
/*Program Example 7.7: Mbed communicates with TMP102 temperature sensor, and scales
and displays readings to screen.
Works for both LPC1768 and F401RE, with pin allocations shown.
                                                                              */
#include "mbed.h"

I2C tempsensor(p9, p10);        //Use D5, D7 for F401RE
const int addr = 0x90;
char config_t[3];
char temp_read[2];
float temp;

int main() {
  config_t[0] = 0x01;                    //set pointer reg to 'config register'
  config_t[1] = 0x60;                    // config data byte1
  config_t[2] = 0xA0;                    // config data byte2
  tempsensor.write(addr, config_t, 3);
  config_t[0] = 0x00;                    //set pointer reg to 'data register'
  tempsensor.write(addr, config_t, 1);   //send to pointer 'read temp'
  while(true) {
    ThisThread::sleep_for(1s);             //wait for one second
    tempsensor.read(addr, temp_read, 2);  //read the two-byte temp data
    temp = 0.0625 * (((temp_read[0] << 8) + temp_read[1]) >> 4);   //convert data
    printf("Temp = %.2f degC\n\r", temp);
  }
}
```

(A)

```
{
   "target_overrides":{
       "*": {
           "target.printf_lib": "std"
       }
   }
}
```

(B)

Program Example 7.7 Communicating by I^{2}C with the TMP102 temperature sensor: (A) the main.cpp file; (B) the Mbed_app.json file (for printing of floating point numbers).

Connect the sensor to your Mbed target according to the information in Table 7.8. External pull-up resistors should not be necessary; to check that onboard resistors are there, check the resistance between SDA/SCL and VCC board terminals with a digital voltmeter. This should give a reading of 4.7 kΩ. Take note of requirements for successfully applying **printf( )** with floating point numbers in Section 5.3. Then compile, download, and run the code of Program Example 7.7.

Test that the displayed temperature increases when you press your finger against the sensor. Try placing the sensor on something warm, for example, a radiator. If you have a calibrated temperature sensor, try to compare readings from both sources.

Exercise 7.5

Rewrite Program Example 7.5 using the I²C Master multi-byte **read( )** and **write( )** functions, as applied in Program Example 7.7. Compile, download, and test that it works as expected.

7.6.2 The SRF08 Ultrasonic Range Finder

The SRF08 ultrasonic range finder, as shown in Fig. 7.11a, can be used to measure the distance between the sensor and an acoustically reflective surface or object in front of it. It makes the measurement by transmitting a pulse of ultrasound from one of its transducers, and then measuring the time for an echo to return to the other. If there is no echo, it times out. The distance to the reflecting object is proportional to the time taken for the echo to return. Given knowledge of the speed of sound in air (343 $m.s^{-1}$), the actual distance can be calculated. The SRF08 has an I²C interface. Data is readily available for it; for example, Reference 7.5, though at the time of writing not as a formalized document.

The SRF08 can be connected to an Mbed target, for example, as shown in Fig. 7.11b. It must be powered from 5 V; with I²C pull-up resistors connected to that voltage. The

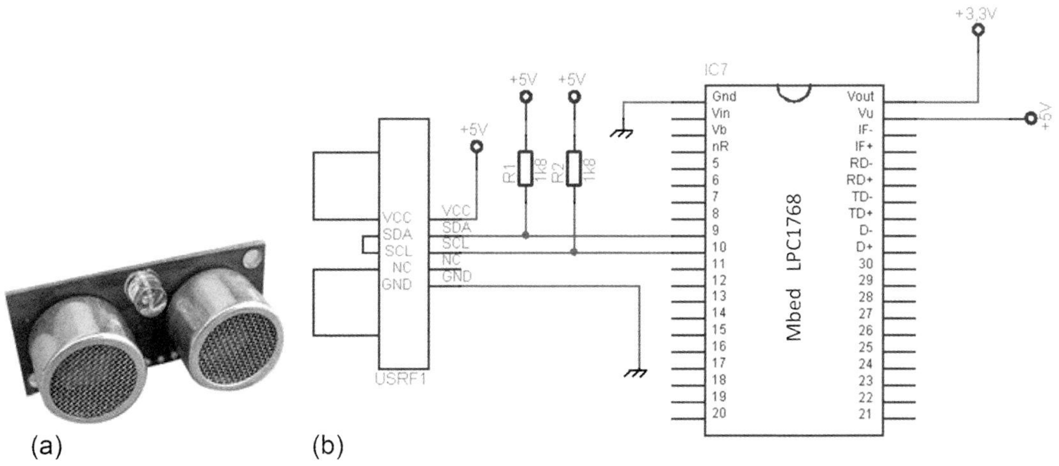

Figure 7.11

The SRF08 ultrasonic range finder: (a) the sensor; (b) connections to the Mbed LPC1768.

internal Mbed circuit remains powered from 3.3 V; but is able to tolerate the higher voltage being presented at its I^2C pins.

The preceding programs having been worked through, it should be easy to grasp how Program Example 7.8 works, noting the following information, taken from the device data:

- The SRF08 (8-bit) I^2C address is 0xE0.
- The pointer value for the command register is 0x00.
- A data value of 0x51 to the command register initializes the range finder to operate and return data in cm.
- A pointer value of 0x02 prepares for 16-bit data (i.e., two bytes) to be read.

The success of the first transmission in the loop is checked, noting that this is denoted by a return value of 0 (see Table 7.6), with LED1 being used to indicate if a transmission error has occurred.

```
/*Program Example 7.8: Configures and takes readings from the SRF08 ultrasonic range
finder, and displays them on screen. A transmission error is indicated on LED1.
Works for both LPC1768 and F401RE, with pin allocations shown.
                                                                                  */
#include "mbed.h"

I2C rangefinder(p9, p10); //Use D5, D7 for F401RE
DigitalOut error_led(LED1);  //will indicate if error in I2C transmission
const int addr = 0xE0;       //SRF08 I2C address
char config_r[2];            //will hold config data
char range_read[2];          //will receive raw data from SRF08
float range;                 //will hold calculated range value

int main() {
  int tr_error;          //will indicate transmission error (1 for error)
  while (true) {
    config_r[0] = 0x00;                      //set pointer reg to `cmd register'
    config_r[1] = 0x51;                      //initialise, result in cm
    //send first message, and test for success
    tr_error = rangefinder.write(addr, config_r, 2);
    error_led = tr_error;        //light LED if transmission not successful.
    ThisThread::sleep_for(70ms); //wait for measurement to complete (around 65 ms)
    config_r[0] = 0x02;                      //set pointer reg to 'data register'
    rangefinder.write(addr, config_r, 1);    //send to pointer 'read range'
    rangefinder.read(addr, range_read, 2);    //read the two-byte range data
    range = ((range_read[0] << 8) + range_read[1]);
    printf("Range = %.2f cm\n\r", range);    //print range on screen
    ThisThread::sleep_for(500ms);          //set this further delay as desired
  }
}
```

(A)

```
{
 "target_overrides":{
  "*": {
    "target.printf_lib": "std"
  }
 }
}
```

(B)

Program Example 7.8 Communicating by I²C with the SRF08 range finder: (A) the main.cpp file; (B) the Mbed_app.json file (for printing of floating point numbers)

Connect the circuit of Fig. 7.11b, or an F401RE equivalent circuit (using connections given in the program). Again, take note of requirements for successfully applying **printf( )** with floating point numbers in Section 5.3. Compile Program Example 7.8, and download to the Mbed target. Verify correct operation by placing the range finder a known distance from a hard flat surface. Then explore its ability to detect irregular surfaces, narrow objects (e.g., a broom handle), and distant objects.

Exercise 7.6

The centimeter readout in Program Example 7.8 gives only modest accuracy. Checking the device data, try triggering a microsecond readout instead. Then deduce a conversion factor from the speed of sound, and multiply the **range** value by this, leading to a new and more precise centimeter readout. Remember that the sound travels to the target and back, so your conversion factor should include division by two.

7.7 Evaluating I^2C

As we have seen, the I^2C protocol is well established and versatile. Like SPI, it is widely applied to short-distance data communication. However, it goes well beyond SPI in its ability to set up more complex networks, and to add and subtract nodes with comparative ease. Although we have not explored it in detail here, it provides for a more reliable system. If an addressed device does not send an acknowledgement, the Master can act upon that fault. Does this mean that I^2C is going to meet all our needs for serial communication, in any application? The answer is a clear No, for at least two reasons.

One is that the bandwidth is comparatively limited, even in the faster versions of I^2C. The second is the security of the data. For instance, while fine in a domestic appliance, I^2C is still susceptible to interference, and does not check for data errors. Therefore, we would not consider using it in a medical, motor vehicle, or other application requiring high reliability.

7.8 Asynchronous Serial Data Communication

The synchronous serial communication protocols that we have seen so far in this chapter, in the form of SPI and I^2C, are extremely useful ways of moving data around. However, the question remains: do we really need to send that clock signal wherever the data goes? Although it allows an easy way of synchronizing the data, it does have these disadvantages:

- An extra (clock) line needs to go to every data node.
- The bandwidth needed for the clock is always twice the bandwidth needed for the data; therefore, it is the demands of the clock which limit the overall data rate.
- Over long distances, clock and data themselves could lose synchronization.

7.8.1 Introducing Asynchronous Serial Data

For the reasons just stated, a number of serial standards have been developed which do not require a clock signal to be sent with the data. This is generally called *asynchronous* serial communication. It is now up to the receiver to extract all timing information directly from the data signal itself. This has the effect of laying new and different demands on the signal, and making transmitter and receiver nodes somewhat more complex than comparable synchronous nodes.

A common approach to achieving asynchronous communication is based on this:

- Data rate is predetermined—both transmitter and receiver are preset to recognize the same data rate. Hence, each node needs an accurate and stable clock source from which the data rate can be generated. However, small variations from the theoretical value can be accommodated.
- Each byte or word is *framed* with a Start and Stop bit. These allow synchronization to be initiated before the data starts to flow.

An asynchronous data format, of the sort used by such standards as RS-232, is shown in Fig. 7.12. There is now only one data line. This idles in a predetermined state, in this example at Logic 1. The start of a data word is initiated by a *Start* bit, which has polarity

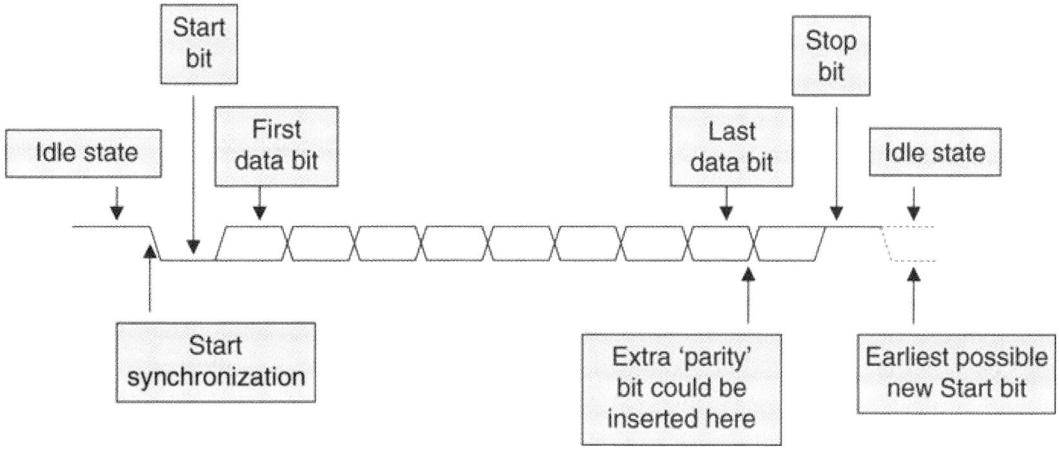

Figure 7.12
A common asynchronous serial data format.

opposite to that of the idle state. The leading edge of the Start bit is used for synchronization. Eight data bits are then clocked in. A ninth bit, for parity checking, is also sometimes used. The line then returns to the idle state, which forms a Stop bit. A new word of data can be sent immediately, following the completion of a single Stop bit, or the line may remain in the idle state until it is needed again.

An asynchronous serial port integrated into a microcontroller or peripheral device is generally called a UART, standing for *universal asynchronous receiver/transmitter*. In simplest form, a UART has one connection for transmitted data, usually called TX, and another for received data, called RX. The port should sense when a start bit has been initiated, and automatically clock in and store the new word. It can initiate a transmission at any time. The data rate at which the receiver and transmitter will operate must be predetermined; this is specified by its *baud rate*. For our purposes, we can view baud rate as being equivalent to bit rate; for more advanced applications, one should check the distinctions between these two terms.

7.8.2 Applying Asynchronous Communication in the Mbed Environment

If we look back at Fig. 2.3, we see that the LPC1768 microcontroller has four UARTs. Three of these appear on the pinout of the Mbed LPC1768, simply labeled "Serial," in Fig. 2.1. The Nucleo F401RE pinout in Fig. 2.6 is a little more complex regarding asynchronous ports. It shows "Serial1," "Serial2," and "Serial6." The latter only has an RX pin, so can receive but not transmit. Serial1 and Serial2 have both RX and TX, while Serial1 has two options for TX. To add further complication, a careful reading of

Reference 2.5 shows that Serial2 is not available to the user without adjustment to on-board solder bridges, as it is used to connect to the interface microcontroller. Therefore, in our use of the board, we will not make use of this serial port.

The Mbed OS has two possible classes to control the UART ports, buffered (**BufferedSerial**) and unbuffered (**UnbufferedSerial**). The buffered serial provides internal data buffering, both for data waiting to be transmitted and for data which has been received. The application code effectively writes to that buffer, without being concerned about whether the data is being transmitted instantaneously, and how many bytes might be in the data queue. The unbuffered serial API is just that; the application must ensure that each data byte has been fully transmitted, before sending a new byte to the port. This can be more efficient of memory usage, but places limitations on the flexibility of programming. Generally, the buffered serial is preferred; this is the one we apply here. The API summary is given in Table 7.9.

We now repeat what we have already done with SPI and I^2C, which is to connect two Mbed targets to demonstrate a serial link, but this time asynchronously. You can view Fig. 7.13, the build we use, as a variation on Figs. 7.5 and 7.9; notice that it is slightly simpler than either of these. We will apply Program Example 7.9. Interestingly, it will be the same program loaded into both Mbeds. It follows a pattern similar to that in Program Examples 7.2 and 7.5, but replaces all SPI or I^2C code with UART code. The code itself applies a number of functions from Table 7.9, and—reading the comments—should not be too difficult to follow. Note that it is essential to create a buffer in order to invoke the **write()** and **read()** functions. Here the buffer is larger than necessary, as we only transfer one byte at a time.

Table 7.9: BufferedSerial API Summary.

Functions	Usage
Constructor	
BufferedSerial(PinName tx, PinName rx, int *baud*)	Create a serial port, connected to the specified transmit and receive pins
Member Functions	
void **set_baud**(int *baud*)	Set the baud rate of the serial port.
void **set_format** (int *bits*, Parity *parity*, int *stop_bits*)	Set the transmission format used by the serial port.
bool **readable**()	Determine if there is a character available to read, in which case return true.
bool **writeable**()	Determine if there is space available to write a character, in which case return true.
ssize_t **read**(void **buffer*, size_t *length*)	Read the contents of a file into a buffer.
ssize_t **write**(const void **buffer*, size_t *length*)	Write the contents of a buffer to a file.

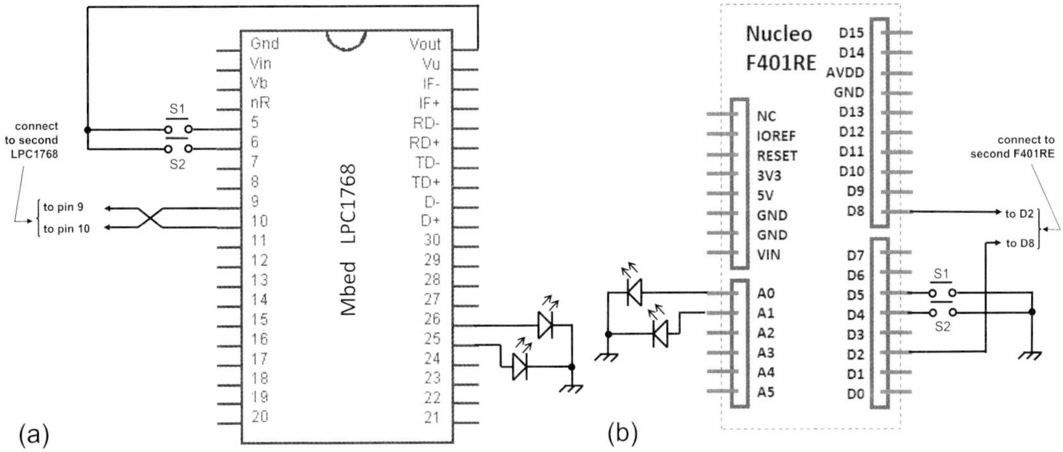

Figure 7.13
Linking two Mbed UARTs: (a) with Mbed LPC1768; (b) with Nucleo F401RE.

```
/*Program Example 7.9: Sets the Mbed up for async communication, and exchanges data
with a similar node, sending its own switch positions, and displaying those of the
other.
Works for both LPC1768 and F401RE, with pin allocations shown.          */

#include "mbed.h"

//set up TX and RX
BufferedSerial async_port(p9,p10);    //TX, RX. Use D8,D2 for F401RE
BusOut leds(p25,p26);                 //LEDs. Use A0, A1 for F401RE
DigitalOut strobe(p7);                //a strobe for `scope. Use D7 for F401RE
DigitalIn switch_ip1(p5,PullUp);      //Use D5 for F401RE
DigitalIn switch_ip2(p6,PullUp);      //Use D4 for F401RE

char switch_word ;                    //the word we will send
char recd_val;                        //the received value
char buf[4] = {0};             //declare 4-byte buffer and initialise to 0

int main() {
  //accept default format, of 9600 Baud, 8 bits, no parity
  while (true){
    //Set up the word to be sent, by testing switch inputs
    switch_word = 0xa0;             //set up a recognisable output pattern
    if (switch_ip1) switch_word = switch_word|0x01;
    if (switch_ip2) switch_word = switch_word|0x02;
    buf[0] = switch_word;
    async_port.write(buf,1);          //transmit switch_word
    if (async_port.readable() == 1){     //is there a character to be read?
      async_port.read(buf,1);         //if yes, then read it
      recd_val = buf[0];
```

```
        leds = recd_val & 0x03;     //set leds according to incoming word
      }
      strobe=1;                     //short strobe pulse.
      wait_us(100);
      strobe=0;
       //At 9600 baud one character transmission requires 10x104us, ie approx. 1ms
       //Adjust loop length for ease of observation on `scope, as required.
       ThisThread::sleep_for(2ms);
    }
  }
```

Program Example 7.9 Bidirectional data transfer between two Mbed UARTs

Connect two Mbed targets as shown in Fig. 7.13, and compile and download Program Example 7.9 into each. As previously, you should find that the switches from one Mbed control the LEDs from the other, and vice versa.

Exercise 7.7

On an oscilloscope, observe the data waveform of one of the TX lines. It helps to trigger the 'scope from the strobe pulse (pin 7/D7). See how the pattern changes as the switches are pressed.

1. What is the time duration of each data bit? How does this relate to the baud rate? Change the baud rate, and measure the new duration.
2. Is the data byte transmitted MSB first, or LSB? How does this compare to the serial protocols seen earlier in this chapter?
3. Are strobe pulse and data word precisely synchronized? Explain.
4. What is the effect of removing one of the data links in Fig. 7.13?

7.8.3 Applying Asynchronous Communication with the Host Computer

As mentioned at the start of Section 7.8.2, each Mbed target has one UART which is committed to communicating back to the USB link. This link can be accessed by the programmer using the identifiers **USBTX** and **USBRX**. The USB UART then acts just like any of the others, in terms of its use of the API in Table 7.9. We have been using it already, but only via the **printf( )** function, for the first time in Program Example 5.4.

```
/*Program Example 7.10:
A simple demonstration of the serial buffer.
Runs on LPC1768 and F401RE without adjustment, no external connection*/
```

```
#include "mbed.h"

BufferedSerial serial_port(USBTX, USBRX);

char buf[14] = "\n\rHello World";      //declare buffer and load text message
int i;
int main() {
  for(i = 2;i < 15;i++){              //i starts at 2, to include control characters
  serial_port.write(buf,i);     //Write i bytes to serial port.
  }
}
```

Program Example 7.10 Writing to the host computer

Program Example 7.10 is a very simple demonstration of using the computer link. It also explores using the buffer in buffered serial mode. The number of characters in the buffer which are written each time is controlled by the variable *i*. Once running, the program should lead to a display similar to Fig. 7.14; one extra character is printed on every line.

7.9 USB

We move now to the USB, which is a huge step in complexity from the serial protocols we have considered so far. Historically, it was introduced later than any of the protocols discussed so far in this chapter, even though it is far from being a recent protocol. Its sophistication arises from many things; notably, it is not just a set of rules to achieve serial data transfer. It contains the capability for devices to identify themselves, establish links,

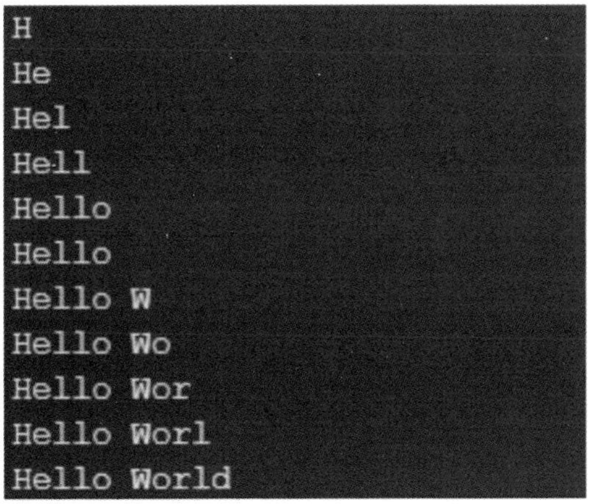

Figure 7.14
Output from Program Example 7.10.

and provide power. Here a quick introduction to the USB is given, followed by an example of USB application with the Mbed LPC1768.

7.9.1 Introducing USB

In the comparatively early days of personal computing, different peripheral devices each came with their own type of connector, and each required software reconfigurations when they were fitted. This was annoying, inefficient, and inflexible. The USB protocol was introduced to provide a more flexible and universal interconnection system, where peripherals could be added or removed without the need for reconfiguring the whole system (i.e., moving to a "plug and play" capability). The USB protocol is now managed by the USB Implementers Forum (USB-IF), Reference 7.6.

The USB is now ubiquitous and very familiar, widely used for its original purpose of connecting a PC to its direct peripherals, but also to connect all manner of devices, for example, digital cameras, mobile phones, webcams, and memory sticks, to the PC. USB version 2.0 was released in April 2000, and prevailed over a number of years. It recognizes three data rates: high speed at 480 Mbps, full speed at 12 Mb/s, and low speed at 1.5 Mb/s. The last of these is for very limited-capability devices, where only a small amount of data will be transferred. USB version 3.0 was published in November 2008, with a view to greatly increase speed. The most recent version, USB 4.0, was launched in 2019, and permits 40 Gbit/s throughput. It is backward compatible with USB 3.2 and 2. Specifications can be found at Reference 7.7.

A USB network has one host, and can have one or many *functions*, that is, USB compatible devices that can interact with the host. It is also possible to include hubs, which can have a number of functions connected to them, and which in turn link back to the host. A USB host has 15 kΩ pull-down resistors connected to each of its data lines. This indicates clearly if no connection has been made.

USB version 2.0 uses a four-wire interconnection. Two wires, labeled D+ and D-, carry the differential signal, and two are for power and earth. Within certain limits, USB functions can draw power from the bus, taking up to 100 mA at a nominal 5 V. This is supplied by the host. The fact that a USB link can supply power is one of this standard's neat features! It is because of this, of course, that we can power Mbed development boards from their USB connectors. A higher power demand can also be requested; alternatively, the functions can be self-powered.

The original USB standard defined a number of connectors; more have been added, and it seems likely that more may be expected. The "A" connector is the one you find on a standard laptop. An example of the type A receptacle is fitted to the Mbed application board, part 14 in Fig. 2.14. USB cables often have a type A plug at one end, to link with the PC, and a "mini-B" type at the other, to link with a peripheral device. There is a reason for this—generally the host, with its type A connector, supplies power. The *function* has a type B connector, and receives power. This is what happens with the LPC1768 or F401RE Mbed targets, which both have mini-B connectors. The application board also has a mini-B type connector, part 10 in Fig. 2.14.

When a device is first attached to the bus, the host resets it, assigns it an address, and interrogates it (a process known as *enumeration*). Thus, it identifies it and gathers basic operating information; for example, device type, power consumption, and data rate. All subsequent data transfers are initiated only by the host. It first sends a data packet which specifies the type and direction of data transfer and the address of the target device. The addressed device responds as appropriate. Generally, there is then a *handshake packet*, to indicate success (or otherwise) of the transfer.

7.9.2 USB Capability in the Mbed Environment

USB capability depends first on the USB capability of the target board microcontroller. Both Nucleo F401RE and LPC1768 Mbed, of course, have a USB connection, which we use for powering, programming, and other interaction with the host computer. In each case, this links to an on-board interface microcontroller. The LPC1768 microcontroller itself has a USB port. This connects to pins 31 and 32 (labeled D+ and D- on the Mbed), as seen in Fig. 2.1; we've barely noticed it so far.

There are a good number of USB APIs available for use, as shown in Table 7.10. Each has its own constructor and set of member functions. Most of them allow the Mbed to emulate a number of external devices, through USB. We will be trialing just one of these, the mouse. Further details on the others can be found on the Mbed OS web site.

7.9.3 Using the Mbed to Emulate a USB Mouse

With **USBMouse,** it is possible to make the Mbed behave like a standard USB mouse, sending position and button press commands to the host. Program Example 7.11 implements a **USBMouse** interface and continuously sends relative position information to move the mouse pointer around four co-ordinates, which make up a square. The co-ordinates are defined by the two arrays **dx** and **dy**; the mouse is moved to these positions within a **for** loop. The mouse, when initialized to its default parameters, uses a relative co-ordinate system, so if $dy = 40$ and $dx = 40$, then the mouse pointer is instructed to move

Table 7.10: Example USB Driver APIs.

USB Driver APIs	Description
USBAudio	Configures the Mbed to be recognized as an audio interface allowing streaming audio to be read, output or analyzed and processed.
USBCDC	Emulates a basic serial port over USB to send or receive data.
USBHID	Allows custom data to be sent and received from a human interface device (HID), allowing custom USB features to be developed without the need for host drivers to be installed.
USBKeyboard	Configures the Mbed to emulate a USB keyboard.
USBMIDI	Allows send and receive of MIDI messages in communication with a host PC using MIDI sequencer software.
USBMouse	Configures the Mbed to emulate a USB mouse.
USBMouseKeyboard	A USB mouse and keyboard feature set combined in a single library.
USBSerial	Emulates an additional standard serial port on the Mbed, through the USB connections.
USBMSD	Emulates a mass storage device over USB, allowing interaction with a USB storage device.

40 pixels right and 40 pixels down. Negative coordinates for x and y planes move the pointer left and up, respectively.

```
/* Program Example 7.11: Emulating a USB mouse
Note this program cannot run on F401RE, as it has no USB port.   */

#include "mbed.h"                       // include mbed library
#include "USBMouse.h"                   // include USB Mouse library
USBMouse mouse;                         // define USBMouse interface

int dx[] = {40,0,-40,0};                  // relative x position co-ordinates
int dy[] = {0,40,0,-40};                  // relative y position co-ordinates

int main() {
  while (true) {
    for (int i = 0; i < 4; i++) {       // scroll through position co-ordinates
      mouse.move(dx[i],dy[i]);    // move mouse to co-ordinate
      ThisThread::sleep_for(200ms);
    }
  }
}
```

Program Example 7.11 Emulating a USB mouse

As mentioned, the Mbed LPC1768 application board has two USB connectors, labeled 10 and 14 in Fig. 2.14. These connect to the Mbed USB port on pins 31 and 32. Program Example 7.11 connects with this port, not the usual Mbed USB connector. It is therefore convenient to run Program Example 7.11 on the app board (though a USB connector *can* be wired to an Mbed sitting in a breadboard). Using the app board approach, compile and download the program to the Mbed, connected in the usual way. Then, with the Mbed plugged into the app board, disconnect your USB cable from the Mbed, and connect it to the app board mini-B connector (item 10 in Fig. 2.14). Notice that this powers the app board again. Work out how by looking at Fig. 2.16. With the program running, your PC cursor should suddenly display a mind of its own, and start making jerky rectangular movements on the screen.

7.9.4 USB On-the-Go

The whole concept of a USB depends on a host device, normally a PC, with other devices connected to it at will. However, as its popularity and technology developed, the need arose to connect portable devices directly to each other, without the need for a host. This is the basis for USB On-the-Go, which is defined as a supplement to USB 2.0. The On-the-Go standard allows the addition of limited host capability to devices which have traditionally been peripheral only. Devices can act as either host or peripheral, or switch between roles. The standard also gives special consideration to low-power requirements. Of course, all of these are of particular interest in the embedded world.

7.10 Using Serial on the ST IoT Discovery Board

We introduced this interesting and complex board in Section 2.4 and will use it extensively in later chapters of the book. Before we place it in an IoT context, let's explore its wide range of sensors, and recognize that all of these depend on serial communication to link with the CPU.

The sensors are seen in the diagram of Fig. 2.7, and summarized in Table 7.11. All communicate by I^2C, and all are made by ST. Notice that two sensors each measure two variables, so overall six sensors measure eight variables. It is interesting to find the data sheets for each sensor on-line.

Program Example 7.12 is an official Mbed example, taken from Reference 7.8, with further comments added. It reads each of the sensors, displaying the result. If you are running the program, you will need to add the Board Support files, found in Reference 7.9.

Table 7.11: Sensors on the ST IoT Discovery Board.

Sensor	Type
Temperature, Humidity	HTS221
Pressure	LPS22HB
3D accelerometer, 3D gyroscope	LSM6DSL
3-axis magnetometer	LIS3MDL
Time of flight ranging and gesture detection	VL53L0X
MEMS omnidirectional microphone	MP34DT01

Do this by using the *Add Library to Active Program* option in Mbed Studio, as described in Section 2.6.2. To enable the printing of floating point numbers, you will also need to add the **Mbed_app.json** file to the program, explained in Section 5.3.

```
#include "mbed.h"

// Sensor drivers present in the Board Support library
#include "stm32l475e_iot01_tsensor.h"   //temperature
#include "stm32l475e_iot01_hsensor.h"   //humidity
#include "stm32l475e_iot01_psensor.h"   //pressure
#include "stm32l475e_iot01_magneto.h"
#include "stm32l475e_iot01_gyro.h"
#include "stm32l475e_iot01_accelero.h"

DigitalOut led(LED1);

int main(){
  float sensor_value = 0;              //used for temp, humidity, pressure
  int16_t pDataXYZ[3] = {0};       //for Magneto and accelerometer
  float pGyroDataXYZ[3] = {0};

  printf("Start sensor init\n");
  //initialise each sensor
  BSP_TSENSOR_Init();    //temperature
  BSP_HSENSOR_Init();    //humidity
  BSP_PSENSOR_Init();    //pressure
  BSP_MAGNETO_Init();
  BSP_GYRO_Init();
  BSP_ACCELERO_Init();

  while(true) {
    printf("\nNew loop, LED1 should blink during sensor read\n");
    led = 1;
    //Read and print sensor values
    sensor_value = BSP_TSENSOR_ReadTemp();
    printf("\nTEMPERATURE = %.2f degC\n", sensor_value);
```

```
        sensor_value = BSP_HSENSOR_ReadHumidity();
        printf("HUMIDITY = %.2f %%\n", sensor_value);
        sensor_value = BSP_PSENSOR_ReadPressure();
        printf("PRESSURE is = %.2f mBar\n", sensor_value);
        led = 0;
        ThisThread::sleep_for(1s);
        led = 1;

        BSP_MAGNETO_GetXYZ(pDataXYZ);
        printf("\nMAGNETO_X = %d\n", pDataXYZ[0]);
        printf("MAGNETO_Y = %d\n", pDataXYZ[1]);
        printf("MAGNETO_Z = %d\n", pDataXYZ[2]);

        BSP_GYRO_GetXYZ(pGyroDataXYZ);
        printf("\nGYRO_X = %.2f\n", pGyroDataXYZ[0]);
        printf("GYRO_Y = %.2f\n", pGyroDataXYZ[1]);
        printf("GYRO_Z = %.2f\n", pGyroDataXYZ[2]);

        BSP_ACCELERO_AccGetXYZ(pDataXYZ);
        printf("\nACCELERO_X = %d\n", pDataXYZ[0]);
        printf("ACCELERO_Y = %d\n", pDataXYZ[1]);
        printf("ACCELERO_Z = %d\n", pDataXYZ[2]);
        led = 0;

        ThisThread::sleep_for(1s);
        }
}
```

(A)

```
{
  "target_overrides":{
    *": {
      "target.printf_lib": "std"
    }
  }
}
```

(B)

Program Example 7.12 Testing the ST IoT discovery board sensors (official Mbed example): (A) the main.cpp file; (B) the Mbed_app.json file

Once running, the program should give a display similar to Fig. 7.15.

```
New loop, LED1 should blink during sensor read

TEMPERATURE = 24.57 degC
HUMIDITY    = 52.85 %
PRESSURE is = 1019.02 mBar

MAGNETO_X = 95
MAGNETO_Y = -514
MAGNETO_Z = 321

GYRO_X = -350.00
GYRO_Y = -2100.00
GYRO_Z = 0.00

ACCELERO_X = 2
ACCELERO_Y = -15
ACCELERO_Z = 1031
```

Figure 7.15
Screen output from Program Example 7.12

Exercise 7.8

Download the data sheets for all the Discovery Board sensors, as itemized in Table 7.11. They are easily found on line. Study also the schematics of the Discovery Board (within Reference 2.8). Draw a block diagram of how each of the sensors in Table 7.11 connect back to the CPU, identifying where possible the address or address range of each.

7.11 Mini-projects

7.11.1 Multi-node I^2C Bus

Design a simple circuit which has a temperature sensor, range finder, and Mbed target connected to the same I^2C bus. Merge Program Examples 7.7 and 7.8, so that measurements from each sensor are displayed in turn on the computer screen. Verify that the I^2C protocol works as expected.

7.11.2 Vibrating Beam Acceleration Threshold Detection

We met the ADXL345 accelerometer earlier in this chapter. It is interesting to note that the device has two interrupt outputs, as seen in Fig. 7.6. These can be connected to an

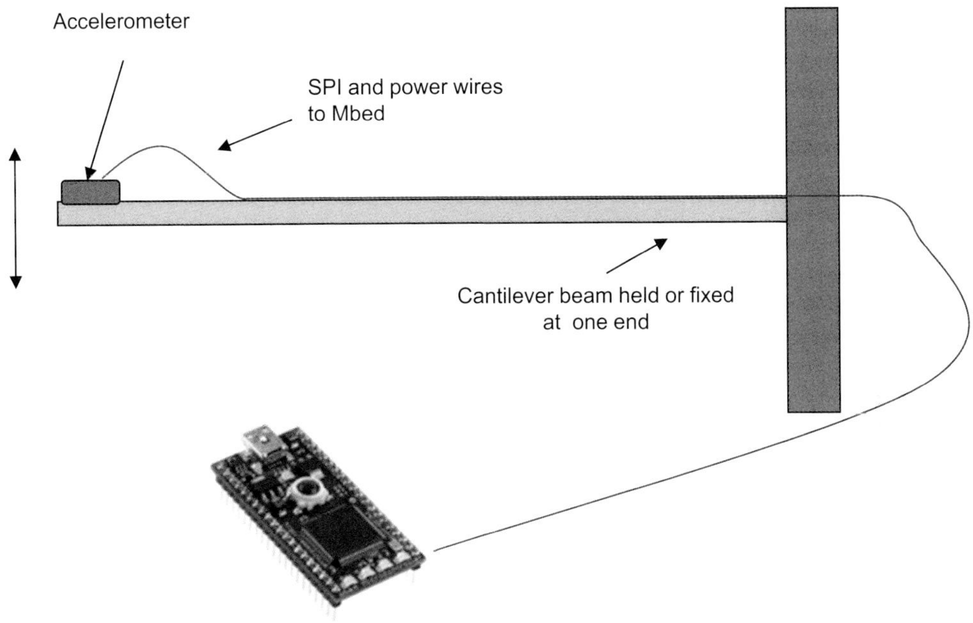

Figure 7.16
The accelerometer setup.

Mbed digital input to run an interrupt routine whenever an acceleration threshold is exceeded. For example, the accelerometer might be a crash detection sensor in a vehicle which, when a specified acceleration value is exceeded, activates an airbag.

Use an accelerometer on a cantilever arm to provide the acceleration data. Fig. 7.16 shows the general construction. A plastic 30-cm or 1-ft ruler, clamped at one end to a table, can be used. Set the accelerometer to generate an interrupt whenever a threshold in the *z*-axis is exceeded. Connect this as an interrupt input to the Mbed target, and program it so a warning LED lights for 1 s whenever the threshold is exceeded. You can experiment with the actual threshold value to alter the sensitivity of the detection system.

Chapter Review

- Serial data links provide a ready means of communication between microcontrollers and peripherals, and/or between microcontrollers.
- SPI is a simple synchronous standard which is still very widely applied. Mbed target boards generally have SPI ports, and the Mbed OS provides a supporting API.
- While a very useful standard, SPI has certain very clear limitations, relating to a lack of flexibility and robustness.

- The I^2C protocol is a more sophisticated serial alternative to SPI; it runs on a two-wire bus, and includes addressing and acknowledgment.
- I^2C is a flexible and versatile standard. Devices can readily be added to or removed from an existing bus, and a Master can detect if a Slave fails to respond. Nevertheless, I^2C has limitations which mean that it cannot be used for high-reliability applications.
- A very wide range of peripheral devices are available, including intelligent sensors, which communicate through SPI and I^2C.
- A useful asynchronous alternative to I^2C and SPI is provided by the UART. Mbed target boards generally have several UARTs, and reserve one for a communication link back to the host computer.
- The USB protocol is designed specifically for allowing plug-and-play communications between a computer and peripheral devices such as a keyboard or mouse. There are a number of Mbed APIs allowing it, for example, to operate as a mouse or keyboard, or as an audio or MIDI interface.
- The ST IoT Discovery Board uses I^2C extensively to link its on-board sensors to the central microcontroller.

Quiz

1. What do the acronyms SPI, I^2C, UART, and USB stand for?
2. Draw up a table comparing the advantages and disadvantages of using SPI versus I^2C for serial communications.
3. What are the limitations for the number of devices which can be connected to a single SPI, I^2C, or UART interconnection?
4. An SPI link is running with a 500-kHz clock. How long does it take for a single message containing one data byte to be transmitted?
5. An Mbed target (choose LPC1768 or F401RE) configured as SPI Master is to be connected to three other similar boards, each configured as Slave. Sketch a circuit which shows how this interconnection could be made. Explain your sketch.
6. An Mbed LPC1768 is to be set up as SPI Master, using pins 11, 12, and 13, running at a frequency of 4 MHz, with 12-bit word length. The clock should idle at Logic 1, and data should be latched on its negative edge. Write the necessary code to set this up.
7. Repeat Question 4, but for I^2C, ensuring that you calculate time for the complete message. Assume for this calculation that start, stop, and acknowledge events each have a duration of one data bit.
8. Repeat Question 5, but for I^2C. Identify carefully the advantages and disadvantages of each connection.
9. You need to set up a serial network, which will have one Master and four Slaves. Either SPI or I^2C can be used. Every second, data has to be distributed so that one byte is sent to Slave 1, four to Slave 2, three to Slave 3, and four to Slave 4. If the

complete data transfer must take not more than 200 μs, estimate the minimum clock frequency which is allowable for SPI and I^2C. Assume there are no other timing overheads.

10. Repeat Question 4, but for asynchronous communication through a UART, assuming a baud rate of 500 kHz. Ensure that you calculate time for the complete message.
11. The Mbed LPC1768 application board has two 15-kΩ pull-down resistors connected to its D+ and D- USB lines, which can be switched in or out. What is their purpose?
12. View the application board power supply circuit in Fig. 2.16. The USB-A and USB-B connectors connect to different points in the board power supply circuit. Explain the connections made, and the role of the diodes in that circuit.

References

7.1. Spasov P. *Microcontroller Technology, the 68HC11*. second ed. Prentice Hall; 1996.
7.2. ADXL345 data sheet, Rev. E. http://www.analog.com/media/en/technical-documentation/data-sheets/ADXL345.pdf.
7.3. The I2C Bus Specification and User Manual, Rev. 7.0 (2021). NXP Semiconductors, Document number UM10204.
7.4. TMP102, Low Power Digital Temperature Sensor, Document number SBOS397B, Texas Instruments, August 2007, rev. October 2008.
7.5. SRF08 data. http://www.robot-electronics.co.uk/htm/srf08tech.shtml .
7.6. USB Implementers Forum Website. www.usb.org.
7.7. USB Specifications. https://usb.org/documents.
7.8. HelloWorld_ST_Sensors, Official Mbed Example. https://os.Mbed.com/teams/ST/code/HelloWorld_ST_Sensors/.
7.9. BSP driver for the B-L475E-IOT01 board. https://os.Mbed.com/teams/ST/code/BSP_B-L475E-IOT01/.

CHAPTER 8

Liquid Crystal Displays

8.1 Display Technologies

We have used light-emitting diodes (LEDs) in previous chapters, particularly individual LEDs and seven-segment displays. LEDs on their own can only inform us of a few different states, for example, indicating if something is on or off, or if a variable is cleared or set. It is possible to be clever in the way LEDs are used—for example, different flashing speeds can be used to represent different states—but there is a limit to how much information a single LED can communicate. With seven-segment displays it is possible to display numbers and a few characters from the alphabet, but we have seen a problem with such displays in that in simple use they require a microcontroller output for each LED segment. The multiplexing technique described in Section 3.7.3 can be used, but many pins are still needed. Further, and importantly, LEDs are relatively power-hungry, which is a major concern for power-conscious designs. Clearly, the use of LEDs for display purposes has limitations.

The *liquid crystal display* (LCD) overcomes many of the difficulties associated with LED displays. In this chapter we introduce LCD principles, and then go on to examine the use of three distinct LCD types—the conventional character display, a monochrome pixel-oriented display, and a color pixel-oriented type. The ability to design with these is an essential skill of the embedded designer.

8.1.1 Introducing Liquid Crystal Technology

LCDs are of great importance these days, both in the electronic world in general, and in the embedded system environment. Their main advantages are their extremely low-power requirements, light weight, and high flexibility. They have been one of the enabling technologies for battery-powered products such as the digital watch, laptop computer, and mobile phone, and are available in a huge range of indicators and displays. LCDs do, however, have some disadvantages; these include limited viewing angle and contrast in some implementations, sensitivity to temperature extremes, and high cost for the more sophisticated graphical display.

LCDs do not emit light, but they can reflect incident light, or transmit or block backlight. The principle of an LCD is illustrated in Fig. 8.1. The liquid crystal is an organic

Fast and Effective Embedded Systems Design. https://doi.org/10.1016/B978-0-323-95197-5.00008-5

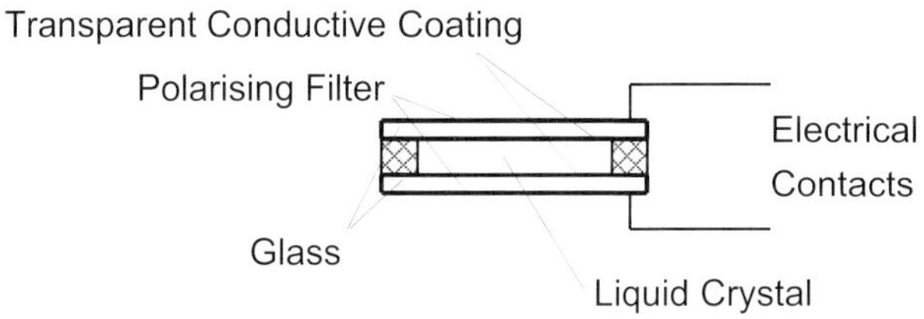

Figure 8.1
A simple liquid crystal structure.

compound which responds to an applied electric field by changing the alignment of its molecules, and hence the light polarization which it introduces. A small quantity of liquid crystal is contained between two parallel glass plates. A suitable field can be applied if transparent electrodes are located on the glass surface. In conjunction with external polarizing light filters, light is either blocked or transmitted by the display cell.

The LCD electrodes can be made in any pattern desired. These may include single digits or symbols, or the standard patterns of bar graph, seven-segment, dot matrix, starburst, and so on. Alternatively, they may be extended to complex graphical displays with addressable pixels. Color LCDs go a big step further in complexity. For each pixel, there are now three subpixels, with red, green, and blue light filtering. Controlling the voltage of each controls its intensity, ultimately giving a high level of control over the resultant combined color.

8.1.2 Liquid Crystal Character Displays

A popular and well-established form of LCD is the character display, seen in Fig. 8.2. These are available from one line of characters to four or more, and are commonly seen on many domestic and office items, such as photocopiers, burglar alarms, or home entertainment systems. Driving this complex array of tiny LCD dots is far from simple, so

Figure 8.2
The Mbed LPC1768 driving an LCD.

such displays always contain a hidden microcontroller customized to drive the display. The first such controller to gain widespread acceptance was the Hitachi HD44780. While this has been superseded by others, they have kept the interface and internal structure of the Hitachi device. It is important to know its main features in order to design with it.

The HD44780 contains an 80-byte RAM to hold the display data and a ROM (i.e., non-volatile memory) for generating the characters. It has a simple instruction set, including instructions for initialization, cursor control (moving, blanking, blinking), and clearing the display.

Communication with the controller is done via an 8-bit data bus, three control lines (RS, $R/\overline{W}$, Busy Flag), and an enable/strobe line (E). These are illustrated in Fig. 8.3. Data written to the controller is interpreted either as instruction or display, depending on the

Signal Name	Signal Function
RS	Register select: 0 = Instruction register 1 = Data register
$R/\overline{W}$	Selects read or write
E	Synchronizes read and write operations
DB4–DB7	Higher-order bits of data bus; DB7 also used as Busy flag
DB0–DB3	Lower-order bits of data bus; not used for 4-bit operation

(a)

RS	$R/\overline{W}$	E	Action
0	0		Write instruction code
0	1		Read busy flag and address counter
1	0		Write data
1	1		Read data

(b)

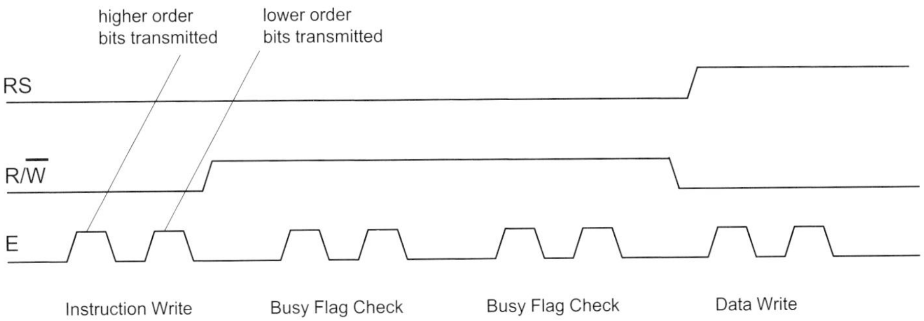

(c)

Figure 8.3
Interfacing with the HD44780: (a) user interface lines; (b) data and instruction transfers; (c) example timing for 4-bit interface.

state of the RS line. This is seen in Fig. 8.3(b). An important use of reading data back from the LCD is to check the controller status via the busy flag. As some instructions take a finite time to implement (for example, a minimum of 40 μs is required to receive one character code), it is useful to be able to read the flag and wait until the LCD controller is ready to receive further data.

The controller can be set up to operate in 8-bit or 4-bit mode. In the latter mode, only 7 instead of 11 interconnections are required; the four most significant bits of the bus are used, and two write cycles are needed to send a single byte. This is shown in Fig. 8.3(c). The convenience of fewer interconnections is paid for with slightly longer data transfer times. In both cases the most significant bit doubles as the busy flag.

For less demanding applications, two important simplifications can be made to the interface described so far. Instead of reading the busy flag, small delays can be inserted in the program after each write to the display, with value greater than the maximum known time needed to process the previous message. If the fastest possible control of the LCD is required, however, then the busy flag should be implemented. Also in some applications, the display can be used in write mode only, in which case the $R/\overline{W}$ line can be tied to ground.

8.2 Using the PC1602F LCD

An Mbed development board can be interfaced with an external LCD. Here the 2×16 character Powertip PC1602F LCD is used as an example; its data sheet can be found at Reference 8.1. A wide variety of displays, with differing manufacturers and sizes but similar functionality, can also be found.

Using LCDs such as these requires some care in implementation. The correct interconnections must be made, the LCD must be correctly initialized at the start of the program, and appropriate control and display data must then be sent. As this results in some complexity in the code, we take the opportunity to use it as an example of modular coding.

8.2.1 Introducing and Connecting the PC1602F Display

The PC1602F display has an on-board microcontroller, based on the HD44780. It has 16 connections, defined in Table 8.1.

In the examples which follow, the LCD is used in 4-bit mode, so only the upper four bits of the data bus (DB4–DB7) are connected. The two simplifications mentioned earlier are also applied—the busy flag is not accessed, and $R/\overline{W}$ is tied permanently to ground.

Table 8.1: PC1602F pin descriptions, with suggested connections to development boards.

PC1602F Pin Number	Pin Name	Function	Mbed LPC1768 Pin Number Used	Nucleo F401RE Pin Number Used
1	V_{SS}	Power supply (GND)	1	GND
2	V_{DD}	Power supply (5 V)	39	5V
3	V_0	Contrast adjust (nc on recent devices)	-	-
4	RS	Register select	19	D9
5	$R/\overline{W}$	Data read / write	1	GND
6	E	Enable	20	D8
7	DB0	Data bus line bit 0	nc	nc
8	DB1	Data bus line bit 1	nc	nc
9	DB2	Data bus line bit 2	nc	nc
10	DB3	Data bus line bit 3	nc	nc
11	DB4	Data bus line bit 4	21	D4
12	DB5	Data bus line bit 5	22	D5
13	DB6	Data bus line bit 6	23	D6
14	DB7	Data bus line bit 7/ busy flag	24	D7
15	A	Power supply for LED back light (5 V)	39	5V, 3V3 or nc
16	K	Power supply for LED back light (GND)	1	GND or nc if 15 is nc

Notes: nc = no connection, A = Anode, K = cathode of the back light LED.

A 1 ms delay between data transfers is used to ensure that all internal processes can complete before the next action.

The display is initialized by sending control instructions to the configuration registers in the LCD. This is done by setting RS and $R/\overline{W}$ low. Once the LCD has been initialized, display data can be sent by setting the RS bit high. For all transfers, the E bit must be pulsed for every nibble of data sent.

The suggested interface for connecting either LPC1768 or F401RE to the PC1602F is shown in Table 8.1, though other digital I/O pins can be chosen. Four outputs are needed to send the 4-bit instruction and display data, and two outputs are needed for the RS and E control lines. LCD pins DB0, DB1, DB2, and DB3 are left unconnected, as these are not required for 4-bit data mode.

8.2.2 Using Modular Coding to Control the LCD

We turn now to developing the modular code. You may wish to check Appendix B, Section B.11 for an overview of how multiple files relate to each other in a C/C++ program. Three files are used for this application:

- a main code file (**main.cpp**), which can call functions defined in the LCD definition file,
- an LCD definition file (**LCD.cpp**), which will include all the functions for initializing and sending data to the LCD,
- and an LCD header file (**LCD.h**), which will be used to declare data and function prototypes.

The following functions will be declared in the LCD header file:

- **toggle_enable( )**: a function to pulse the E bit,
- **LCD_init( )**: a function to initialize the LCD,
- **display_to_LCD( )**: a function to display characters on the LCD.

The **LCD.h** header file, Program Example 8.1, defines the function prototypes.

```
/* Program Example 8.1: LCD.h header file
   Runs on either LPC1768 or F401RE with pins shown in Table 8.1
                                                           */
#ifndef LCD_H
#define LCD_H

#include "mbed.h"

void toggle_enable(void);           //function to toggle/pulse the enable bit
void LCD_init(void);                //function to initialise the LCD
void display_to_LCD(char value);    //function to display characters

#endif
```

Program Example 8.1 The LCD.h header file

8.2.3 Initializing the Display

A specific initialization procedure must be programmed in order for the PC1602F display to operate correctly. Full details are provided in the datasheet, Reference 8.1. This shows that we first need to wait a short period (approximately 20 ms), and then set the RS and E lines to zero and send a number of configuration messages to set up the LCD. Configuration data is then sent to the Function Mode, Display Mode, and Clear Display registers. These are now introduced.

Function Mode

The function mode determines the basic operating conditions of the display. To set it, the RS and R/$\overline{W}$ bits should be set as shown in Fig. 8.4, and the data bus values then sent as two nibbles.

If, for example, a binary value of 00101000 (0x28 hex) is sent to the LCD data pins, this defines 4-bit mode, 2-line display, and 5×7 dot characters. In this case, we send the value 0x2, pulse E, then send 0x8, then pulse E again. This command appears in the **LCD_init( )** function detailed in Program Example 8.2.

Display Mode

Here we need to send a command to switch the display on, which also determines the cursor function. The Display Mode register is shown in Fig. 8.5.

To switch the display on with a blinking cursor, the value 0x0F (in two 4-bit nibbles) is needed. This command is used in the **LCD_init( )** function in Program Example 8.2.

Clear Display

Before data can be written to the display, it must be cleared, and the cursor reset to the first character in the first row, or any other location that you wish to write data to. The Clear Display command is shown in Fig. 8.6.

RS	R/$\overline{W}$
0	0

DB7	DB6	DB5	DB4	DB3	DB2	DB1	DB0
0	0	1	BW	N	F	X	X

BW = 0 → 4 bit mode
BW = 1 → 8 bit mode
X = Don't care bits (can be 0 or 1)

N = 0 → 1 line mode
N = 1 → 2 line mode

F = 0 → 5×7 pixels
F = 1 → 5×10 pixels

Figure 8.4
Function Mode control register.

RS	R/$\overline{W}$
0	0

DB7	DB6	DB5	DB4	DB3	DB2	DB1	DB0
0	0	0	0	1	P	C	B

P = 0 → display off
P = 1 → display on

C = 0 → cursor off
C = 1 → cursor on

B = 0 → cursor no blink
B = 1 → cursor blinking

Figure 8.5
Display Mode control register.

RS	R/W
0	0

DB7	DB6	DB5	DB4	DB3	DB2	DB1	DB0
0	0	0	0	0	0	0	1

Figure 8.6
Clear Display command.

8.2.4 Sending Display Data to the LCD

The American Standard Code for Information Interchange (ASCII) is a method for defining alphanumeric characters as 8-bit values. On receiving a single ASCII byte, the display recognizes which character should be shown. The complete ASCII table is included with the LCD datasheet, but for interest some common ASCII values are shown in Table 8.2. It can be seen, for example, that if we send the data value 0x48 to the display, the character "H" will be displayed.

The LCD definition file (**LCD.cpp**) contains the three C functions **toggle_enable( ), LCD_init( )**, and **display_to_LCD( )** as described above. It appears as Program Example 8.2. Note that the **toggle_enable( )** function has two 1 ms delays, removing the need to monitor the busy flag; the downside to this is that we have introduced a timing delay into the program.

The **display_to_LCD( )** function transmits characters to the LCD. Characters are transmitted by setting the RS flag to 1 (data setting), and then sending a data byte containing the required ASCII character. The function accepts an 8-bit value as a data input, using the **char** data type. As we are using 4-bit mode, the most significant bits of the ASCII byte must be shifted right in order to be output on the 4-bit bus created through **BusOut**. The lower four bits can then be output directly.

Table 8.2: Common ASCII values.

		Less significant bits (lower nibble)															
		0x0	**0x1**	**0x2**	**0x3**	**0x4**	**0x5**	**0x6**	**0x7**	**0x8**	**0x9**	**0xA**	**0xB**	**0xC**	**0xD**	**0xE**	**0xF**
More significant bits (upper nibble)	0x2		!	"	#	$	%	&	'	(	)	*	+	,	-	.	/
	0x3	0	1	2	3	4	5	6	7	8	9	:	;	<	=	>	?
	0x4	@	A	B	C	D	E	F	G	H	I	J	K	L	M	N	O
	0x5	P	Q	R	S	T	U	V	W	X	Y	Z	[	\	]	^	_
	0x6	'	a	b	c	d	e	f	g	h	i	j	k	l	m	n	o
	0x7	p	q	r	s	t	u	v	w	x	y	z	{	|	}	~	

```
/* Program Example 8.2: Declaration of objects and functions in LCD.cpp file
  Runs on either LPC1768 or F401RE with pins shown below
                                                                            */
#include "LCD.h"
DigitalOut RS(p19);                   //Use D9 for F401RE
DigitalOut E(p20);                    //Use D8 for F401RE
BusOut data(p21, p22, p23, p24);      //Use D4,D5,D6,D7 for F401RE

void toggle_enable(void){
  E = 1;
  wait_us(1000);
  E = 0;
  wait_us(1000);
}

//initialise LCD function
void LCD_init(void){
  wait_us(20000);             // pause for 20 ms
  RS = 0;                       // set low to write control data
  E = 0;                        // set low

  //function mode
  data = 0x2;                 // 4 bit mode (data packet 1, DB4-DB7)
  toggle_enable();
  data = 0x8;                 // 2-line, 7 dot char (data packet 2, DB0-DB3)
  toggle_enable();
  //display mode
  data = 0x0;                 // 4 bit mode (data packet 1, DB4-DB7)
  toggle_enable();
  data = 0xF;                 // display on, cursor on, blink on
  toggle_enable();

  //clear display
  data = 0x0;                 //
  toggle_enable();
  data = 0x1;                 // clear
  toggle_enable();
}

//display function
void display_to_LCD(char value){
  RS = 1;                     // set high to write character data
  data = (value >> 4);        // value shifted right 4 = upper nibble
  toggle_enable();
  data = value;               // value bitmask with 0x0F = lower nibble
  toggle_enable();
}
```

Program Example 8.2 The LCD.cpp file

8.2.5 Calling the LCD Functions from main( )

We can now develop a main control file (**main.cpp**) to use the LCD functions described above. The simple example of Program Example 8.3 initializes the LCD, displays the word "HELLO," and then displays the numerical characters from 0 to 9.

```
/* Program Example 8.3 Utilising LCD functions in the main.cpp file
Runs on either LPC1768 or F401RE with pins shown in Example 8.2
                                                                    */
#include "LCD.h"

int main() {
  LCD_init();                   // call the initialise function
  display_to_LCD(0x48);         // 'H'
  display_to_LCD(0x45);         // 'E'
  display_to_LCD(0x4C);         // 'L'
  display_to_LCD(0x4C);         // 'L'
  display_to_LCD(0x4F);         // 'O'
  for(char x = 0x30;x <= 0x39;x++){
    display_to_LCD(x);          // display numbers 0-9
  }
}
```

Program Example 8.3 The main.cpp file

Applying Table 8.1, connect a Powertip PC1602F LCD to an LPC1768 or F401RE development board, and construct a new program with the files from Program Examples 8.1 to 8.3, putting each in its own file. Compile and run the program, and verify that the word "HELLO" and the numerical characters are correctly displayed, with a flashing cursor in the final digit.

■ Exercise 8.1

Change the program described above so that your name appears on the display, after the word "HELLO," replacing the numerical characters 0-9

■

■ Exercise 8.2

Modify the **LCD_init( )** function to disable the flashing cursor. Here you will need to modify the value sent to the Display Mode register.

■

8.2.6 Adding Data to a Specified Location

The display has a memory which is mapped so that each display digit has a unique memory address, as shown in Fig. 8.7. A display pointer can be set before data is transferred; the data sent (in the form of an ASCII character) will then appear in the position specified. For example, if the display pointer is set to address 0x40, data will be displayed at the first position on the second line. To change the pointer address, the desired 6-bit address value must be sent in a control byte with bit 7 also set, as shown in Fig. 8.8.

We can create a new function to set the value of the display pointer. We will call this function **set_location( ),** as shown in Program Example 8.4. Notice that bit DB7 is set by ORing the location value with 0x80.

```
/* Program Example 8.4 function to set the display location. Parameter
"location" holds address of display unit to be selected
                                                                        */

void set_location(char location){
    RS = 0;
    data = (location|0x80) >> 4;        // upper nibble
    toggle_enable();
    data = location&0x0F;             // lower nibble
    toggle_enable();
}
```

Program Example 8.4 Function to change the display pointer position

Display Position		1	2	3	4	5	6	7	8	9	10	11	12	13	14	15	16
Display Pointer Address	Row 1	00	01	02	03	04	05	06	07	08	09	0A	0B	0C	0D	0E	0F
	Row 2	40	41	42	43	44	45	46	47	48	49	4A	4B	4C	4D	4E	4F

Figure 8.7
Screen display address values.

RS	R/$\overline{W}$
0	0

DB7	DB6	DB5	DB4	DB3	DB2	DB1	DB0
1	AC6	AC5	AC4	AC3	AC2	AC1	AC0

Figure 8.8
Display pointer control. AC6—AC0 contain the 6-bit display pointer address.

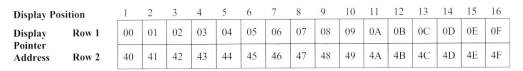

Exercise 8.3

Add the **set_location( )** function shown in Program Example 8.4 to the **LCD.cpp** definition, also declaring the function prototype in **LCD.h**. Now add **set_location( )** function calls to **main.cpp** so that the word "HELLO" appears in the center of the first line and the numerical characters 0–9 appear in the center of the second line of the display.

Exercise 8.4

Try writing a revised program, so that the delays in **toggle_enable( )** are removed, and the busy flag is tested instead. You will need to activate $R/\overline{W}$ and set it to 1 in order to read the flag, indicated by data bit DB7. Once Busy is clear, the program should proceed. Estimate how much time is saved by your new program. You can test this with an oscilloscope by making your program write a digit continuously to the display, and measuring on the oscilloscope the time between the E pulses.

8.3 Using the Mbed TextLCD Library

There are a number of software libraries available that make an alphanumeric LCD much simpler and quicker to program. For example, the Mbed **TextLCD** library, Reference 8.2, is more advanced than the simple functions we have just created. Simple **printf( )** statements are used to display characters and strings on the LCD screen, so a function call is not required for each individual character, as was the case in Program Example 8.3. A summary of **TextLCD** is given in Table 8.3.

Table 8.3: Example functions, TextLCD API.

Functions	Usage
Constructor	
TextLCD *name*(int *rs*, int *e*, int *d0*, int *d1*, int *d2*, int *d3*)	Create object, choosing *name*, specifying pins as indicated, and initialize.
Member Functions	
void **character**(int *column*, int *row*, int *c*)	Write character c, to specified column and row.
void **locate**(int *column*, int *row*)	Locate cursor at specified column and row.
void **cls**()	Clear screen, and set cursor to 0.
void **writeByte**(int *value*)	Write *value* with unchanged setting of RS.
void **writeCommand**(int *command*)	Set RS to 0 and write *command*.
void **writeData**(int *data*)	Set RS to 1 and write *data*.

Program Example 8.5 is a simple "Hello World" example using the **TextLCD** library.

```
/*Program Example 8.5:  TextLCD library example
  Runs on either LPC1768 or F401RE with pins shown below
                                                          */
#include "mbed.h"
#include "TextLCD.h"

TextLCD lcd(p19, p20, p21, p22, p23, p24); //rs,e,d0,d1,d2,d3
          //Use D9,D8,D4,D5,D6,D7 or own choice for F401RE

int main() {
  lcd.printf("Hello World!");
}
```

Program Example 8.5 TextLCD Hello World

Create a new program around Program Example 8.5 and import the **TextLCD** library file (Reference 8.2) using the procedure described in Section 2.6.2. At the time of writing, there were still some deprecated **wait()** functions in this library. It is easy to find and replace these, for example, replacing **wait_ms(5)** with **wait_us(5000)**. Compile and run, verifying that the program displays the characters "Hello World" correctly.

Exercise 8.5

Applying Fig. 8.7, experiment with the **locate()** function and verify that the Hello World string can be positioned to any desired location on the display. The row value can be 0 or 1, the column value 0 to 15.

Program Example 8.6 displays a count variable, incrementing every second, on the LCD display.

```
/* Program Example 8.6: LCD Counter example
 Runs on either LPC1768 or F401RE with pins shown below
                                                  */
#include "mbed.h"
#include "TextLCD.h"

TextLCD lcd(p19, p20, p21, p22, p23, p24); // rs,e,d0,d1,d2,d3
          //Use D9,D8,D4,D5,D6,D7 or own choice for F401RE
int x = 0;

int main() {
  lcd.printf("LCD Counter");
  while (true) {
    lcd.locate(5,1);     //locate the cursor on the 2nd row, 6th digit
```

```
      lcd.printf("%i",x);  //print the value of x as an integer
      wait_us(1000000);     //wait 1 second
      x++;
   }
 }
```

Program Example 8.6 LCD counter

Implement Program Example 8.6 as a new program. Don't forget to import the TextLCD library again. Run the program, and observe the count on the display.

Exercise 8.6

Increase the speed of the counter in Program Example 8.6 and investigate how the apparent cursor position changes as the count value increases.

8.4 Displaying Analog Input Data on the LCD

In Section 5.3, we read in an analog input and displayed its value on the computer screen. We can now do a similar thing, but displaying on the LCD, as shown in Program Example 8.7. Using the Mbed **AnalogIn** API, the analog input variable is assigned a floating point value between 0 and 1, where 0 is 0 V and 1 represents 3.3 V. The input value is multiplied here by 100 to display a percentage between 0% and 100%. The program loops and clears the screen every time before a new value is displayed.

```
/*Program Example 8.7: Display analog input data
Runs on either LPC1768 or F401RE with pins shown below              */

#include "mbed.h"
#include "TextLCD.h"
TextLCD lcd(p19, p20, p21, p22, p23, p24); //rs,e,d0,d1,d2,d3
         //Use D9,D8,D4,D5,D6,D7 or own choice for F401RE
AnalogIn Ain(p17);  //Use A5 for F401RE
float percentage;

int main() {
  while(true){
    percentage = Ain*100;
    lcd.printf("%1.2f",percentage);
    ThisThread::sleep_for(20ms);
    lcd.cls();
  }
}
```

(a)

```
{
  "target_overrides":{
    "*": {
      "target.printf_lib": "std"
      }
    }
}
```

(**b**)

Program Example 8.7 Display analog input data: (a) the main.cpp file; (b) the Mbed_app.json file

Implement Program Example 8.7. You will again need to import the TextLCD library, and also (because we're going to display floating point numbers) create an Mbed_app.json file, as described in Section 5.3. Connect a potentiometer between 0 V and 3.3 V, with wiper connected to chosen analog input pin. Verify that readings between 0% and 100% can be obtained.

Exercise 8.7

Modify the **wait_us( )** statement in Program Example 8.7 to smaller and larger values and evaluate the change in performance. It is interesting to make a mental note that a certain range of update rates appear irritating to view (too fast), while others may be perceived as too slow.

Exercise 8.8

Create a program to make the Mbed target and display act as a voltmeter, with a display layout similar to that shown in Figure 8.9. Potential difference should be measured between 0 V and 3.3 V and displayed to the screen. Note the following:

- You will need to convert the 0.0–1.0 analog input value to a value which represents 0 V–3.3 Volts.
- Check the display with the reading from an actual voltmeter—is it accurate?
- Increase the number of decimal places that the voltmeter displays. Evaluate the noise and accuracy of the voltmeter readings with respect to the target system's ADC resolution.
- Consider adding averaging, as described in Section 5.3.3, to stabilize the display reading.

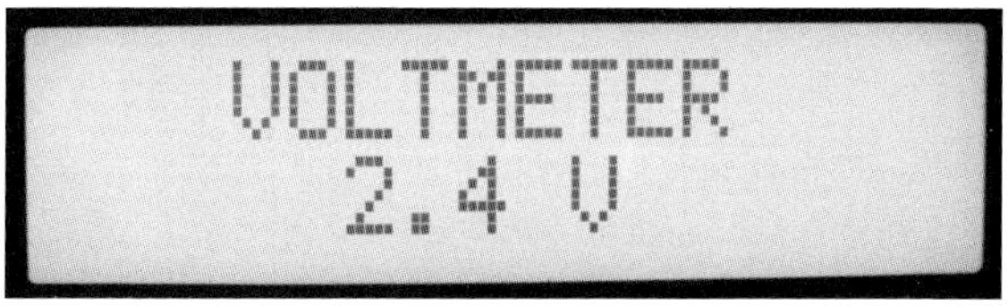

Figure 8.9
Voltmeter display.

8.5 Pixel Graphics—Implementing the NHD-C12832 Display

A more advanced LCD display allows the programmer to set or clear each individual pixel on the screen. For example, the NHD-C12832, as shown in Fig. 8.10, is designed as an LCD matrix of 128×32 pixels, allowing more intricate images and text messages to be displayed. The datasheet for this display can be found at Reference 8.3. Interconnection is through SPI, plus a few further control lines.

App Board Conveniently, the Mbed app board has a C12832 included, making it easy to test and verify graphical displays. While the app board is of course designed to operate with an Mbed LPC1768 plugged into it, it is not difficult to wire across to a Nucleo F401RE, as seen (for a different circuit) in Fig. 3.9(b). In this case the USB connection to the F401RE board is retained to provide power, with its 3.3-V output powering the app board. In the absence of an app board, a standalone display can be wired across to the Mbed target board, using connections shown in the C12832 data.

The C12832 can be controlled by importing and using the C12832 library by developer Kevin Anderson, Reference 8.4. The main features of this API are shown in Table 8.4.

Figure 8.10
The NHD-C12832 LCD display on the Mbed app board.

Table 8.4: NHD-C12832 summary API table.

Functions	Usage
C12832 (mosi,sck,rst,a0,cs1)	Create a C12832 LCD object with Mbed pin definitions. Pins are serial data, serial clock, reset, register select (0 = instruction, 1 = data), chip select.
void **setmode**(int *mode*)	mode: NORMAL = standard operation mode: XOR = write toggles pixel status
void **invert**(unsigned int *x*)	x = 1 inverts the pixel "color," black to white, etc. x = 0 standard
void **cls**(void)	Clear the LCD.
void **set_contrast**(unsigned int *x*)	x = contrast value (value of 10–35 will be visible)
printf(string)	Prints formatted string to LCD display.
void **locate**(int *x*, int *y*)	x, y sets display cursor position.
void **set_font**(unsigned char* *font*)	font = Small_6, Small_7, Arial_9, Arial12x12, Arial24x23
int **_putc**(int *c*);	Print the char c on the actual cursor position. Returns this value.
void **character**(int *x*, int *y*, int *c*)	Print the char c at position x, y.
void **line**(int *x0*, int *y0*, int *x1*, int *y1*, int *c*)	Draw a single pixel line from x0, y0 to x1, y1 (c = color 0 or 1).
void **rect**(int *x0*, int *y0*, int *x1*, int *y1*, int *c*)	Draw a rectangle from x0, y0 to x1, y1 (c = color 0 or 1).
void **fillrect**(int *x0*, int *y0*, int *x1*, int *y1*, int *c*)	Draw a filled rectangle from x0, y0 to x1, y1 (c = color 0 or 1).
void **circle**(int *x*, int *y*, int *r*, int *c*)	Draw a circle with x, y center, and radius r (c = color 0 or 1).
void **fillcircle**(int *x*, int *y*, int *r*, int *c*)	Draw a filled circle with x, y center, and radius r (c = color 0 or 1).
void **pixel**(int *x*, int *y*,int *c*)	Set a single pixel at x, y (c = color 0 or 1). Note that this function does not update the screen until **copy_to_lcd()** is called.
void **print_bm**(Bitmap bm, int *x*, int *y*)	Print bitmap object. Screen is only updated when **copy_to_lcd()** is called.
void **copy_to_lcd**(void)	Copy buffer contents to the screen.

Note: x, y are pixel coordinates, where (0,0) is top left, (127,31) is bottom right.

Program Example 8.8 prints a simple formatted string to the display that continuously counts up. Notice the use of the **cls( )** function to clear the screen at the start of the program, and the **locate( )** function to set the text start position.

```
/*Program Example 8.8: Displaying a formatted string on the NHD-C12832
  Runs on either LPC1768 or F401RE with pins shown below           */

#include "mbed.h"
#include "C12832.h"

C12832 lcd(p5, p7, p6, p8, p11); //use D11,D13,D12,D10,D9 for F401RE
//For F401RE, connect also GND pins, and 3.3V output to app board pin 40

int main(){
  int j = 0;
  lcd.cls();                  // clear screen
  while(true){
    lcd.locate(10,10);             // set location to x=10, y=10
    lcd.printf("Counter : %d",j); // print counter value
    j++;                           // increment j
    ThisThread::sleep_for(500ms); // wait 0.5 seconds
  }
}
```

Program Example 8.8 Displaying a formatted string on the NHD-C12832

Implement the program example in the usual way, and add the library from the Reference link, following the procedure outlined in Section 2.6.2. Use the Mbed application board or standalone display. Note that, at the time of writing, there was one instance of the use of a deprecated function, **wait_ms(5)**, appearing in the **c12832.cpp** file. If this is still the case, simply go into that file and replace the line with **ThisThread::sleep_for(5ms),** or **wait_us(5000).** You will need to do this for every new program for which you download the library (or simply copy the corrected **c12832.cpp** file from the previous project).

Exercise 8.9

For every **while** loop iteration in Program Example 8.8, draw a horizontal line across the screen, with variable j as the x coordinate. The screen will slowly fill. What happens as the lines start to cover the Counter reading? Do this so the screen fills from the top and then from the bottom.

■

It is also possible on the C12832 display to set pixels individually. Program Example 8.9 draws a small cross with a center point at location **x = 10**, **y = 10**.

```
/*Program Example 8.9: Setting individual pixels on the C12832 LCD
 Runs on either LPC1768 or F401RE with pins shown below           */

#include "mbed.h"
#include "C12832.h"

C12832 lcd(p5, p7, p6, p8, p11); //use D11,D13,D12,D10,D9 for F401RE
//For F401RE board, connect also GND pins, and 3.3V output to app board pin 40
```

```
int main(){
  lcd.cls();           // clear screen
  lcd.pixel(10,9,1);   // set pixel 1
  lcd.pixel(10,10,1);  // set pixel 2
  lcd.pixel(10,11,1);  // set pixel 3
  lcd.pixel(9,10,1);   // set pixel 4
  lcd.pixel(11,10,1);  // set pixel 5
  lcd.copy_to_lcd();   // Send pixel data to screen
}
```

Program Example 8.9 Setting individual pixels on the C12832

We can now create more dynamic displays, as we can use data and variables to dictate the display output. For example, we could modify Program Example 8.9 to use two potentiometer readings to define the center point for the cross displayed. The Mbed application board has potentiometers connected to pins 19 and 20, so this is fairly easy to implement. Program Example 8.10 defines the two potentiometer inputs as **AnalogIn** objects and uses those input values to define the position in the display for the center point of the cross. This is done by multiplying one analog input value by 128 (the display's pixel range on the x axis), and the second analog input value by 32 (the display's pixel range on the y axis).

```
/*Program Example 8.10: Dynamically drawing pixels based on analog data
      Runs on either LPC1768 or F401RE with pins shown below     */

#include "mbed.h"
#include "C12832.h"

C12832 lcd(p5, p7, p6, p8, p11); // use D11,D13,D12,D10,D9 for F401RE
//For F401RE, connect also GND pins, and 3.3V output to app board pin 40

AnalogIn pot1(p19);     // potentiometer 1. For F401 select 2 analog inputs
AnalogIn pot2(p20);     // potentiometer 2. and connect potentiometers.

int main(){
  int x,y;              // initialise x, y variables
  while(true) {
    x = pot1 * 128;          // set pot 1 data as x screen coordinate
    y = pot2 * 32;           // set pot 2 data as y screen coordinate
    lcd.cls();               // clear LCD
    lcd.pixel(x,y - 1,1);    // set pixel 1
    lcd.pixel(x,y,1);        // set pixel 2
    lcd.pixel(x,y + 1,1);    // set pixel 3
    lcd.pixel(x - 1,y,1);    // set pixel 4
    lcd.pixel(x + 1,y,1);    // set pixel 5
    lcd.copy_to_lcd();       // send pixel data to screen
  }
}
```

Program Example 8.10 Dynamically drawing pixels based on analog data

Implement Program Example 8.10 on an Mbed application board or standalone display. Check that the cross moves in response to changes in the potentiometer settings.

Exercise 8.10

You will see that the display sometimes goes a little dim as it is clearing and redrawing the screen thousands of times per second. Add a short wait function to the while loop and see how that affects program performance. Experiment with different delay times and find a setting that seems to work best.

With the ability to set and clear individual pixels, it is possible to start displaying graphics and images. Fig. 8.11 shows a 32×32 pixel image that represents a flower, with the lines numbered on the left-hand side. When this array of data is displayed on an LCD, pixels represented with an X can be set while pixels represented with a dash will be left clear.

```
01 _ _ _ _ _ _ _ _ _ _ _ _ _ _ X X X X X _ _ _ _ _ _ _ _ _ _ _ _ _
02 _ _ _ _ _ _ _ _ _ _ _ _ _ X X X X X X X _ _ _ _ _ _ _ _ _ _ _ _
03 _ _ _ _ _ _ _ _ _ _ _ _ _ X X _ _ _ X X _ _ _ _ _ _ _ _ _ _ _ _
04 _ _ _ _ _ _ X X X X _ _ _ X _ _ _ _ _ X X _ _ _ _ _ _ _ _ _ _ _
05 _ _ _ _ _ X X X X X X X _ X _ _ _ _ _ _ X X _ X X X X _ _ _ _ _
06 _ _ _ _ X X X _ _ _ _ X X X _ _ _ _ _ _ X X X X X X X X X _ _ _
07 _ _ _ _ X X _ _ _ _ _ _ X X _ _ _ _ _ _ X X X _ _ _ _ X X X _ _
08 _ _ _ _ X X _ _ _ _ _ _ _ _ _ _ _ _ _ _ _ _ _ _ _ _ _ _ X X _ _
09 _ _ _ _ X X _ _ _ _ _ _ _ _ _ X X X X _ _ _ _ _ _ _ _ _ X X _ _
10 _ _ _ _ _ X X _ _ _ _ _ _ X X X X X X X X _ _ _ _ _ _ _ X X _ _
11 _ _ _ _ _ _ X _ _ _ _ X X X _ _ _ _ _ _ X X X _ _ _ _ X X X _ _
12 _ _ _ X X X X X _ X X X _ _ _ _ _ _ _ _ _ _ X X _ _ X X X _ _ _
13 _ X X X X _ _ _ X X X _ X X _ _ _ X X _ _ _ X X _ X X X X X _ _
14 X X X _ _ _ _ _ X X _ _ X X _ _ _ X X _ _ _ X X _ _ _ _ _ X X _
15 X X _ _ _ _ _ _ X X _ _ X X _ _ _ X X _ _ _ _ X X _ _ _ _ _ X X
16 X X _ _ _ _ _ X X _ _ _ _ _ _ _ _ _ _ _ _ _ _ X X _ _ _ _ _ X X
17 X X X _ _ _ _ X X _ _ _ X _ _ _ _ _ _ X _ _ _ X X _ _ _ _ _ X X
18 _ X X X X X _ X X _ _ X X _ _ _ _ _ _ X X _ _ X X _ _ _ _ X X X
19 _ _ _ _ X X _ X X _ _ _ X X _ _ _ _ X X _ _ _ X X _ X X X X X _
20 _ _ _ X X _ _ _ X X _ _ _ X X X X X X _ _ _ X X _ _ _ _ X X _ _
21 _ _ X X _ _ _ _ X X X _ _ _ _ _ _ _ _ _ _ _ X X _ _ _ _ _ X X _
22 _ _ X X _ _ _ _ _ X X X X _ _ _ _ _ _ _ X X X _ _ _ _ _ _ X X _
23 _ _ X X _ _ _ _ _ _ _ X X X X X X X X X X X _ _ _ X _ _ X X X _
24 _ _ X X _ _ _ _ _ _ X X _ X X X X X X X _ _ _ _ _ X X X X X _ _
25 _ _ X X X _ _ _ X X X X _ _ _ _ _ _ _ _ _ _ _ _ _ _ X X X _ _ _
26 _ _ _ X X X X X X X X X _ _ _ _ _ _ _ _ _ _ _ _ _ _ X X _ _ _ _
27 _ _ _ _ X X X X _ _ X X _ _ _ _ _ _ X X X _ _ _ _ _ X X _ _ _ _
28 _ _ _ _ _ _ _ _ _ _ X X _ _ _ _ _ _ X X X X _ _ _ X X X _ _ _ _
29 _ _ _ _ _ _ _ _ _ _ X X _ _ _ _ _ X X X _ X X X X X X _ _ _ _ _
30 _ _ _ _ _ _ _ _ _ _ X X X _ _ _ X X X _ _ _ X X X X _ _ _ _ _ _
31 _ _ _ _ _ _ _ _ _ _ _ X X X X X X X _ _ _ _ _ _ _ _ _ _ _ _ _ _
32 _ _ _ _ _ _ _ _ _ _ _ _ X X X X X _ _ _ _ _ _ _ _ _ _ _ _ _ _ _
```

Figure 8.11
Bitmap image of a flower.

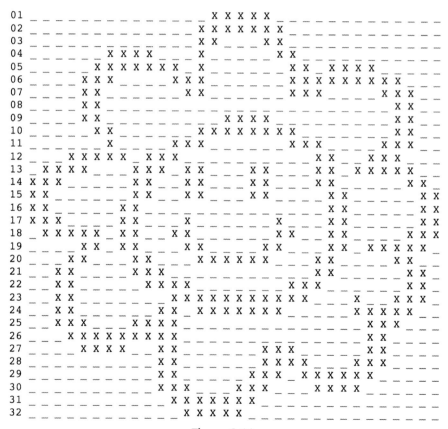

Table 8.5: Representing bitmap image data as 8-bit values.

Row 18 pixel pattern	_XXXXX_X	X__XX___	___XX__X	X____XXX
Row 18 Hex. data values	0x7D	0x98	0x19	0x87

The image shown in both Figs. 8.10 and 8.11 can be defined by an array of binary values that represent the status of each pixel. This type of binary image data is often described as a *bitmap*. Table 8.5 shows how the bitmap data is constructed in binary form, considering row 18 as an example.

It can be seen from Table 8.5 that each row of data is defined as four consecutive 8-bit values. The first eight pixel values in row 18 represent the binary value b01111101 or 0x7D hexadecimal. Program Example 8.11 shows the entire data array within a header file called **flower.h**.

```
/*Program Example 8.11: Bitmap header file flower.h
                                                            */
#ifndef flower_H
#define flower_H

#include "C12832.h"

static char Flower[] = {
  0x00, 0x03, 0xE0, 0x00, // ____ ____ ____ __XX XXX_ ____ ____ ____
  0x00, 0x07, 0xF0, 0x00, // ____ ____ ____ _XXX XXXX ____ ____ ____
  0x00, 0x06, 0x30, 0x00, // ____ ____ ____ _XX_ __XX ____ ____ ____
  0x03, 0xC4, 0x18, 0x00, // ____ __XX XX__ _X__ ___X X___ ____ ____
  0x07, 0xF4, 0x0D, 0xE0, // ____ _XXX XXXX _X__ ____ XX_X XXX_ ____
  0x0E, 0x1C, 0x0F, 0xF8, // ____ XXX_ ___X XX__ ____ XXXX XXXX X___
  0x0C, 0x0C, 0x0E, 0x1C, // ____ XX__ ____ XX__ ____ XXX_ ___X XX__
  0x0C, 0x00, 0x00, 0x0C, // ____ XX__ ____ ____ ____ ____ ____ XX__
  0x0C, 0x01, 0xE0, 0x0C, // ____ XX__ ____ ___X XXX_ ____ ____ XX__
  0x06, 0x07, 0xF8, 0x0C, // ____ _XX_ ____ _XXX XXXX X___ ____ XX__
  0x02, 0x1C, 0x0E, 0x1C, // ____ __X_ ___X XX__ ____ XXX_ ___X XX__
  0x1F, 0x70, 0x03, 0x38, // ___X XXXX _XXX ____ ____ __XX __XX X___
  0x78, 0xEC, 0x63, 0x7C, // _XXX X___ XXX_ XX__ _XX_ __XX _XXX XX__
  0xE0, 0xCC, 0x63, 0x06, // XXX_ ____ XX__ XX__ _XX_ __XX ____ _XX_
  0xC0, 0xCC, 0x61, 0x83, // XX__ ____ XX__ XX__ _XX_ ___X X___ __XX
  0xC1, 0x80, 0x01, 0x83, // XX__ ___X X___ ____ ____ ___X X___ __XX
  0xE1, 0x88, 0x11, 0x83, // XXX_ ___X X___ X___ ___X ___X X___ __XX
  0x7D, 0x98, 0x19, 0x87, // _XXX XX_X X__X X___ ___X X__X X___ _XXX
  0x0D, 0x8C, 0x31, 0xBE, // ____ XX_X X___ XX__ __XX ___X X_XX XXX_
```

```
  0x18, 0xC7, 0xE3, 0x0C, // ___X X___ XX__ _XXX XXX_ __XX ____ XX__
  0x30, 0xE0, 0x03, 0x06, // __XX ____ XXX_ ____ ____ __XX ____ _XX_
  0x30, 0x78, 0x0E, 0x06, // __XX ____ _XXX X___ ____ XXX_ ____ _XX_
  0x30, 0x1F, 0xFC, 0x4E, // __XX ____ ___X XXXX XXXX XX__ _X__ XXX_
  0x30, 0x37, 0xF0, 0x7C, // __XX ____ __XX _XXX XXXX ____ _XXX XX__
  0x38, 0xF0, 0x00, 0x38, // __XX X___ XXXX ____ ____ ____ __XX X___
  0x1F, 0xF0, 0x00, 0x30, // ___X XXXX XXXX ____ ____ ____ __XX ____
  0x0F, 0x30, 0x38, 0x30, // ____ XXXX __XX ____ __XX X___ __XX ____
  0x00, 0x30, 0x3C, 0x70, // ____ ____ __XX ____ __XX XX__ _XXX ____
  0x00, 0x30, 0x77, 0xE0, // ____ ____ __XX ____ _XXX _XXX XXX_ ____
  0x00, 0x38, 0xE3, 0xC0, // ____ ____ __XX X___ XXX_ __XX XX__ ____
  0x00, 0x1F, 0xC0, 0x00, // ____ ____ ___X XXXX XX__ ____ ____ ____
  0x00, 0x0F, 0x80, 0x00, // ____ ____ ____ XXXX X___ ____ ____ ____
};

Bitmap bitmFlower = {
  32, // XSize
  32, // YSize
  4, // Bytes in each line
  Flower,  // Pointer to picture data
};

#endif
```

Program Example 8.11 Bitmap header file flower.h

Program Example 8.12 prints the flower data to the display, using the **print_bm()** function.

```
/*Program Example 8.12: Displaying a bitmap image on the C12832 display
  Runs on either LPC1768 or F401RE with pins shown below        */

#include "mbed.h"
#include "C12832.h"
#include "flower.h"

C12832 lcd(p5, p7, p6, p8, p11); // use D11,D13,D12,D10,D9 for F401RE
//For F401RE, connect also GND pins, and 3.3V output to app board pin 40

int main(){
  lcd.cls();
  lcd.print_bm(bitmFlower,50,0);  // print flower at location x=50, y=0
  lcd.copy_to_lcd();
}
```

Program Example 8.12 Displaying a bitmap image on the C12832 display

Implement Program Example 8.12, creating a separate file for **flower.h**, to show the flower image on the C12832 display. This should appear as seen in Fig. 8.10.

Exercise 8.11

Extend Program Example 8.12 to:

1. make the flower float across the screen; you will need to implement a program loop that continuously prints and clears the display, while incrementing the y-axis print location on each iteration;
2. draw a rectangular frame around the perimeter of the display, and draw a circle symmetrically on either side of the flower.

■

8.6 Color LCDs and the uLCD-144-G2

Nowadays, LCD technology is used in advanced displays for mobile phones, PC monitors, and televisions. For color displays, each pixel is made up of three subpixels for red, green, and blue. Each subpixel can be set to 256 different shades of its color, so it is possible for a single LCD pixel to display $256 \times 256 \times 256 = 16.8$ million different shades! The pixel color is usually referred to by a 24-bit binary value where the highest 8 bits define the red shade, the middle 8 bits the green, and the lower 8 bits the blue. Example values are shown in Table 8.6.

Given that each pixel needs to be assigned a 24-bit value and a 1280×1024 LCD computer display has over 1 million pixels ($1280 \times 1024 = 1{,}310{,}720$), we clearly need to send a lot of data to a color LCD display. Standard color LCDs are set to refresh the display at a frequency of 60 Hz, so the digital input requirements are much greater than those associated with the displays seen earlier in this chapter.

Table 8.6: 24-bit color values.

Color	24-bit Value
Red	0xFF0000
Green	0x00FF00
Blue	0x0000FF
Yellow	0xFFFF00
Orange	0xFF8000
Purple	0x800080
Black	0x000000
White	0xFFFFFF

The uLCD-144-G2 display module is a compact 1.44" LCD color screen, as shown in Fig. 8.12. It has very wide-ranging capability, which we only glimpse in the introduction given here. Further useful description can be found on the manufacturer's website, Reference 8.5, and on the Mbed site itself, in Reference 8.6.

The display uses a UART serial communications interface to receive display data from a connected microprocessor. To limit data transfer required, the display has an optional SD card holder, as can be seen. Video clips, special fonts, or images can be held on the card and called up by the program. Connections on the back of the display are by two rows of five pins each, and only the odd-numbered pins are required for simple applications. Recommended connections to either of our chosen development boards are shown in Table 8.7. These rely on UART capability, so take care if you wish to change them. For ease of prototyping, it is a good idea to buy the ribbon cable and accompanying connector that are offered with the display.

The **4DGL-uLCD-SE** library, by developer Jim Hamblen and found at Reference 8.7, can be used to interface with the display module. This library is a modification of an earlier one that was developed for the **uLCD_4DGL** display, and retains that naming for its

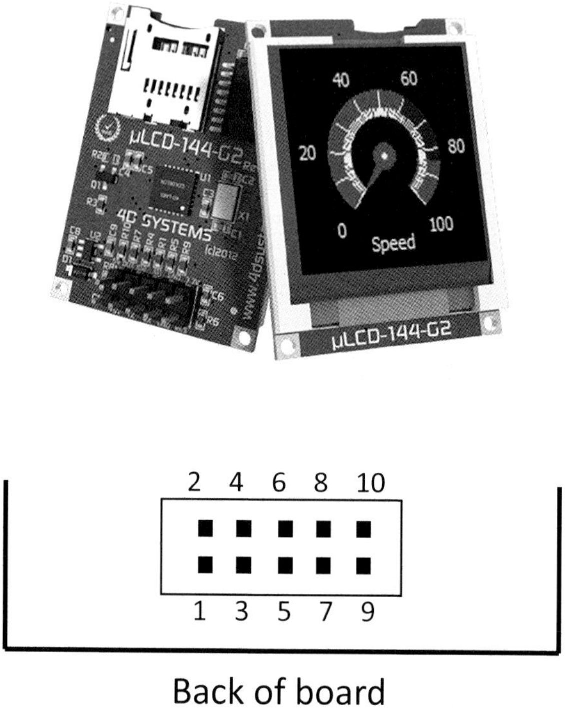

Figure 8.12
uLCD-144-G2 color display module. *Image courtesy of 4D Systems.*

Table 8.7: Pin connections for the uLCD-144-G2 display module.

Function	uLCD-144-G2 Pin	Mbed LPC1768 Pin	Nucleo F401RE Pin
Power (5 V)	1	39	5V
TX/RX	3	10	D2
RX/TX	5	9	D8
Ground	7	1	GND
Reset	9	11	D9

header file. The library shares many similar API functions with the C12832 library used previously. Colors are defined as 24-bit hexadecimal numbers, as detailed in Table 8.6. At the time of writing, the library had not been updated to eliminate a significant number of deprecated functions. If this remains the case, you can download a revised version from the book website.

Program Example 8.13 prints formatted text in red, green, and blue using the library's **color( )** and **printf( )** functions.

```
/*Program Example 8.13: Displaying color text on the uLCD-144-G2
Works on both LPC1768 and F401RE, using pins shown.              */

#include "mbed.h"
#include "uLCD_4DGL.h"

//Create the uLCD_4DGL object, defining respectively tx, rx, reset.
uLCD_4DGL uLCD(p9,p10,p11); //Use D8, D2, D9 for F401RE

int main(){
  uLCD.color(0xFF0000);              // set text color to red
  uLCD.printf("Text in RED\n");
  uLCD.color(0x00FF00);              // set text color to green
  uLCD.printf("Text in GREEN\n");
  uLCD.color(0x0000FF);              // set text color to blue
  uLCD.printf("Text in BLUE\n");
}
```

Program Example 8.13 Displaying color text on the uLCD-144-G2

Connect a uLCD-144-G2 color display to an Mbed target, using pin connections from Table 8.7. Create a new project around Program Example 8.13. Import the library and compile and download. Verify that the three lines of text are displayed in the chosen colors.

Exercise 8.12

Modify the text colors in Program Example 8.13 to test different 24-bit color values. You can try all the color values shown in Table 8.7 and some other 24-bit values of your own choosing.

Program Example 8.14 draws a series of concentric blue circles with increasing radii. The circles are drawn with a shared center at pixel (64,64), the center pixel of the display. The circles increment in radius by three pixels on each iteration until the maximum radius of 64 is reached.

```
/*Program Example 8.14: drawing concentric color circles on the uLCD_4DGL
Works on both LPC1768 and F401RE, using pins shown.                    */

#include "mbed.h"
#include "uLCD_4DGL.h"     // library also supports uLCD-144-G2 variant

//Create the uLCD_4DGL object, defining respectively tx, rx, reset.
uLCD_4DGL uLCD(p9,p10,p11); //Use D8, D2, D9 for F401RE

int main(){
  while(true) {
    for (int r = 0; r <= 64; r += 3) { // increment r by 3 each time
      uLCD.circle(64, 64, r, 0x0000FF); // draw blue circle of radius r
      ThisThread::sleep_for(100ms);
    }
    uLCD.cls();
  }
}
```

Program Example 8.14 Drawing concentric circles on the uLCD-144-G2

Compile Program Example 8.14, along with the uLCD library, and download to your target board. Verify that blue concentric circles are repeatedly drawn.

Exercise 8.13

Try these modifications to Program Example 8.14:

- Modify the color of the circles to test different 24-bit color values.
- Modify the radius increment value and observe the changes. You can modify the maximum increment value and the wait time too.

- Make the circles shrink back down in size, after they have reached maximum size, rather than reset each time.
- Make the color change with each increment.

■

8.7 Mini-projects

8.7.1 Digital Spirit Level

Design, build, and test a digital spirit level using an Mbed development board. Use an ADXL345 accelerometer (Section 7.3) to measure the angle of orientation in two planes, a digital push-to-make switch to allow calibration and zeroing of the orientation, and a color LCD to output the measured orientation data, in degrees.

To help you proceed, consider the following:

1. Design your display to show a pixel or image moving around the LCD screen with respect to the orientation of the accelerometer. The spirit level should be able to respond to movements in two axes (sometimes referred to as *tilt* and *roll*); we can refer to these as the x and y axes for simplicity.
2. Add the digital switch to allow simple calibration and zeroing of the data.
3. Improve your display output to give measurements of x and y axis angles in degrees from the horizontal. For example, a perfectly flat spirit level will read 0° in both the x and y axes. Tilting the spirit level will cause a positive or negative reading in the x axis, whereas rolling the spirit level to either the left or right will give a positive or negative reading in the y axis.
4. How accurate are your x and y readings? Set up a number of know angles and test your spirit level; continuously improve your code until accurate readings are achieved at all angles in both axes.

8.7.2 A Self-contained Metronome

The metronome described in Section 6.8.1 is interesting, but it doesn't result in something that a musician would really want to use. So try revising the program, and its associated build, to make a self-contained battery-powered unit, using an LCD instead of the host computer screen to display beat rate. Experiment also with getting a loudspeaker to "tick" along with the LED. If you succeed in this, then try including the facility to play "concert A" (440 Hz), or another pitch, to allow the musicians to tune their instruments. This project will work on either a breadboard build or the application board. It is attractive to do it on the latter, with its built-in speaker, LCD, and joystick.

Chapter Review

- Liquid crystal displays (LCDs) use an organic crystal which can polarize and block light when subjected to an electric field.
- Many types of LCDs are available and, when interfaced with a microcontroller, they allow digital control of alphanumeric character displays, or pixel-oriented high-resolution monochrome and color displays.
- The PC1602F is a 16-column by 2-row character LCD which can be controlled by an Mbed development board.
- Character data can be defined using the 8-bit ASCII table.
- The Mbed **TextLCD** library can be used to simplify working with alphanumeric LCDs; it allows the display of formatted data using the **printf( )** function.
- The **NHD-C12832** display, which is installed on the Mbed application board, has 128×32 pixels, allowing graphics and images, as well as alphanumeric text, to be displayed.
- Color LCDs frequently allocate each pixel a 24-bit color setting, with consequent high data-transfer demands.
- The **uLCD-144-G2** display module is a color LCD screen that uses a UART serial communications interface to receive display data from a microcontroller, such as an Mbed development board.

Quiz

1. What are the advantages and disadvantages of using an alphanumeric liquid crystal display in an embedded system?
2. What types of cursor control are commonly available on alphanumeric LCDs?
3. How does the Mbed **BusOut** object help to simplify interfacing an alphanumeric display?
4. What is the function of the E input on an alphanumeric display such as the PC1602F?
5. What does the term ASCII refer to?
6. What are the ASCII values associated with the numerical characters from 0 to 9?
7. Referring to the **TextLCD** library, write the C code required to display the value of a floating point variable called "ratio" to 2 decimal places in the middle of the second row of a 2×16 character alphanumeric display?
8. What is a bitmap and how can it be used to display images on an LCD display?
9. List and describe five practical examples of a color LCD used in an embedded system.
10. A color LCD is filled in turn with a single background color, with 24-bit codes as shown. What colors will be shown?
    a. 0x00FFFF
    b. 0x00007F
    c. 0x7F7F7F

References

8.1. Powertip PC1602F extended datasheet (available from Rapid Electronics). http://www.rapidonline.com/pdf/57-0913.pdf.
8.2. TextLCD library. https://os.Mbed.com/users/simon/code/TextLCD/docs/tip/TextLCD_8cpp_source.html.
8.3. NHD-C12832A1Z-FSW-FBW-3V3 COG (Chip-On-Glass) Liquid Crystal Display Module, Newhaven Display International, June 2013. https://www.newhavendisplay.com/specs/NHD-C12832A1Z-FSW-FBW-3V3.pdf.
8.4. C12832 library by Kevin Anderson. https://developer.Mbed.org/users/askksa12543/code/C12832/.
8.5. uLCD-144-G2 1.44 Intelligent LCD Module. https://4dsystems.com.au/products/ulcd-144-g2.
8.6. uLCD-144-G2 128 by 128 Smart Color LCD. https://developer.Mbed.org/users/4180_1/notebook/ulcd-144-g2-128-by-128-color-lcd/.
8.7. 4DGL-uLCD-SE library. Jim Hamblen. https://developer.Mbed.org/users/4180_1/code/4DGL-uLCD-SE/.

CHAPTER 9

Programming in Real Time

9.1 Multitasking and Real Time

Almost every embedded system has more than one activity that it needs to perform. A program for a simple autonomous vehicle, such as seen in Fig. 9.1, may need to sense its environment through proximity and light sensors, measure the distance it has moved, keep track of its battery charge, move its manipulator arm, and calculate and implement drive values for its motors. As a system becomes more complicated, it becomes increasingly difficult to balance the needs of the different things it does. Each will compete for CPU time and may therefore cause delays in other areas of the system. The program needs a way of dividing its time "fairly" between the different demands laid upon it. We call these different activities *tasks*, and the process of keeping them all going *multitasking*.

Figure 9.1
An experimental multitasking autonomous vehicle. *Photo courtesy of Richard Richards photography.*

Fast and Effective Embedded Systems Design. https://doi.org/10.1016/B978-0-323-95197-5.00009-7

An important parallel aspect of the need to multitask is the need to ensure that things are happening at the right time. This is important in almost every embedded system, and the problem just gets worse when there are multiple activities competing for CPU attention.

Many of us in this busy modern world feel we spend our lives multitasking. A parent may need to get two or three children ready for school—one has lost a sock, one feels sick, and the other has spilled the milk; and the dog needs feeding, the saucepan is boiling over, the postman is at the door, and the phone is ringing. Many things need to be done, but we can only do one thing at a time. The microcontroller in an embedded system can feel as harassed as this parent. It can be surrounded by many things, each demanding its attention. It will need to decide what to do first and what can be left until later. Common to both situations is the idea of deadlines: the child *must* be ready for the school bus at a certain time; the gripper *must* close as the component passes on the conveyor belt.

Systems which have to meet critical deadlines in order to function properly are said to be operating in *real time*. A simple but completely effective definition of real time, often used by veteran real-time programmer David Kalinsky, is as follows:

> *A system operating in real time must be able to provide the correct results at the required time deadlines.*

Notice that this definition carries no implication that working in real time implies high speed, although this can often help. It simply states that what is needed must be ready at the time when it is needed. This definition has wide-ranging applications, from providing a result correctly for a single event to delivering an ongoing stream of data in applications such as real-time digital signal processing.

The *real-time operating system*, or RTOS (pronounced "Arr-Toss"), offers an effective mechanism for programming in a multitasking real-time environment. It requires a completely different approach to program development, and takes us far from the assumptions of traditional sequential programming. With the RTOS, we hand over control of the CPU and all system resources to the OS. It is the OS which now determines which section of the program is to run and for how long, and how it accesses system resources. The application program itself is subservient to the OS, and is written in a way that recognizes the requirements of the operating system.

Multitasking describes a situation where there are many tasks which need to be performed, ideally simultaneously. A program written for an RTOS is structured into *tasks* or *threads*. While there are subtle differences between the use of the words task and thread, let's simplify by saying that the task is the thing that needs to be done, and the thread is the code that delivers it. A thread is *a program strand or section which has a clear and distinct purpose and outcome.* With the RTOS, each thread is written as a self-contained program module, a bit like a function. The threads can be prioritized, though this is not always the case. The RTOS performs three main functions:

- it decides which thread should run and for how long,
- it provides communication and synchronization between threads,
- and it controls the use of resources shared between the threads, such as memory and hardware peripherals.

An RTOS itself is a general-purpose program framework. It is adapted for a particular application by the programmer writing threads for it and by customizing it in other ways. While you can write your own RTOS, it is pretty much a specialist activity and generally best done by specialists.

We introduced the very basics of the Mbed OS in Chapter 2, and made brief mention there of the RTOS. In this chapter we at last get into its detail and consider the challenges and opportunities of writing multitasking programs in the real-time environment. You may ask: Haven't we already been using the OS, with the RTOS apparently an integral part? The answer is yes, but as we have not been invoking any of the RTOS APIs, the OS has been acting in a conventional, not real-time manner.

While we attempt a working overview of the RTOS, it's a big topic, and it's a good idea to read further. There are many good sources available. The underlying concepts of the RTOS have been stable for several decades, so we recommend a very readable old classic here, Reference 9.1.

9.2 Scheduling

9.2.1 Scheduling and Context Switching

A central part of the RTOS is the scheduler. This determines which thread is allowed to run at any particular moment. Among other things, the scheduler must be aware of what threads are ready to run and their priorities (if any). There are a number of fundamentally different scheduling strategies, which we consider now. In the several diagrams which follow, the horizontal strand represents program execution; the thread that is being executed at any one time is indicated by a shaded, numbered box on that strand.

Cyclic scheduling is the simplest scheduling form, and is represented in Fig. 9.2. It more or less reflects the superloop type of programming we have been doing. Each thread runs

Figure 9.2
Cyclic scheduling.

to completion before it hands over to the next, and cannot be discontinued as it runs. The threads may take a different duration each time they run, so the following thread may have to wait an unpredictable time before it can execute.

In round-robin scheduling the operating system is driven by a regular interrupt (the "clock tick"). Threads are selected in a fixed sequence for execution. On each clock tick, the current thread is discontinued and the next is allowed to start execution. All threads are treated with equal importance and wait in turn for their slot, or "time slice" of CPU time. Threads are not allowed to run to completion, but are pre-empted; that is, their execution is discontinued mid-flight. This is an example of a pre-emptive scheduler.

The implications of this pre-emptive thread switching and its overheads are not insignificant and should not be overlooked. When the thread is allowed to run again, it must be able to pick up operation seamlessly, with no side effect from the pre-emption. Therefore, complete context saving (all flags, registers, and other memory locations) must be undertaken as the thread switches.

A diagrammatic example of round-robin scheduling is shown in Fig. 9.3. The numbered blocks once more represent the threads as they execute, but there is a major difference from Fig. 9.2. Now each thread gets a slot of CPU time, which has a fixed length. The clock tick, which causes this thread switch, is represented in the diagram by a vertical arrow. When that time is up, the next thread takes over, whether the current one has completed or not. At one stage Thread 2 completes and does not need CPU time for several time slices. It then becomes ready for action again and takes its turn in the cycle.

As the thread and context are switched, there is an inevitable time overhead, which is represented by the black bars. This is the time taken serving the requirements of the RTOS, which is lost to the application program.

In round-robin scheduling, threads become subservient to a higher power, the operating system, as we have seen. Yet all threads are of equal priority, so an unimportant thread

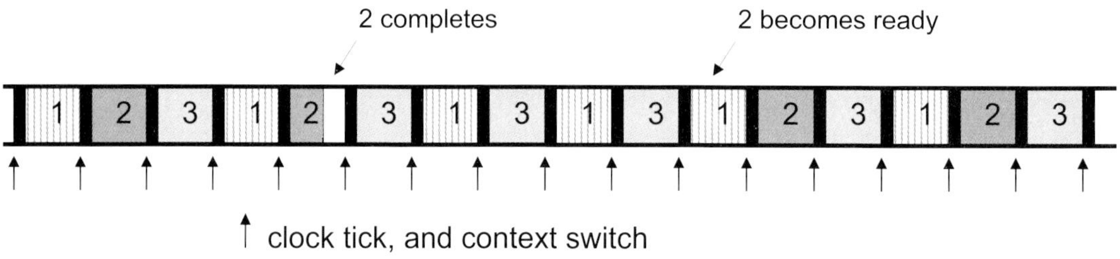

Figure 9.3
Round-robin scheduling.

gets just as much access to the CPU as one of tip-top priority. We can change this by prioritizing threads.

In the prioritized pre-emptive scheduler, threads are given priorities. High-priority threads are now allowed to complete before any time whatsoever is given to threads of lower priority. The scheduler is still run by a clock tick. On every tick it checks which ready thread has the highest priority. Whichever that is gets access to the CPU. An executing thread which still needs CPU time, and is highest-priority, keeps the CPU. A low-priority thread which is executing is replaced by one of higher priority, if that has become ready. The high-priority thread becomes the "bully in the playground." In almost every case it gets its way.

The way this scheduling strategy works is illustrated in the example of Fig. 9.4. This contains a number of the key concepts of the RTOS and is worth understanding well. The diagram shows three threads, each of different priority and different execution duration. At the beginning, all are ready to run. Because Thread 1 has the highest priority, the scheduler selects it to run. At the next clock tick, the scheduler recognizes that Thread 1 still needs to run, so it is allowed to continue. The same happens at the next clock tick and the thread completes during the following time slice. Thread 1 does not need CPU time now and becomes suspended. At the next clock tick the scheduler selects the ready thread which has the highest priority, which is now Thread 3. This also runs to completion.

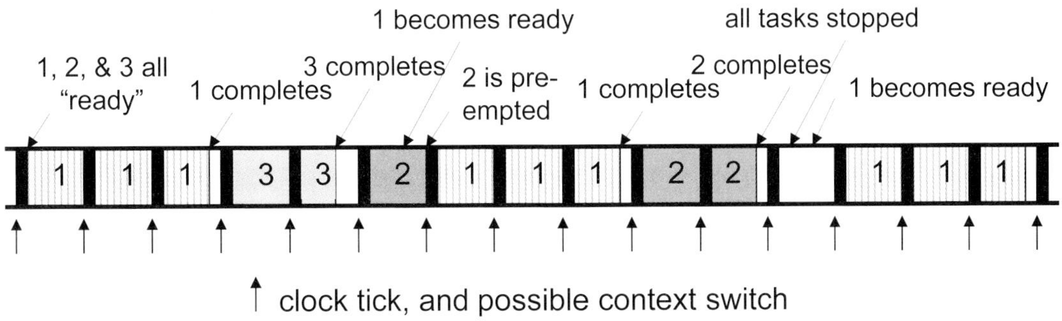

Thread	Priority	Duration (in time slices)
1	1 (highest)	2.7
2	3	2.8
3	2	1.5

Figure 9.4
Prioritized pre-emptive scheduling.

At last Thread 2 gets a chance to run! Unfortunately for it, however, during its first time slice Thread 1 becomes ready again. At the next clock tick the scheduler therefore selects Thread 1 to run again. Once more, this is allowed to run to completion. When it has, and only because no other thread is ready, Thread 2 can re-enter the arena and finally complete. Following this, for one time slice, there is no active task and no CPU activity. Thread 1 then becomes ready one more time and starts to run again to completion.

A final scheduling strategy which should be mentioned is *cooperative*. While pre-emptive scheduling is very effective, it does have the demands on time and memory already mentioned. In cooperative scheduling the tasks themselves relinquish CPU access, and thus manage their own context saving, if needed. Cooperative scheduling is unlikely to be as responsive to tight deadlines as prioritized pre-emptive, but for a small system of limited resources it may prove to be the most effective option.

9.2.2 Thread States

It is worth pausing at this moment to consider what is happening to the threads now that they are being controlled by a scheduler. Clearly, only one thread can run at any one time. Others may need to run, but at any one instant do not have the chance. Others may just need to respond to a particular set of circumstances and only be active at certain times during program execution.

Therefore, it is important to recognize that threads move between different states. A possible state diagram for this is shown in Fig. 9.5. The states are described below. Note, however, that the terminology used and the way the state is managed vary to some extent from one RTOS to another. Therefore, in some cases, different terms are used to describe a certain state, and different versions of the diagram can be found in the literature.

Ready (or eligible). The thread is ready to run and will do so as soon as it is allocated CPU time. The thread leaves this state and enters the running state when it is started by the scheduler.

Running (or active). The thread has been allocated CPU time and is executing. A number of things can cause the thread to leave this state. Maybe it simply completes and no longer needs CPU time. Alternatively, the scheduler may pre-empt it, so that another thread can run. Finally, it may enter a blocked or waiting state for one of the reasons described below.

Blocked/waiting. This state represents a thread which is ready to run, but for one reason or another is not allowed to. There are a number of reasons why this may be the case, some will be seen later in the chapter. The thread could be waiting for some data to arrive

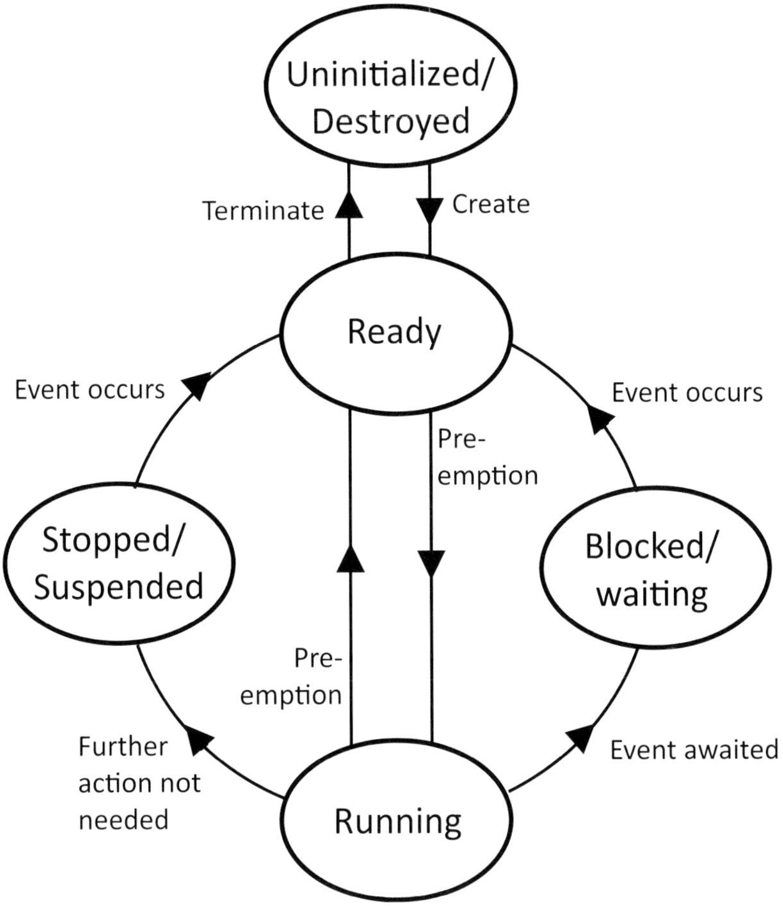

Figure 9.5
Thread states.

or for a resource that it needs, which is currently being used by another thread, or it could be waiting for a period of time to be up. The state is left when the thread is released from the condition which is holding it there.

Stopped/suspended. The thread does not at present need CPU time. A thread leaves this state and enters the ready state when it is activated again, for whatever reason.

Uninitialized/destroyed. In this state the thread no longer exists as far as the RTOS execution is concerned. An implication of this is that a thread does not need to have continuous existence throughout the course of program execution. Generally, it has to be created or initialized in the program before it can run. If necessary, it can later be destroyed and possibly another created instead. Removing unneeded threads from the thread list simplifies scheduler operation and reduces demands on memory.

9.3 The Mbed RTOS

As shown in Fig. 2.9, the Mbed RTOS combines features of CMSIS (ARM's "Common Microcontroller Software Interface Standard") and Keil's RTOS core, RTX. While references are available for each, they do not make easy reading for the learner, and it is not always easy to recognize which elements they contribute. We suggest that it is better to keep as main reference the website introduction to the Mbed OS, Reference 9.2, and the many pages to which it leads.

The Mbed RTOS applies the principles which have just been introduced in this chapter. Its underlying scheduling strategy is prioritized pre-emptive scheduling. However, for threads of equal priority, it applies round-robin scheduling. The default time slice is 1 ms.

A program always has at least one user-managed thread, which is **main( );** this is allocated 4 KB of stack memory. The user can create further threads, which are also allocated (by default) 4 KB of stack. Other threads are added by the system; for example, the **Idle** thread is run by the scheduler when there is no other activity in the system, while the **Timer** thread handles system and user-timer objects. Some drivers also create additional threads.

Numerous configuration settings are available. Some can be viewed in Reference 9.3, though most need not concern us at this stage.

The scheduling capabilities of the Mbed OS give its RTOS qualities. The APIs fall into the categories shown (in reverse alphabetical order) in Table 9.1. We meet examples from these APIs in the rest of this chapter.

Fundamental to successful use of the RTOS is the creation and control of threads. This is done using the APIs summarized in Table 9.2. We use some of these features immediately, and others in later sections.

As well as the thread class, it will also be necessary to start using functions from the **ThisThread** namespace, with examples shown in Table 9.3. Here at last we see the **ThisThread::sleep_for( )** function that we have been using since the earliest programs in this book.

Tables 9.2 and 9.3 also contain some features that we don't develop in this chapter, but are included as consideration for further reading. The **yield( )** function is applied in cooperative scheduling. Four examples of functions which access thread flags are also given in Table 9.3; these match the **flags_set( )** function in Table 9.2.

Table 9.1: Scheduling APIs.

API Classes/Features	Summary
Thread	Defines, creates, and controls parallel threads.
ThisThread	Provides features for control of the current thread.
Semaphore	Manages thread access to a pool of shared resources of a certain type.
Queue	Enables queueing of pointers to data from producer threads to consumer threads.
Mutex	Protects access to a shared resource and synchronizes the execution of threads.
Mail	Provides a queue combined with a memory pool for allocating messages.
MemoryPool	Defines and manages memory pools of fixed size.
IdleLoop	Background system thread, executed when no other threads are ready to run.
EventFlags	An event channel that provides a way of notifying other threads about condition changes or events.
ConditionVariable	Provides a mechanism to wait for or signal state changes, for example a mutex change.

Table 9.2: API for threads.

Name/Format	Description
Constructor	
`Thread (osPriority priority, uint32_t stack_size, unsigned char *stack_mem, const char *name)`	Allocate a new thread without starting execution, setting (as needed) priority, stack size, stack pointer, and thread name.
Member Functions	
`uint_32 flags_set(uint_32 flags)`	Set the specified flags for the thread, as indicated by *flags*.
`osPriority get_priority()`	Get priority of an active thread.
`osStatus set_priority (osPriority priority)`	Set priority of an active thread.
`osStatus join()`	Wait for thread to terminate (then calling function can continue).
`osStatus start (mbed::Callback<void()> thread)`	Start a thread executing the function specified by the thread. Returns status code that indicates the execution status of the function.
`State get_state()`	Get the state of a thread. Returns code indicating thread status.
`osStatus terminate()`	Terminate execution of a thread and remove it from active threads.

Table 9.3: Selected functions: ThisThread namespace.

Name/Format	Description
uint32_t **flags_clear** (uint32_t *flags*)	Clears the specified thread flags of the currently running thread.
uint32_t **flags_get** ()	Returns the thread flags currently set for the currently running thread.
uint32_t **flags_wait_all** (uint32_t *flags*, bool *clear*)	Waits for all of the specified thread flags to become signaled for the current thread, clear the flags if *clear* = 1.
uint32_t **flags_wait_any** (uint32_t *flags*, bool *clear*)	Wait for any of the specified thread flags to become signaled for the current thread; clears the flags if *clear* = 1.
void **sleep_for** (chrono duration)	Sleeps for the specified time period.
void **sleep_until** (chrono duration)	Sleeps until the specified time.
void **yield** ()	Passes control to next equal-priority thread that is in ready state.

9.4 Writing Multi-threaded Programs

We now turn to developing simple programs, applying the Mbed RTOS.

9.4.1 Defining and Writing Threads

Threads should be written as if they are to run continuously, as self-contained and semi-autonomous functions, even though they will be discontinued by the scheduler. They cannot call on a section of another's code, but can access common code, for example, C libraries. They may depend on services provided by each other and may need to be synchronized with each other. In either case, the RTOS will have special services to allow this to happen.

In the earliest programs which follow, we will use a thread to simply flash an LED, or something equally mundane, simply to show it is running. However, for practical applications, we will need to make decisions about which activities are allocated to which thread. This becomes an interesting undertaking for the programmer. The number of threads created should not be too great. More threads generally imply more programming complexity, and for every thread switch there is a time and memory overhead.

A useful starting point is to consider what the deadlines are and then to allocate one thread per deadline. A set of activities which are closely related in time are likely to serve similar deadlines and should therefore be grouped together into a single thread. A set of activities which are closely related in function and interchange a large amount of data may also be grouped into a single thread.

9.4.2 Two-threaded Programs

Program Example 9.1 shows a simple two-threaded program. The first thread is **main( ).** A second thread, called **thread2**, is created with the **Thread** constructor. It is immediately started from the first thread, that is, within **main( ),** using the **start( )** member function. The two continue switching their LEDs indefinitely. The initial sleep durations specified are deliberately not multiples of each other; any other values can be trialed.

```
/* Program Example 9.1. Two threads, of which one is main(), run indefinitely.
Runs on LPC1768 or F401RE (with LEDs shown in Fig 3.1) */

#include "mbed.h"

DigitalOut led1(LED1); //Use A0 for F401RE
DigitalOut led2(LED2); //Use A1 for F401RE

Thread thread2;    //Create thread2

void led2_thread(){
   while (true) {
     led1 = !led1;
     ThisThread::sleep_for(1100ms);
   }
}

int main(){
  thread2.start(led2_thread);     //Launch thread2.
  while (true) {
    led2 = !led2;
    ThisThread::sleep_for(500ms);
  }
}
```

Program Example 9.1 A two-threaded program

Program Example 9.2 shows how one thread can both start and terminate another, with **thread2** being started from the **main( )** thread. This, we will see, is standard practice for all subsequent programs. However, after a wait of 10 s, **thread2** is stopped by the **main( )** thread, by setting the **running** variable to false.

Notice that **running** has been declared as **volatile**. This is invoked because the compiler might incorrectly deduce that **running** cannot change, because it is not changed within the **blink( )** function. In this case, the compiler might optimize it out. Declaring it as **volatile** signals that a change may occur external to the loop, and so the compiler keeps the test.

```
/* Program Example 9.2. A program in which one thread starts and stops another.
Runs on LPC1768 or F401RE */

#include "mbed.h"
Thread thread2;    //name thread2 chosen for this thread
DigitalOut led1(LED1);
volatile bool running = true;

// Blink function toggles the led in a long running loop
void blink() {
  while (running) {
    led1 = !led1;
    ThisThread::sleep_for(1s); //set blink rate
  }
}

int main() {                    // Launches a thread which then runs for 10 seconds
  thread2.start(blink);         //Starts thread2 executing.
  //Set how long thread2 will be able to run
  ThisThread::sleep_for(10s);
  running = false;
  thread2.join();         //Wait for thread2 to terminate.
}
```

Program Example 9.2 One thread controlling another

9.4.3 A Multi-threaded Program

Program Example 9.3 provides a more practical example than the previous two. It uses features of the Mbed app board (although standalone components could also be used), notably the temperature sensor, the speaker, two potentiometers, the LCD, and the multicolored LED. The program contains three threads:

main( ), which calls a diagnostic function, launches the other threads, and then sleeps;
temp_read_Th, which reads the on-board temperature sensor;
display_Th, which displays values read and controls the LED and speaker.

It should not be difficult to follow the program through. Each thread is moderately self-contained in structure. A striking thing about the program is that it contains no less than three **while(true)** loops, one per thread; in conventional programming this would not be possible.

```
/* Program Example 9.3. An RTOS-based program applying the Mbed App Board (or
separate components).
Works with LPC1768 or F401RE, with pin connections shown.                */
```

```
#include "mbed.h"
#include "C12832.h"
#include "LM75B.h"    //note: more than one version of LM75V available on Mbed web

C12832 lcd(p5, p7, p6, p8, p11); //use D11,D13,D12,D10,D9 for F401RE
LM75B tmp(p28,p27);    //I2C Temperature Sensor. Use D8,D2 for F401RE
PwmOut red(p23);       //Use D2 for F401RE
PwmOut grn(p24);       //Use D3 for F401RE
PwmOut blu(p25);       //Use D4 for F401RE
PwmOut speaker(p26);   //Speaker with PWM driver. Use D7 for F401RE
AnalogIn pot1(p19);    //Pot 1 (near LCD. Sets alarm threshold. Use A4 for F401RE
AnalogIn pot2(p20);    //Pot 2 (near RGB LED. Sets LED intensity. Use A5 for F401RE

float board_temp;
float alarm_temp;

//Function prototypes
void diag(void);
void temp_read (void);
void display_fn (void);

//Threads
Thread Temp_read_Th;
void temp_read (void){
  while(true){
    board_temp = tmp.temp();    //read temperature from sensor
  }
}
//This thread deals with both display and alarm
Thread Display_Th;
void display_fn (void){
  speaker.period(1.0/800.0); // 800 Hz period
  red = grn = blu = 1.0; //PWM 100% duty cycle, all LEDs off. Blue remains off throughout
  while(true){
    lcd.cls();        //clear LCD
    lcd.locate(0,0);
    lcd.printf("Board Temperature = %0.2f\n\r", board_temp);
    alarm_temp = 50.0 * pot1;
    lcd.printf("Temp Alarm Setting = %0.2f\n\r", alarm_temp);
    if(board_temp > alarm_temp) {    //check temp for alarm
      red = 1.0 - pot2;              //RGB LED red
      grn = 1.0;
      speaker = 0.5;                 // PWM is 50%, alarm tone sounds
    }
    else{
      grn = 1.0 - pot2;              //RGB LED green
      red = 1.0;
      speaker = 0.0;
    }
    ThisThread::sleep_for(2s);
  }
}
```

```
int main(){
  diag();
  Temp_read_Th.start(temp_read);   //initialise threads
  Display_Th.start(display_fn);
  while (true) {
    ThisThread::sleep_for(2s);
  }
}
//Diagnostic; flash LED colors to indicate successful program launch
void diag(void){
  red = 0, grn = 1, blu = 1;
  ThisThread::sleep_for(500ms);
  red = 1, grn = 0, blu = 1;
  ThisThread::sleep_for(500ms);
  red = 1, grn = 1, blu = 0;
  ThisThread::sleep_for(500ms);
  red = 0, grn = 1, blu = 0;
  ThisThread::sleep_for(500ms);
}
```

(A)

```
{
  "target_overrides":{
      "*": {
          "target.printf_lib": "std"
      }
  }
}
```

(B)

Program Example 9.3 A multi-threaded program: (A) the main.cpp file; (B) the Mbed_app.json file (for printing of floating point numbers)

Run Program Example 9.3 on your chosen target system. If wiring the F401RE to the app board, check the connection guidance in Section 3.5. The program should run, continuously displaying temperature reading on the screen and sounding the alarm if the temperature is high, with corresponding changes in LED color. Notice at this stage that data is transferred between threads simply by using global variables. We will meet another form of inter-thread data transfer when we consider the **Queue** and **MemoryPool** APIs.

Exercise 9.1

In Program Example 9.3, try splitting the **Display_Th** thread in two, one thread managing the display, and the other the alarm function.

9.4.4 Setting Thread Priority

In most programs, threads are not all of equal importance. A high-priority thread should have the right to execute before a low-priority thread. Each thread also has a deadline, or could have one estimated for it. Generally, an RTOS allows the programmer to set thread priorities. In the case of *static* priority, priorities are fixed. In the case of *dynamic* priority, priorities may be changed as the program runs. One way of determining priority is to consider how important a thread is to the operation and well-being of the system, its user, and its environment. A simple three-level distinction is as follows:

- Highest priority: threads essential for system survival
- Middle priority: threads essential for correct system operation
- Low priority: threads needed for adequate system operation—these threads might occasionally be expendable or a delay in their completion might be accepted.

Priorities can also be considered by evaluating the thread deadlines. In this case, high priority is given to threads which have very tight time deadlines. However, if a thread has a demanding deadline, but just isn't very important in the overall scheme of things, then it may still end up with a low priority.

Table 9.2 has shown that there are two mechanisms in the Mbed OS to set a thread priority, either as the thread is created, or by using the **set_priority( )** member function. Table 9.4 shows a selection of the priority levels available. A name and value are defined for each, with the higher value implying a higher priority. The default **osPriorityNormal,** unless specified otherwise, is applied to every new thread. It is also the priority of the **main( )** thread.

Program Example 9.4 explores the behavior of three threads as their priorities are changed (these are dynamic priorities). Each thread simply blinks an LED. The blink rate of

Table 9.4: Thread priorities.

Priority Level	Value
osPriorityIdle	1
osPriorityLow	8
osPriorityBelowNormal	16
osPriorityNormal	24
osPriorityAboveNormal	32
osPriorityHigh	40
osPriorityRealtime	48
osPriorityError	−1

thread3 is determined, as we might expect, by invoking a **sleep_for()** function. However, the rates of **thread1** and **thread2** are determined by timing loops inserted into the thread. Thus, each of these threads is permanently "busy". The **main( )** function first launches the three threads, which then run for 10 s. Priorities are then changed, and the program executes for a further 10 s, and so on. Each thread is promoted to top priority for one 10 s period. Priorities are selected from Table 9.4. These are kept equal to or below *Normal*. In this program structure, if a priority is set higher than *Normal*, then the high-priority task may execute continuously, not allowing **main( )** to execute again and reset the priorities.

```
/* Program Example 9.4. Demonstrates the setting of thread priorities.
Runs on LPC1768 or F401RE (with LEDs shown in Fig 3.1) */

#include "mbed.h"
DigitalOut led1(LED1);
DigitalOut led2(LED2);    //Use A0 for F401RE
DigitalOut led3(LED3);    //Use A1 for F401RE
DigitalOut led4(LED4);    // Use A2 for F401RE, the LED is not essential, but
                          //useful to indicate timing loops are running

Thread thread1; //create three threads
Thread thread2;
Thread thread3;

void thread1_fn(){
  while(true){
    led1 = !led1;
    for (int j = 0;j < 1000000;j++) {
    led4 = !led4; //do something, or compiler can "optimise out" this loop
    }
  }
}
void thread2_fn(){
  while(true){
    led2 = !led2;
    for (int i = 0;i < 1000000;i++) {
    led4 = !led4; //do something, or compiler can "optimise out" this loop
    }
  }
}
void thread3_fn(){
  while (true){
    led3 = !led3;
    ThisThread::sleep_for(1s);
  }
}
int main(){
  thread1.start(thread1_fn); //Starts thread1 executing.
  thread2.start(thread2_fn); //Starts thread2 executing.
  thread3.start(thread3_fn); //Starts thread3 executing.
```

```
    //run for 10s equal priorities
    ThisThread::sleep_for(10s);
    while(true){            //changes thread priorities every 10 seconds
      thread1.set_priority(osPriorityNormal);
      thread2.set_priority(osPriorityBelowNormal);
      thread3.set_priority(osPriorityLow);
      ThisThread::sleep_for(10s); //run for 10s new priorities
      thread1.set_priority(osPriorityLow);
      thread2.set_priority(osPriorityNormal);
      thread3.set_priority(osPriorityBelowNormal);
      ThisThread::sleep_for(10s); //run for 10s new priorities
      thread1.set_priority(osPriorityLow);
      thread2.set_priority(osPriorityBelowNormal);
      thread3.set_priority(osPriorityNormal);
      ThisThread::sleep_for(10s);
    }
  }
```

Program Example 9.4 Setting thread priority

Open Program Example 9.4 in a new project, and download it to your target system. Observe that during the first 10 s of execution, all LEDs blink. In the second period, when **thread1** has top priority, LED1 is the only active LED. This is because this thread is permanently active, and will not yield to any other thread. There is similar behavior in the next 10-s period, except that **thread2** is the only one executing. In the next period, however, although **thread3** has highest priority, both it and **thread2** (second priority) are in action. This is because when **thread3** enters sleep, it allows other threads to execute. However, **thread2** grabs all available remaining time, and **thread1** is forced to be inactive.

Exercise 9.2

Having gained familiarity with Program Example 9.4, try putting one of the threads at a priority level above *Normal*, and observe the effect.

9.5 The Mutex

9.5.1 The Mutex Concept

With all that thread switching going on in an RTOS, new issues arise. One thread, using a certain resource, may be pre-empted before that use is complete. The next thread may then try to use the same resource, before the first has finished with it. This could be damaging or catastrophic to overall program performance. Imagine, for example, if two tasks were writing to a screen using **printf( )**, and mid-message one thread was suspended and the other took over.

Mutex (from *mutual exclusion*) allows controlled access to shared resources. Mutex action is illustrated symbolically in Fig. 9.6, which shows five phases in the execution of a program having two threads and a mutex. A single resource is shared between two threads. Thread 1 aims to use it first, so it locks the mutex, denying any other thread the chance to access. When it has completed use of that resource, it unlocks the mutex. Thread 2 can then make use of the resource as needed, locking the mutex in a similar way, so that it has sole access during the time of its use. Unsurprisingly, only the thread which has locked the mutex can unlock it.

It is interesting here to refer back to Fig. 9.5, and consider the states that the threads move between, in relation to that diagram.

Because one thread can in this way influence the timing of a certain action by another thread, the mutex also then provides a means of synchronization between threads. A block of code protected by a mutex is called a *critical section*. Use of mutex *may* introduce the issue of deadlock, a situation when two threads are each waiting for the other to release a mutex.

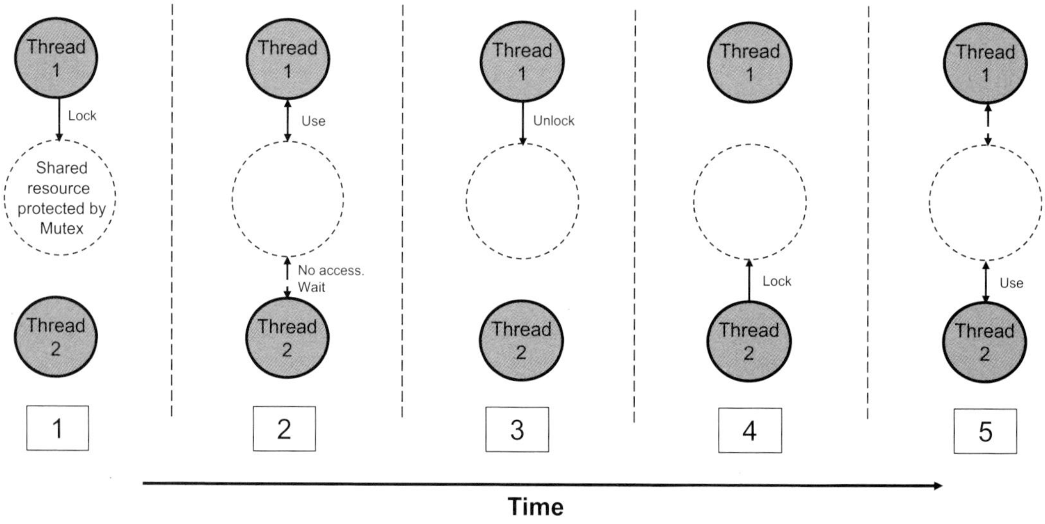

Figure 9.6
The mutex concept.

9.5.2 A Mutex Program

Selections from the mutex API are shown in Table 9.5. A C++ constructor can create a mutex, which can then be locked, tried, and unlocked.

It would be simple to demonstrate the need for a mutex by using **printf( ),** but the reality is that in this RTOS **printf( )** is already protected by an inbuilt mutex. Program Example 9.5 simulates a simple control system, in which two processes are being monitored, with status information output to a display. It makes use of **printf( ),** placing it in a function which is used by two threads, that is, a shared resource. However, each access to the **printf( )** function prints only a single character, making the overall printing of a message vulnerable to corruption due to multiple access. The program has three threads. One is **main( ).** Two further threads are called **Process1** and **Process2.** Each of these accesses a display function called **printing( ).** A mutex, called **print_mutex,** is available to protect the **printing( )** function. The mutex is locked and then unlocked by each thread before and after access to **printing( ).**

Table 9.5: API summary for mutex.

Name/Format	Description
Constructor	
`Mutex (const char *name)`	Create and Initialize a mutex object.
Member Functions	
`void lock()`	Wait until a mutex becomes available.
`bool trylock()`	Try to lock the mutex, and return immediately, returns true if mutex acquired.
`void unlock()`	Unlock the mutex that has previously been locked by the same thread.

```
/* Program Example 9.5. Mutex applied to a printing application.
Works on LPC1768 and F401RE with no external connections
                                            */
#include "mbed.h"

DigitalOut led1(LED1);
Thread Process1;
Thread Process2;
Mutex print_mutex;     //Create and name Mutex

void printing (int);
//2 possible messages for display
char messages[] = "System operational Boiler overheating";
```

```
void thread1_fn(){          //Thread for Process1
  while(true){
    print_mutex.lock();
    printing(0);            //pass pointer to "System operational" message
    print_mutex.unlock();
    ThisThread::sleep_for(800ms);
  }
}

void thread2_fn(){          //Thread for Process2
  while(true){
    print_mutex.lock();
    printing(20);           //pass pointer to "Boiler overheating" message
    print_mutex.unlock();
    ThisThread::sleep_for(800ms);
  }
}

void printing (int msg_ptr){    //print the message passed with msg_ptr
  for (char i = 0; i < 18; i++){
    printf("%c",messages[msg_ptr+i]); //print a single character from the string
    wait_us(2000);                    //represents other processing commitments
  }
  printf("\n");
}

int main(){
  Process1.start(thread1_fn); //Start Process1 executing.
  Process2.start(thread2_fn); //Start Process2 executing.
  while(true){
    led1 = !led1;
    ThisThread::sleep_for(1s);
  }
}
```

Program Example 9.5 Applying mutex

Compile, download, and run Program Example 9.5 to your chosen target system. With the running system connected to a serial terminal, you should see the display of Fig. 9.7(a). Although each thread is trying to access the **printing( )** function, it is the first to reach it that locks the mutex and completes its message. The second must wait until the mutex is unlocked. This then locks the mutex and writes its message. If you "comment out" the mutex locks and unlocks, then each thread accesses the **printing( )** function alongside the other. A display similar to Fig. 9.7(b), where the two messages are scrambled together, follows.

It is useful to notice that the "scrambling" effect is particularly bad, but visible, because the two process threads require the **printing( )** function at more or less the same moment. A more realistic, and more difficult to detect, clash in a practical system might occur, for example, only every few hours or even days, possibly with catastrophic consequences.

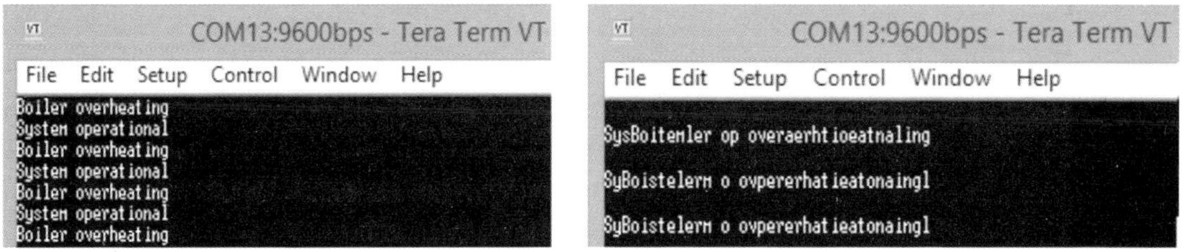

(a) (b)

Figure 9.7
Serial terminal output from Program Example 9.5: (a) with mutex in place; (b) with mutex disabled.

Exercise 9.3

Try the following variations on the mutex example program.

1. "Comment out" the **unlock( )** function calls, but leave the **lock( )** calls in, first for just one call to the **printing( )** function, and then for both. Can you explain the effect on the program behavior?
2. Try having one process thread sleep for exactly half the time of the other, so that some calls to **printing( )** cause a clash, and others don't. What is the effect on the terminal display?
3. Can you adjust the program so that the mutex is used explicitly to synchronize two activities?

9.6 The Semaphore

9.6.1 The Semaphore Concept

The mutex is a very useful concept, and it turns out that it is a special case of a more general one. A semaphore, long before the age of radio, was a long-distance signaling device used where line of sight was possible, for example from hilltop to hilltop, castle tower to ship, and so on. Now we take over the semaphore name to do a bit of remote signalling in the software world.

A semaphore, in the RTOS context, is simply an integer created in the RTOS, with an initial value determined by the programmer. This may be used to manage access to a certain shared resource or group of resources. The number can be thought of as being a number of tokens, which can be taken or returned within the program, as the resource is accessed. The semaphore can be incremented (but not above its initial value) or decremented (but not below zero) by one. If a thread tries to decrement the semaphore and the result would be negative, then that thread enters a blocked state. It needs another thread to increment the semaphore, whereupon it can proceed. The value of the semaphore represents the number of threads that can access the semaphore, without being blocked. The value is not generally known to the program, so a thread accessing the semaphore will not "know" if it is about to be blocked or not. Similarly, it does not know if by incrementing the semaphore it will release another thread from the blocked state. A zero value of the semaphore implies that the next thread to decrement it will become blocked.

A wide variety of terminology and imagery is used for this simple process of incrementing and decrementing. We have already mentioned the idea of the semaphore holding tokens. Others explain it in terms of train signaling, or other stop/go mechanisms.

The semaphore concept is illustrated in Fig. 9.8, which shows five stages of an example program execution. Let's think in terms of tokens, and use the descriptive terminology of acquire/release, as used in the Mbed OS. A semaphore is initialized with two tokens in a program which has four threads running. Early in program execution, Threads 1 and 4 each take a token, leaving none. A little later, Thread 3 attempts to take a token. As none are there, the thread is required to wait, entering a blocked state. Thread 1 then releases a token. This is acquired by the waiting Thread 3, which can then proceed. In the final phase shown, Thread 4 releases a token, available to the next thread which needs it.

A semaphore can be allocated to a shared resource, which could be hardware (including memory or peripheral) or a common software module. Several threads may need to access this same shared resource.

The semaphore structure described above is called a *counting semaphore,* for obvious reasons. If only one token is made available, it is called a *binary semaphore*. The first

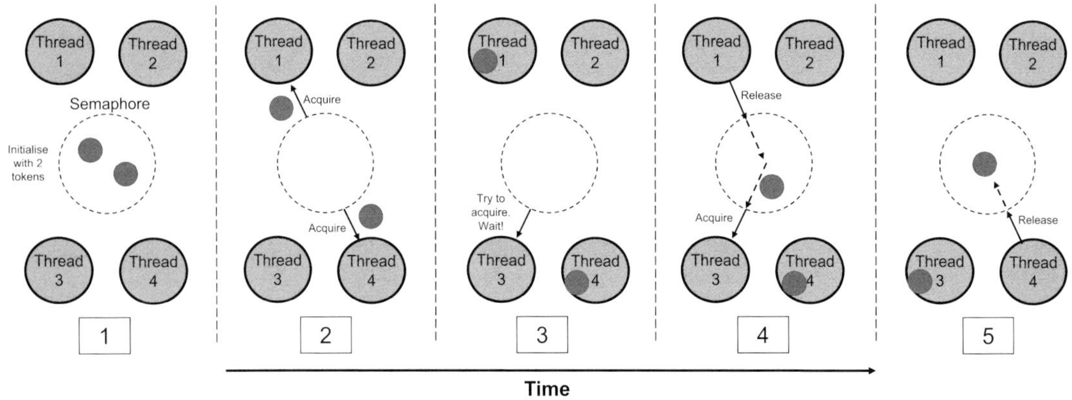

Figure 9.8
A simple counting semaphore in action.

thread taking a token can proceed, but any other will be blocked. This is similar to the mutex—when one thread is accessing the resource, all others are excluded.

The semaphore can be used to control access to a set of resources. In this case, it is initially set to the number of units that are available. As any thread uses one of the units, it decrements the semaphore by one, incrementing it again on completion of use.

An effect of taking the last token is that another thread becomes blocked, semaphores—like the mutex—can be used as a means of providing synchronization and signaling between the threads. One thread can block another by taking the last token, and can release the blocked thread at a time of its choosing by releasing the token.

If a low-priority thread claims a semaphore token for a resource that the high-priority thread needs, it can block that thread. This leads to a dangerous condition known as *priority inversion*.

If you want to get deeper into the world of semaphores, and enjoy some of the beauty and fascination of software engineering, then try "The Little Book of Semaphores," Reference 9.4.

9.6.2 Programming with Semaphores

The main features of the Mbed Semaphore API are shown in Table 9.6. It is not difficult to relate these to what is going on in Fig. 9.8.

Program Example 9.6 demonstrates simple semaphore action, behaving in a way similar to Fig. 9.8. It can be thought of as three similar processes running in an industrial system; from time to time each requires use of a certain resource, of which there are two. Three threads, including **main( ),** are created. The threads have nearly identical, but non-synchronous

Table 9.6: API summary for Semaphore.

Name/Format	Description
Constructor	
`Semaphore (int32_t count, uint16_t max_count)`	Create and initialize a Semaphore object used for managing resources, where *count* is number of available resources.
Member Functions	
`void acquire()`	Acquire a Semaphore resource; wait till available if necessary.
`bool try_acquire()`	Try to acquire a Semaphore resource, and return immediately. Returns True if a resource was acquired, False otherwise.
`osStatus release (void)`	Release a Semaphore resource that was obtained with acquire().

activity. Each flashes an LED, first slowly, and then fast; this should run continuously. To proceed to the fast-flashing phase, where the shared resource is required, they must acquire a semaphore token. A semaphore, **demo_sem**, loaded with two tokens, is thus created. The token is released by the thread when the fast-flashing phase is complete. However, as there are only two items of resource and two tokens, there can only be two threads fast flashing at any one time. If a third wants to enter this phase, then it must wait until one of the others has completed and returned its token.

```
/* Program Example 9.6. Demonstrate Semaphore principles.
Works on LPC1768 and F401RE (with 2 extra LEDs)
                                          */
#include "mbed.h"

DigitalOut led1(LED1);
DigitalOut led2(LED2);
DigitalOut led3(LED3);

Thread Thread1;
Thread Thread2;
Semaphore demo_sem(2);     //Create Semaphore with 2 tokens

void thread1_fn(){         //Thread 1
  while(true){
    for (char i = 0;i < 8;i++){ //Phase 1 - Slow flashing
      led1 = !led1;
      ThisThread::sleep_for(1s);
    }
    demo_sem.acquire(); //Acquire a token
    for (char i = 0;i < 120;i++){ //Phase 2 - Use resource, fast flashing
      led1 = !led1;
      ThisThread::sleep_for(150ms);
    }
    demo_sem.release(); //Release the token
  }
}

void thread2_fn(){     //Thread 2
  while(true){
    for (char j = 0;j < 12;j++){ //Phase 1 - Slow flashing
      led2 = !led2;
      ThisThread::sleep_for(750ms);
    }
    demo_sem.acquire(); //Acquire a token
      for (int j = 0;j < 400;j++){ //Phase 2 - Use resource, fast flashing
      led2 = !led2;
      ThisThread::sleep_for(120ms);
    }
    demo_sem.release(); //Release the token
  }
}
```

```
int main(){                        //Let's call this Thread 3
  Thread1.start(thread1_fn); //Start Thread 1 executing.
  Thread2.start(thread2_fn); //Start Thread 2 executing.
  while(true){
    for (char k = 0;k < 20;k++){ //Phase 1 - Slow flashing
      led3 = !led3;
      ThisThread::sleep_for(500ms);
    }
    demo_sem.acquire(); //Acquire a token
    for (char k = 0;k < 100;k++){ //Phase 2 - Use resource, fast flashing
      led3 = !led3;
      ThisThread::sleep_for(200ms);
    }
    demo_sem.release(); //Release the token
  }
}
```

Program Example 9.6 Applying the Semaphore

Download and run this program on your chosen development board. The threads start simultaneously, flashing their LEDs at a slow pace, with slightly different rates and overall durations. At the end of its opening bout of slow flashing, each thread attempts to acquire a semaphore token, in order to access the imagined resource and proceed to Phase 2 of its operation. Thread 3 goes on with this flashing just a little longer, so it is the last to try to get a token; but all have gone! While the other threads enter their fast-flashing phase, Thread 3 is now blocked. When the first of Thread 1 or 2 completes this phase, it returns its token, and Thread 3 can take it and proceed. Program execution continues, with different threads being blocked from time to time.

The principle of Program Example 9.8 is further illustrated in the simple timing diagram of Fig. 9.9. Three threads are labeled A, B, and C, and each has a Phase 1 and a Phase 2 of operation. To execute Phase 2, the thread must acquire a Semaphore token, of which there are two. Tokens are symbolized by asterisks, with arrow directions indicating acquisition and release. Task C is the first to be blocked, with both tokens being taken by the time it needs to execute Phase 2. Having acquired a token it can proceed. Later Thread B becomes blocked, for a similar reason.

Exercise 9.4

Run Program Example 9.6 again, but now giving the semaphore one and then three tokens. Explain program behavior in each case.

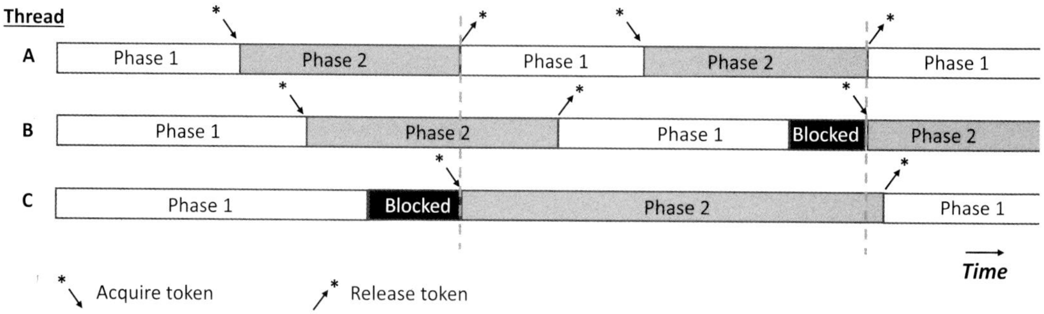

Figure 9.9
Simplified timing diagram for Program Example 9.6.

9.7 Using Interrupts with the RTOS

Interrupts form an essential part of RTOS programming, but they must be used with care. The first use of interrupts, unseen to the programmer, is almost always to provide the clock tick, provided by a timer interrupt on overflow. Incorrectly used interrupt service routines (ISRs) can upset the correct timing of the RTOS scheduling.

An ISR should not disturb the flow of the RTOS, so it should execute in the shortest possible time. Hence, there are greater restrictions when working within an RTOS environment. The best use of the ISR is to supply urgent information to the threads or scheduler. It could, for example, be set to signal that a certain event had occurred, thereby releasing a thread from a blocked state. The ISRs themselves are not normally used as threads. Any function which is the least bit time-consuming should be avoided, including wait/delay loops, lengthy while loops, and calls to time-hungry library functions.

Aside from the need to complete the ISR at the highest speed possible, many RTOS features simply are not available in the ISR. Table 9.7 gives examples of these. Unsurprisingly, no constructors listed are usable. The functions which can be used tend to be those which can complete simply and fast. For example, in the semaphore class, **acquire( )** is not available, as it might force a delay. However, **try_acquire( )** can be used, as it returns immediately. Check the Mbed OS website for functions not listed here.

Program Example 9.7 extends Example 9.2 to become a simple interrupt example. There are two threads, each of which simply flashes an LED. One, however, is dependent on the status of the variable **running**, which the ISR reverses every time it is called. Reflecting good practice, the ISR is as short as possible, with all the action within the threads.

Table 9.7: Example API restrictions in ISRs.

Function Name	Can it be used in an ISR?	Function Name	Can it be used in an ISR?
Thread		**Semaphore**	
constructor	X	constructor	X
get_priority()	X	acquire()	X
set_priority()	X	try_acquire()	√
start()	X	release()	√
join()	X	**Queue**	
Mutex		constructor	X
constructor	X	empty()	√
member functions	X	full()	√

```
/* Program Example 9.7. An interrupt controls the action of a thread.
   Runs on LPC1768 or F401RE (with LEDs shown in Fig 3.1) */

#include "mbed.h"
Thread thread2; //name thread2 chosen for this thread
DigitalOut led1(LED1);
DigitalOut led2(A0);          //Replace A0 with LED2 for LPC1768
InterruptIn button(BUTTON1); //Replace BUTTON1 with p5 for LPC1768
volatile bool running = true;

// Blink function toggles the led in a long running loop
void blink(void) {
  while (true){
    if (running) {
     led1 = !led1;
     ThisThread::sleep_for(1s); //set blink rate
    }
  }
}
void ISR1(void){
  running = !running;
}

int main() {                  // Launches a thread which runs for 10 seconds
  thread2.start(blink);       //Starts thread2 executing.
  button.rise(&ISR1);
  while(true){
    ThisThread::sleep_for(800ms);
    led2 = !led2;
  }
}
```

Program Example 9.7 A simple interrupt application

9.8 Queues and Memory Pools

9.8.1 The Queue Concept

A queue allows you to queue pointers to data—for example, to pass data from threads which produce it to threads which apply the data. The queue acts as a data pipeline, with buffering, between two threads. The data items stored in the queue can be prioritized. In this case, they are retrieved in order of descending priority. If all priorities are equal, they are withdrawn on a FIFO (first-in, first-out) basis. Data stored can be integer, pointer, or a type specified by a user-provided template. The Mbed API is summarized in Table 9.8.

Queues are also used with the MemoryPool class, Table 9.9, which allows you to define and manage fixed-size memory pools of objects of a given type.

9.8.2 Programming with Queues and Memory Pools

Program Example 9.8 demonstrates use of a queue and memory pool. A set of data, containing simulated voltage and current readings and a counter, is transferred from one thread to another using a queue. There are two user-defined threads running, **main( )** and **thread2.** This latter thread, which generates the data and puts it in the queue, runs every second. The **main( )** thread extracts data from the queue and displays it. However, it is set to run only once every 4 s. Thus, the queue is required to fill up before it is emptied.

Table 9.8: Selected Mbed API functions for queues.

Name/Format	Description
Constructor	
`Queue <T, queue_sz>`	Create and initialize a message queue of objects, of data type T and maximum capacity queue_sz.
Member Functions	
`bool empty() const`	Check if the queue is empty. Returns True if queue is empty, False otherwise.
`bool full() const`	Check if the queue is full. Returns True if queue is full, False otherwise.
`bool try_get(T** dataout)`	Retrieve a message from the queue, storing message in location pointed to by *dataout*. Returns immediately, with True for success.
`bool try_put(T* data, uint8_t prio)`	Insert the message pointed to by *data* into the end of the queue, with priority *prio* (higher number for higher priority). Returns immediately, with True for success.

Table 9.9: Selected Mbed API functions for MemoryPool.

Name/Format	Description
Constructor	
MemoryPool<*T, pool_sz*>	Create and Initialize a memory pool, of data type *T* and number of objects *pool_sz*
Member Functions	
T* try_alloc()	Allocate a memory block from a memory pool, without blocking. Returns address of allocated memory block. or **nullptr** if no memory available.
osStatus free(T* *block*)	Free a memory block. Returns **osOK** for success.

Using **typedef**, a **struct** data block is defined as a new data type, **message_t.** In **thread2**, the simulated data is calculated, based on a cycle counter **i**. The memory pool is populated with these calculated values and the data placed in the queue. To illustrate application of priority, every third data item is prioritized.

In the **main( )** thread the values in the queue are read back and printed. Higher-priority items are automatically read first, followed by the others (of equal priority) thereafter. Because we are printing floating point numbers, the usual addition of the little **Mbed_app.json** file (as described in Section 5.3.1) is needed.

```
/* Program Example 9.8. Demonstrates memory pools and queue, including prioritisation.
   Developed from https://os.Mbed.com/docs/Mbed-os/v5.15/apis/memorypool.html
   Runs on LPC1768 or F401RE with no external connection needed */

#include "mbed.h"

// Define this struct as new type, named message.
typedef struct {
  float voltage;     // ADC result of measured voltage
  float current;     // ADC result of measured current
  uint32_t counter; // A counter value
} message_t;

// Declare a MemoryPool object of 16 elements, with type message_t.
MemoryPool<message_t, 16> mpool;

// Create queue, with same capacity as the memory pool.
Queue<message_t, 16> queue;

Thread thread2;            // Declare a thread object for the send_thread.

void send_thread(void) {
   uint32_t i = 0;
```

```
    while (true) {
      i++;                    // cycle counter
      // Allocate a message object
      message_t *message = mpool.try_alloc();
      if (message == nullptr) {
        printf("error: unable to allocate memory\n");
        continue;
      }
      // Use i to calculate simulated data
      message->voltage = (i * 3.3);
      message->current = (i * 1.1);
      message->counter = i;
      // Set priority to 5, whenever i is divisible by 3
      int prio = (i % 3) ? 0 : 5;
      if (!queue.try_put(message, prio)) {
        printf("error: unable to insert message into queue\n");
        mpool.free(message);
        continue;
      }
      ThisThread::sleep_for(1s);
    }
  }

  int main(void) {                         // viewed as Thread 1
    thread2.start(callback(send_thread)); // launch thread2
    printf("New set of readings:\n");     //Indicate new readings
    printf("--------------------\n");
    while (true) {
      message_t *message;
      if (queue.try_get(&message)) {
        // Print the data
        printf("Voltage: %.2f V\n", message->voltage);
        printf("Current: %.2f A\n", message->current);
        printf("Number of cycles: %u\n\n", message->counter);
        mpool.free(message);           // Free the message object
      } else {
        printf("Waiting for readings...\n");  //indicate queue is now empty
        ThisThread::sleep_for(4s);
        printf("\nNew set of readings:\n");   //Indicate new readings
        printf("--------------------\n");
      }
    }
```

(A)

```
{
  "target_overrides":{
    "*": {
        "target.printf_lib": "std"
         }
    }
}
```

(B)

Program Example 9.8 Applying queues and memory pools: (A) the main.cpp file; (B) the Mbed_app.json file (for printing of floating point numbers)

Compile and run Program Example 9.8 on your chosen development board. When this is connected to a virtual terminal, you should get a display similar to Fig. 9.10. Every batch of readings is started with a new heading. In each batch, notice how any cycle which is divisible by 3 is prioritized, and appears first in the list.

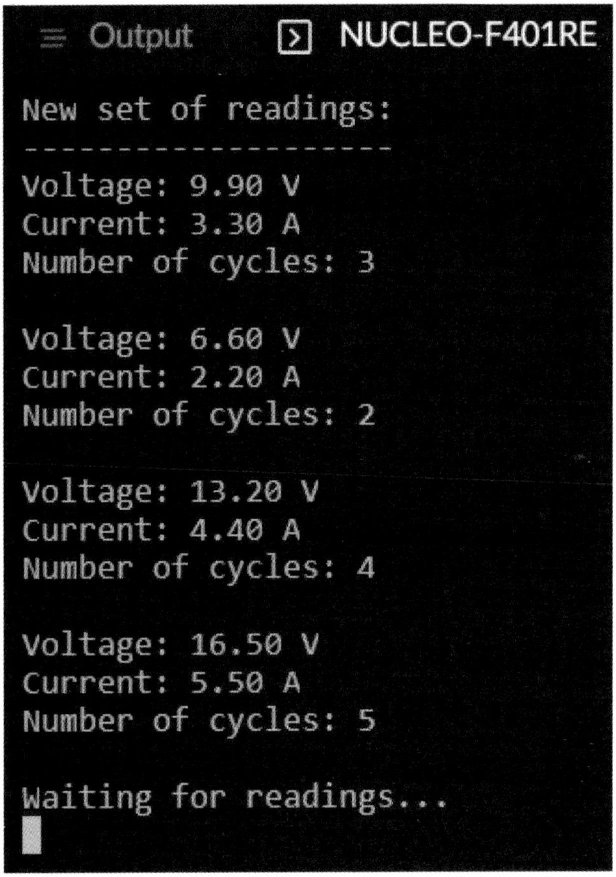

Figure 9.10
Example output of Program Example 9.8.

Exercise 9.5

i) Identify and remove the piece of code in Program Example 9.8 which prioritizes those samples whose counter value is divisible by 3. Rerun the program, checking that samples are now displayed in order. Explore using a different criterion to set priority, and include more than one priority level.

ii) Change the sleep duration at the end of **main( )** from 4 s to 500 ms, and explain the new serial output.

9.9 The "Bare Metal" Profile

Developing an application with the full Mbed RTOS does have resource implications; think for example of the 4 KB stack memory allocated per thread. For simple programs, or in resource-constrained situations, it is not worth using the RTOS; a clear alternative is to adopt the "bare metal profile," mentioned already in passing in Section 2.5.4.

The bare metal profile is a subset of the Mbed OS, which minimizes the size of the final application. It is intended for simple programs and/or constrained hardware, and results in efficient and compact code. Only the set of APIs that applications require are used. Compared with the full OS profile, this gives better control of the application's final size.

The bare metal profile uses only a subset of the OS APIs. It is important to know which these are when using it! Examples are shown in Table 9.10. Drivers, like **DigitalIn** or **AnalogOut**, are mostly available, but most of the RTOS API of course are not. Note that you can manually add missing APIs if your application requires it.

A project must have its **Mbed_app.json** file correctly configured for this implementation. The simplest way to launch a bare metal project is to select the bare metal template in Mbed Studio or Studio Cloud. In this the **Mbed_app.json** file has already been configured.

Chapter Review

- A system operating in real time is one that is able to meet its deadlines.
- The requirement of multitasking, common to almost every embedded system, carries with it the concepts of threads, deadlines, and priorities.

Table 9.10: RTOS API availability in bare metal profile

API	Available in Bare Metal
Thread	No
ThisThread	Yes
Semaphore	Yes
Queue	No
Mutex	Yes
Mail	No
MemoryPool	No
IdleLoop	No
EventFlags	Yes
ConditionVariable	No

- Use of a real-time operating system (RTOS) requires that programs be structured in a different way, with the programmer clearly understanding the underlying principles of the operating system.
- The RTOS schedules threads, with cyclic, round-robin, prioritized pre-emptive, or cooperative scheduling.
- Threads can interact with each other, applying carefully managed techniques.
- Access to shared resources and synchronization can be provided by the mutex and the semaphore.
- Transfer of data between threads can be managed with queues and memory pools.
- Interrupts remain available in the RTOS environment, but must be used with particular care.

Quiz

1. An RTOS takes the concepts of a general OS, but applies certain further requirements. In connection with this, identify the correct statements below.
   i. Real time implies that all actions of the RTOS are linked to "real times" of the day; for example, an event must happen at 5:30 pm.
   ii. A system operating in real time must be able to provide the correct results by the required time deadlines.
   iii. The RTOS works in the embedded environment; this leads to relaxed operational demands, as there is usually no need to manage a keyboard or display.
   iv. The RTOS needs to result in very compact code; hence, the embedded system needs to optimize processing power and memory resources.
   v. The terminology "real time" is used to emphasize that everything in the computing world takes time; it is similar to the US expression "real estate."
2. An important part of the RTOS is its scheduler. In connection with the scheduler, identify the correct statements below.
   i. The scheduler shares use of the CPU between the different threads, taking account of their priority if needed.
   ii. In round-robin scheduling, thread priority is emphasized; the highest-priority task completes before any other.
   iii. In cooperative scheduling, each thread relinquishes access to the CPU of its own accord.
   iv. Cyclic scheduling means that each thread executes until it has completed; the CPU then starts executing the next thread in the list.
   v. Threads experience two different states, called running and pending.
3. In any two (or more) of Program Examples 9.2, 9.4, 9.5, or 9.6, describe the states that the threads go through, with reference to Fig. 9.5.

4. At a certain moment that an RTOS runs, three threads are ready to start. Their subsequent execution is shown in Fig. 9.11. Here the larger numbers indicate the task being executed in each time slice, while the smaller numbers provide a time slice number. Each time slice has a duration of 2 ms, and no task is executing in time slice 12 or 13. During the time shown, threads run to completion, and just one of them is made ready to run again. From the diagram, determine the scheduling method applied and the relative priorities of each thread and their duration.
5. Considering the mutex, identify the correct statements in the list below.
    i. A benefit of the mutex is that it allows synchronization between threads.
    ii. A mutex applies three conditions: lock, unlock, and half-open, to manage resource access.
    iii. In a poorly thought-through program, a deadlock condition can arise through the use of mutex.
    iv. If a thread has locked a mutex, no other thread can access that resource.
    v. A mutex applies two conditions, lock and unlock, to manage resource access.
6. Considering the semaphore, identify the correct statements in the list below.
    i. There are three types of semaphore: *counting*, *binary*, and *decimal*.
    ii. A binary semaphore is very similar to a mutex.
    iii. The number of tokens in any semaphore starts with a positive number; it can, however, go negative—threads can simply owe the semaphore a token.
    iv. Semaphore tokens can be passed directly from one thread to another, as well as to and from the semaphore itself.
    v. A semaphore can be used to signal to a thread that data has become available in a queue.
7. An inexperienced programmer writes the following ISR while applying the Mbed RTOS, but then experiences problems. Identify its weaknesses, and explain why.

```
void ISR1(void){          //ISR called to display pressure
  wait_us(5000);          //delay to let switch settle
  op_mutex.unlock(); //can now release mutex
  printf("Pressure= %.2f V\n\r", ADC_op);
}
```

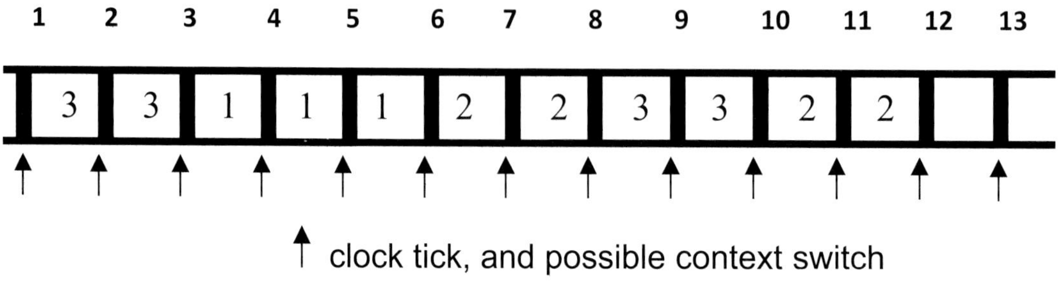

Figure 9.11

8. In connection with queues, identify the correct statements below.
   i. The Mbed queue API uses **empty( )** to check if the queue is empty, and **remind( )** to trigger a request to periodically flag data availability.
   ii. A queue acts as a data pipeline between two threads.
   iii. Data in the queue is always treated as having equal priority.
   iv. If one thread puts some data into a queue it must then be immediately removed by another; otherwise it will be lost.
   v. The Mbed queue API uses **try_put( )** and **try_get( )** functions to enter and remove data from the queue.
9. A system has three threads, as shown in Table 9.11.

Table 9.11

Task	Priority	Execution Time (ms)	Period (ms)
A	1	2	16
B	2	9	intermittent
C	3	4	intermittent

A multi-tasking real-time program is developed, using a time slice of 2 ms. Switching takes place as each time slice completes. Task A needs to run every 16 ms. Tasks B and C are re-launched only as indicated below.

(a) Show in diagrammatic form how the system executes the tasks, over the first 20 ms of operation, if round-robin scheduling is applied. Assume that a non-ready task is automatically skipped.
(b) Repeat a(), using prioritized pre-emptive scheduling.
(c) Repeat (b) if Task A runs every 8 ms.
(d) Repeat (c) if Task B runs again at $t = 17$ ms, for the first 32 ms.

10. The company you work for is developing an energy management system for remote houses which do not have access to the electricity grid. Energy will be sourced from a wind turbine and a solar panel and stored in a battery bank. You are designing the control system for this, using a single microcontroller. Your system must undertake the following:
    - Monitor voltage and current supplied from solar panel
    - Monitor voltage and current supplied from wind turbine
    - Select energy sources to charge battery, or disconnect battery from sources
    - Estimate charge state of the battery
    - Monitor voltage and current supplied from battery to house
    - Display main status information to user

- Shut down system if excess current is being supplied to house
- Issue audible warning if battery voltage is low

The microcontroller software is to be structured using a small RTOS, which allows a maximum of four threads and three priority levels.

Identify the threads you would create, indicate their priority, and explain your answers.

References

9.1. Simon David. *An Embedded Software Primer*. Addison Wesley; 1999.
9.2. An introduction to ARM Mbed OS6. https://os.Mbed.com/docs/Mbed-os/v6.15/introduction/index.html.
9.3. Scheduling Options and Config. https://os.Mbed.com/docs/Mbed-os/v6.15/apis/scheduling-options-and-config.html.
9.4. A.B. Downey, The Little Book of Semaphores, second ed., Open Textbook Library. https://open.umn.edu/opentextbooks/textbooks/83.

CHAPTER 10

Memory and Data Management

10.1 A Memory Review

10.1.1 An Overview of Memory Technologies

Broadly speaking, a microprocessor needs memory for two reasons: to hold its program, and to hold the data that it is working with. We often call these memories *program memory* and *data memory.*

To meet these needs, there are a number of different semiconductor memory technologies available, which can be embedded on the microcontroller chip. Memory technology, summarized in Fig. 10.1, is divided broadly into two types, *volatile* and *non-volatile*.

Non-volatile memory retains its data when power is removed, but tends to be more complex to write to in the first place. For historical reasons it is still often called ROM (read only memory). Non-volatile memory is generally required for program memory, so that the program data is there and ready when the processor is powered up.

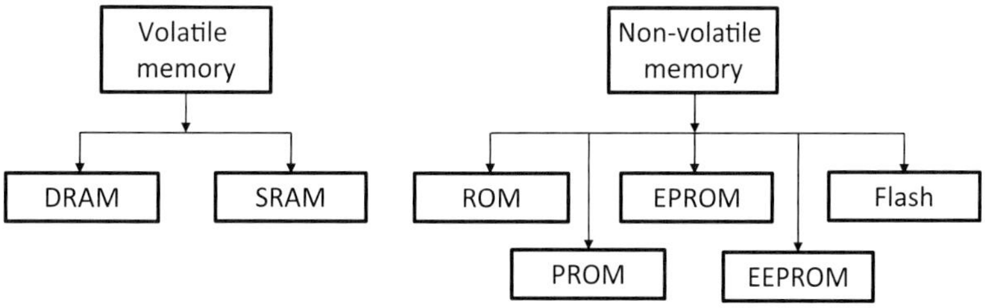

Key

DRAM: Dynamic random access memory

EPROM: Erasable programmable read only memory

EEPROM: Electrically erasable programmable read only memory

PROM: Programmable read only memory

ROM: Read only memory

SRAM: Static random access memory

Figure 10.1
Electronic memory types.

Fast and Effective Embedded Systems Design. https://doi.org/10.1016/B978-0-323-95197-5.00010-3

Volatile memory loses all data when power is removed, but is easy to write to. Volatile memory is traditionally used for data memory; it is essential to be able to write to memory easily, and there is little expectation for data to be retained when the product is switched off. For historical reasons it is often called RAM (random access memory), although this terminology tells us nothing that is useful. These categorizations of memory, however, give an over-simplified picture. It can be useful to change the contents of program memory, and there are times when we want to save data long-term. Moreover, new memory technologies now provide non-volatile memory which is easy to write to.

In any electronic memory we want to be able to store all the 1s and 0s which make up the data. There are several ways that this can be done; a few essential ones are outlined here. A simple 1-bit memory is a coin. It is stable in two positions, with either "heads" facing up, or "tails". We can try to balance the coin on its edge, but it will soon fall over. We recognize that the coin is stable in two states; we call this *bistable*. It could be said that "heads" represents logic 1, and "tails" logic 0. With 8 coins, an 8-bit number can be represented and stored. If we had 10 million coins, we could store the data that makes up one photograph of good resolution, but that would take up a lot of space indeed!

There are of course a number of electronic alternatives to the coin, which take up much less space. The most common are illustrated in Fig. 10.2(b)–(e), generally in simplified form. Any block of memory is made up of a vast array of such cells; if you understand the characteristics of one cell, then broadly you understand the characteristics of that memory type.

Let's start with the electronic bistable (or "flip-flop") circuit, as shown in Figs. 10.2(b) and 10.2(c). These circuits are stable in only two states, and each can be used to store one bit of data. Circuits like these have been the bedrock of volatile memory. Static RAM (SRAM) consists of a vast array of memory cells based on the circuit of Fig. 10.2(b). These have to be addressable, so that just the right group of cells is written to, or read from, at any one time. To make this possible, two extra transistors are added to the two outputs. To reduce the power consumption, the two resistors are usually replaced by two transistors also. That means six transistors per memory cell. Each transistor takes up a certain area on the IC, so when the circuit is replicated thousands or millions of times, it can be seen that this memory technology is not actually very space-efficient. Despite this, it is of great importance; it is low-power, can be written to and read from with ease, can be embedded onto a microcontroller, and forms the standard way of implementing data memory in most embedded systems. Of course, all data is lost when power is removed. Fig. 10.2(c) is a neat circuit, but requires even more transistors, so we will not consider it as a candidate for microcontroller memory here.

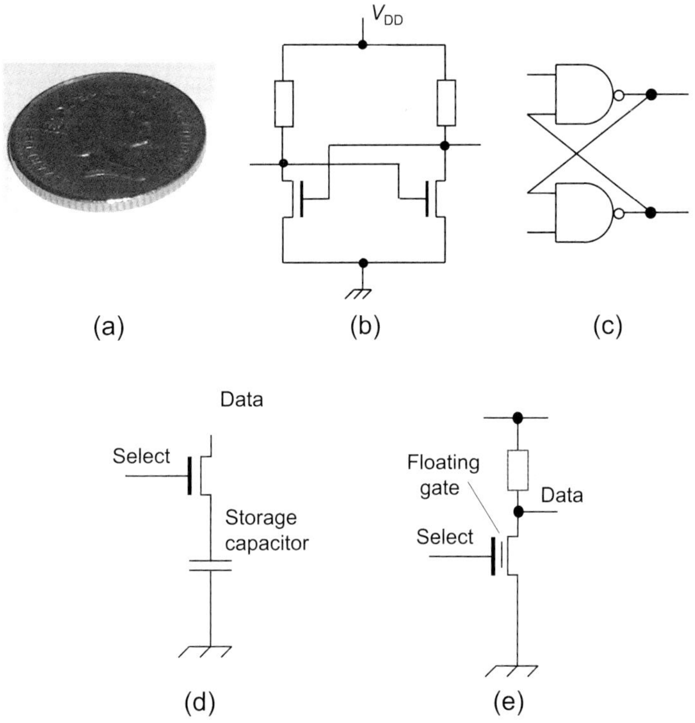

Figure 10.2
Five ways of implementing a one-bit memory: (a) a coin; (b) two transistors and two resistors; (c) two NAND gates; (d) charged capacitor; (e) floating gate.

Dynamic RAM (DRAM) is intended to do the same thing as SRAM with a reduced component count, and reduced silicon area. It is represented in Fig. 10.2(d). Instead of using a number of transistors, one bit of information is stored in a tiny capacitor, like a small rechargeable battery. A charged capacitor can represent Logic 1, and a discharged capacitor Logic 0. Such capacitors can be fabricated in large quantity on an IC. To select the capacitor for reading or writing, a simple transistor switch is required. So far, so good. Unfortunately, due to the small capacitor values and unavoidable leakage currents on the chip, the memory loses its charge over a rather short period of time (around 10 ms to 100 ms). So the DRAM needs to be accessed every few milliseconds to refresh the charges; otherwise the information is lost. DRAM has about four times better storage capacity than SRAM at about the same cost and chip size, with a compromise of the extra work involved in regular refreshing. Moreover, it is power-hungry and inappropriate for any battery-powered device. DRAM has found wide application as high-density data memory in mains powered computers, such as desktop PCs.

The original ROMs and PROMs (programmable ROMs) could only be programmed once (hence in subsequent use "read *only*"), and have now completely disappeared in normal usage. The first type of non-volatile reprogrammable semiconductor memory, the EPROM (electrically-programmable ROM), represented a huge step forward—a non-volatile memory could now be reprogrammed. With a process called *hot electron injection* (HEI), electrons can be forced through a very thin layer of insulator onto a tiny conductor embedded within the insulator, and can be trapped there almost indefinitely. This conductor is placed so that it interferes with the action of a field effect transistor (FET); it is called a *floating gate*, and is represented in Fig. 10.2(e). When it is charged/discharged, the action of the FET is disabled/enabled. This modified FET effectively becomes a single memory cell, far denser than the SRAM discussed previously. Moreover, the memory effect is non-volatile; trapped charge is trapped charge! This programming does require a comparatively high voltage (around 25 V), so it generally needs a specialized piece of equipment. The memory is erased by exposing it to intense ultraviolet light; EPROMs can always be recognized by the quartz window on the IC, which allows this to happen.

The next step beyond HEI was *Nordheim Fowler tunnelling*. This requires even finer memory cell dimensions, and gives a mechanism for the trapped charge to be retrieved electrically, which had not previously been possible. With electrically erasable and programmable ROM (EEPROM), words of data are individually writeable, readable, and erasable and are non-volatile. The cell structure is still broadly the same as in Fig. 10.2(e). The downside of this is that more transistors are needed to select each word. In many cases we don't need this flexibility. A revised internal memory structure led to *flash* memory; in this, the ability to erase individual words is not available. Whole blocks have to be erased at any one time, "in a flash." This compromise leads to a huge advantage: Flash memory is very high-density indeed, more or less the highest we can get. This memory type has been a key feature of many recent products which we have now become used to, like smart phones, digital cameras, memory sticks, or solid-state drives.

A curious feature of Flash and EEPROM, unlike most electronics, is that they exhibit a wear-out mechanism. Electrons can get trapped in the insulator through which they are forced when a write operation takes place. Therefore, this limitation is often mentioned in data sheets, for example, a maximum of 100,000 write-erase cycles. This is of course a very large number, and is unlikely to be experienced in normal use.

Although EPROM had become very widely used, and had been integrated onto microcontrollers, it was rapidly and completely replaced by Flash memory. Now in embedded systems, the two pervasive memory technologies are Flash and SRAM. A glance back at Figs. 2.2 and 2.3 shows how important they are, as an example, to the

Mbed LPC1768. Program memory on the LPC1768 is Flash, and data memory is SRAM. On the Mbed board, the so-called "USB disk" is a Flash IC. The LPC1768 microcontroller has 512 Kbytes of Flash program memory and 64 Kbytes of SRAM; the F401RE microcontroller also has 512 Kbytes of Flash program memory, but 96 Kbytes of SRAM. The other memory technology that we are likely to meet at times is EEPROM; this is still used where the ability to rewrite single words of data remains essential.

10.1.2 Memory Mapping

When that mass of memory is embedded onto the microcontroller IC, each memory location, typically of one byte, is allocated a binary address. This is a unique number by which that location is identified, and hence accessed while a program runs. Different parts of memory are allocated different places in the resulting "memory map." To these are added the registers which control the peripherals, and possibly other features of the microcontroller. A simplified version of the F401RE microcontroller memory map, itself an application of the ARM Cortex-M memory map (see Reference 1.1), is shown in Fig. 10.3. To the left are the addresses, expressed in hexadecimal, from 0x0000 0000 to

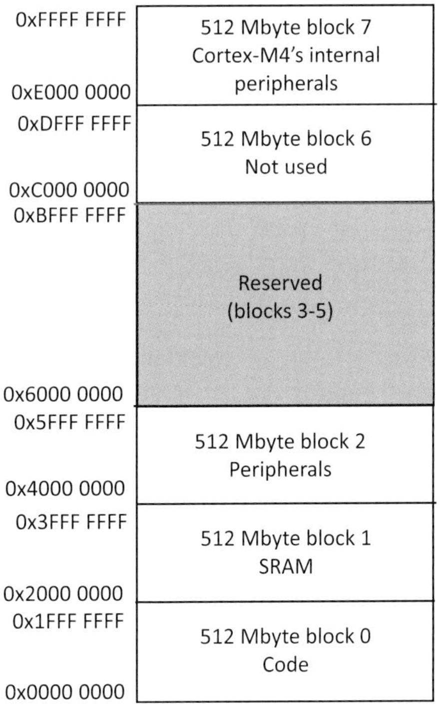

Figure 10.3
Simplified memory map, STM32F401RE microcontroller.

0xFFFF FFFF. This address range gives the possibility of 2^{32} memory locations, or 4,294,967,296, or 4 Gbytes! This is more than a microcontroller of this complexity usually needs. It can be seen that eight memory blocks, each of 512 Mbytes, are identified. These are allocated to different memory types, including flash, SRAM, and peripherals. Three blocks are not used at all, and a closer look at the data would reveal that even the blocks which are used are not fully populated. A similar memory map can be found for the LPC1768 in Table 3 of its User Manual (Reference 2.4).

Happily for regular C/C++ programming, we are shielded from having to think about these numerical addresses; the compiler makes memory allocations on our behalf. However, the next section explains how we can usefully exploit the concept of memory addressing. Individual memory and register locations can also be accessed using the debugger. We do make use of the numerical addresses in Chapter 14, where we take a deeper look at how peripherals are addressed and set up.

10.2 Some Programming Techniques for Data and File Management

10.2.1 Using Addresses and Pointers

C/C++ code feature

Let's see how we apply the memory address concept practically in C/C++ programming, linking addresses to the related concept of pointers. We will illustrate this with a specific Mbed example using addresses and pointers. These are also reviewed in Section B.8.2.

Pointers are used to indicate where a particular element or block of data is stored in memory. When a pointer is defined it can be set to point at a particular memory location; C/C++ syntax then allows access to the data at that address. Pointers are required for a number of reasons; one is because the C/C++ standard does not allow arrays of data to be passed to and from functions—in this case pointers are used instead. For example, we may wish to pass an array of 10 data values to a function in order to perform a simple averaging calculation, but in C/C++ this is not possible. Instead, it is necessary to pass a single pointer value as an input argument to the function. In this instance, the pointer gives the memory address of the first element of the data array and is usually accompanied by a single argument that defines the size of the data array in question. Pointers can also be used to improve programming efficiency and speed by accessing memory locations and data directly, though they do bring some extra programming complexity.

Pointers are defined in a similar way to variables, but by additionally using the * operator. For example, the following declaration defines a pointer called **ptr** which points to data of type **int**:

```
int *ptr;            // define a pointer which points to data of type int
```

The specific address of a data variable can also be assigned to a pointer by using the & operator, for example:

```
int datavariable = 7; // define a variable called datavariable with value 7
int *ptr;             // define a pointer which points to data of type int
ptr = &datavariable;  // assign the pointer to the address of datavariable
```

In program code we can also use the * operator to get the data from the given pointer address, for example:

```
int x = *ptr;         // get the contents of location pointed to by ptr and
                      // assign to x (in this case x will equal 7)
```

We can also use pointers with arrays, because an array is a number of data values stored at consecutive memory locations. So if the following is defined:

```
int dataarray[] = {3,4,6,2,8,9,1,4,6}; //define an array, arbitrary values
int *ptr;                              // define a pointer
ptr = &dataarray[0];                   // assign pointer to the address of
                                       // the first element of the data array
```

the following statements will be true:

```
*ptr == 3;      // the first element of the array pointed to
*(ptr+1) == 4; // the second element of the array pointed to
*(ptr+2) == 6; // the third element of the array pointed to
```

So array searching can be done by moving the pointer value to the correct array offset. To illustrate, Program Example 10.1 implements a function for analyzing an array of data and returns the average of that data. It requires no external connections, and can be run on any Mbed platform, returning display data to a serial monitor.

```
/* Program Example 10.1: Pointers example for an array averaging function
                                                                         */
#include "mbed.h"

char data[] = {5,7,5,8,9,1,7,8,2,5,1,4,6,2,1,4,3,8,7,9}; //define some input data
char *dataptr;            // define a pointer for the input data
float average;            // floating point average variable

float CalculateAverage(char *ptr, char size); // function prototype

int main() {
  dataptr = &data[0]; // point pointer to address of the first array element
```

```
  average = CalculateAverage(dataptr, sizeof(data)); // call function
  printf("\n\rdata = ");
  for (char i = 0; i < sizeof(data); i++) {     // loop for each data value
    printf("%d ",data[i]);                      // display all the data values
  }
  printf("\n\raverage = %.3f",average);         // display average value
  printf("\n\rfinished");
}
// CalculateAverage function definition and code
float CalculateAverage(char *ptr, char size) {
  int sum = 0; //define variable for calculating the sum of the data
  float mean; //define variable for floating point mean value
  for (char i = 0; i < size; i++) {
    sum = sum + *(ptr+i); //add all data elements together
  }
  mean = (float)sum/size; //divide by size and cast to floating point
  return mean;
}
```

(A)

```
{
  "target_overrides":{
    "*": {
      "target.printf_lib": "std"
      }
    }
}
```

(B)

Program Example 10.1 Averaging function using pointers: (a) the main.cpp file; (b) the Mbed_app.json file (for printing of floating point numbers)

C/C++ code feature

Looking at some key elements of Program Example 10.1, we can see that the pointer **dataptr** is assigned to the address of the first element of the **data** array. The **CalculateAverage( )** function takes in a pointer value which points to the first value of the data array and a second value which defines the size of the array. The function returns the floating point mean value. There is also an additional C/C++ keyword used here, **sizeof,** which deduces the size of a particular array. This gets the size (i.e., the number of data elements) in the array **data**. Additionally, note that the calculation of the mean value is in the form of an integer divided by a char, yet we want the answer to be a floating point value. To implement this, we *cast* the equation as floating point by using **(float)**; this ensures that the resultant mean value is to floating point precision. Because we are printing floating point numbers, we need the **Mbed_app.json** file added, as explained in Section 5.3.

Implement Program Example 10.1 and check that the data average is calculated correctly.

Exercise 10.1

In Program Example 10.1, modify the size and values within the data array and check that the **CalculateAverage( )** function still performs as expected.

10.2.2 File Management in C/C++

C/C++ code feature

Simple arrays, strings, or other data groupings are useful in their own way, but we often need to hold much larger data collections, maybe of diverse forms and uncertain size. Here we turn to files and filing systems. In C/C++ we can create files, open and close them, read and write data from or to them, and scan them in search of particular data.

Functions for input and output operations are defined by the C Standard Input and Output Library **stdio**. See Reference 10.1 for further detail. Using **stdio**, we can store data in files (as **chars**) or as strings of text. For data storage examples given in the book, we use the functions summarized in Table 10.1 (this is effectively Table B.8, repeated here for convenience). In the table there is repeated reference to data streams. These are flows of data, essentially sequences of bytes, linked to an input or output. Streams are represented in the **stdio** library as pointers to FILE objects, which uniquely identify the stream.

10.2.3 Using Data Files with the Mbed OS

The Mbed OS has a range of file handling APIs belonging to the data storage API group seen in Table 2.2. These are far more extensive than can be covered in this book. They can be viewed in Reference 10.2, with a subset shown in Table 10.3. The OS file systems make some use of the **stdio** functions described.

A long-standing file management mechanism is the *file allocation table* (FAT) file system. This was originally developed as a file format with an 8-bit addressing system, for use with floppy disks. The file allocation table itself, with a second copy for security, is placed in a fixed location at the beginning of the memory space. The size of the address table dictates the amount of data that can be held on a single memory device, so the standard has since been updated to include FAT16 and FAT32 variants. These allow much larger memory devices, such as SD cards, to be managed.

In later sections we also meet the concept of the block device. This is a mass storage non-volatile device that stores data in blocks whose location it allocates. Block devices can include floppy discs, CD ROMs, solid state storage devices, and so on. The block device can manage some of its functions, like block allocations, or data buffering.

Table 10.1: Useful stdio library functions.

Function	Format	Summary Action
fopen	`FILE *fopen(const char *filename, const char *mode)`	Opens the file of name *filename* with chosen mode (Table 10.2).
fclose	`int fclose(FILE *stream)`	Closes a file.
fflush	`int fflush(FILE *stream)`	Writes any unwritten data in file buffer to the file.
fgetc	`int fgetc(FILE *stream)`	Gets a character from a stream.
fgets	`char *fgets(char *str, int n, FILE *stream)`	Gets a string of **n** chars from a stream.
fputc	`int fputc(int character, FILE *stream)`	Writes *character* to a stream.
fputs	`int fputs(const char *str, FILE *stream)`	Writes a string to a stream.
fprintf	`int fprintf(FILE *stream, const char *format,... )`	Writes formatted data to a stream.
fseek	`int fseek(FILE *stream, long int offset, int origin)`	Moves file pointer to specified location, applying: SEEK_CUR: current pointer position SEEK_END: end of file SEEK_SET: start of file

Notes:
str An array containing the null-terminated sequence of characters to be written.
stream Pointer to a FILE object that identifies the stream where the data is to be written.
... Indicates that additional formatted arguments may be specified in a list.

Table 10.2: Example access modes for fopen( ).

Access mode	Meaning	Action
"r"	Read	Open an existing file for reading.
"w"	Write	Create a new empty file for writing. If a file of the same name already exists it will be deleted and replaced with a blank file.
"a"	Append	Append to a file. Write operations result in data being appended to the end of the file. If the file does not exist a new blank file will be created.

Of the APIs seen in Table 10.3, we start with the local file system, peculiar to the LPC1768 and LPC11U24 Mbeds. We then move on to explore the FAT file system when using SD cards.

Table 10.3: Example Mbed OS file systems.

Mbed OS File System	Summary Description
FATFileSystem	A file system originally for disc storage. Uses File Allocation Table (FAT). Main use with Mbed is for interfacing with SD cards.
File	Provides generic access to a file system.
FileSystem	Implements a file system on a block-based storage device.
LittleFileSystem	A fail-safe file system designed for embedded systems, specifically for microcontrollers that use external flash storage.
LocalFileSystem	A file system for use only with LPC1768 and LPC11U24 Mbeds, with files written onto the "USB disk."

10.3 The Mbed LPC1768 Local File System

As Fig. 2.2 shows, the Mbed LPC1768 has a standalone 16 Mbit flash memory, sometimes called the USB disk drive, linked by serial interface to the interface microcontroller. Its contents can be viewed from a host PC when the Mbed is connected by a USB cable. Although it holds the program binary files which are downloaded to the board, it also contains plenty of space which can be used for data storage.

The **LocalFileSystem** library allows the programmer to set up a local file system for accessing the USB disk. It allows programs to read and write files in this memory, which can then be accessed from the host computer. Once the system has been set up, the standard C/C++ **stdio** file access functions—seen in Table 10.1—can be used to open, read, and write files.

The **LocalFileSystem** declaration sets up the Mbed device as an accessible storage unit and defines a directory for storing local files. To implement, add the following line to the declarations section of a program:

```
LocalFileSystem local("local"); //Create local file system named "local"
```

Note that the name **local** is given twice in the declaration; this is because we need to define first a C/C++ object that can be used in the code, and second a file directory path (in quotation marks) that can be referred to when using **stdio** library functions. For convenience we give both the same name here.

10.3.1 Opening and Closing Files

With a **LocalFileSystem** object defined, a file (in this example called **datafile.txt**) can be created on the Mbed with the following command:

```
FILE* pFile = fopen("/local/datafile.txt","w");
```

C/C++ code feature

The **fopen( )** function call uses the * operator to assign a pointer with name **pFile** to the file at the specific location given. From here on, the file is accessed by referring to its pointer rather than having to use the specific filename.

We also need to specify the *access* mode; three common ones were given in Table 10.2. In the above example the "w" specifier is used, denoting write access. A number of other access modes are shown in Table B.9.

When a file is opened, an *internal position indicator* is created for the file, defining the position at which the next data within the file will be either read from or written to. The position indicator can be modified with the **stdio fseek( )** function.

When we have finished using a file for reading or writing it is essential to close it, for example using:

```
fclose(pFile);
```

If you fail to do this you might lose all access to the Mbed; see the following section!

10.3.2 Recovering a "Lost" Mbed LPC1768

When the microcontroller program opens a file on the local drive, the Mbed will be marked as "removed" on a host PC. This means the PC will often display a message such as "insert a disk into drive" if you try to access the Mbed at this time; this is normal and stops both the Mbed and the PC trying to access the USB disk at the same time. The USB drive will only re-appear when all file pointers are closed in your program, or the microcontroller program exits. If a running program on the Mbed does not correctly close an open file, you will no longer be able to see the USB drive when you plug the Mbed into your PC. It is therefore important for a programmer to take care when using files to ensure that all files are closed when they are not being used.

If a running program on the Mbed LPC1768 does not exit correctly, use the following procedure to allow you to see and download to the Mbed again:

1. Unplug the Mbed.
2. Hold the Mbed reset button down.
3. While still holding the button, plug in the Mbed. The Mbed USB drive should appear on the host computer screen.
4. Keep holding the button until the new program is saved onto the USB drive.

10.3.3 Simple File Data Transfers

If the intention is to store numerical data, this can be done in a simple way by storing individual 8-bit data values. The **fputc( )** function allows this, for example as follows:

```
char write_var=0x0F;
fputc(write_var, pFile);
```

This stores the 8-bit variable **write_var** to the data file at the position indicated by the file's internal position indicator. When the data is written with the **fputc( )** command, the internal position indicator automatically increments by one.

Data can also be read from a file to a variable as follows:

```
read_var = fgetc(pFile);
```

The **fgetc()** function returns the character currently pointed to by the file's internal file position indicator. The internal file position indicator is then automatically advanced to the next character.

Program Example 10.2 creates a data file and writes the arbitrary value 0x23 to that file. The file is saved on the Mbed USB disk. The program then opens and reads back the data value and displays it to the screen in a host terminal application.

```
/* Program Example 10.2: read and write data bytes to Mbed USB disk
   Runs only on LPC1768   */

#include "mbed.h"
LocalFileSystem local("local");    // define local file system
int write_var, read_var;           // create data variables

int main (){
  // open file with name datafile.txt, in write mode
  FILE* File1 = fopen("/local/datafile.txt","w");
  write_var = 0x23;            // example data, 0x23 is ASCII code for #
  fputc(write_var, File1);     // put char (data value) into file
  fclose(File1);               // close file

  File1 = fopen ("/local/datafile.txt","r"); // open file for reading
  read_var = fgetc(File1);                   // read first data value
  fclose(File1);                             // close file
  printf("input value=%i \n",read_var);      // display read data value
```

Program Example 10.2 Saving data to a file

Create a new project and add the code in Program Example 10.2. Run the program, and verify that the data file is created on the Mbed and read back correctly. View the Mbed files on your computer. Most will be program binary files, but you should see the **datafile.txt** file. If you open this in a standard text editor program (such as Microsoft Notepad), you should see a hash character "#" in the top left corner, the ASCII character for code 0x23 (check Table 8.3).

Exercise 10.2

Change Program Example 10.2 to experiment with other data values; check these against their associated ASCII codes. Write the numbers 0 to 9, and view them in the saved file.

10.3.4 String File Access

Using the **stdio** functions, it is possible to read and write words and strings and search or move through files looking for particular data elements. Program Example 10.3 creates a file and writes a string of text data to that file. The file is saved on the Mbed LPC1768 target. The program then opens and reads back the text data and displays it to the screen in a host terminal application.

```
/* Program Example 10.3: Read and write text string data to Mbed USB disk
 Runs only on LPC1768    */

#include "mbed.h"
LocalFileSystem local("local"); //define local file system
char write_string[64];          //character array up to 64 chars
char read_string[64];           //character array up to 64 chars)

int main (){
  FILE* File1 = fopen("/local/textfile.txt","w");      //open file access
  fputs("lots and lots of words and letters", File1);  //put text into file
  fclose(File1);                                       //close file
  File1 = fopen ("/local/textfile.txt","r");   //open file for reading
  fgets(read_string,256,File1);                //read 256 chars of data
  fclose(File1);                               //close file

  printf("text data: %s \n",read_string);       //display read data string
}
```

Program Example 10.3 Saving a string to a file

Compile and run Program Example 10.3. Verify that the text file is created and read back correctly to a host terminal. If you open the file **textfile.txt**, found on the Mbed LPC1768, the correct text data should be found within.

Exercise 10.3

Taking Program Example 10.3, modify the size and format of the text string data and verify that it is always read back correctly and outputs to the host terminal application.

When data is read from a file, the file pointer can be moved with the fseek( ) function. For example, the following command will reposition the file pointer to byte 8 in the text file:

```
fseek(File2, 8, SEEK_SET); // move file pointer to byte 8 from the start
```

The **fseek( )** function needs three input terms; first the name of the file pointer, second the value to offset the file pointer to, and third an "origin" term which tells the function where to apply the offset. The term **SEEK_SET** is a predefined origin term (defined in the stdio library) which ensures that the 8-byte offset is applied from the start of the file.

Exercise 10.4

Add the following **fseek( )** statement to Program Example 10.2 just prior to the data being read back:

```
fseek (File2, 8, SEEK_SET); // move file pointer to byte 8
```

Verify that a host terminal only displays the data after byte 8 of the data file. Remember that byte values increment from zero, so it will actually be after the ninth character in the file.

10.3.5 Using Formatted Data

C/C++ code feature

It is possible to store formatted data in a file. We may want, for example, to log specific events to a data file and include variable data values such as time, or sensor input values, in a form which will be readily readable.

Logging of formatted data can be done with the **fprintf( )** function, which has syntax very similar to that of **printf( )**, except that the filename pointer is also required. Program Example 10.4 uses the **fprintf( )** function in a simple interrupt-controlled data logging project. Each time the pushbutton is pressed, an LED toggles. On each button press the file **log.txt** is updated to include the time elapsed since the previous button press, the current LED state, and a time stamp from the Real-time Clock (taking some lines of code from Program Example 6.10).

```
/* Program Example 10.4: Formatted data logging to text file triggered by button
push, with time stamp.
Runs on LPC1768 only
                                                                              */

#include "mbed.h"
InterruptIn button(p5);            //Interrupt on digital input p5
DigitalOut led1(LED1);             //digital out to onboard LED1
Timer main_timer;                  //define elapsed time timer
LocalFileSystem local("local");    //define local file system
bool button_pressed;          //flags button has been pressed
void toggle(void);            //function prototype
using namespace std::chrono;

int main() {
  set_time(1666697756); //Set RTC time (this is Tues, 25 Oct 2022 11:35:56)
  button.rise(&toggle); //attach the toggle function to the rising edge
  button_pressed = 0;
  main_timer.start();
  while(true){
    if(button_pressed==1){
      //open a text file named log, append mode
      FILE* Logfile = fopen ("/local/log.txt","a");
      //print elapsed time from Timer, and state of led
      fprintf(Logfile,"Elapsed time=%u ms, ",duration_cast<milliseconds>
                                  (main_timer.elapsed_time()).count());
      if(led1.read() == 1)fprintf(Logfile,"led is 1\n\r");
      else fprintf(Logfile,"led is 0\n\r");
      time_t seconds = time(NULL); //return elapsed seconds
      fprintf(Logfile,"Time of reading = %s\n\n", ctime(&seconds));
      fclose(Logfile); // close file
      button_pressed = 0;
    }          //end of if
  }            //end of while
}
//this is ISR
void toggle() {
   wait_us(5000); //simple debounce to settle
   if (button == 1){
     led1 = !led1; // toggle LED
     button_pressed = 1;
   }
}
```

Program Example 10.4 Logging data and time on button press

Try running this program example, using the circuit of Fig. 6.3. Press the button a few times, at for instance 5 s intervals; then view the **log.txt** file on the Mbed. Verify that the correct elapsed time and LED state have been written to the file.

Exercise 10.5

Try one or more of these developments of Program Example 10.4, checking each time the record in the **log.txt** file.

1. Enhance the usefulness and readability of the program by including an opening title in the text file. Then add a record counter, so that each file entry is individually numbered.
2. Connect a temperature sensor or potentiometer to an analog input, and record its value on every button press.
3. Replace the pushbutton with a Ticker, so that a truly periodic record can be made.

■

10.4 Using External SD Card Memory

Flash SD (secure digital) cards are a very familiar and useful form of storage. They have been in use for over 20 years, and—as may be expected—have shown considerable evolution during that time. Thus, their physical size has decreased in successive versions, while their capacity has dramatically increased. Their specifications are complex. "Simplified" specifications, still very detailed, can be found via Reference 10.3. Fortunately, again as we may expect, software libraries are available to allow comparatively easy access to this important technology.

10.4.1 Simple SD Card Block Transfers

As the SD card is a block device, it makes use of a number of features which are shared with other such devices. A summary of the **SDBlockDevice** API is given in Table 10.4.

SD cards can be linked to an Mbed target via the SPI protocol. Using a micro SD card with a card holder cradle (such as shown in Fig. 10.4(a)), it is possible to access the SD card as an external memory. The card is linked by SPI, applying—for either of our target boards—the connections given in Table 10.5.

Program Example 10.5 is adapted from Reference 10.4. It detects card data and prints it to the PC screen. It then writes a test text file to the card, reads this back, and displays. It uses the pin allocations given in Table 10.5. The **Mbed_app.json** file must also be created; this works for both targets, as the **SDBlockDevice** initialization within the program overwrites pin allocations in the **json** file.

Table 10.4: API summary for SDBlockDevice.

Functions	Usage
Constructor	
SDBlockDevice (PinNames mosi, miso, sclk, cs, uint64_t hz=1000000, bool crc_on=0)	Creates an SD Block Device on an SPI bus specified by pins.
Member Functions	
virtual int **deinit**()	Deinitialize a block device.
virtual int **erase**(bd_addr_t *addr*, bd_size_t *size*)	Erase block and prepare for writing; *addr* is start of erase block, of size *size*.
virtual int **frequency**(uint64_t *freq*)	Set transfer frequency, *freq*.
virtual bd_size_t **get_erase_size** (bd_addr_t *addr*)	Get the size (by return) of an erasable block.
virtual mbed::bd_size_t**get_program_size**()	Get the size (by return) of programmable block.
virtual mbed::bd_size_t **get_read_size**()	Get the size (by return) of readable block.
virtual int **init**()	Initialize a block device.
virtual int **program**(const void* *buffer*, mbed::bd_addr_t *addr*, mbed::bd_size_t *size*)	Program blocks to a block device. The blocks must be erased prior to programming. *buffer* = buffer to write to, *addr* = address of block, *size* = size to read.
virtual int **read**(void* *buffer*, mbed::bd_addr_t *addr*, mbed::bd_size_t *size*)	Read block and parameters as above.

```
/*Program Example 10.5. Writing data to/from SD card.
Works on LPC1768 and Nucleo F401RE, with pin connections shown
*/

#include "mbed.h"
#include "SDBlockDevice.h"

SDBlockDevice serdev(D4, D5, D3, D6);      //use p5,p6,p7,p8 resp. for LPC1768

uint8_t block[512] = "Hello World!\n";     //the string to be stored
uint8_t retrieve[512];                     //data block to store retrieved data

int main(){
  // Initialise the SDBlockDevice
  if (0 != serdev.init()) {
    printf("Init failed \n");
    return -1;
  }
```

```
    printf("serdev size: %llu\n", serdev.size());
    printf("serdev read size: %llu\n", serdev.get_read_size());
    printf("serdev program size: %llu\n", serdev.get_program_size());
    printf("serdev erase size: %llu\n", serdev.get_erase_size());
    // Set the frequency to 5 MHz
    if (0 != serdev.frequency(5000000)) {
      printf("Error setting frequency \n");
    }
    if (0 != serdev.erase(0,serdev.get_erase_size())){
      printf("Error Erasing block \n");
    }
    // Write data block to the device
    if (0 == serdev.program(block, 0, 512)) {
      // Read the data block from the device
      if (0 == serdev.read(retrieve, 0, 512)) { //read to new array
        // Print the contents of the block
        printf("%s", retrieve);
      }
    }
    serdev.deinit(); // de-initialise
}
```

(A)

```
{
  "target_overrides":{
    "*": {
          "target.components_add" : ["SD"],
          "sd.SPI_MOSI" : "D4",
          "sd.SPI_MISO" : "D5",
          "sd.SPI_CLK" : "D3",
          "sd.SPI_CS" : "D6"
    }
  }
}
```

(B)

Program Example 10.5 Writing data to an SD card: (a) the main.cpp program; (b) the Mbed_app.json file

Construct a circuit using the connections shown in Table 10.5. Compile and download Program Example 10.5. If all wiring is correct, and the SD card is in place, you will get an output similar to Fig. 10.5. The overall storage size shown here aligns with the 2-Mbyte size of the card in use. Unsurprisingly, the three block sizes, of read, program, and erase, are all the same. If wiring is incorrect, or there is no SD card in place, you will get the "Init failed" error message.

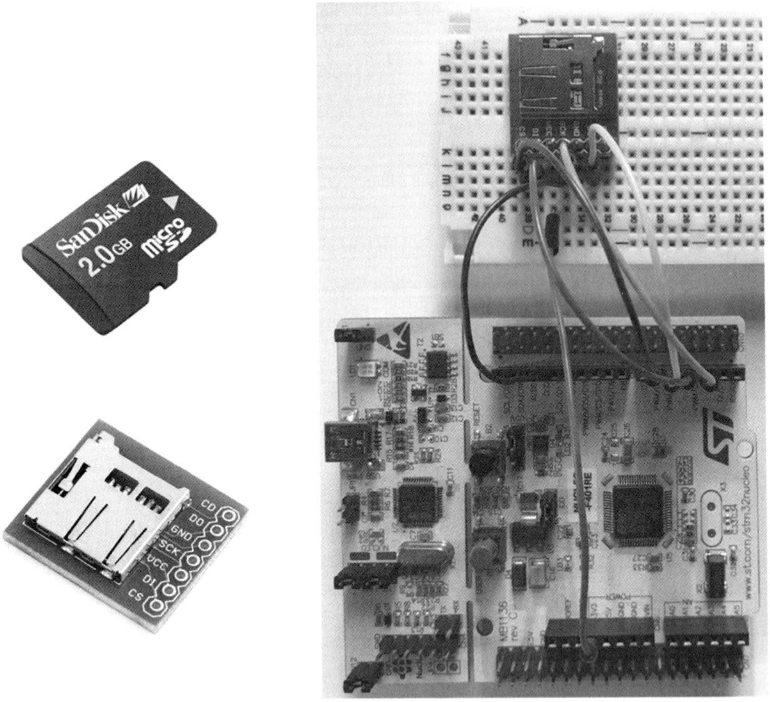

(a) (b)

Figure 10.4
Applying a Micro SD card: (a) Micro SD card with holder (Images courtesy of Sparkfun); (b) connection to Nucleo F401RE.

Table 10.5: Connections for SPI access to the SD card.

MicroSD Breakout	Mbed LPC1768 Pin	Nucleo F401RE Pin
CS	8 (DigitalOut)	D6
DI	5 (SPI MOSI)	D4
Vcc	40 (Vout)	+3V3
SCK	7 (SPI SCLK)	D3
GND	1 (GND)	GND
DO	6 (SPI MISO)	D5
CD	No connection	No connection

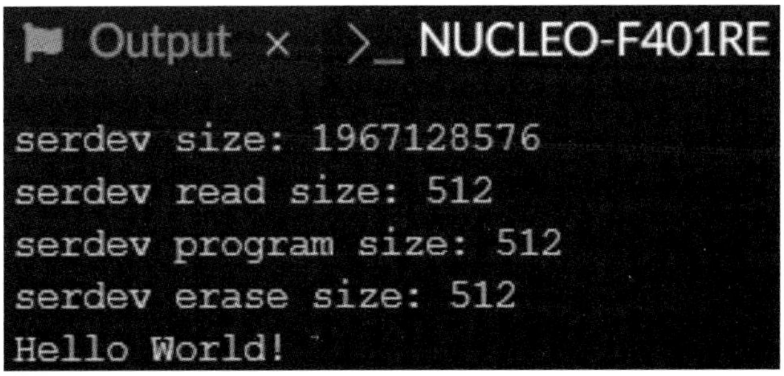

Figure 10.5
Output from Program Example 10.5, using 2-Mbyte SD card.

10.4.2 Creating Files with the FAT File System

The previous program example examined the size of the SD card and wrote and read blocks of data. It did not, however, create a file, so you couldn't easily read the card in a card reader on your PC. We now apply the FAT file system to the SD card, making use of a few of the API functions shown in Table 10.6.

Table 10.6: API summary for FATFileSystem.

Functions	Usage
Constructor	
`FATFileSystem (const char* name, BlockDevice* bd)`	Creates a file system called *name,* on block device *bd.*
Member Functions	
`virtual int mount (BlockDevice *bd)`	Mount a file system to a block device; returns 0 on success, negative error code on failure.
`virtual int reformat (BlockDevice *bd, int allocation_unit)`	Reformat a file system. Results in an empty and mounted file system; *allocation_unit* is number of bytes in a cluster; returns 0 on success, negative error code on failure.
`virtual int unmount( )`	Unmount a file system from block device; returns 0 on success, negative error code on failure.
`virtual int mkdir(const char *path, mode_t mode)`	Create a directory in the file system; returns 0 on success, negative error code on failure.

Program Example 10.6 performs effectively the same function as Program Example 10.4, but this time on the SD card. Therefore, most lines of code are the same, and—to save space—you are asked to copy certain sections from the earlier program. Distinctive features within **main( )** are the functions to mount the file system (i.e., to locate it on the SD card and make it accessible), and to reformat it if mounting fails. Again, the **Mbed_app.json** file is essential, adjusted for the target board in use, as this both identifies the SD card as the target block device and indicates pin connections.

```
/* Program Example 10.6 Using FATfile system to write create a file on the SD card
Works on LPC1768 and Nucleo F401RE, with pin connections shown
*/

#include "mbed.h"
#include <stdio.h>
#include <errno.h>

#include "BlockDevice.h"
#include "FATFileSystem.h"

//Take the system's default block device (from json file)
BlockDevice *bd = BlockDevice::get_default_instance();
FATFileSystem fs("fs");

InterruptIn button(BUTTON1);       //Interrupt input, use p9 for PC1768
DigitalOut led1(LED1);             //digital out to onboard LED1
Timer main_timer;                  //define elapsed time timer
bool button_pressed;               //flags button has been pressed

void toggle(void);                 //function prototype

using namespace std::chrono;

int main() {
  printf("--- Let's log some data ---\n");
  set_time(1666697756); // Set RTC time (this is Tues, 25 Oct 2022 11:35:56)
  button.rise(&toggle); // attach the toggle function to the rising edge
  button_pressed = 0;
  main_timer.start();
  // Try to mount the filesystem
printf("Mounting the filesystem");
fflush(stdout);
int err = fs.mount(bd); //non-zero return indicates mount failure
printf("%s\n", (err ? "Fail :(" : "OK")); //Print OK, or Fail :( on value of err
if (err) {
  //Reformat if we can't mount the filesystem
  printf("formatting  ");
  fflush(stdout);
  err = fs.reformat(bd);
  printf("%s\n", (err ? "Fail :(" : "OK"));
  if (err) {
```

```
        error("error: %s (%d)\n", strerror(-err), err);
    }
  }
  while(true){
  ***Use same while(true) loop as Program Example 10.4***

  }

  //this is ISR
  ***Use same ISR as Program Example 10.4***
```

(A)

```
{
  "target_overrides":{
    "*": {
        "target.components_add" : ["SD"],
        "sd.SPI_MOSI" : "D4",
        "sd.SPI_MISO" : "D5",
        "sd.SPI_CLK" : "D3",
        "sd.SPI_CS" : "D6"
    }
  }
}
```

replace D4,D5,D3,D6 with p5,p6,p7,p8 for LPC1768

(B)

Program Example 10.6 Using the FATFileSystem on an SD card: (a) the main.cpp program; (b) the Mbed_app.json file

Try running this program example. Use the SD card connections of Table 10.5; for the Mbed LPC1768 add an external pushbutton linked to pin 9, following the pattern of Fig. 6.3. Press the button a few times, at timed intervals. Then transfer the SD card to a USB card reader, and view the **log.txt** file on your PC. Verify that the correct elapsed time and LED state have been written to the file.

■ Exercise 10.6

Repeat Exercise 10.5, but applying one or more of the suggested enhancements to Program Example 10.6.

■

10.5 Mini Project: Accelerometer Data Logging on Exceeding Threshold

This project develops from the mini-project in Section 7.11. Create a program which records the acceleration profile encountered in a simple vibrating cantilever, and plots the acceleration data as shown in Fig. 10.6. Record the data on an SD card. Implement the project with the following specifications:

1. Use an LPC1768 or Nucleo F401RE target board.
2. Attach an SPI accelerometer to a simple plastic cantilever with a flying lead to the Mbed target.
3. Program the accelerometer to cause an interrupt trigger when a certain acceleration is encountered.
4. Create an interrupt routine which triggers the recording of 100 data samples to a file on the external memory device. You may wish to use the **fprintf( )** function to format accelerometer data in the text file.
5. The resultant text file can be opened using an SD card reader plugged into a PC. This can be transferred to a spreadsheet program, for example Microsoft Excel, to plot the recorded acceleration waveform.

Note: to achieve a suitable sample period, you may want to apply an Mbed Ticker or Timer to ensure that the accelerometer logs data regularly. A sampling frequency of around 50 Hz should be sufficient to record a detailed acceleration waveform.

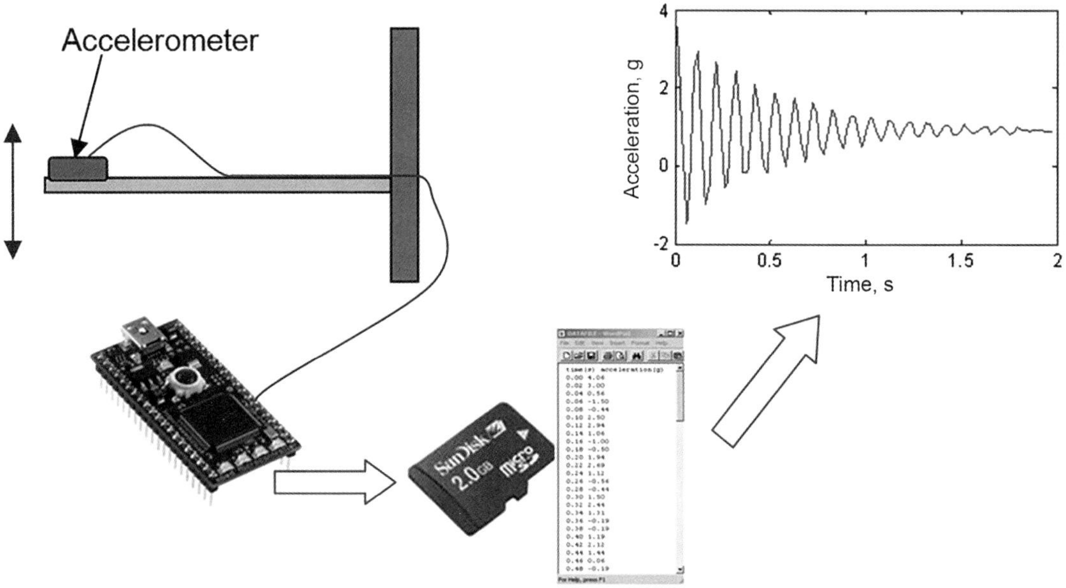

Figure 10.6
Accelerometer data-logging mini-project.

Chapter Review

- Microcontrollers use memory for holding the program code (program memory) and the working data (data memory) in an embedded system.
- Volatile memory loses its data once power is removed, whereas non-volatile memory can retain data with no power. A number of different technologies are used to realize these memory types, including SRAM and DRAM (volatile) and EEPROM and Flash (non-volatile).
- The modern microcontroller has large on-chip blocks of both Flash memory and SRAM. These are organized into a single memory map, along with the microcontroller control registers.
- Pointers point to memory locations to give a means of access to the data stored at the pointed location; they provide easy access to arrays, strings, and other data structures.
- The **stdio** C library contains functions that allow us to create, open, and close files, as well as read data from and write data to files.
- The Mbed operating system has a wide selection of APIs used to access memory and create files and filing systems. These include the local file system for the Mbed LPC1768 and the widely used FAT file system.
- An external SD memory card can be interfaced with an Mbed target to provide extended memory capability.

Quiz

1. What does the term *bistable* mean?
2. How many bistables would you expect to find in the LPC1768's SRAM?
3. What are the fundamental similarities and differences between SRAM and DRAM type memory?
4. What are the fundamental similarities and differences between EEPROM and Flash type memory?
5. A block of memory seen in the memory map of a microcontroller has a start address of 0x8000 0000 and end address of 0x9FFF FFFF.
    a. How many address lines are required to address this memory?
    b. How many memory locations are within this block?
6. Describe the purpose of pointers and explain how they are used to access the different elements of a data array.
7. What C/C++ function would open a text file for adding additional text to the end of the current file?
8. What C/C++ function should be used to open a text file called "data.txt" and read the 12th character?

9. An automatic weather monitoring station records air temperature, soil temperature, light intensity, wind speed, and wind direction once per minute. Each is recorded as a 16-bit number, along with a time and date stamp which occupies 6 bytes. The system must hold up to 90 days of data before it is downloaded. What memory capacity is required?
10. Give one reason that pointers are used for direct manipulation of memory data.
11. Write the C/C++ code that defines an empty five-element array called **dataarray** and a pointer called **datapointer** that is assigned to the first memory address of the data array.

References

10.1. C++ stdio reference. https://cplusplus.com/reference/cstdio/.
10.2. Storage APIs (Mbed OS). https://os.Mbed.com/docs/Mbed-os/v6.15/apis/data-storage.html.
10.3. SD Association. Notice of SD Simplified Specifications. https://www.sdcard.org/downloads/pls/.
10.4. Mbed OS: SDBlockDevice. https://os.Mbed.com/docs/Mbed-os/v6.15/apis/sdblockdevice.html.

PART 2

Wireless and the Internet of Things

CHAPTER 11

Wireless Communication

11.1 Introducing Wireless Data Communication

For many years, flying radio-controlled model aircraft has been a popular pastime. The aircraft "pilot" holds the radio-control unit, which sends simple data-control messages to the plane soaring above. The control unit is configured to transmit at a particular radio frequency. The pilot may be displaying a colored tag showing on which frequency the radio is operating. If someone else arrives and wants to fly a plane at the same place, then they must set their radio to a different frequency, or there will be interference between the two, possibly leading to a spectacular plane crash. If more people come, then even greater care must be taken to make sure that each has a unique radio frequency. Contrast this simple image of radio data communication with a more recent setting. Picture instead a busy airport lounge: hundreds of travellers are waiting for their flights. Many are on their mobile phones; others have laptops or tablets in use, maybe with wireless-linked mice, keyboards, or headphones. Potentially thousands of data messages are flying through the air. All must get through reliably; none should interfere with any other.

This chapter explores some of the issues and mysteries relating to wireless data communications. It is a big topic, so we focus mainly on two important and well-known protocols for shorter-range communication, Bluetooth LE and Zigbee; we close with a mention of LoRaWAN, for longer range. We start with a very swift review of some important aspects of wireless communication. There is much of the theoretical background that cannot be covered, so do check Reference 11.1 for further information on any topic where you want to see the detail.

11.1.1 Some Wireless Preliminaries

The traditional way of transferring data between devices or subsystems has been through electrical connections: wires, cables, or PCB (printed circuit board) tracks. Yet this physical connection is in many situations inconvenient, annoying, or just plain impossible to implement. This is particularly true for applications where things are distant from each other, or need to move around, or engage and disengage. It is, of course, possible to make a data connection without any physical link. Alternatives to wired connection are very familiar, and include infrared, radio, and visible light. All are examples of electromagnetic

Fast and Effective Embedded Systems Design. https://doi.org/10.1016/B978-0-323-95197-5.00011-5

waves; they can be found in the electromagnetic spectrum, seen in Fig. 11.1. This shows a very wide range of frequencies, where frequency and wavelength are related by Formula 11.1. Here f is the frequency (in Hz), λ is the wavelength (in meters), and c is the speed of light, known to be approximately 3 x 10^8 m/s.

$$c = f\lambda \tag{11.1}$$

The spectrum usefully shows where various wireless activities sit. Any frequency on the spectrum, from the lowest frequency up to visible light, can be used for data communication. Almost all of this is very strictly regulated by national and international agencies—you cannot just start broadcasting a radio signal at any frequency you like! As more and more wireless activity has come along, the spectrum has become increasingly crowded. The International Telecommunication Union, a United Nations agency, manages the allocation of the radio spectrum between different broadcasters and applications. It reserves certain frequency bands for industrial, scientific, and medical (ISM) applications. These bands are *unlicensed*, and some vary between countries. The 2.4 GHz band is, however, reserved for unlicensed use in all regions. It has become widely used, mainly for short-range, low-power applications.

While the spectrum represents the "pure" radio frequencies, they only become useful once they are carrying information. This is done by the process of *modulation*; the information to be carried is imprinted onto the carrier frequency through one of a number of different techniques. Amplitude modulation (AM) and frequency modulation (FM) are the old favorites of the broadcast industry. One effect of modulation is to cause fluctuations around the base frequency; thus, if we say that a certain radio station can be found at the frequency of 103 MHz, in fact it is in a narrow band of frequencies centered on 103 MHz. The word *bandwidth* is used to define a range of frequencies within which a particular transmission may be taking place.

The relationship between the information a signal can convey and its bandwidth is direct—higher data rates demand wider bandwidth. In general, higher-frequency bands

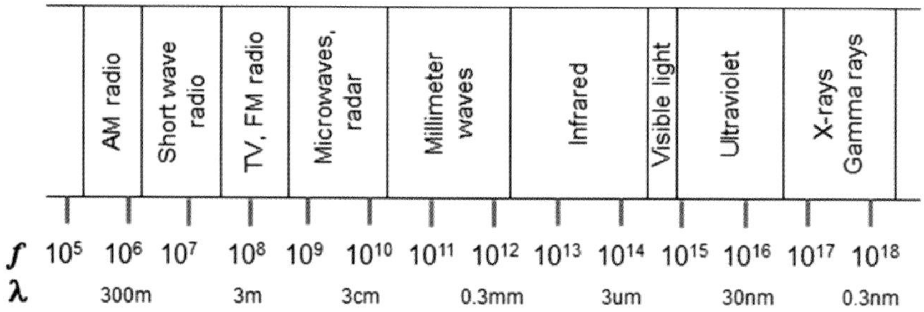

Figure 11.1
The electromagnetic spectrum.

offer more channels and bandwidth, so they are used to provide larger networks and higher data throughput. However, lower frequencies propagate better than higher frequencies, so they can achieve better ranges, for example, for neighborhoods and within buildings.

The basics of a simple radio link are shown in Fig. 11.2. Data is acquired, conditioned, coded, and transmitted by the radio transceiver. The data is used in some way to modulate the carrier frequency. This is implied in the waveform at the center of the figure. In this example, the amplitude of a carrier wave is modulated by the signal; this is hence an example of AM. The electrical signal produced flows to the antenna, which causes an electromagnetic wave of the same frequency to be radiated. The range of this radio signal will depend on many things, including the power of the transmitter, the efficiency of the antenna, and the signal frequency itself. If there is another antenna within range and of appropriate dimension, then the signal can be received and detected by a receiver circuit. Of course, if another radio is transmitting at the same frequency in the same range, then the signals will interfere, and confusion will reign.

The performance and efficiency of the antenna at the frequency of operation are determined by its physical shape and size. The length of the antenna should be a multiple or fraction of the wavelength. One of the big challenges in low-cost and small-size wireless communication has been to scale the antenna to the rest of the product, whether that is a mobile phone or a wearable health monitor. In doing so, trade-offs are made between physical size and efficiency. For example, a wireless computer mouse transmitting over a short distance with a low data rate can operate with a less-efficient antenna than a high data-rate link operating over a greater distance. With new and miniature devices, it is now common to see antennas formed from a trace of PCB track, within a chip, or a tiny wire antenna.

Returning to our comparison of the radio-controlled aircraft and the airport lounge full of people on mobile devices—do these people run around and agree between themselves who is going to transmit on which frequency? This of course is absurd and impossible. One technique applied is the use of *spread spectrum* transmission. In this strategy, the

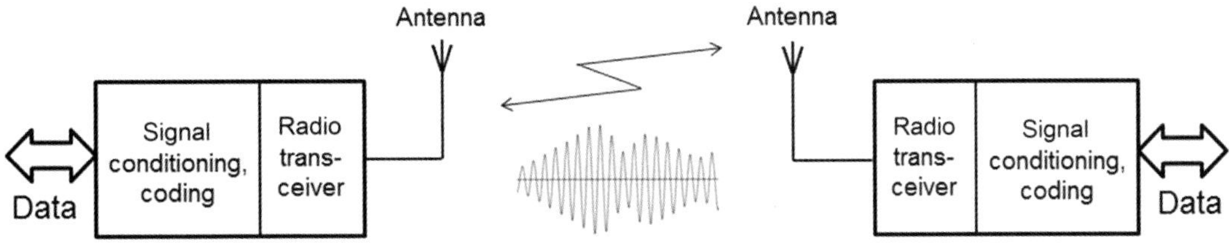

Figure 11.2
A basic radio data link.

transmitting frequency keeps changing, or hopping, within a certain bandwidth. The receiver needs to know the transmitter's frequency hopping pattern in order to receive the signal properly. This technique was originally applied in secure systems—if you can't keep track of the frequency hopping, then you can't snoop on the signal. However, it's also useful when space and bandwidth are shared. Suppose several data links are all in action close to each other. If two are at the same continuous frequency, then we know they will interfere. However, if they are continuously switching frequency—having first agreed within their pairs or networks on a frequency-hopping pattern—then if they do occasionally clash, this can be detected and corrected. Most of the time the probability is that different networks will select different frequencies and will not interfere.

11.1.2 Wireless Networks

There are many situations where we need to provide connections between different systems or subsystems. In the domestic environment, the automated household is now a reality—different household appliances and gadgets are increasingly being connected together, and to the wider internet in the Internet of Things (more on this in the next chapters). Elsewhere, there are other needs for networking. The modern motor vehicle may contain dozens of embedded systems, all engaged in very specific activity, but all interconnected. In the home or car situation, connections may be long-term and stable. However, other networks or connections are transitory—for example, when data is downloaded from a smartphone to a laptop over a wireless link.

Providing a network is about much more than just providing connectivity, as important as this is. In a complex system it is also essential to deal in depth with how data is formatted and interpreted, how addressing is achieved, and how error correction can be implemented. All of this is independent of the physical interconnection itself. For different nodes to communicate on a network, there must be very clear rules about how they create and interpret messages. We have already seen aspects of this with definitions of standards like I^2C and USB. This set of rules is called a protocol, taking the word from its diplomatic and legal origins.

To establish terminology, networks are sometimes divided into four categories, as shown in Fig. 11.3. The personal area network (PAN) usually relates to devices worn on the person, such as smart watches, personal entertainment, or health or performance monitoring. The local area network (LAN) typically applies to a single building, for example, a network of computers in a home, company, school, or office. The neighborhood area network (NAN) reflects a need to network in a wider area still, such as a smart transport or smart energy system; this brings us towards the realization of the smart city. The wide area network (WAN) effectively includes national or global systems, most notably the Internet. Each has distinct demands, based both on technical issues and

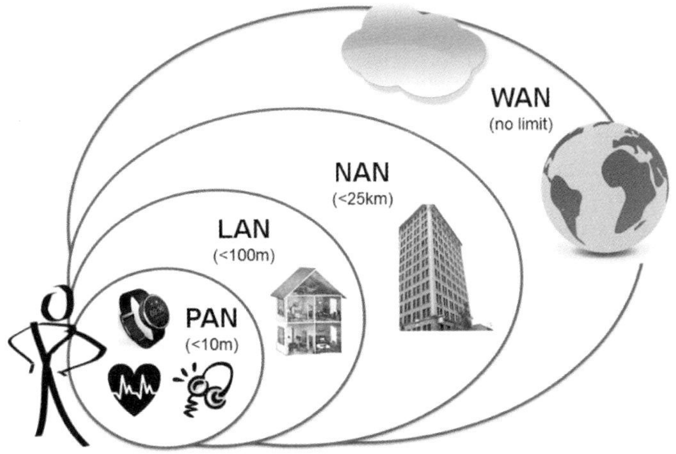

PAN: Personal area network LAN: Local area network

NAN: Neighbourhood area network WAN: Wide area network

Figure 11.3
Network ranges.

on the type of data generated. Of course, as is often the case, one type of network can link to another. Internet communications on wireless LANs/WANs will be introduced in Chapter 13; this chapter will predominantly discuss wireless connectivity over short ranges, applicable to a PAN or possibly LAN setup, but also will introduce long-range low-power wireless communication, which has seen a recent surge of development.

11.1.3 A Word on Protocols

With large networked systems, protocols can become very complicated, defining every aspect of the communication link. Some of these aspects are obvious, but others are not. To aid in the process of defining a protocol, the International Organization for Standardization (ISO) devised a "protocol for protocols," called the open systems interconnect (OSI) model. This is shown in Fig. 11.4. Each layer of the OSI model provides a defined set of services to the layer above, and each depends on the services of the layer below. The lowest three layers depend on the network itself and are sometimes called the media layers. The physical layer defines the physical and electrical link, specifying, for example, what sort of connector is used and how the data is represented electrically. The link layer is meant to provide reliable data flow and includes activities

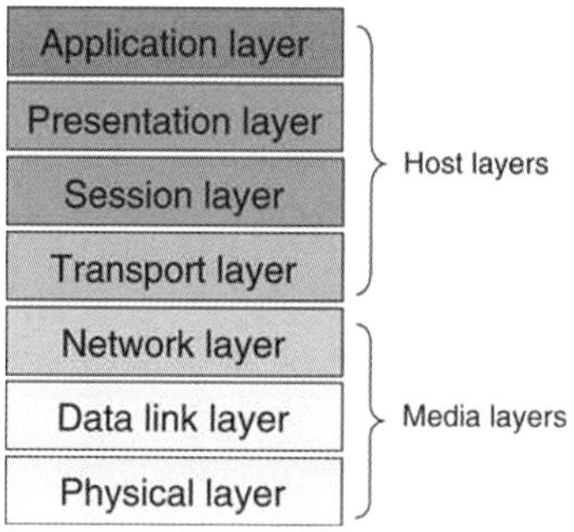

Figure 11.4
The ISO OSI model.

such as error checking and correcting. The network layer places the data within the context of the network and includes activities such as node addressing.

The upper layers of the OSI model are all implemented in software, although some recent innovations have introduced hardware acceleration that allows "offloading" of network processes. This takes place on a host computer, and these layers are called the *host layers*. The software implementation is often called a *protocol stack*. For a given protocol and hardware environment, it can be supplied as a standard software package. A designer adopting a protocol stack may need to interface with it at the bottom end, providing physical interconnection, and at the top end, providing a software interface for the application.

This model forms a framework against which new protocols can be defined and a useful point of reference when studying the various protocols already available. In practice, any one protocol is unlikely to prescribe for every layer of the OSI model, or it may only follow it in an approximate way.

The IEEE (the Institute of Electrical and Electronic Engineers) plays a major role in defining standards and protocols. Unsurprisingly, it is very active in the field of networked communications and maintains a set of standards for LANs, allocated the number 802. A small number of these which are relevant to this and the next chapter are shown in Table 11.1. More can be found in Reference 11.2.

11.2 Bluetooth Low Energy

In this section we look at the short-range Bluetooth low-energy (BLE) protocol, which is a ubiquitous wireless communication technology found in a wide variety of embedded systems. There is a lot of support for BLE in hardware ecosystems, and the various programming interfaces and applications available make it easy to get going.

Table 11.1: Example IEEE 802 working groups.

IEEE Working Group	Description
802.3	Ethernet
802.11	Wireless LAN, including Wi-Fi
802.15	Wireless PAN
802.15.1	Bluetooth
802.15.3	High-rate wireless PAN
802.15.4	Low-rate wireless PAN, e.g., Zigbee

11.2.1 Classic Bluetooth

Bluetooth is a digital radio protocol, intended primarily for PAN applications, and operates in the 2.4-GHz radio band. It was developed by the Swedish phone company Ericsson, which took the name of a tenth-century Viking king to name their communication protocol. Bluetooth provides wireless data links between such devices as mobile phones, wireless audio headsets, computer interface devices like mice and keyboards, and systems requiring the use of remote sensors. Bluetooth standards are now controlled by the Bluetooth Special Interest Group (Reference 11.3), though the IEEE has made an important contribution. It is a formidably complex protocol. There is a core specification (the version 5.2 specification sitting at over 3000 pages!), and then over 40 different *profiles*, which specify different Bluetooth applications, such as for audio, printers, file transfer, and simple cable replacement.

In this chapter, however, we will be focusing on BLE, which takes inspiration from the classic Bluetooth protocol, but is aimed at (as the name suggests) tiny, embedded devices that are power-constrained. The BLE protocol is still designed and maintained by the Bluetooth Special Interest Group, but it is not compatible with the classic Bluetooth protocol, although the two can co-exist on the same device.

11.2.2 Bluetooth Low Energy

BLE was developed as a response to the emerging low-energy device market, which includes applications such as fitness trackers, sport, healthcare, and smart homes. It first appeared in the Bluetooth 4.0 standard in 2011 and has been extended and improved over the years. Classic Bluetooth already boasted a relatively low power profile, with power requirements sitting around the 1 W mark. However, BLE takes this to the extreme, with power requirements in the 0.01 W to 0.5 W range.

A wide range of factors can affect transmission range, including physical constraints and the device vendor's implementation, but typically BLE can achieve ranges from 10 m to 100 m. To achieve greater distances, more power is required, so the trade-off is power versus transmission range.

The specification dictates a transmission rate of one symbol per microsecond, and with one bit encoded per symbol, this gives an over-the-air data rate of 1 Mbps. However, there are some key constraints that affect the actual achievable throughput. Notably, there is an inter-frame delay of 150 μs, and every packet transmitted must be acknowledged by the receiver. An acknowledgement packet is 80 bits, and therefore takes 80 μs to send. Additionally, the upper-level protocols add overhead to the data packets in the form of "headers," which take up space in the packet, limiting the maximum amount of data that can be transmitted per frame to 27 bytes.

Taking these factors into account, sending 27 bytes of data will add 14 bytes of overhead (from the packet headers), giving a total of 41 bytes in the frame, and taking a total of 328 μs (41 bytes $\times$ 8 bits) to send. Adding to these the inter-frame delay of 150 μs, the acknowledgment packet of 80 μs, and further inter-frame delay of 150 μs means it will take 708 μs to send 27 bytes of useful data, giving an effective data rate of around 305 kbps.

Although the above example is based on a data rate of 1 Mbps, the specification allows several different bit rates, namely 125 kbps, 500 kbps, 1 Mbps, and 2 Mbps.

A BLE system comprises so-called "central" devices and "peripheral" devices. A central device can connect to many peripheral devices, forming a piconet, and a peripheral device can be connected to more than one central device, effectively connecting multiple piconets together and forming a scatternet. This topology is illustrated in Fig. 11.5.

BLE uses the same radio frequency as classic Bluetooth (2.4-GHz ISM band), but employs 40 2-MHz channels, where classic Bluetooth employs 79 1-MHz channels. Fig. 11.6 shows how the channel numbers are allocated. Note that 37, 38, and 39 appear out of order—these channels are reserved for advertisement data packets, explained in the next section. The specification dedicates three channels for this purpose to help reduce interference. Advertisement channels are selected arbitrarily by the advertiser.

11.2.3 Establishing a Connection

The Generic Access Profile (GAP) specification dictates how BLE devices can discover each other and establish connections.

BLE devices that want to be found ("advertisers") periodically broadcast an "advertisement" packet, to let other devices in range ("scanners") know of their presence.

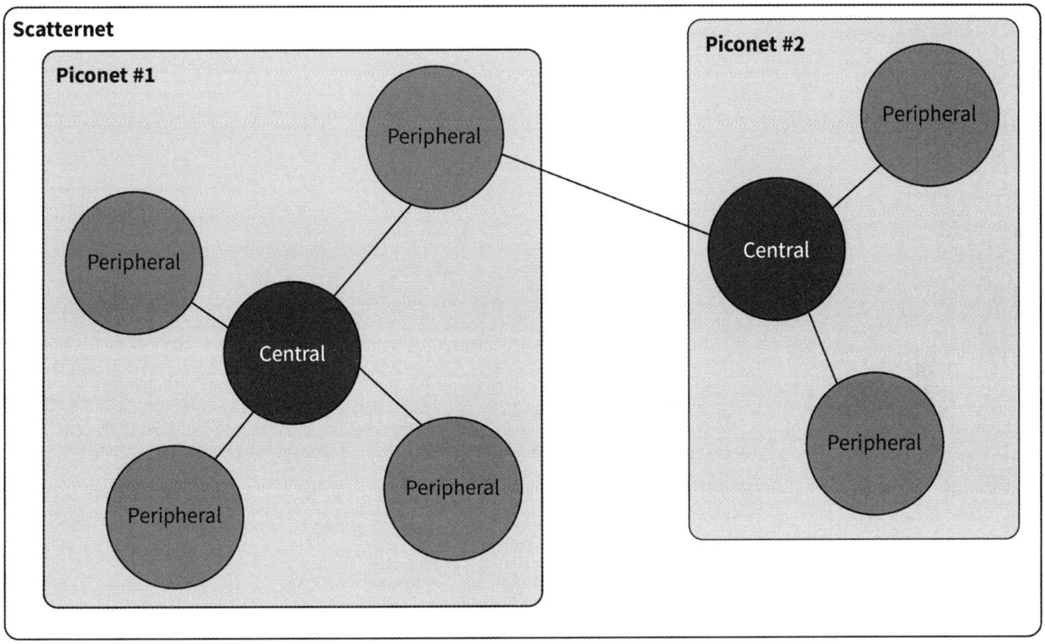

Figure 11.5
BLE network topology.

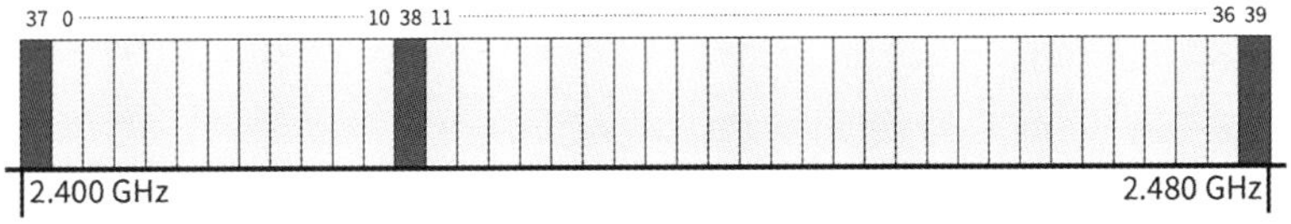

Figure 11.6
BLE channels in the frequency domain.

The period of broadcast is called the "advertising interval," and a random delay of up to 10 ms is added to avoid interference between multiple advertisers.

The actual connection process between two devices takes place in three phases. These phases are outlined in Fig. 11.7.

Phase 1 – Discovery:

During this phase, a scanner looks for advertisers by monitoring the three advertisement frequencies and looking for advertisement packets. The scanning window dictates how long the scanner waits for an advertisement, and the scanning interval dictates how often it initiates a scan.

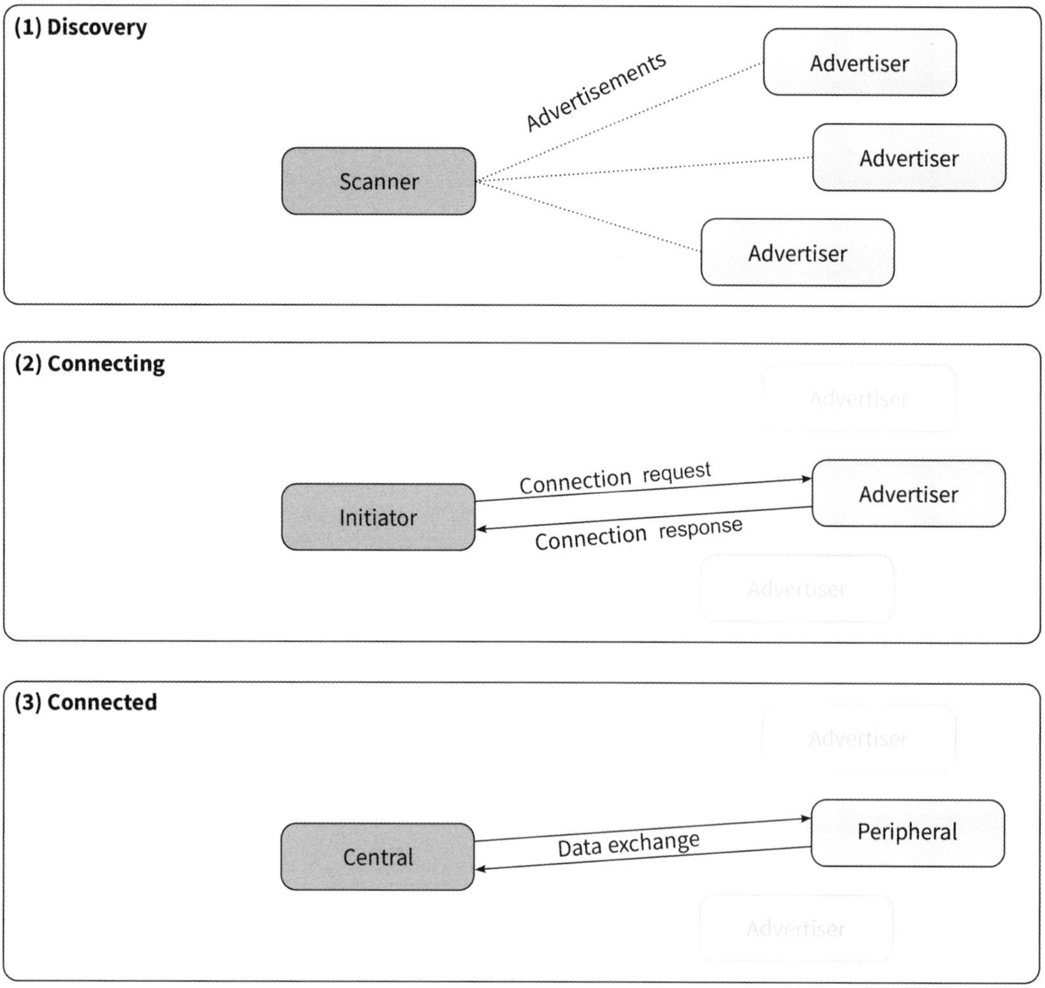

Figure 11.7
BLE connection procedure.

Phase 2 – Connecting:

When the scanner decides it wishes to connect to a particular device, it becomes the "initiator" and sends a connection request to the advertiser. If the advertiser is agreeable to the connection request, it sends back a connection response.

Phase 3 – Connected:

Upon agreement, the connection is established, the initiator assumes the "central" role, and the advertiser assumes the "peripheral" role. Data can now be exchanged between the two devices.

11.2.4 Exchanging Data

While the GAP specification defines how to establish a link between devices, the generic attribute (GATT) profile specification is all about exchanging data, including the type and meaning of that data. We will be using the GATT profile in the forthcoming examples to build a system for transferring environmental sensor data.

The GATT profile is used to describe one or more services offered by a device—such as a "blood pressure monitoring service" or a "device information service." A service is a collection of characteristics that represent the information offered and accepted by that service. This structure is shown in Fig. 11.8.

Characteristics are attribute types that have a name, a uniform type identifier, and an assigned universally unique identifier (UUID) number. These semantics are defined in the GATT Specification Supplement, available from the Bluetooth Assigned Numbers data. A characteristic offered by a service can be readable, writable, and/or notifiable. A notifiable characteristic tells connected devices when the value of the characteristic changes—for example, a temperature characteristic might indicate a new temperature value when it has changed.

11.2.5 BLE on the ST IoT Discovery Board

We will now revisit the ST IoT Discovery board, as introduced in Section 2.4 and demonstrated in Section 7.10. We have already seen that it carries an interesting range of sensors, and we will now explore its wireless connectivity options, which are summarized in Table 11.2. In this chapter, we will be using the Bluetooth V4.1 module, which implements BLE.

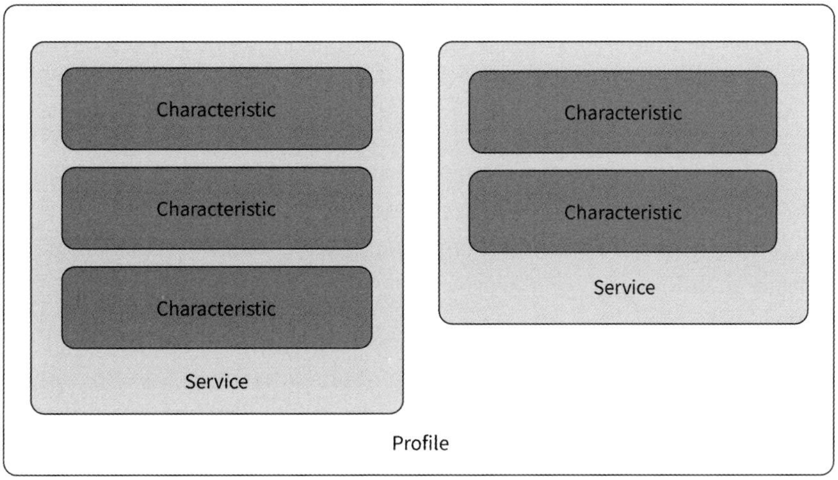

Figure 11.8
Relationship between BLE profiles, services, and characteristics.

Table 11.2: Communication modules on the ST IoT Discovery Board.

Module	Type
Bluetooth V4.1 module	SBTLE-RF
Sub-GHz low-power programmable RF module	SPSGRF-868 or SPSGRF-915
802.11 b/g/n Wi-Fi module	ISM43362-M3G-L44
Dynamic NFC tag	M24SR

Detailed information is available on the Mbed OS BLE APIs in Reference 11.4, along with a range of examples in Reference 11.5. For the following examples, you will require the ST IoT Discovery board and a smart phone with a BLE scanning app. A popular choice is "nRF Connect," used in the example below. Other BLE scanning apps are also available; for iPhone consider "BLE Scanner."

11.2.6 Simple BLE: Sending Data to a Smart Phone

C/C++ code feature

In this example, we will see how to initialize the BLE service on the IoT Discovery board and use it periodically to transmit sensor data to any interested device. By the end, you should have a simple working example, which you can build on for future projects. It is worth mentioning at this stage that the complexity of the C++ code we are using takes a step up here, as we engage with the BLE APIs, and more closely with the creation and use of classes for various applications. This goes beyond the outline support we can offer in Appendix B, though its references remain very useful. Don't worry if you don't grasp some of the more advanced programming. We will break the code down into chunks and tackle each BLE concept separately, and you will see how the various code modules can be applied.

You should start by creating a new Mbed OS project, and include the board support files for the ST IoT Discovery board from Reference 7.9 (as we did for Program Example 7.12) and the BLE utility library from Reference 11.6.

Starting from an empty **main.cpp** program file, we need to include some header files and create the required infrastructure for interfacing with the Bluetooth library. Program Example 11.1 shows how to begin. It contains the necessary **#include** statements to bring in the appropriate definitions.

The only class in this code section, **ServerProcess**, manages the interface to the BLE API. It can run multiple BLE-related tasks, so that communications can operate asynchronously. It extends the **BLEProcess** superclass and allows the BLE subsystem to run the various BLE activities (e.g., advertisement and data transmission) on an event queue.

```
#include "mbed.h"
#include "ble/BLE.h"
#include "ble_process.h"
#include "stm32l475e_iot01.h"
#include "stm32l475e_iot01_hsensor.h"
#include "stm32l475e_iot01_psensor.h"
#include "stm32l475e_iot01_tsensor.h"
#include <events/mbed_events.h>

// Declare the ServerProcess class
class ServerProcess : public BLEProcess {
public:
  ServerProcess(events::EventQueue &event_queue, BLE &ble_interface)
    : BLEProcess(event_queue, ble_interface) {}

  // Return a friendly name for the BLE device
  const char *get_device_name() override {
    static const char name[] = "Sensor Server";
    return name;
  }
};
```

Program Example 11.1: ServerProcess, a class for running multiple BLE activities

Next, we add a class that will manage the BLE service and characteristics that we want to support and expose on the BLE link. Program Example 11.2 gives an example of a Server class that implements the necessary routines for advertising an "Environmental Service."

```
class Server : ble::GattServer::EventHandler {
public:
  void start(BLE &ble, events::EventQueue &event_queue) {
    // The UUID of the service we wish to advertise.
    const UUID uuid = GattService::UUID_ENVIRONMENTAL_SERVICE;

    // We will fill this array in later to expose sensor data.
    GattCharacteristic *charTable[] = {};

    // Initialise and register the service object.
    GattService sensorService(uuid, charTable,
                              sizeof(charTable) / sizeof(charTable[0]));

    ble.gattServer().addService(sensorService);
    ble.gattServer().setEventHandler(this);

    printf("Service started.\n");
  }
};
```

Program Example 11.2: A Server class that specifies which BLE services and characteristics are available

This Server class creates a GATT service of the "Environmental" type and registers it with the BLE subsystem. At this point, we have just about enough code to advertise our presence over BLE, and detect it on a smart phone—so let's try it out by initializing everything in the **main( )** function, as shown in Program Example 11.3.

```
// Declare an event queue for dealing with various BLE activities.
static EventQueue event_queue(10 * EVENTS_EVENT_SIZE);

// This function initialises the sensors we want to use.
static void initialiseSensors() {
  BSP_TSENSOR_Init();
  BSP_HSENSOR_Init();
  BSP_PSENSOR_Init();
}

int main() {
  // Acquire a reference to the BLE management object.
  BLE &ble = BLE::Instance();

  // Initialise the on-board sensors
  initialiseSensors();

  // Create the process that will handle all BLE events.
  ServerProcess ble_process(event_queue, ble);
  Server server;
  ble_process.on_init(callback(&server, &Server::start));

  // Start the BLE process.
  ble_process.start();
  return 0;
}
```

Program Example 11.3: Main function that initializes sensors and starts the BLE subsystem

The **main( )** function initializes the sensors we are going to use, and then starts the various BLE processes. Once this code is in place, you should be able to build and run this application on the board. When the board is running, if you open the BLE scanning app on your mobile device, you should see something similar to the display in Fig. 11.9. You will probably see more devices that are broadcasting in the range—especially if you are in a busy area—but crucially, you should see "Sensor Server," which is the name of the BLE device defined in Program Example 11.1.

If everything went as planned, your device should be advertising, and you should see it in the scanner app. But we are not actually sending any data over the connection yet—at this point,

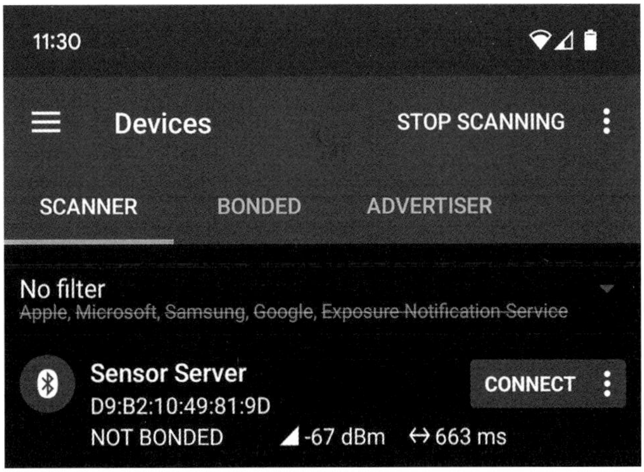

Figure 11.9
A BLE scanning app on a mobile phone showing the IoT discovery board advertising its presence.

we are just checking that we have the BLE subsystem working. To proceed, we will initialize a few of the on-board sensors and then create BLE characteristics for those sensors. This requires us to create another class that represents a characteristic for hosting a sensor value.

```
template <typename T>
class SensorReadingCharacteristic : public GattCharacteristic {
public:
  SensorReadingCharacteristic(const UUID &uuid)
    : GattCharacteristic(
        uuid, (uint8_t *)&internalValue_, sizeof(internalValue_),
        sizeof(internalValue_),
        GattCharacteristic::BLE_GATT_CHAR_PROPERTIES_READ |
                  GattCharacteristic::BLE_GATT_CHAR_PROPERTIES_NOTIFY),
    internalValue_(0) {}

// Called to update the value being broadcasted by this characteristic.
void updateValue(BLE &ble, T newValue) {
  // Update the internal value
  internalValue_ = newValue;
  // Trigger a value change in the BLE subsystem.
  ble.gattServer().write(getValueHandle(), (uint8_t *)&internalValue_,
                         sizeof(internalValue_));
}

private:
  T internalValue_;
};
```

Program Example 11.4: A class to represent a sensor reading characteristic

Program Example 11.4 defines a class, **SensorReadingCharacteristic,** to help track the current value of a sensor, and to let this value be represented as a BLE characteristic. In our Serverclass, which we are about to modify, we will construct the **SensorReadingCharacteristic** for each sensor, allowing other devices to read the value and be notified when the value changes. This lets us update the value repeatedly, and the client will see those changes when it is connected. In the class, the **internalValue_** field tracks the current value, and calling **updateValue** allows us to record a new sensor value and tell the BLE subsystem that the value has changed.

The **"template<typename T>"** syntax at the beginning of the class lets us represent different number types internally—which we will see in use when constructing instances of this class. So let's do that by modifying the Server class we created previously, applying Program Example 11.5.

```
class Server : ble::GattServer::EventHandler {
public:
  Server()
    : temperature_(GattCharacteristic::UUID_TEMPERATURE_CHAR),
      humidity_(GattCharacteristic::UUID_HUMIDITY_CHAR),
      pressure_(GattCharacteristic::UUID_PRESSURE_CHAR) {}

// ... existing code ... //

private:
  SensorReadingCharacteristic<short> temperature_;
  SensorReadingCharacteristic<short> humidity_;
  SensorReadingCharacteristic<unsigned int> pressure_;
};
```

Program Example 11.5: Modifications to the Server class to construct the sensor characteristics

In Program Example 11.5 we've created three fields in the Serverclass to represent the three characteristics we want to broadcast—one each for temperature, humidity, and pressure. We've also specified the data type that represents the particular characteristic, which is defined in the GATT specification. In this case, both temperature and humidity are signed 16-bit integers (indicated by the "short" datatype), and pressure is an unsigned 32-bit integer (represented by the "unsigned int" datatype). We've also initialized them (in the constructor) with the correct identifiers, so that BLE clients understand how to interpret the values.

Finally, we have to tell our service to advertise these characteristics, by modifying the charTable array in the **start** method:

```
GattCharacteristic *charTable[] = {&temperature_, &humidity_, &pressure_};
```

Now, rebuild and run the application on the board. Check the BLE scanner on your mobile device, and click on "Connect" to connect to "Sensor Server." Select "Environmental Sensing" to expand the list, and you should see something similar to Fig. 11.10 .

We are now seeing the three characteristics that were defined, but they are not showing any values. This is because we haven't yet told the Bluetooth process to update its values with the real sensor data and to actually send the data. To do this, we need to create a routine that updates the sensor data values, and then we need to call that routine periodically (e.g., every second). Add the method in Program Example 11.6 anywhere inside the Server class.

```
void updateSensors(BLE &ble) {
  printf("Updating sensors...\n");
  temperature_.updateValue(ble, BSP_TSENSOR_ReadTemp() * 100);
  humidity_.updateValue(ble, BSP_HSENSOR_ReadHumidity() * 100);
  pressure_.updateValue(ble, BSP_PSENSOR_ReadPressure() * 10);
}
```

Program Example 11.6: Update sensors method that reads sensor data and puts the values into the characteristics

This routine calls **updateValue( )** on each characteristic to provide a new data value for the sensor. You will notice that we have to manipulate the values a little bit (multiplying by 10 or 100), because the GATT specification dictates the resolution of the data being sent, and thus the multiplier that we need to use. The final piece of the puzzle is to modify the **start** method (again, in the **Server** class) to include the following line at the end:

```
event_queue.call_every(1000ms, [this, &ble] { updateSensors(ble); });
```

This triggers **updateSensors( )** to be called every second. Now, you can rebuild and rerun your application, and you should be able to see the sensor values appearing on your mobile device, in the appropriate section, as shown in Fig. 11.11.

Congratulations! You have successfully created a BLE device that periodically broadcasts sensor information to any device that is interested.

■ Exercise 11.1

Using a serial terminal (e.g., Tera Term), monitor the sequence of messages coming from the development board as the program launches and runs. Make sure that you understand the sequence you see.

■

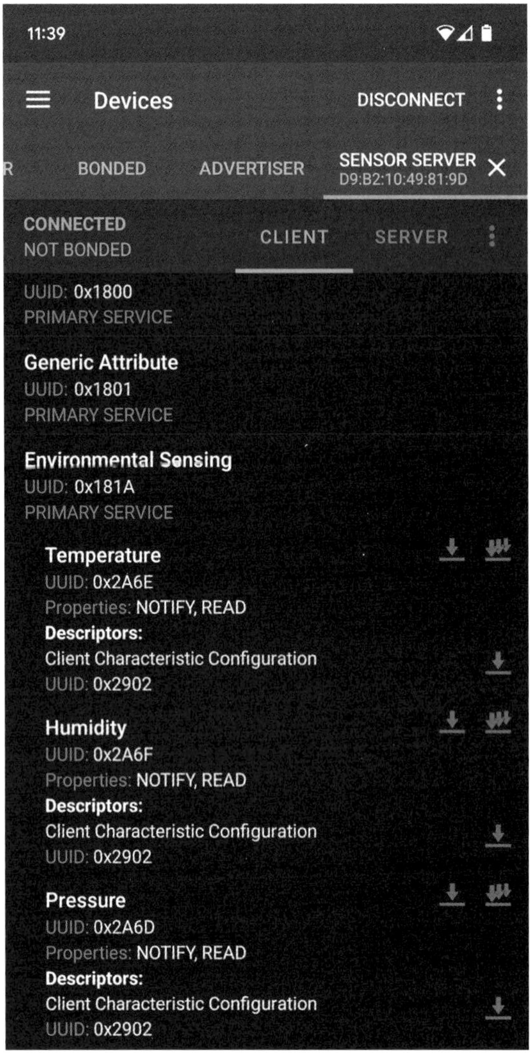

Figure 11.10
Expanded view of "Sensor Server," showing the "Environmental Sensing" service and the three sensor characteristics.

11.2.7 Evaluating BLE

BLE is an exciting technology that enables short-range wireless communication. This has many valuable applications where wires are intrusive, expensive, or difficult/impossible to install. The various software frameworks that are available (especially in the Mbed ecosystem) make interacting with BLE modules much easier than trying to develop software from scratch.

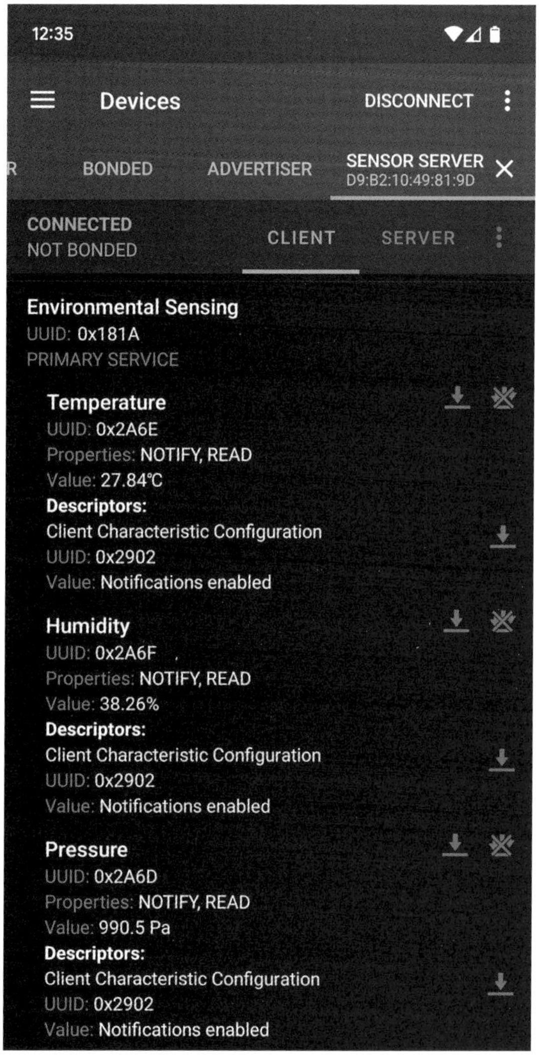

Figure 11.11
Sensor values being displayed in the BLE scanner application.

Like any technology, care should be taken to evaluate if it is suitable for the application, and so it is important to note that BLE has limitations. First, as the name suggests, with lower energy comes a reduced transmission range and a reduced data rate. This may or may not be a problem, as typically embedded systems have limited data to transfer anyway. As packet sizes get bigger, however, the overhead of the BLE stack can quickly add up and increase packet latency. Second, operating in the 2.4-GHz ISM band means that the signal itself is subject to a lot of interference (given the relative "crowding" of this RF band) and that it is subject to physical limitations too. Higher frequencies are absorbed more easily by objects such as walls and the human body.

11.3 Zigbee

11.3.1 Introducing Zigbee

BLE has been very usefully positioned as being of lower power, lower data rate than classic Bluetooth, retaining the ability to form and reform piconets with great flexibility. Hence, it has a very distinct and useful range of applications. However, a set of applications exists where even lower power demand is needed—maybe with continuous operation from a single battery over months or years—and data rates might be very low indeed.

Zigbee is intended for extreme low-power systems, with low-data rates. It applies and builds on the IEEE 802.15.4 Low-Rate WPAN standard (Table 11.1). Like BLE, it operates in the ISM bands of the radio spectrum. The Zigbee protocol is managed by the Connectivity Standards Alliance (Reference 11.7). The Zigbee name refers to the waggle dance of bees when they return to the hive after seeking nectar. (If this name seems completely odd to you, remember this—the bees are themselves little data carriers, communicating through their dance information about the location of pollen they have found.) Zigbee has some similarities to BLE, but aims to be simpler and cheaper, with smaller software overhead and different target applications. Like BLE, Zigbee devices apply spread spectrum communication.

There are three Zigbee device types:

The end device: This is the simplest device, with just enough capability to undertake simple measurement actions and pass back the data. It can only do this to its "parent," the router or coordinator which allows it to join. It is likely to spend a large part of its time in sleep mode. It wakes up briefly, just to confirm it is still part of the network.
The router: After joining a Zigbee network, it can receive and transmit data, and can allow further routers and end devices to join the network. It may need to buffer data if an end device is asleep. It can also exercise a useful function (e.g., measurement). It cannot sleep, so it may need to be mains powered.
The coordinator: This is the most capable Zigbee device; there can only be one in a network. It launches a network by selecting a PAN ID and a channel to communicate over. It can allow routers and end devices to join the network. The coordinator cannot sleep, so it must normally be mains powered.

Zigbee is particularly appropriate for home automation and other measurement and control systems, with the ability to use small, cheap microcontrollers. Data rates vary from 20 kbps (in the 858-MHz band) to 250 kbps (in the 2.4-GHz band). Because of this low data-rate expectation, Zigbee end devices are able to implement sleep modes; hence, power consumption can be minimal.

Once established, each network is defined by a 64-bit (previously 16-bit) PAN ID (personal area network identifier). All Zigbee devices have a 16-bit and a 64-bit address. The latter—sometimes called the extended address—is assigned at manufacture and is unique. However, the device receives a 16-bit address as it joins a network; hence, this address is sometimes called the *network address*. This address is not permanent. If the device left the network and then rejoined, it would probably be assigned a different address. Zigbee data can be sent unicast (from one device to a specific target device) or broadcast (throughout the entire network).

Fig. 11.12 shows example Zigbee networks of increasing complexity. It shows the three device types and how they might connect. The simplest connection is just a pair, of which one must be a coordinator, and the other can be a router or end device. A step up in complexity is the star, with a single coordinator linking to a surrounding set of end devices. Finally, there is an advanced mesh structure, fully exploiting the Zigbee capability. Here routers take on an important role. Think of this diagram installed in a smart multi-story building. The coordinator might be placed in a central location. Each

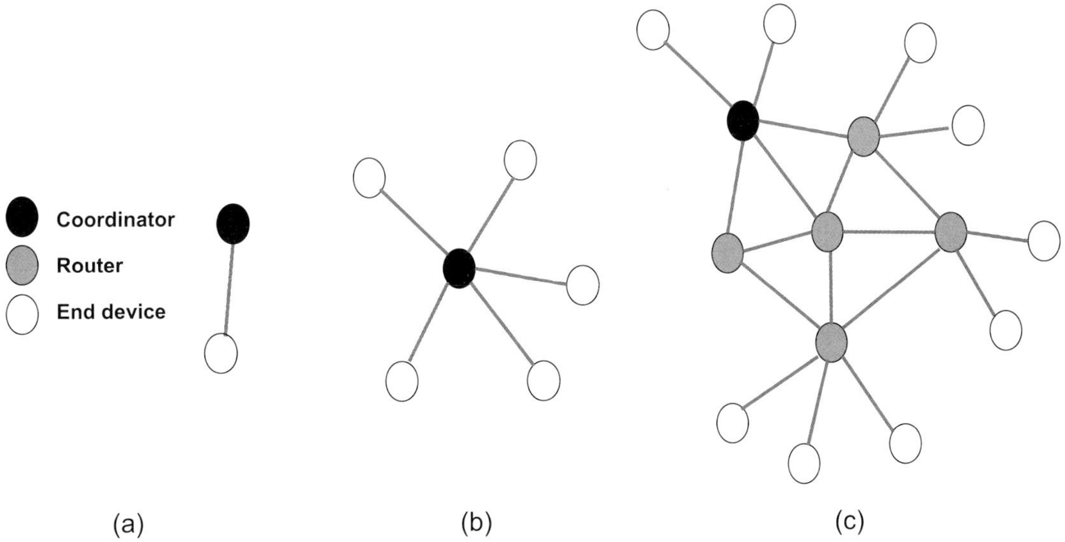

Figure 11.12
Zigbee networks: (a) pair; (b) star; (c) mesh.

floor has a router, and distributed end devices monitor temperature, air quality, and light conditions. Because Zigbee devices can transfer data, a wider physical range is achieved.

11.3.2 Introducing XBee Wireless Modules

The XBee wireless modules, made by Digi International (Reference 11.8), can be used to configure Zigbee networks rapidly. Different firmware versions allow them to take on coordinator, router, or end-device roles. There are a variety of these modules, but be warned, some are superseded, and not all communicate with each other. There are standard versions and PRO versions, which tend to be higher-power. Two XBee modules are pictured in Fig. 11.13, one with a whip, and one with a PCB antenna. We apply here the types labeled S2C; these are configurable and flexible, and should work with Zigbee-compliant devices from other makers. A detailed user manual is given in Reference 11.9, while summary data appears in Table 11.3.

Each XBee device has a media access control (MAC) address which communicating devices can recognize and initialize interaction with, if required. The MAC address (a component of the data link layer seen in Fig. 11.4) is unique to the device, and is embedded in its hardware by the manufacturer.

The XBees operate in either AT (application transparent) or API mode. In AT mode the radio link is effectively transparent; data sent to one will immediately be transmitted to the module whose destination address is held by the transmitter. This is the simplest, and default, configuration. AT mode can operate in transparent or command forms. As we shall

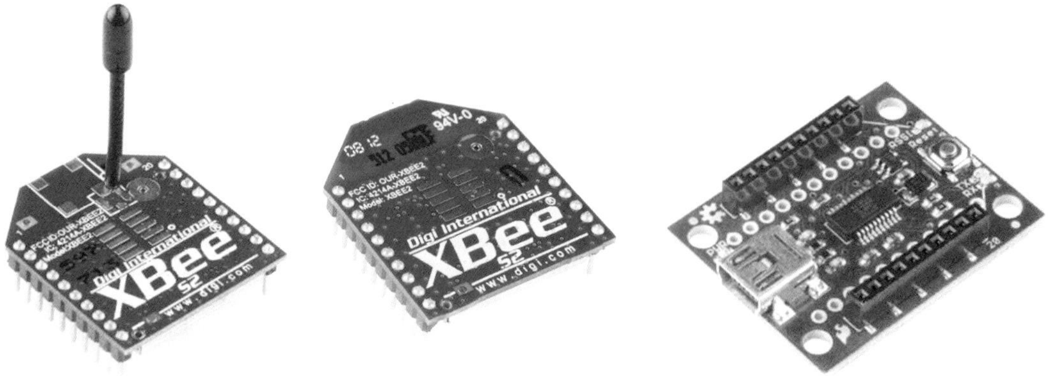

(a) (b) (c)

Figure 11.13
XBee modules: (a) XBee with whip antenna, (b) XBee with PCB antenna, (c) XBee Explorer USB interface. *Images courtesy of Cool Components.*

Table 11.3: XBee S2C module characteristics.

Variable	XBee	XBee-PRO
Range, indoor/urban	Up to 60 m	Up to 90 m
Range, outdoor/line of sight	Up to 1200 m	Up to 3200 m
Transmit power output	3.1 mW (normal)	63 mW
Maximum data throughput	Up to 96,000 bps	Up to 96,000 bps
Transmit current	33 mA (normal)	120 mA
Idle/receive current	28 mA (normal)	31 mA
Power-down current	< 1 uA	< 1 uA

see, the API mode allows considerably more sophistication. A network can contain a mix of modules operating in both modes, configured according to the roles they play. Importantly, XBees are not just conduits of data. They have input and output capability with, for example, pins that can be configured to read analog or digital data.

11.3.3 Introducing the XCTU Software

When setting up BLE links earlier in this chapter, we had the distinct advantage that modern smartphones are generally equipped with BLE capability, so we were able to make quick and easy connections to an experimental device. However, smartphones and PCs do not have Zigbee capability, so it is important to create that link. This requires two extra pieces of kit. The first is a USB interface, such as the Explorer, shown in Fig. 11.13(c). This contains a USB-to-serial converter, which allows the XBee to link with a PC USB port. The second is the official interface software from Digi, known as XCTU (XBee configuration and test utility). This can be downloaded free from the Digi website, Reference 11.8. This brings the Explorer interface to life and allows diagnostic testing of an XBee and downloading of new firmware, plus access to remote XBee devices by radio link. Using XCTU, the XBee device can be configured as coordinator, router, or end device.

To follow the practical work for yourself, you will need to download XCTU and have the Zigbee-related parts listed in Appendix D. Here we use XCTU version 6.5.12. Recognizing that software is continuously revised, it is useful to view the excellent introductory video on the Digi XCTU page linked from Reference 11.8.

To proceed, carefully mount an XBee in the USB Explorer and connect to your PC. If at any time you experience unreliable behavior from the Explorer device, check the guidance given in Reference 11.10. Launch XCTU and click on the "Discover Radio Modules ..." icon, which appears at the top left of the screen. This is the symbol with the magnifying glass sign on a little XBee outline, seen in Fig. 11.14. Accept the default offerings in the pop-ups which follow, and the software will then proceed to identify the XBee you've plugged in, displaying a panel as shown at the left of Fig. 11.14. You will be invited to add this to the XCTU display. Clicking on this panel also calls up a much more detailed screen, as seen to the right of the figure. You can scroll down to view a very wide range of details about the XBee, or search for a particular parameter from the search box. Parameters displayed can be read and written to. The buttons across the top of this block, Read—Write—Default—Update—Profile, provide access to the main features of XCTU.

Label and number your XBees and write down the MAC address of this first one; you will need it later. Note that all XBee radios have 0013 A200 as their higher bytes. For example, the experiments which follow were done with XBees with these addresses:

Coordinator: 0013 A200 4172 4F77 Router: 0013 A200 4153 1CD2

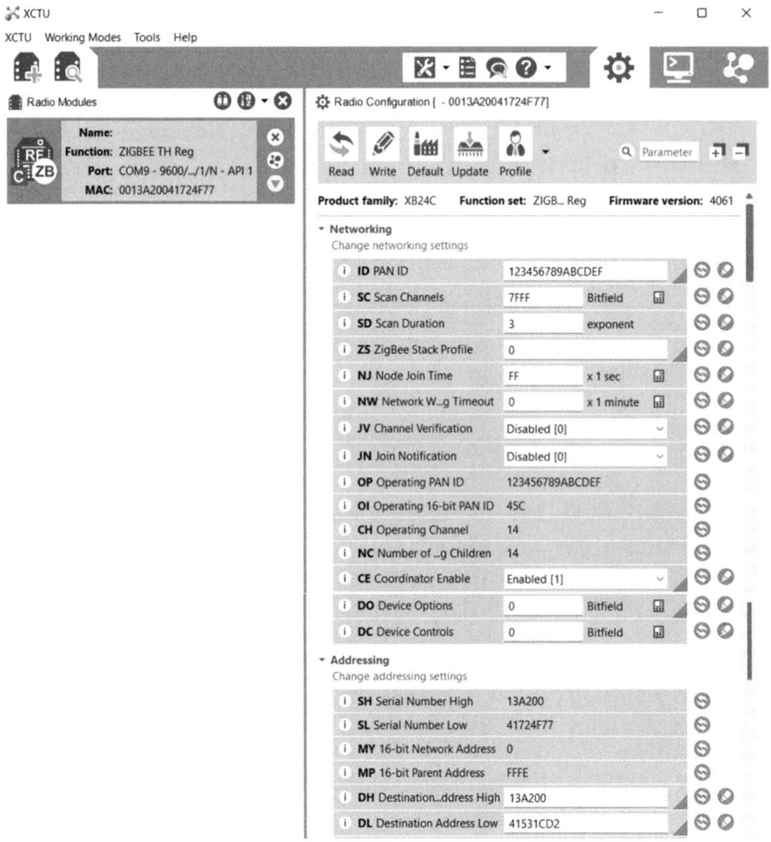

Figure 11.14
XCTU main screen, showing a coordinator module having been discovered.

Disconnect the USB Explorer, carefully take out the first device, and put in the second. Reconnect "discover" it again through XCTU, and get it displayed on the screen, as in Fig. 11.14. When prompted, you will need to "remove" the previous device from the display. Record the new MAC address.

11.3.4 Configuring an XBee Pair

In the simple experiments that follow, we do not attempt to explore or exploit differences between coordinator, router, and end-device behavior. We will refer to a "router," but will not attempt any routing action. With one XBee in the USB Explorer and "discovered" by XCTU, click the "Update" button on the XCTU screen. You will be offered a choice of firmware that can be transferred to the XBee module, similar to Fig. 11.15. Make the selection shown, and then click "Update." The updating process (if needed) takes a minute or more, with interesting information appearing on the screen as it does so. At the end, the original front screen of Fig. 11.14 may be updated to show any change in module status.

For our tests, each XBee must also have the same PAN ID selected, in the range 0 to 0xFFFFFFFFFFFFFFFF, and each must have the address of the other as its destination address. Complete this through XCTU. Fig. 11.14 shows a PAN ID of 0123456789ABCDEF that has been arbitrarily chosen (a simpler one might have been better!). To select the coordinator mode for this XBee, simply enable it in the **CE** line, as seen in the figure. Then set up the destination address, seen in the last two lines of the figure, applying the address of the router XBee.

Now disconnect the USB Explorer and put in the second XBee, which will be the router. Check that the firmware is up-to-date, via the screen of Fig. 11.15. Configure it with the

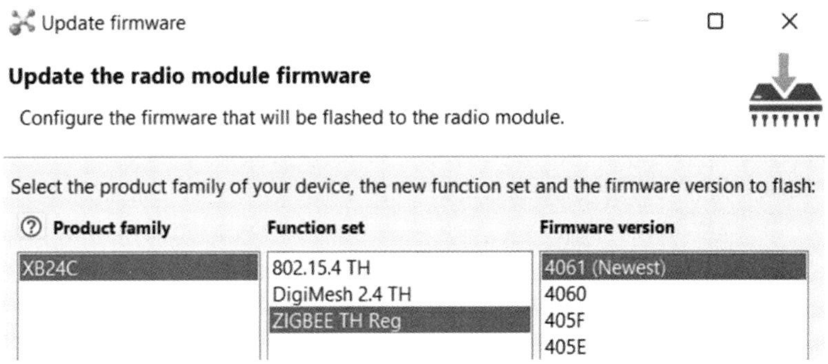

Figure 11.15: Updating the XBee firmware.

same PAN address as the coordinator and set the coordinator MAC address as its destination.

11.3.5 Implementing Zigbee Links

App Board We now move to setting up a Zigbee-to-Zigbee link, with one end linked to a PC. Connect the router XBee in the circuit of Fig. 11.16. An easy way to do this is to mount the XBee in a "breakout board," as listed in Appendix D, and solder jumper wires from this; these can then be connected into a breadboard connecting the Mbed target. Alternatively, the Mbed application board has a very useful XBee socket, item 12 on Fig. 2.14, with connections shown in Table 2.3. Only the TX and RX lines are needed. An Mbed LPC1768 can simply be plugged into the board, or an F401RE (with extra LEDs linked in) can be wired across. Compile and download Program Example 11.7 into the Mbed target.

Put the coordinator XBee in the USB Explorer, plugged into the host PC. Set up Tera Term or CoolTerm, and let things run. The two XBees should link and the terminal count continuously from 0 to 9, with the Mbed LEDs counting up in synchrony. Your first Zigbee link is up and running! You will be even more convinced if you power the router circuit from a battery.

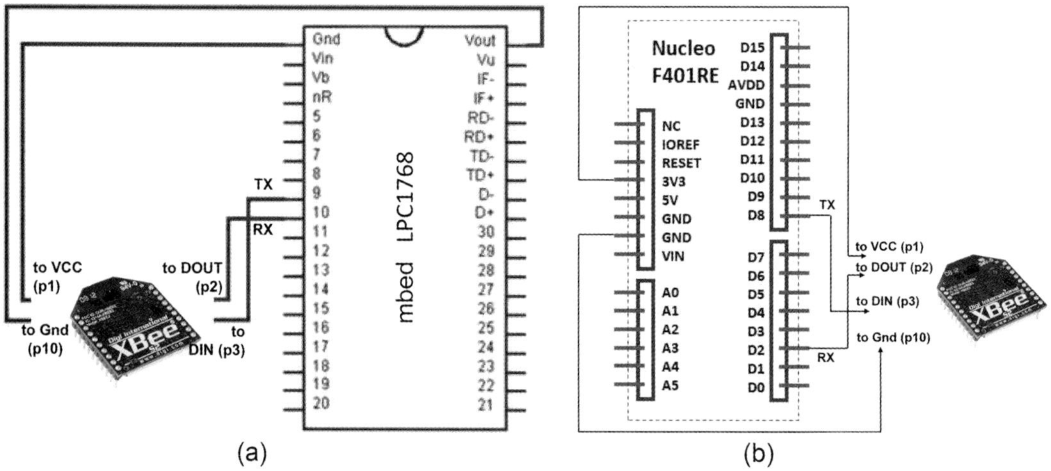

Figure 11.16
Alternatives for connecting an Mbed target to the Xbee: (a) Mbed LPC1768; (b) Nucleo F401RE.

```
/* Program Example 11.7: Zigbee serial test data program.
Data is transferred from Mbed to PC via Zigbee to Zigbee link.
Requires a set of "paired" XBee modules. The Zigbee coordinator connects with
the PC, the router with the Mbed.
Works with both LPC1768 and F401RE, with pin connections shown */

#include "mbed.h"

//name the serial port xbee
BufferedSerial xbee(p9,p10);     //Use D8,D2 for F401RE
BusOut leds(LED4,LED3,LED2,LED1); //Use A1,A0,D12,D13 or own choice for F401RE
char buf[8]= {0};              //declare 8-byte buffer, initialise to 0

int main() {
  xbee.set_baud(9600);     // set baud rate for xbee
  while (true) {
    for (char x = 0x30;x <= 0x39;x++){ // ASCII numerical characters 0-9
      buf[0] = x;
      xbee.write(buf,1);    //transmit current value of x
      leds = x&0x0F;        // set LEDs to count in binary
      ThisThread::sleep_for(500ms);
    }
  }
}
```

Program Example 11.7: Zigbee serial test, Mbed to PC.

The XBees can transfer data in either direction, and it's a simple step now to demonstrate this. Retain the hardware configuration used directly above. Program Example 11.8 shows the router requesting data from the PC-connected XBee. This is provided by the user from the PC keyboard, via Tera Term or similar. The router Mbed echoes each data byte back, which is then displayed on the terminal screen. It also displays the lower 4 bits on the Mbed LEDs. The exchange is terminated after 10 bytes have been received and a message sent.

Compile and download Program Example 11.8 to the Mbed target. With Tera Term (or equivalent) active, the PC keyboard number presses should be transmitted by the coordinator in the USB explorer, and reflected in the Mbed LEDs linked to the router. The router echoes each byte back, to be displayed on the Tera Term screen.

```
/* Program Example 11.8:
Data is transferred bidirectionally between Mbed and PC via Zigbee.
The Zigbee coordinator connects with the PC, the router with the Mbed target.
Works with both LPC1768 and F401RE, with pin connections shown */

#include "mbed.h"

//set up the serial port and name xbee
```

```
BufferedSerial xbee(p9,p10);       //Use D8,D2 for F401RE
BusOut leds(LED4,LED3,LED2,LED1); //Use A1,A0,D12,D13 or own choice for F401RE
char bufdata[8]= {0};             //declare 8-byte buffer and initialise to 0
char buftxt1[24] = "Send 10 data samples\n\r";  //declare buffer and load text message
char buftxt2[24] = "\n\rNo more samples\n\r";   //declare buffer and load text message
char i = 0;

int main() {
  xbee.set_baud(9600);          //set up baud rate
  xbee.write(buftxt1,24);       //Send the message in buftxt1
  while (i < 10) {              //read in 10 bytes
    if (xbee.readable()) {      //check if data available
       xbee.read(bufdata,1);    //read a data byte into the buffer
       xbee.write(bufdata,1);   //echo that byte back, it will display on screen
       leds = bufdata[0];       //display LS Byte to LEDs
       i++;
    }
  }
  xbee.write(buftxt2,20);       //Send the message in buftxt2
}
```

Program Example 11.8: Zigbee bidirectional data test

It is a further simple step to set up an Xbee-to-XBee link, with no PC intervention. Two Mbed targets are now needed, one for each XBee, and each with a different program. The one linked to the coordinator XBee will need a C12832 LCD display, such as we used in Section 8.5 of Chapter 8. This is also found on the Mbed app board; one or two of these could usefully be applied here. A possible overall system is represented in Fig. 11.17. Although this shows the Mbed LPC1768, implementations based on the F401RE are equally possible, applying Fig. 11.16(b).

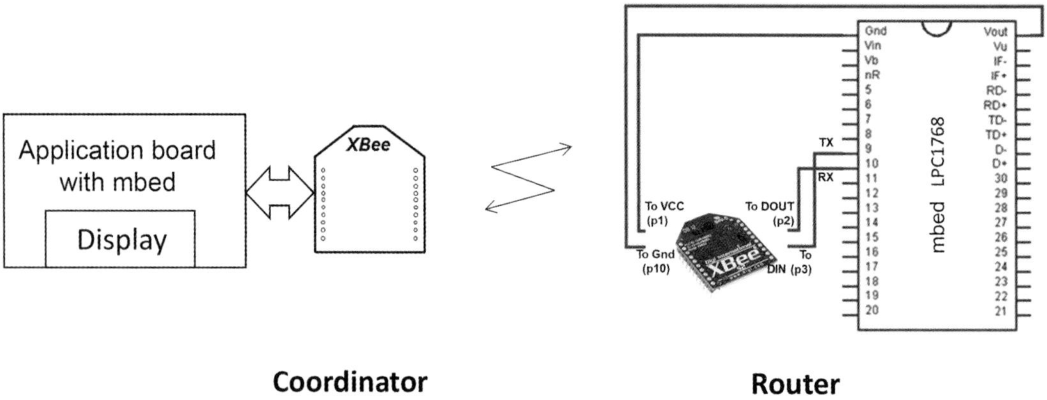

Figure 11.17
Zigbee link applying the app board.

Program Examples 11.9 and 11.10 should be downloaded and run in their respective Mbed targets. All program features should be familiar. The library of Table 8.4 is used for the C12832 LCD. The coordinator should either detect and display the incoming "data" from the router, or display "Wireless link lost!"

```
/* Program Example 11.9
Requires a paired set of XBees, configured in XCTU. This is coordinator program.
Works with both LPC1768 and F401RE, with pin connections shown        */

#include "mbed.h"
#include "C12832.h"

C12832 lcd(p5,p7,p6,p8,p11);     //Use D11,D13,D12,D10,D9 for F401RE
//name the serial port xbee
BufferedSerial xbee(p9,p10);     //Use D8,D2 for F401RE

BusOut leds(LED4,LED3,LED2,LED1); //Use A1,A0,D12,D7 or own choice for F401RE

char x,j;
char bufdata[8]= {0};

int main(){
  lcd.cls();          //clear lcd screen
  lcd.locate(0,3);    //locate the cursor
  lcd.printf("Zigbee Test Program");  //Print title to lcd
  ThisThread::sleep_for(500ms);
  while(true) {
    if (xbee.readable()){    // if data available
      j = 0;
      xbee.read(bufdata,1);
      leds = bufdata[0];     // output LSByte to LEDs
      lcd.locate(0,15);
      lcd.printf("Remote data = %d",bufdata[0]);
    }
    else {        //count no of times there is no data
      j++;
      ThisThread::sleep_for(10ms);
    }
    if (j > 250){
      lcd.locate(0,15);
      lcd.printf("Wireless link lost!");
      j = 0;      //reset counter
    }
  }
}
```

Program Example 11.9: XBee to Xbee link—coordinator

```
/* Program Example 11.10
Paired Zigbees – this is router program
The router generates data and sends to coordinator
Works with both LPC1768 and F401RE, with pin connections shown */

#include "mbed.h"

//name the serial port xbee
BufferedSerial xbee(p9,p10);      //Use D8,D2 for F401RE
BusOut leds(LED4,LED3,LED2,LED1); //Use A1,A0,D12,D7 or own choice for F401RE

char x;
char bufdata[8]= {0};

int main() {
  xbee.set_baud(9600);
  while (true) {
    x++;                    // increment x
    if (x > 0x0F)           // limit to 4 bits
      x = 0;
    bufdata[0] = x;
    xbee.write(bufdata,1);
    leds = x;                // set LEDs to count in binary
    ThisThread::sleep_for(500ms);
  }
}
```

Program Example 11.10: XBee to Xbee link—router.

11.3.6 Introducing the XBee API

So far we've done some neat things with the XBees, but in truth we have done little more than replace a length of wire with a radio link. We have not approached anywhere near the flexibility that the introductory section on Zigbee seemed to imply. This section is intended to provide a glimpse of how the real power of the Zigbee/XBee combination can be implemented. To avoid this becoming a book on Zigbee (and that would be a very interesting thing to do!), this remains a glimpse. However, it should give the motivation, confidence, and pointers to go much further.

We have already mentioned the distinction between operating in AT and API modes. It is the API mode which unlocks the power of Zigbee through the XBee. Using the API, the XBee can, for example, change the destination address dynamically, perform error checking, reconfigure remote radios, and exploit the remote XBee I/O capability. It is for this last reason that the next example uses the API.

In API mode all data is packaged into *frames,* which carry both the data itself, and a range of ID, addressing, and error-checking capability. The general frame structure is shown in Fig. 11.18. The frame always starts with the identifier 0x7E, then two bytes which indicate the length of the following "frame data" section. The first byte of the frame data is a single-byte command identifier (also called an API identifier). This indicates the purpose of the frame, and thus determines the structure of the frame data section which follows. The final byte is always a *checksum*, which provides a simple means of error checking.

There are 18 API identifiers possible, which allow commands, data, or status information to be transferred. Each has its own distinct format within the frame data. In this simple demonstration, we will use the remote XBee to make readings on just one of its analog inputs. This requires use of the "Zigbee IO data sample Rx indicator," with frame structure shown in Table 11.4.

11.3.7 Applying the XBee API

We now set up a simple Zigbee link, in which one XBee is set up as a standalone device, making analog measurements. It periodically transmits data values to an XBee coordinator. This can run on an Mbed app board, as pictured in Fig. 11.19. Alternative arrangements have already been described. The coordinator now runs in API mode. The other XBee is configured as a router. Its pin 20, labeled ADO/DIO0, will be configured as an analog input; it can remain in AT mode to play this role. The XBee ADC is a 10-bit device, and has an input range of 1.2 V. A potentiometer has been chosen to provide an analog source, as shown. This has itself been placed in a potential divider, through the addition of a 20 kΩ resistor, to match the ADC input range. The 3.3 V can be supplied by a battery (e.g., a pair of AA cells) or a voltage regulator or bench supply. A simple solution if you have a spare Mbed available is to use it just for its 3.3 V regulated output, powering it through a USB or battery pack.

As before, the XBees need to be configured in turn, using an XBee USB explorer connected to a laptop running XCTU. The router is then configured through XCTU, in the screen shown in Fig. 11.14, with settings given in Table 11.5. Nothing else needs to be changed.

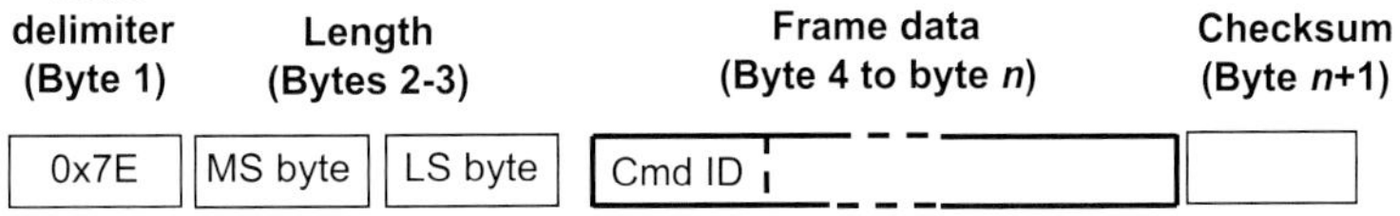

Figure 11.18
General XBee API data frame structure.

Table 11.4: Frame Structure for "Zigbee I/O data sample Rx indicator."

Byte	Purpose	Description
1	Start delimiter	0x7E
2	Length MS byte	Number of bytes between Byte 4 and Checksum, MS byte
3	Length LS byte	Number of bytes between Byte 4 and Checksum, LS byte
4	Command ID	0x92
5–12	64-bit address of sender	MS byte first
13–14	16-bit address of sender	MS byte first
15	Receive options	01: Packet acknowledged 02: Packet was broadcast
16	Number of sample sets	Number of sample sets in payload (always set to 1)
17–18	Digital Channel Mask	Bitmask field indicating which digital IO lines on the sending device have sampling enabled (if any)
19	Analog Channel Mask	Bitmask field indicating which analogue IO lines on the sending device have sampling enabled (if any)
20–21	Digital samples (if any)	All enabled digital inputs are mapped within these two bytes
22-	Analog samples	Each enabled analog input returns a 2-byte value indicating the ADC output
	Checksum	Single byte: disregarding first three bytes, add all other bytes, keep only the lowest 8 bits of the result and subtract the result from 0xFF

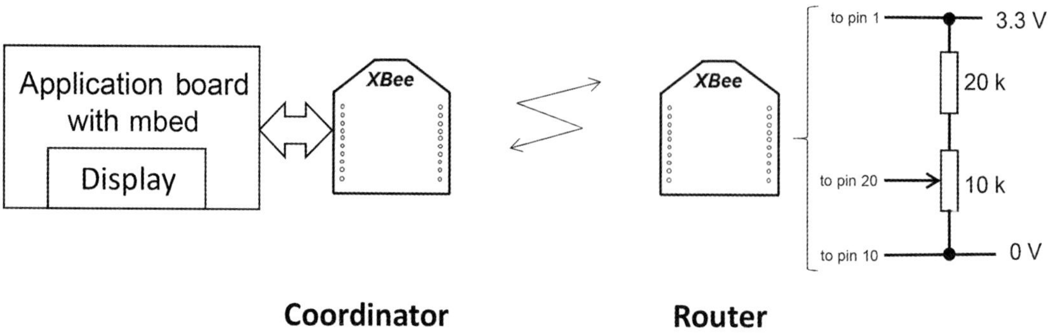

Figure 11.19
Diagnostic circuit for API trial.

Table 11.5: Router configuration for API trial.

Code	Name	Select this Value	Comment
ID	PAN ID	A 16-digit value of your choice; keep it simple	Must be the same as the coordinator; we chose 0123456789ABCDEF earlier
JV	Channel verification	Enabled [1]	Router attempts to join coordinator when powered up
D0	AD0/DIO0 configuration	ADC [2]	Selects this pin as analog input
IR	IO sampling rate	200 × 1 ms	The input will be sampled and transmitted with a period of 200 msh

The coordinator should be enabled as such (we have done this for earlier programs) and should be set in API mode. This is done through line **A0** in the settings menu of Fig. 11.14. As before, it should be configured to have the same PAN address as the router. Nothing else needs to be changed.

Program Example 11.11 runs on the Mbed target linked to the coordinator XBee. The coordinator receives transmissions from the router, determined by the internal 200 ms timer that has been set up. These are then communicated to the Mbed target. It should not be too difficult to work out the program features. Broadly speaking, it unpicks the data frame of Table 11.4 and displays chosen values from the frame. For this simple demo, certain important items within the data frame are discarded. It is easy to change what is selected and displayed to inspect other parts of the frame. Compile and download the program.

```
/* Simple XBee API application
Requires XBee and Mbed in app board, plus remote XBee, or equivalent
The coordinator receives data from router, and displays on lcd
Works with both LPC1768 and F401RE, with pin connections shown. */

#include "mbed.h"
#include "C12832.h"

C12832 lcd(p5,p7,p6,p8,p11);     //Use D11,D13,D12,D10,D9 for F401RE

BufferedSerial xbee(p9,p10);     //Use D8,D2 for F401RE
DigitalOut led (LED4);           //Use A0 for F401RE
char x,xhi,xlo,j,len,ftype;      //some useful internal variables
int result;
char bufdata[8]= {0};

int main(){
  lcd.cls();                     //clear lcd screen
```

```
    lcd.locate(0,3);              //locate the cursor
    lcd.printf("Zigbee API Test");
    ThisThread::sleep_for(500ms);
    while(true) {
      if (xbee.readable())        // if data is available
        xbee.read(bufdata,1);     // get data
      x = bufdata[0];
      if (x == 0x7E){             //test for start of frame
        led = 1;                  //New frame detected, set diagnostic LED
        while (xbee.readable() == 0);  //wait for next byte
        xbee.read(bufdata,1);        //discard length msb, assume zero
        while (xbee.readable()==0);
        xbee.read(bufdata,1);        //save length lsb
        len = bufdata[0];
        while (xbee.readable() == 0);
        xbee.read(bufdata,1);        //save frame type
        ftype = bufdata[0];
        j = 1;
        //now discard 15 bytes: i.e. 64 bit address, 16-bit address, et al
        while(j < 16){
          while (xbee.readable() == 0); //wait for next byte
          xbee.read(bufdata,1);         //read, discard
          j++;
        }
        while (xbee.readable() == 0);
        xbee.read(bufdata,1);        //get ms ADC byte
        xhi = bufdata[0];
        while (xbee.readable() == 0);
        xbee.read(bufdata,1);        //get ls ADC byte
        xlo = bufdata[0];
        result = xhi*256 + xlo;      //convert to 16 bit number
        lcd.locate(0,15);
        lcd.printf("length = %d",len);     //include values as desired
        lcd.printf(" data = %d",result);   //display result
        led = 0;
      }                              //end of if
    }                                //end of while()
  }
```

Program Example 11.11: Applying the XBee API—coordinator program

With both ends of the Fig. 11.19 data link powered, the XBees should find each other; the LED on the Mbed target then starts flashing five times a second, each time the data frame is detected. The C12832 display should show the text message "Zigbee API Test," followed by values for frame length and data. If the potentiometer is adjusted, the value on the screen should be updated immediately, giving a numerical value in the range from 0 to 1023.

This example, though still simple, gets you into the world of the XBee API, through which the true power of Zigbee, and the XBee modules, can be exploited. Clearly the Program

Example used and the XBee settings made neglect many important Zigbee features, which a more sophisticated program would exploit.

Exercise 11.2

Note the value of **len** displayed when Program Example 11.11 runs, and consider Quiz question 10.

i) Now display variable **ftype** instead of **len**, and explain the value found.

ii) Display the least significant two bytes of the 64-bit sender address. Do they give the expected value?

■

Exercise 11.3

Replace the potentiometer in Fig. 11.19 with a light sensor, as seen in Fig. 5.7. Transmit light data over the Zigbee link.

■

11.3.8 Conclusion on Zigbee and Further Work

There is huge scope to go much further with Zigbee and these wonderful little XBee devices. However, this book is about the Mbed environment, and quite a bit more XBee-specific knowledge would be needed. To progress further, try other API data frames, set up simple networks with router and end devices, and explore Sleep mode. Check Reference 11.11 also. Though now a little dated (and no new edition has so far been forthcoming), this detailed guide to Zigbee and the use of XBees contains many useful ideas, and projects can be adapted to the Mbed environment.

11.4 LoRa and LoRaWAN

LoRa (a contraction of "long range") is one of a series of long-range wireless communication technologies designed for power-constrained embedded systems, especially those running on battery power. The trade-off here is the low data rate, with LoRa supporting data transfer rates in the range of 0.3 to 5.5 kilobits per second. It claims a transmission distance of up to 16 km line of sight, or 5 km in built-up areas. It also claims good communication in difficult settings—for example, through concrete and in tunnels.

LoRa is a proprietary protocol based on a chirp spread spectrum technique that defines the physical layer for long-range communication, and on its own does not dictate the actual communication protocol. Its specification is managed by the LoRa Alliance and can be seen on their site in Reference 11.12.

LoRaWAN is a protocol that defines an overall architecture and specifies how devices communicate. As the name suggests, it uses LoRa for the physical layer and overlays wide area networking capabilities.

A typical LoRaWAN architecture is shown in Fig. 11.20 and comprises end nodes, gateways, and network servers. The end node is the actual embedded device deployed in the field, which establishes an RF link with the gateway. This forwards data to the network server, usually over a WAN (such as the internet), which can then transfer it as needed. We explore this style of system architecture in more detail in the following chapters.

LoRa is geared toward end nodes initiating a communication transfer, that is, end nodes transmitting information to the gateway, rather than network servers transmitting information to end nodes. However, there is an opportunity for end nodes to receive data too. This happens during a period of time called the "receive window."

The LoRaWAN specification defines three classes of device, as illustrated in Fig. 11.21:

- **Class A:** An end node initiates a transmission, which is immediately followed by two short receive windows where data can be received. All LoRaWAN devices must support Class A operation.
- **Class B:** Operates as in Class A, but additional receive windows are available at scheduled times for non-device-initiated receives.

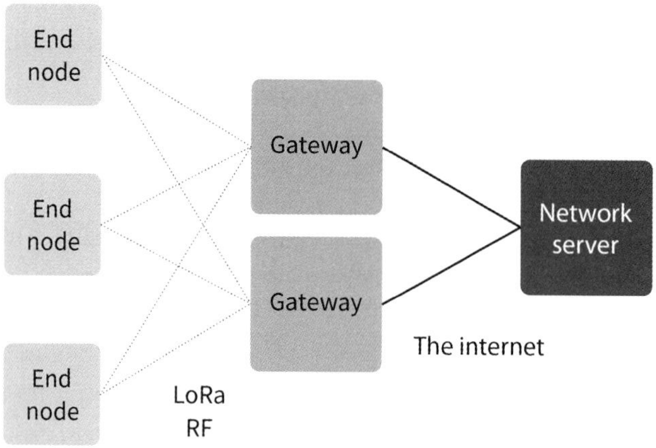

Figure 11.20
A typical LoRaWAN architecture.

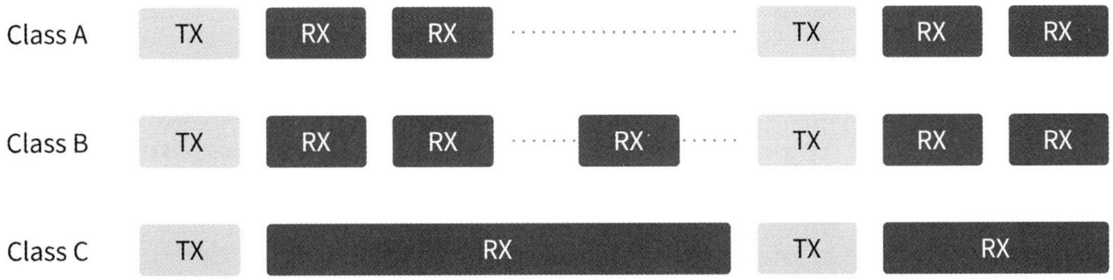

Figure 11.21
Transmit (TX) and receive (RX) windows for different LoRa device classes.

- **Class C:** The receive window is continuously open, except when transmitting. This imposes additional power requirements: the device must be continuously powered.

The Mbed OS offers a LoRaWAN interface API, Reference 11.13, along with an extended example. Clearly its use is not without complexity, but it can form a useful launchpad for experimentation in this interesting field. ST offers an Mbed-enabled development board, Reference 11.14, with LoRaWAN capability.

LoRaWAN appears to offer the impossible—long range communication at extremely low power, albeit at the price of a low data rate. It opens many doors to remote sensing opportunities, for example, in areas without mobile phone, cable, or fiber coverage. It is enjoying very rapid take-up with many reported uses, for example, Reference 11.15.

11.5 Mini Projects

11.5.1 BLE Mini Project

Building on the implementation of the BLE sensor example, implement a characteristic that can be written to so that you can toggle the LED on the board. You can use the already existing **ReadWriteGattCharacteristic** from the BLE API and implement the logic for toggling the LED by overriding the **onDataWritten( )** function in the **Server** class.

11.5.2 Zigbee Mini Project

Depending on how many XBees you have, set up a network with multiple nodes. The routers and/or end devices should be fitted with temperature sensors. Place one in each of several rooms in a house or other building, with one selected as coordinator. Display the temperature in each room.

Chapter Review

- Wireless links exploit the characteristics of the electromagnetic spectrum, notably in radio, infrared, or visible light.
- A wide range of protocols and technologies exist to implement wireless data links across personal, local, neighborhood, and wide area networks.
- BLE is a complex, yet effective, protocol defined within the IEEE 802 group, which allows Bluetooth-enabled devices to connect and transfer data wirelessly with potentially high data rates.
- Zigbee is another important protocol defined within the IEEE 802 group, targeted toward low data rate, distributed measurement systems, and extremely low power.
- The XBee module provides Zigbee capability, which can be linked to an Mbed target. The XBee has its own on-board processing power and input/out capability, so it can also act as a standalone device.
- LoRaWAN brings long transmission ranges to low-power communication, with the trade-off of lower data rates.

Quiz

1. An FM signal has a frequency of 103.0 MHz. What is its wavelength?
2. The antenna for a BLE device operating at 2.4 GHz is to have a "quarter-wavelength antenna", meaning that the length of the antenna should be one-fourth of the radio signal wavelength. How long should it be?
3. Explain briefly the term *spread spectrum*. Why does this reduce interference between channels?
4. How many layers are there in the OSI model? Using BLE and Zigbee as examples, justify the statement, "The IEEE 802 standards tend to link to the … data link layer and physical layer." You may need to do a little more background reading to do this.
5. What is the effective data rate of a typical BLE connection?
6. What does the term "MAC address" refer to?
7. Briefly describe the three types of Zigbee device and the roles they play.
8. What is the nominal range of an XBee PRO device, when operating out side?
9. Think of a moderate-sized building that you know well. Sketch a Zigbee network for it which monitors temperature, light, and air quality in each room. Show where you would place the coordinator, routers, and end devices.
10. When Program Example 11.11 runs, the app board display shows a frame length of 18. Explain carefully why this is so.
11. A rural hospital in Africa draws its water from a borehole 8 km from the main hospital site. The borehole pump, which is not in mobile phone range, is powered by a diesel electric generator. Sensor signals available within 10 m of each other detect

borehole water depth, diesel fuel level, and supply voltage and current to the pump. A technician at the hospital (which has internet and mobile phone connectivity) needs to be able to access all this sensor data by mobile phone. Sketch a wireless network which could be installed to meet this need.

References

11.1. Bear Cory, Stallings William. *Wireless Communication Networks and Systems*. Pearson; 2016.
11.2. IEEE 802 LAN/MAN Standards Committee. http://www.ieee802.org/.
11.3. The official Bluetooth website. http://www.bluetooth.com.
11.4. Mbed OS BLE API. https://os.Mbed.com/docs/Mbed-os/v6.15/apis/bluetooth-apis.html.
11.5. Mbed OS BLE Examples. https://github.com/ARMMbed/Mbed-os-example-ble/.
11.6. ARMMbed/Mbed-os-ble-utils. https://github.com/ARMMbed/Mbed-os-ble-utils.
11.7. The Connectivity Standards Alliance web site. https://csa-iot.org/all-solutions/zigbee/.
11.8. The Digi Web site. http://www.digi.com.
11.9. *XBee/XBee-PRO S2C 802.15.4 Radio Frequency Module User Guide*. Digi International; 2022. https://www.digi.com/resources/documentation/digidocs/pdfs/90001500.pdf.
11.10. Optimizing USB to serial port settings, DIGI. Optimizing USB to serial port settings | Digi International.
11.11. Faludi Robert. *Building Wireless Sensor Networks*. O'Reilly; 2011.
11.12. LoRaWAN Specifications https://resources.lora-alliance.org/technical-specifications.
11.13. LoRaWAN Interface https://os.Mbed.com/docs/Mbed-os/v6.15/apis/lorawan-apis.html.
11.14. The B-L072Z-LRWAN1 LoRa®Discovery kit https://os.Mbed.com/platforms/ST-Discovery-LRWAN1/.
11.15. Laveyne J, Van Eetvelde G, Vendevelde L. *Application of LoRaWAN for Smart Metering: An Experimental Verification*; 2018. https://www.researchgate.net/publication/337285895_Application_of_LoRaWAN_for_Smart_Metering_An_Experimental_Verification.

CHAPTER 12

Towards the Internet of Things

12.1 What is the Internet of Things?

The IoT is a growing network of physical objects that are connected to the internet, and collect and exchange data. These "things," also known as "smart" or "connected" devices, range from everyday household items such as smart thermostats and appliances to industrial or office equipment, and they work together in a larger system to enable intelligent transportation systems, civil infrastructure, and even entire cities.

The IoT allows the seamless exchange of information between devices and systems, enabling them to communicate and coordinate their actions, often without human intervention. This has the potential to revolutionize how we live, work, and interact with the world around us.

12.1.1 An IoT Overview

At its core, the IoT is made up of four main types of components: sensors and actuators, connectivity, processing, and applications. As earlier chapters have shown, sensors are devices that can detect and measure various aspects of the physical world, such as temperature, humidity, or motion. These sensors are embedded in or attached to physical objects and collect data about their surroundings. Actuators are devices that can perform physical actions in the real world, such as emitting light, making a sound, or turning equipment on and off.

Connectivity refers to the ability of these devices to communicate with each other and with other systems, often over the internet. This can be achieved through a variety of technologies, such as Wi-Fi, Bluetooth, cellular networks, and long-range communication protocols.

Processing refers to the ability of the IoT system to analyze and act on the data collected by the sensors. This can happen on the device itself, which involves processing data at the source rather than sending it to a centralized location, or by using cloud computing, which involves storing and processing data on remote servers. Often, IoT systems are deployed using a public cloud service, such as Google Cloud, Amazon AWS, or Microsoft Azure. However, this is not always the case. For example, a company may wish to keep its IoT

Fast and Effective Embedded Systems Design. https://doi.org/10.1016/B978-0-323-95197-5.00012-7

deployment completely internal and utilize its own private servers for the processing aspect of the system.

Finally, the applications are the software that allows users to interact with the IoT system, often implemented as mobile or web apps. These applications allow the data collected to be visualized and enable operations to be performed on the embedded systems hosting the sensors—including making things happen in the "real world."

Fig. 12.1 gives a high-level overview of a variety of IoT systems. Here, the sensors and actuators are present on the IoT devices themselves. Connectivity is implemented with a variety of technologies, with gateways being devices that can propagate data between the small embedded devices and the large data centers over the internet. Connectivity into the cloud itself can take place over digital subscriber lines (DSL), which is the technology used in the UK for copper-based home broadband, cable internet (data over cable service interface specification, or DOCSIS), which is the most prevalent technology in the US, or sometimes with direct fiber-optic connections, as is the case for a cell tower. Mobile devices connect to cell towers using LTE (long-term evolution, or 4G) technology. Processing is done in the cloud, potentially involving big data and machine-learning systems. Applications are deployed to mobile devices which can communicate with the cloud and interact with these IoT systems.

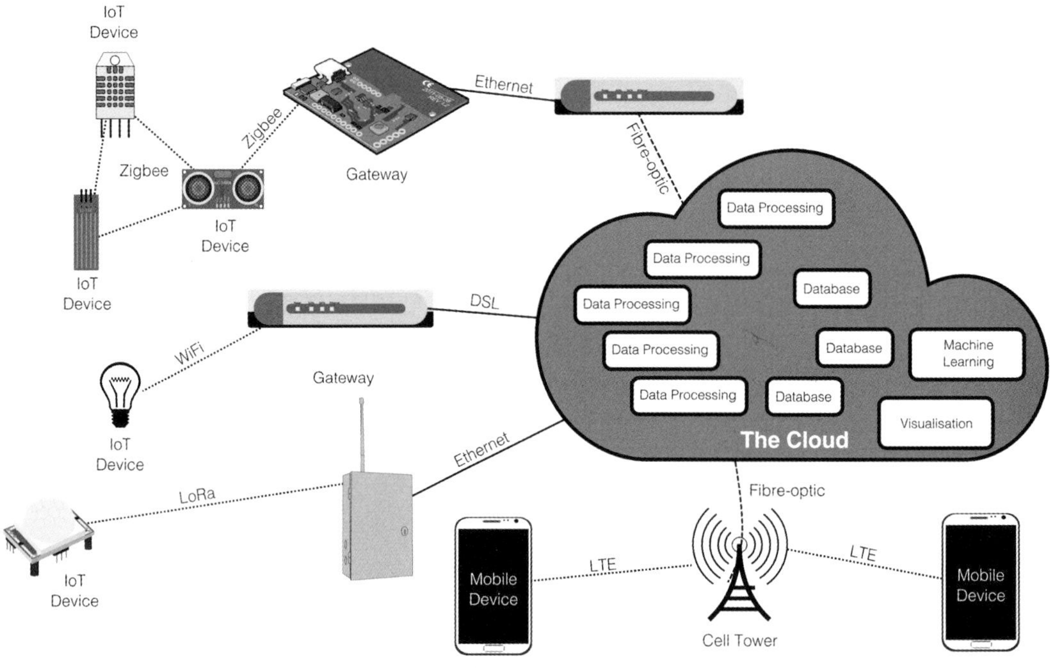

Figure 12.1
A high-level overview of how the Internet of Things might be put together.

12.1.2 Applications of the IoT

The IoT has the potential to transform a wide range of industries and applications. Some examples include the following:

Smart homes: The IoT can be used to automate and control various aspects of a home, such as lighting, temperature, and security. An example of this is shown in Fig. 12.2, where a home can have a multitude of different sensors and actuators that work together and provide various benefits to the users. For example, a smart thermostat can learn a homeowner's preferred temperature settings and automatically adjust the heating and cooling based on the time of day and the presence of people in the home.

Smart cities: The IoT can be used to improve the efficiency and sustainability of urban areas. For example, connected streetlights can dim or brighten based on the presence of pedestrians or vehicles, and smart waste management systems can optimize rubbish-bin collection routes.

Industrial automation: The IoT can be used to improve the efficiency and productivity of industrial processes. For example, connected machines can alert maintenance workers when they need servicing, and smart inventory systems can automatically reorder supplies when they run low.

Transportation: The IoT can be used to improve the efficiency and safety of transportation systems. For example, connected vehicles can communicate with each other

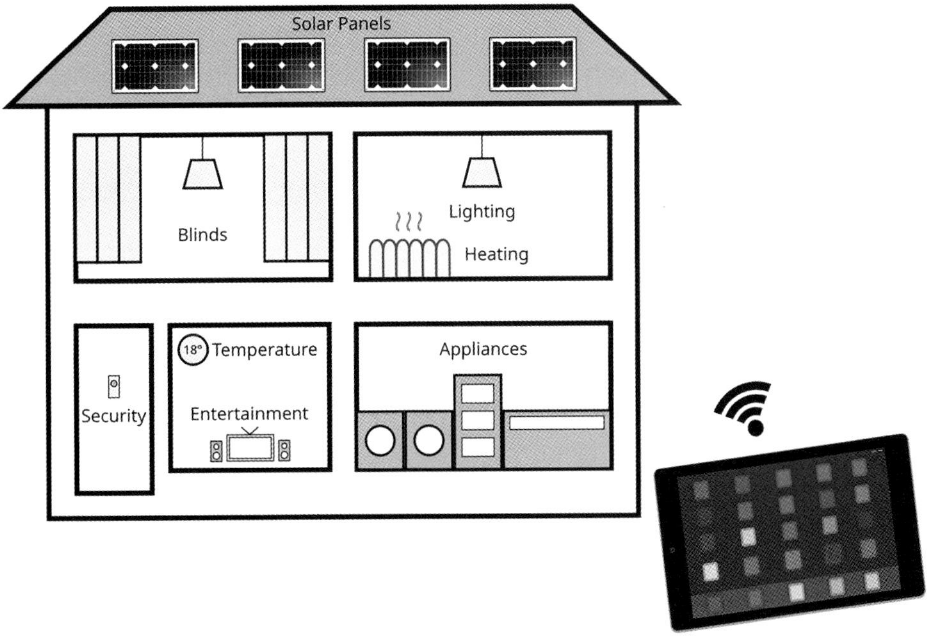

Figure 12.2
A connected smart home.

and with infrastructure to avoid accidents and reduce congestion, and smart public transportation systems can optimize routes and schedules based on real-time demand.

Medicine: The IoT can be used to provide medical services, such as remote monitoring of patients' vital signs, medication management to remind people to take their medication and report back to the healthcare provider if they miss a dose, in surgery to help guide surgeons during complex procedures, and to support clinical decisions in conditions that are difficult to diagnose. See Reference 12.1 for a research paper describing the use of Zigbee and the IoT in remote patient monitoring.

12.1.3 A Little IoT History

The concept of the IoT can be traced back to the 1980s, when researchers and engineers began exploring the idea of connecting everyday objects to the internet. Some associate this with a group of students at Carnegie Mellon University, who connected a vending machine to what was then the early version of the internet—ARPANET (Reference 12.2). The vending machine could report on its stock levels and temperature, so that students could decide whether or not to visit the machine for a beverage.

However, it was not until the widespread adoption of the internet and the development of cheap, low-power sensors and connectivity technologies in the early 21st century that the IoT began to gain traction.

In 1999, Kevin Ashton, a British technology pioneer, coined the term "internet of things" in a presentation he gave to Procter & Gamble. He argued that the potential for connecting everyday objects to the internet was vast and could revolutionize how we live and work.

Since then, the IoT has continued to grow and evolve. In the 2010s, the widespread adoption of smartphones and the development of cloud computing and big data analytics made it easier for businesses and individuals to connect devices to the internet and extract value from the data they generated. In fact, it was estimated at this point that there were more devices than people connected to the internet (Reference 12.3).

In the 2020s, the IoT has become an increasingly integral part of our lives, with billions of connected devices in use around the world. However, the rapid growth of the IoT has also raised concerns about privacy, security, and the impact on society and the environment.

As the IoT continues to evolve, researchers and policymakers are working to address these challenges and ensure that the technology is used in a responsible and ethical manner. Despite this, the IoT is expected to play a major role in shaping the future of our world and revolutionizing how we live, work, and interact with the world around us.

12.2 Core Components in the IoT

In this section, we will visit the core components of IoT deployments in more detail and see how embedded systems facilitate this exciting ecosystem.

12.2.1 Sensors and Actuators

Sensors and actuators are key components of the IoT. Sensors allow IoT systems to gather data about the environment, enabling them to make decisions and take actions based on this data. In the context of the IoT, actuators are often used to control devices or systems based on the data collected by sensors. For example, an actuator might be used to open or close a valve in a smart irrigation system based on a combination of climatic data, stored water availability, the time of day, and the moisture levels measured by sensors in the soil.

Sensors and actuators work together to create a feedback loop in an IoT system. Sensors gather data which is processed and analyzed by the system. Based on this analysis, the system sends a signal to the actuators to perform an action. The action taken by the actuators can then be measured by the sensors, completing the feedback loop.

Here are some examples of sensors and actuators that are commonly seen in the IoT:

Sensors:

- **Temperature sensors:** These can measure temperature in a range of environments, from extremely cold to extremely hot. They are often used in smart thermostats, industrial equipment monitoring, weather stations, and other such devices.
- **Humidity sensors:** These can measure the amount of water vapor in the air. They are often used in domestic or office environments, weather stations, greenhouses, and other environments where humidity is important.
- **Pressure sensors:** These can measure the pressure of gases or liquids. They are often used in industrial processes, weather stations, and other applications where pressure is important.
- **Motion sensors:** These can detect movement or changes in position. They are often used in security systems, industrial control, smart home automation systems, and other devices that need to detect the presence of people or objects.
- **Light sensors:** These are used to measure the intensity of light, often in the form of a photometer or photodiode. They can be used in applications such as adjusting the brightness of a display or detecting the presence of an object.

Actuators:

- **Motors:** These are used to drive mechanical equipment, often in the form of a servomotor or stepper motor. They can be used in applications such as controlling the movement of a robot or the position of a valve.

- **Solenoids:** These are used to control the flow of a fluid, often in the form of a valve or pump. They can be used in applications such as controlling the flow of water in a smart irrigation system or the flow of fuel in a car.
- **Relay actuators:** These are used to control electrical circuits, often in the form of a switch or contactor. They can be used in applications such as turning a light or heater on or off, or activating a motor.
- **LEDs:** These can be used as actuators to provide visual feedback or an indication of the status of a system. For example, an IoT device might use an LED to indicate that it is connected to the internet, or to show the current status of a system such as a door lock. LEDs can also be used as part of more complex systems, such as lighting systems.

As we have seen in previous chapters, it is the embedded system itself that interfaces with the sensors and actuators, providing the processing and control capabilities that enable the IoT system to function. It is sometimes also responsible for processing the data that is collected by the sensors and acting upon it by controlling the actuators in response. The embedded system is typically programmed to perform tasks such as reading sensor data, converting it into a usable format, and sending commands to the actuators based on this data, or from commands that come from a remote server.

Overall, sensors and actuators are essential components of the IoT, and are used in a wide variety of applications to enable intelligent and automated decision-making. There are many more examples of sensors and actuators than listed above, and it is up to the designer of the IoT system to decide which ones are appropriate.

12.2.2 Connectivity

Connectivity refers to the ability of devices to communicate with each other and with other systems, often over the internet. This is what allows the IoT to function as a network, enabling the seamless exchange of information between devices. There are many different technologies that can be used to achieve connectivity in the IoT, and different ways for internet access to be supported. Some common ones include the following:

- **Wi-Fi:** Wi-Fi is the ubiquitous wireless networking technology that allows devices to connect to the internet over short distances. It is commonly used in the IoT because it is widely available, well supported, and easy to use.
- **BLE:** As we saw in Chapter 11, BLE is a wireless technology that allows devices to communicate over short distances, typically within 10 m. It is often used in the IoT to connect devices that are close to each other, such as smart home devices or wearable devices.

- **Near-Field Communication (NFC):** NFC is a short-range wireless communication technology that is often used for device-to-device communication in applications such as mobile payments and access control. It is well suited for applications that require high security and low data rates.
- **Cellular networks:** Cellular networks are used to connect devices to the internet over long distances, typically using mobile phone towers. They are often used in the IoT to connect devices that are not in range of a Wi-Fi or Bluetooth network, such as within smart cities or connected vehicles.
- **LoRaWAN:** As we saw in Chapter 11, LoRaWAN allows devices to communicate over long distances, typically several kilometers, using low-power radio waves with low data rates. LoRaWAN is used in a variety of IoT applications, such as smart cities, connected agriculture, and industrial automation. It is particularly well suited to these types of applications because it can support many devices. It is low-cost and has extremely low power requirements.
- **Zigbee:** As described in Chapter 11, Zigbee is a wireless networking protocol that is designed for short-range low-power applications, fitting nicely into the IoT's remit. It allows devices to communicate over short distances, typically within 100 m, and is often used in smart-home and industrial automation applications.
- **Z-Wave:** Z-Wave is another wireless networking protocol that is designed specifically for the IoT. It is similar to Zigbee in that it allows devices to communicate over short distances and is often used in smart home applications.

There are two main communication strategies to be considered here: device-to-device communication, and device-to-server communication.

Device-to-device communication refers to the direct exchange of data between two or more devices in a given IoT system. This type of communication is often used to enable the devices to collaborate and work together to achieve a common goal, but also as a way for devices to get upstream connectivity to the internet, if a direct connection does not exist. Fig. 11.12 has already shown how devices can group themselves into linear, star, or mesh networks to enable co-operative working between them. There are several technologies that can be used for device-to-device communication, including Bluetooth and Zigbee from the examples above.

Device-to-server communication refers to the exchange of data between an IoT device and a centralized server or network of servers. This type of communication is often used to enable a device to access services and resources that are not available locally, such as data storage, analytics, or security. This kind of communication can be direct, when the device can access the internet directly (such as Wi-Fi or cellular), or indirect, if it must access the internet through a gateway (such as LoRaWAN).

There are several networking protocols that can be used in device-to-server communication, including the following:

- **HTTP:** Hypertext Transfer Protocol is the most popular protocol for transmitting data over the internet. It is often used for device-to-server communication in the IoT, as it is well established, flexible, and widely supported.
- **MQTT:** Message Queue Telemetry Transport is a lightweight messaging protocol that is designed for use in low-bandwidth, high-latency networks such as those used in the IoT. It is often used for device-to-server communication in applications such as remote monitoring and control.
- **CoAP:** Constrained Application Protocol is another lightweight messaging protocol similar to MQTT, but is specifically designed for use in resource-constrained devices and networks. It is often used for device-to-server communication in applications such as smart energy and building automation.

12.2.3 Processing: Edge Computing

Processing refers to the ability of the IoT system to analyze and act on the data collected by the sensors. This is what allows the IoT to be not only a collection of connected devices, but also a network of intelligent systems that can make decisions and take actions based on the data they receive.

There are two main approaches to processing in the IoT: edge computing and cloud computing.

Edge computing involves processing data at or near the source, rather than sending it to a centralized location. This can be useful in the IoT because it allows devices to make decisions and take actions based on local conditions, without the need for a constant connection to the internet. For example, consider a smart irrigation system that uses sensors to measure soil moisture levels in a field. If the soil is too dry, and rain is not forecast, the system could use edge computing to turn on the irrigation system without the need to send data to the cloud and wait for a response. This would allow the system to respond more quickly to changing conditions and would also reduce the amount of data that needs to be transmitted over the internet.

Edge computing is often used in applications where low latency or high data volumes make it impractical to send data to the cloud for processing, an extreme example being a self-driving car. It can also be useful in situations where a constant connection to the internet is not available, such as in remote or rural areas.

Edge computing would still usually have a connection to a remote server somewhere, but it wouldn't depend on this for its real-time operation. Instead, it would send periodic, brief

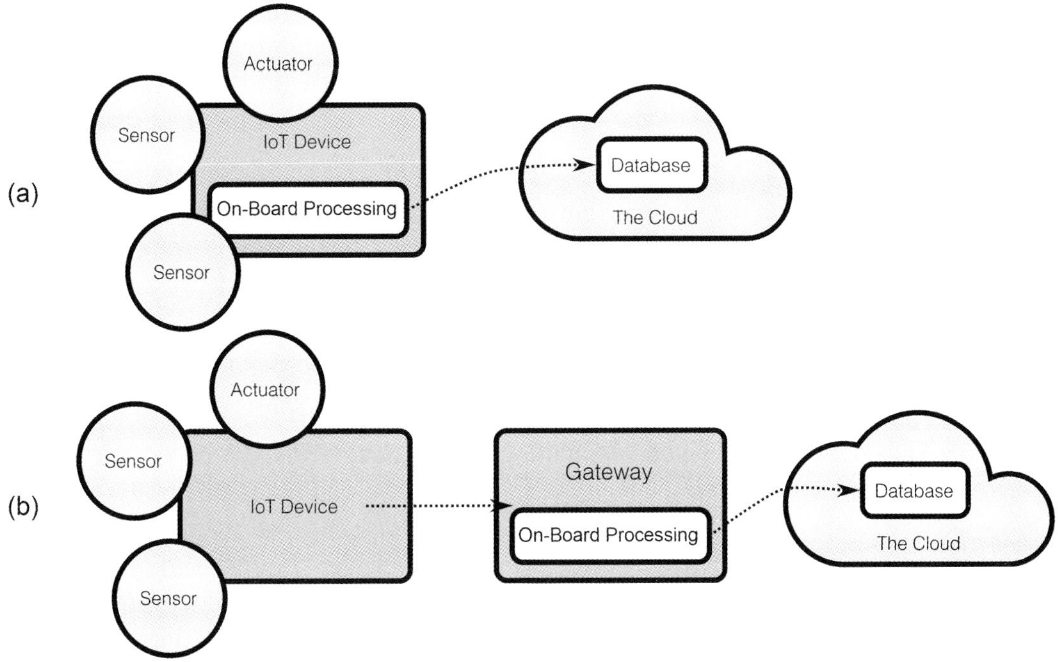

Figure 12.3
Edge computing: Processing is done locally on the device (a) or on a gateway (b), with minimal amounts of data sent to the cloud.

updates to the cloud, allowing remote users to monitor the system, but not actually depending on it for its operation.

Fig. 12.3 gives an example of an edge computing architecture. Fig. 12.3a shows data processing happening on the IoT device itself, with a direct connection to a simple database in the Cloud. This connection could, for example, be through Wi-Fi, giving the IoT device direct access to the Internet, or with another communication technology that requires a gateway. Fig. 12.3b shows data processing happening on a gateway. In this example, the IoT device might communicate with the gateway via BLE, and then the gateway might communicate with the cloud over Wi-Fi. Imagine a smart doorbell IoT system—you'd want your doorbell to work even if your internet connection had gone down, and edge computing is one way to solve this. The doorbell push button could communicate directly with the bell system, telling it to ring when pushed. If the internet connection is working, then a notification can be pushed to your mobile phone.

12.2.4 Processing: Cloud Computing

Cloud computing involves storing and processing data on remote servers, accessed over the internet. This allows devices in the IoT to leverage the processing power and storage capacity of these servers, without the need for expensive hardware in the device itself.

For example, consider a security camera that is part of a smart home system. The camera might transmit video data to the cloud for storage and analysis, allowing the system to identify patterns or anomalies that might indicate a security threat. This would allow the overall system to perform complex tasks, such as facial recognition or motion detection, without the need for expensive hardware in the device itself.

Cloud computing is often used in applications where the data volumes or processing requirements are too high to be handled by the device itself. It can also be useful in situations where it is not practical or cost-effective to install and maintain local processing infrastructure. An additional benefit is that software running in the cloud can be updated quickly and easily, so that everyone benefits immediately from the update, without changes having to be sent to every individual IoT device.

Fig. 12.4 gives an example of a cloud computing architecture. In Fig. 12.4a, the IoT device has a direct connection to the cloud (e.g., over Wi-Fi) and sends data from its sensors directly to cloud applications that perform the processing. The cloud might send instructions back to the IoT device to cause its outputs to change. In Fig. 12.4b, a gateway is involved, but simply acts as a data forwarder. The IoT device generates data and forwards it to the cloud through the gateway. Similarly, the gateway will pass on any messages intended for the device from the cloud.

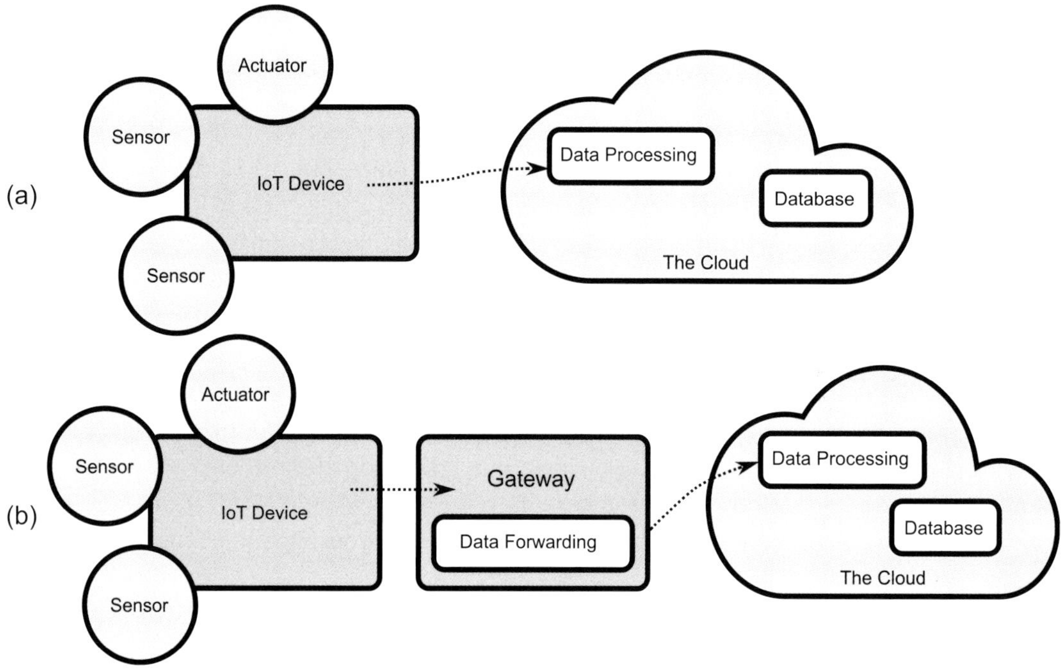

Figure 12.4

Cloud computing: Processing is done remotely in the cloud. The device sends its data directly to the cloud (a) or forwards it through a gateway (b).

12.2.5 Mobile and Web Applications

Mobile and web apps can be used to control and configure IoT systems. These apps allow users to access and manage their IoT devices remotely, often through a smartphone or tablet. Fig. 1.4 from our opening example shows such an app in action.

Mobile apps are particularly useful for controlling IoT systems because they provide a convenient and user-friendly interface that can be accessed from anywhere. For example, a smart home app might allow a user to control their lighting, heating, and appliances from their phone, even if they are not at home.

In addition to providing control, mobile apps can also provide a range of other features and functions, such as monitoring and tracking, alerts and notifications, and automation. For example, a smart irrigation app might allow a user to monitor the moisture levels in their plants and receive notifications if the plants need watering. It is important to note that a mobile app could also act as the gateway for an IoT device, as well as the user interface. See Section 12.3.2 for an example of this style of architecture.

In addition to mobile apps, web applications are commonly used to control and configure IoT systems. Like mobile apps, these web-based applications allow users to remotely access and manage their IoT devices through a web browser on a computer or other device.

Web applications are particularly useful for controlling IoT systems because they can provide a more comprehensive and feature-rich interface than mobile apps. They can also be accessed from any device with a web browser, which can be convenient for users who need to manage their IoT devices from multiple locations.

Web applications can provide many of the same features and functions as mobile apps, such as monitoring and tracking, alerts and notifications, and automation. For example, a smart building web application might allow a facility manager to monitor and control the lighting, heating, and security systems for a building from their computer.

Overall, both mobile apps and web applications are important tools for controlling and configuring IoT systems, and each has its own strengths and capabilities. Many IoT systems use a combination of both, providing users with a range of options for accessing and managing their devices.

12.3 Case Studies

12.3.1 A Smart Lightbulb

Smart lightbulbs are a common example of an IoT system. These bulbs are equipped with sensors and connectivity technologies that allow them to be controlled and configured remotely, often through a smart phone app or voice assistant. As an IoT "thing," a smart lightbulb comprises several key components:

- **Sensors:** Some smart lightbulbs include sensors that are used to gather data about the environment, such as a light sensor to measure the ambient light level or a temperature sensor to measure the ambient temperature, adding value to a more comprehensive home automation system.
- **Processing:** The smart lightbulb includes a processor that is responsible for collecting data generated by the sensors and adjusting the light output of the bulb accordingly, along with forwarding this data to the cloud and receiving any control instructions.
- **Connectivity:** The smart lightbulb includes a connectivity layer that is responsible for transmitting the data collected by the sensors and receiving commands from the user or other devices. This connectivity layer may use a variety of technologics, such as Wi-Fi, Bluetooth, or Zigbee, to transmit and receive data.
- **Actuation:** The smart lightbulb's actuator is the actual light bulb, almost always high-powered LEDs. Often, these LEDs are capable of changing color, so that the user can set a "mood" or adjust the color temperature to suit the environment.

One example of a smart lightbulb is the Philips Hue (Reference 12.4). This bulb can be controlled through a smart phone app or a voice assistant, such as Amazon Alexa or Google Assistant. The app allows users to adjust the color and brightness of the bulb, set schedules for when the bulb should turn on or off, and even create "scenes" that can change the lighting based on the time of day or the user's mood.

In addition to the convenience of being able to control the lighting remotely, smart lightbulbs can also help save energy. For example, the Philips Hue has a "sleep" function that gradually dims the light over a set period of time, helping users fall asleep more easily. It also has a "wake-up" function that gradually increases the light in the morning, helping users wake up more naturally.

Fig. 12.5 shows an example of a smart lightbulb system, in which the smart lightbulb has a direct connection to the cloud, using a home Wi-Fi connection. This means that little additional hardware is required by the end user—for example, no additional equipment other than a wireless home router, which most people have anyway.

One of the challenges of this style of architecture is how to support the initial configuration of the device (in particular, joining the device to a Wi-Fi network) when there is no physical user interface. Often, and as shown in Fig. 12.5, smart devices will include BLE technology to facilitate local configuration, allowing a smart phone to input the Wi-Fi details before the device becomes fully connected to the Cloud.

Once configured, a smart phone can access the system either internally, when the user is on the home Wi-Fi, or remotely, over a cellular connection. Crucially, however, all

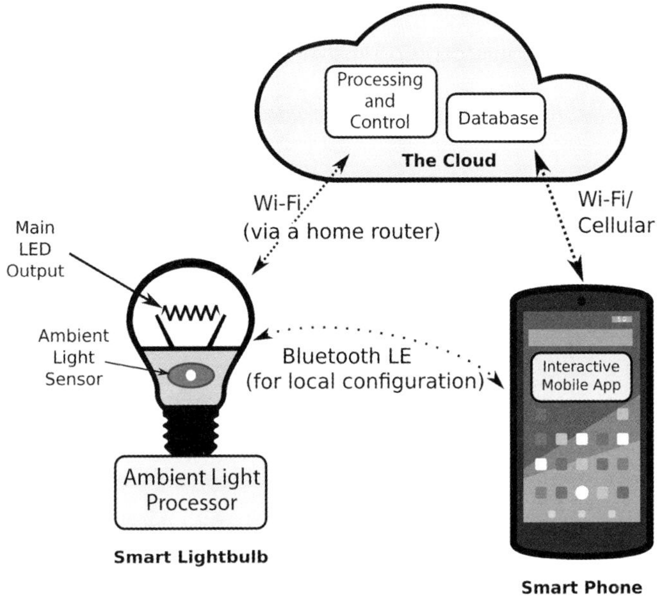

Figure 12.5
An example of a smart lightbulb IoT system.

commands are processed by the cloud—so if the user's home internet connection goes down, the lightbulb will cease to function, even if they are still connected to the local Wi-Fi. In this example, the lightbulb also has an on-board processor, which can dynamically adjust the light output in response to changes in the ambient light, as measured by the ambient light sensor.

12.3.2 A Wearable Fitness Tracker

A wearable fitness tracker is a device that is worn on the body and is used to track physical activity, such as steps taken, calories burned, and distance traveled (e.g., Reference 12.5). Many fitness trackers also have additional features, such as heart rate monitoring, sleep tracking, and GPS tracking. As an IoT system, a wearable fitness tracker consists of several key components:

- **Sensors:** The fitness tracker includes a variety of sensors that are used to collect data about the user's activity, such as an accelerometer to track movement, a heart rate sensor to measure the user's heart rate, and a GPS sensor to track the user's location.
- **Processing:** The fitness tracker includes a processor that is responsible for analyzing the data collected by the sensors and providing insights and feedback to the user. The processor may also be responsible for other tasks such as storing data, managing the device's power, and interacting with other devices.

- **Connectivity:** The fitness tracker includes a connectivity layer that is responsible for transmitting the data collected by the sensors to a central server or to other devices. This connectivity layer may use a variety of technologies, such as Bluetooth or Wi-Fi, to transmit the data.
- **Actuation:** Some fitness trackers may also include actuators that are used to perform actions in response to the data collected by the sensors. For example, a fitness tracker with a vibrating motor might use this actuator to provide haptic feedback to the user.

Fig. 12.6 shows an example of a wearable fitness tracker using a hybrid cloud/edge style architecture. In this example, the fitness tracker contains some sensors for measuring activity and an organic LED display for giving feedback. The tracker communicates with a mobile app over a BLE connection, and transmits the collected sensor data. It also performs a minimal amount of local processing, so that it can show the user's measured heart rate. The mobile app forwards this data to the cloud, which analyzes and processes it, returning a summary to the mobile app to show the user. There is also an interactive web application that the user can access from a desktop or laptop to visualize their data.

12.4 Building a Full-stack IoT System: An Activity Tracker

We turn now to actually developing an IoT system for ourselves. In this comprehensive example, we're going to build a full-stack IoT system from the ground up. Full-stack

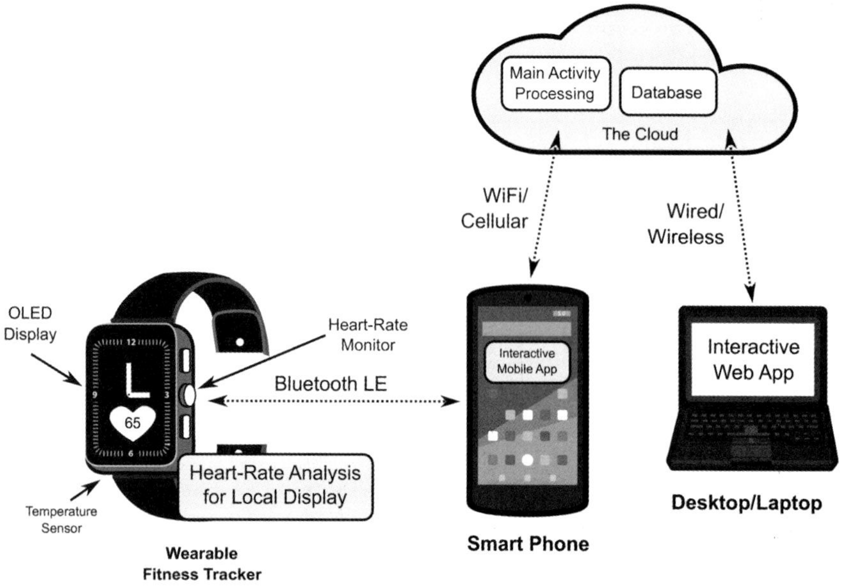

Figure 12.6
An example of a fitness-tracking IoT system that uses a mobile phone as a gateway and as a means of interacting with the system.

means everything from the embedded device through the software running in the cloud to a mobile app to visualize the data. The resources you will need for this are as follows:

- The ST IoT discovery board and a suitable Mbed development environment.
- Access to a Wi-Fi network with an internet connection—note that standard university networks may not be suitable, due to the type of security they employ.
- A Google Cloud account.
- An Android mobile device in developer mode.
- The Android Studio development environment.

We will be using the Android ecosystem in this example, as it has a lower barrier to entry than other mobile development platforms. There are many tutorials available online for Android development, so if you do find yourself stuck, chances are someone has had the same problem and a solution will exist.

Regarding programming, we will be departing from C/C++ and using Java and JavaScript instead. Don't worry if you are unfamiliar with these; all code examples are provided in full, and you won't need to write any original code. Moreover, your knowledge of C/C++ will help you grasp many of the code features. However, if you are familiar with one or both, you will enjoy recognizing the code details and trying your own developments.

12.4.1 Overview

The IoT system we will be building will be the essence of an activity tracker. This will involve periodically sending sensor data from an onboard inertial measurement unit (IMU) sensor to the cloud, processing it, and making the results available to an Android app in close to real-time. We are going to tackle this project in three phases:

1. The cloud
2. The embedded device
3. The mobile app

Fig. 12.7 gives a high-level overview of how the various components you are about to build fit together. The ST IoT discovery board will be programmed to connect to Wi-Fi and record sensor readings from the on-board IMU every second. Every 5 s, the previous five sensor readings will be sent to Google Cloud over an HTTP connection via a Google Cloud function. This function will process the sensor readings and make a prediction of the type of movement activity (whether it was sitting, walking, or running) that occurred. The result will be stored in a Cloud Firestore database. The Android App will periodically (every second) call a different Google Cloud function to retrieve the list of predicted movement activities from the database and display an aggregated version of these in a list on the screen.

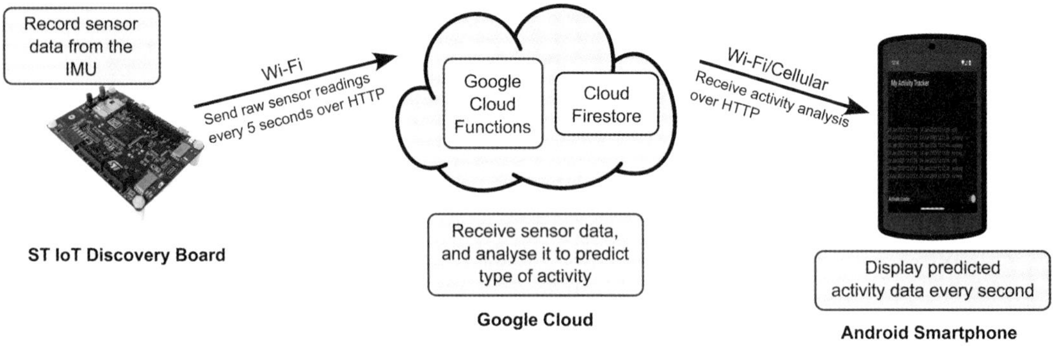

Figure 12.7
A high-level overview of the activity tracker project.

12.4.2 Getting Started with Google Cloud

The first phase of this project is to set up the cloud environment and create the functions necessary for receiving, processing, and accessing data, along with some basic data storage. A "cloud function" is a short snippet of code that lies dormant in the cloud, and is run in response to a triggering event. In this case, we will write our cloud function in the JavaScript programming language and trigger it with an HTTP request. These cloud functions will store and retrieve data from the "Firestore" database service.

Before we can start writing any code, however, you will need to register for a Google Cloud account and create a new Cloud project. If you have a Google account already, you're halfway there—otherwise you will need to register for one. To access the Google Cloud developer's console, navigate to https://console.cloud.google.com, and log in or create a new account. Unless you are using a corporate/university Google Cloud account, you will most likely be asked to enter billing information. This will be required when activating resources such as Firestore, but it is likely you will fall within the free tier for this project. However, please take note of the terms and conditions and the charges that may be incurred if you exceed the daily usage allowance.

From the cloud dashboard, create a new project by selecting "Create Project" from the top navigation bar, calling it, for example, "My Activity Tracker." This new cloud project will contain everything related to this example. Take note of the automatically assigned ID that Google Cloud created for the project, usually the lowercase name of your project, with dashes for spaces and a number at the end. This will be required for some of the code later.

For our cloud functions to be able to access data, we need to create a secret key that gives the functions the necessary permissions to access the database service. To do this, we need to create a service account, which will act as the authenticated user when storing and retrieving data. On the left-hand pull-down menu, navigate to the "Service accounts"

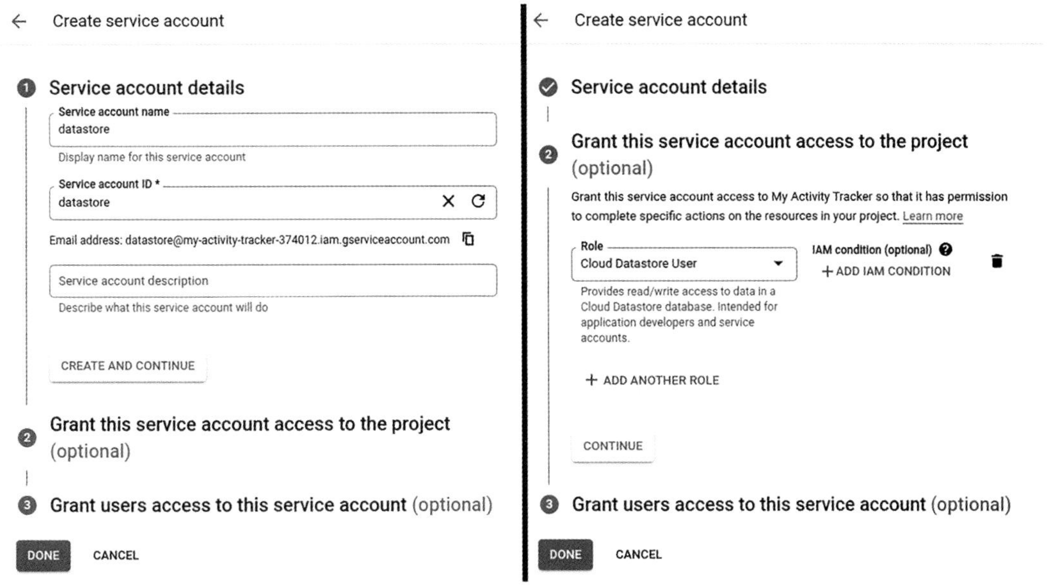

(a) (b)

Figure 12.8
Stages in creating a service account: (a) opening screen; (b) permissions screen.

section of the "IAM and admin" section (where IAM stands for "identity and access management") and choose "Create Service Account." Here, enter a name for the service account, for example, "datastore," and give it the "Cloud Datastore User" permission. You may need to search for this role in the dropdown box. Fig. 12.8 shows further the two successive screens you will need to complete. Clicking "Create and Continue" on the first leads to the second screen. When this is complete, click "Continue."

Once this account is created, and on the same screen, navigate to the "Keys" tab. This is where we will create the secret key. Click "Add Key" and choose "Create new key." Then select the JSON key type (Fig. 12.9). Clicking "Create" will now create this key and download the JSON file to your computer. Keep this JSON secret key file safe (and secure!).

12.4.3 Firestore

Our IoT project will require a place to store the various bits of data that are being passed around the system. We will set this up now, so that our cloud functions can access it.

Navigate to the "Firestore" section of the Google Cloud console. You'll find this in the menu on the left, underneath the "Databases" heading (Fig. 12.10). You might find it convenient to pin this service to the list, as we will be visiting it frequently.

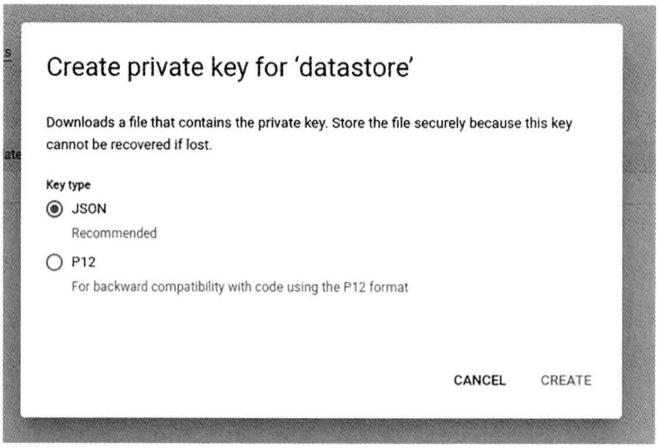

Figure 12.9
Creating a JSON private key for the service account.

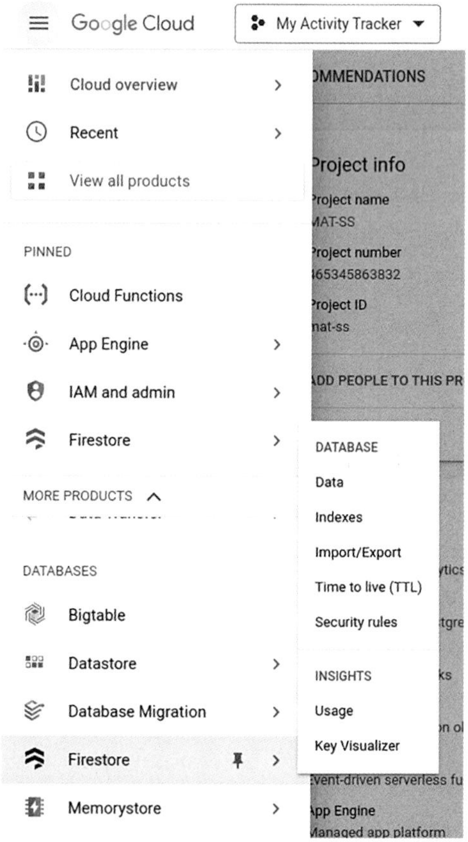

Figure 12.10
Locating Firestore in the Google Cloud developer console.

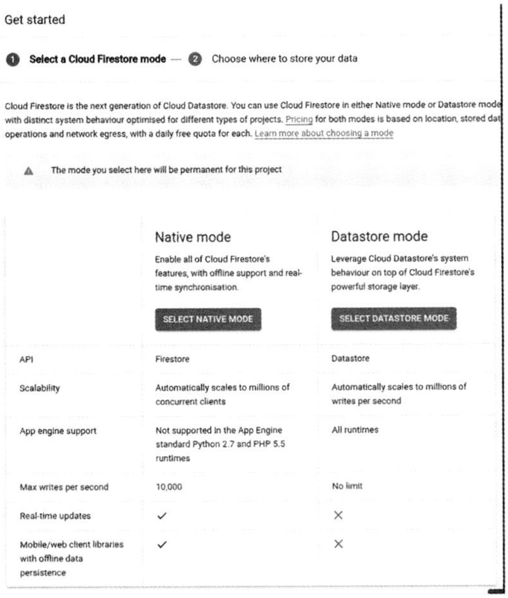

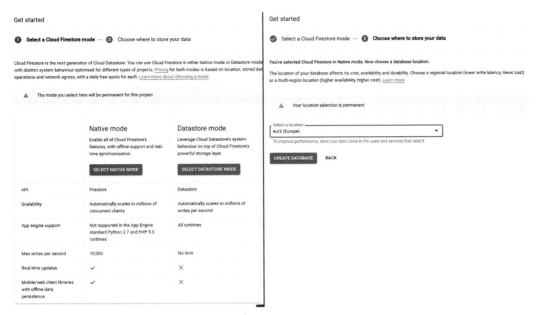

Figure 12.11
Activating Firestore.

When you first access this section, you will be prompted to activate the Firestore service. Choose to activate the service in "Native Mode" (Fig. 12.11) and choose a location geographically close to you for your data storage. Choosing a nearby location is beneficial for two main reasons: (1) the data is physically closer to you, and so faster to access, and (2) there may be legal requirements that prohibit you from storing data in locations outside your own country. This location cannot be changed later, and you should use the same one for any other services you activate. Click "Create Database."

After a few minutes, the Firestore database service will be active and ready to use.

12.4.4 Cloud Functions

We are going to develop two cloud functions for dealing with data. Both of these functions will be used by some part of the IoT system. The **processSensorData( )** function will be called by the IoT discovery board when the embedded device is uploading sensor data to the cloud. The **getActivityData( )** function will be called by the Android app to retrieve the processed activity data.

Navigate to the "cloud functions" section of the Google Cloud console. You'll find this in the menu on the left, underneath the "Serverless" heading (Fig. 12.12). Again, you might find it convenient to pin this service to the list.

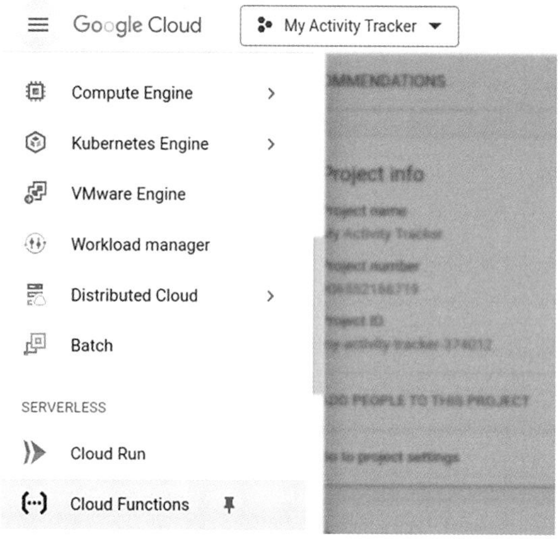

Figure 12.12
Locating cloud functions in the Google Cloud developer console.

Function 1 processSensorData( )

Click "Create Function" to create a new cloud function. You will be presented with several configuration options. The defaults will mostly suffice, but you should modify the following settings:

- **Function name:** "processSensorData"
- **Trigger - Authentication:** Choose "Allow unauthenticated invocations."
- **Trigger - Require HTTPS:** De-select this option.

HTTPS is an encrypted form of HTTP, which requires cryptography libraries and certificates to be correctly installed on the device. We return to it in Chapter 13. Because this cloud function will be triggered by the embedded device, it needs to permit unauthenticated invocations, so that the device does not have to log in to use the function. Additionally, it makes the coding on the device much more straightforward if HTTPS is not enforced.

Save the Trigger setup and click "Next" to continue creating the function. You will be presented with an editor that allows you to enter code. There are a few things we need to change here. Firstly, enter "process" in the "Entry point," and then take the code from Program Example 12.1 and put it into the editor for the **index.js** file, overwriting the sample code you find there. In the code you will need to change **'YOUR-PROJECT-ID'** to the ID of your project, which was generated when you created it. This should also be contained within the single quotation marks.

Next, modify the **package.json** file and replace it with the contents of Program Example 12.2. Finally, create a new file called **key.json** (click the + button), and copy in the contents of the multi-digit JSON secret key that you created and downloaded in Section 12.4.2. This should be exactly in the form "———BEGIN PRIVATE KEY———\n [PRIVATE-KEY]\n———END PRIVATE KEY———\n" – nothing more, nothing less.

Program Example 12.1 Contents of index.js in the processSensorData( ) Cloud Function

```
// Import the Firestore module
const Firestore = require('@google-cloud/firestore');

// Connect to the Firestore database
const db = new Firestore({
  projectId: 'YOUR-PROJECT-ID',
  keyFilename: 'key.json'
});

// Function to compute arithmetic mean over values in arr
function computeMean(arr) {
  return arr.reduce((a, b) => a + b) / arr.length;
}

// Function to compute standard deviation over values in arr
function computeStdDev(arr) {
  const mean = computeMean(arr);
  const variance = arr
    .map(x => Math.pow(x - mean, 2))
    .reduce((a, b) => a + b) / arr.length;
  return Math.sqrt(variance);
}

// Function to make an activity prediction, based on
// aggregated sensor data
function predict(data) {
  if (data.acceleration_x_std < 10 &&
      data.acceleration_y_std < 10 &&
      data.acceleration_z_std < 10) {
    return {
      label: 'still'
    };
  } else if (data.acceleration_x_std < 100 &&
             data.acceleration_y_std < 100 &&
             data.acceleration_z_std < 100) {
    return {
      label: 'walking'
    };
  } else {
    return {
      label: 'running'
    };
  }
}
```

```
// Cloud Function entry point
exports.process = async (req, res) => {
  // Read and validate incoming sensor data message
  const messageData = req.body;
  if (messageData === undefined || messageData === null) {
    console.error('Unable to read message');
    res.status(400).send();
    return;
  }

  // Extract individual readings for accelerometer
  // and gyro axes into arrays.
  const accelX = messageData.readings.map(r => r.accel.x);
  const accelY = messageData.readings.map(r => r.accel.y);
  const accelZ = messageData.readings.map(r => r.accel.z);
  const gyroX = messageData.readings.map(r => r.gyro.x);
  const gyroY = messageData.readings.map(r => r.gyro.y);
  const gyroZ = messageData.readings.map(r => r.gyro.z);

  // Compute mean and standard deviation over readings
  const inputData = {
    acceleration_x_mean: computeMean(accelX).toFixed (6),
    acceleration_x_std: computeStdDev(accelX).toFixed (6),
    acceleration_y_mean: computeMean(accelY).toFixed (6),
    acceleration_y_std: computeStdDev(accelY).toFixed (6),
    acceleration_z_mean: computeMean(accelZ).toFixed (6),
    acceleration_z_std: computeStdDev(accelZ).toFixed (6),
    gyro_x_mean: computeMean(gyroX).toFixed (6),
    gyro_x_std: computeStdDev(gyroX).toFixed (6),
    gyro_y_mean: computeMean(gyroY).toFixed (6),
    gyro_y_std: computeStdDev(gyroY).toFixed (6),
    gyro_z_mean: computeMean(gyroZ).toFixed (6),
    gyro_z_std: computeStdDev(gyroZ).toFixed (6)
  };

  // Make the activity prediction
  const prediction = predict(inputData);
  console.log('Activity data processed: ', prediction);

  // Insert the prediction into the database
  const timestamp = Date.now()
  await db.collection('activity-data').add({
    timestamp,
    activity: prediction.label
  });

  // Return a successful result
  res.status(200).send();
};
```

Program Example 12.2 Contents of package.json in the processSensorData( ) Cloud Function

```
{
  "name": "processSensorData",
  "version": "0.0.1",
  "dependencies": {
    "@google-cloud/firestore": "~6.4.2"
  }
}
```

There seems to be a lot going on in **index.js**, so let's break it down. At the top of the code are some definitions that give the rest of the code access to the datastore, where we will be storing data. Then we define several helper functions to compute the mean and standard deviation of a list of numbers. The **predict( )** function makes a simple prediction of activity by looking at the standard deviation of the accelerometer in the sensor data that came in from the embedded device. In the function entry point itself, the data is received from the device and split up into its constituent parts. Then that extracted data is passed into **predict( )**, which returns either "still," "walking," or "running," based on its simple heuristic. Once the prediction is made, it is stored in the database.

You can finally click "Deploy" to save the function—it will take a couple of minutes to activate. Once it is ready, you can run a test invocation from the "Testing" tab. Here, you can enter some simulated activity data into the testing input (the section on the left called "Configure triggering event") and run the function by clicking "Test the function." If the function succeeds, you should see a prediction stored in the database if you navigate back to the Firestore section. To try this, use the following test data:

```
{
    "readings":
    [
        {
            "accel": {"x": 1, "y": 2, "z": 3},
            "gyro": {"x": 1, "y": 2, "z": 3}
        }
    ]
}
```

To be able to run this function externally, that is, from the Discovery board, you will need to add invocation permissions for all users. In the "Permissions" tab of the cloud function, click "Grant Access" and set "allUsers" as the principal. Choose "Cloud Functions Invoker" for the role and save this permission (Fig. 12.13).

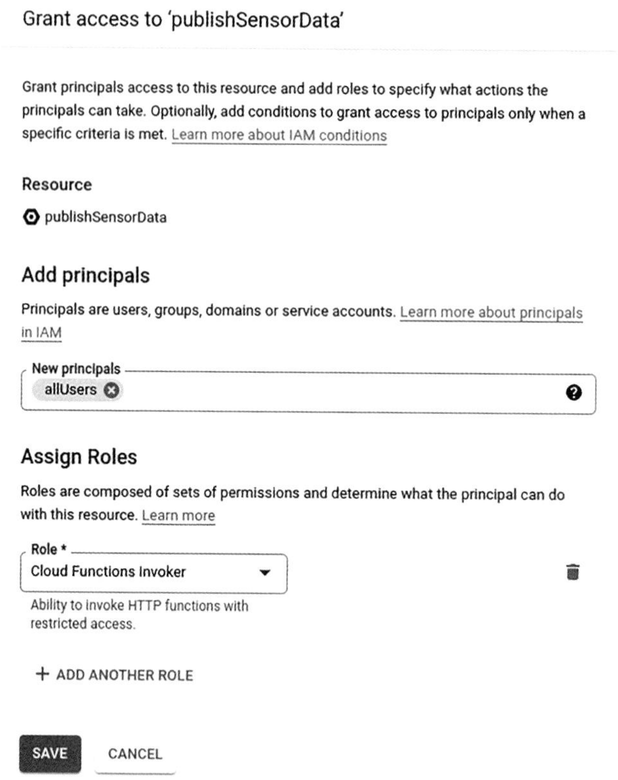
Grant access to 'publishSensorData'
Grant principals access to this resource and add roles to specify what actions the principals can take. Optionally, add conditions to grant access to principals only when a specific criteria is met. Learn more about IAM conditions
Resource
publishSensorData
Add principals
Principals are users, groups, domains or service accounts. Learn more about principals in IAM
New principals
allUsers
Assign Roles
Roles are composed of sets of permissions and determine what the principal can do with this resource. Learn more
Role *
Cloud Functions Invoker
Ability to invoke HTTP functions with restricted access.
+ ADD ANOTHER ROLE
SAVE
CANCEL

Figure 12.13
Adding permissions for all users to invoke the cloud function.

Function 2 getActivityData( )

This is a very straightforward function that simply accesses the prediction data and returns it. It will be used by the Android App to access the predictions made by the **processSensorData( )** function. Create a function as before, using the same settings but instead calling it "getActivityData".

In the code section, you'll need to modify **package.json** and add the Firestore dependency, as in Program Example 12.2. For the entry point, use the function name "getData," and use the code from Program Example 12.3 for the contents of **index.js**. Finally, upload the contents of your secret key to a new file called **key.json**, as before.

```
// Import the Firestore module.
const Firestore = require('@google-cloud/firestore');

// Connect to the Firestore database.
const db = new Firestore({
  projectId: 'YOUR-PROJECT-ID',
  keyFilename: 'key.json'
});
```

```
exports.getData = async (req, res) => {
  // Permit cross-site invocation.
  res.set('Access-Control-Allow-Origin', '*');

  if (req.method === 'OPTIONS') {
    // Settings for cross-site invocation.
    res.set('Access-Control-Allow-Methods', 'GET');
    res.set('Access-Control-Allow-Headers', 'Content-Type');
    res.set('Access-Control-Max-Age', '3600');
    res.status(204).send('');
  } else {
    // Retrieve activity data from the DB.
    const coll = db.collection('activity-data');
    const activityDataSnapshot = await coll.orderBy('timestamp').get();
    const activityData = activityDataSnapshot.docs.map(doc => doc.data());

    // Return the activity data as a JSON object.
    res.json(activityData);
  }
};
```

Program Example 12.3 Contents of index.js in the getActivityData( ) Cloud Function

Here, the function checks to see if the request type is OPTIONS and sends back information that allows the browser to make a "Cross-Site Request". If the request type is anything else, the database is accessed and the list of activity data is sent back as a JSON object.

Once you have edited the code, click "Deploy" to publish the function. After a few minutes, the function will be available, and you should be able to test it. You will also need to add the "Cloud Functions Invoker" permission to this function to enable external access from the Android app.

At this point, we've created (and hopefully tested!) all the necessary cloud infrastructure for our IoT system and can move on to the next phase.

12.4.5 Preparing the Embedded Device

It's now on to programming the embedded device and getting it to communicate with the cloud. For this to work on the IoT Discovery board, you will need a Wi-Fi connection. Your home Wi-Fi connection is probably suitable for this, but university networks generally are not, as they often use WPA Enterprise encryption, which is not easily supported by the device. Some universities do, however, maintain separate networks for experimentation such as this.

To get started, create a new Mbed OS 6 project in your IDE, and make sure you have the following libraries downloaded to the project folder (see Section 2.6.2):

- The Board Support files for the IoT Discovery board: https://os.Mbed.com/teams/ST/code/BSP_B-L475E-IOT01/ (Reference 7.9)
- The ISM43362 Wi-Fi component: https://github.com/ARMMbed/wifi-ism43362/ (select the 'master' branch)
- An Mbed OS 6 HTTP client: https://github.com/rasmus0201/Mbed-http-client.git (select the 'master' branch)

The main application will take a sensor reading every second, so start by using the code from Program Example 12.4, in **main.cpp**.

```
#include "http_request.h"
#include "mbed.h"
#include "stm32l475e_iot01.h"
#include "stm32l475e_iot01_accelero.h"
#include "stm32l475e_iot01_gyro.h"
#include <sstream>
#include <vector>

static void initialiseSensors() {
  BSP_ACCELERO_Init();  // Initialise the acceleromter
  BSP_GYRO_Init();  // Initialise the gyro
}

// A struct to record sensor readings
struct SensorData {
  float gyro[3];
  int16_t accel[3];
};

// A list of sensor readings
static std::vector<SensorData> sensorReadings;

static void recordSensors() {
  SensorData data;

  // Read the sensor data
  printf("Recording sensors...\n");
  BSP_GYRO_GetXYZ(data.gyro);
  BSP_ACCELERO_AccGetXYZ(data.accel);

  // Add the sensor data to the list
  sensorReadings.push_back(data);

  printf("Accel: %d %d %d, Gyro: %f %f %f\n", data.accel[0], data.accel[1],
         data.accel[2], data.gyro[0], data.gyro[1], data.gyro[2]);
}

static void tick(NetworkInterface *net) { recordSensors(); }
```

```
int main() {
  initialiseSensors();

  auto net = NetworkInterface::get_default_instance();

  printf("Connecting to the network...\r\n");

  // Connect to the network
  nsapi_size_or_error_t result = net->connect();
  if (result != 0) {
    printf("Error! net->connect() returned: %d\r\n", result);
    return -1;
  }

  SocketAddress ipaddr;
  net->get_ip_address(&ipaddr);
  printf("Connected with IP address: %s\r\n",
         ipaddr.get_ip_address() ? ipaddr.get_ip_address() : "(none)");

  // Collect sensor readings every second
  while (true) {
    tick(net);
    ThisThread::sleep_for(1s);
  }

  return 0;
}
```

Program Example 12.4 A skeleton application that takes sensor readings every second

To get this working, you will need to create an **Mbed_app.json** file to include your Wi-Fi connection details. Program Example 12.5 shows what the contents of this file should be. Make sure you replace **<YOUR-SSID>** and **<YOUR-PASSWORD>** with the appropriate values, taken from your Wi-Fi hub.

```
{
    "target_overrides": {
        "*": {
            "nsapi.default-wifi-security": "WPA_WPA2",
            "nsapi.default-wifi-ssid": "\"<YOUR-SSID>\"",
            "nsapi.default-wifi-password": "\"<YOUR-PASSWORD>\"",
            "platform.stdio-baud-rate": 115200,
            "rtos.main-thread-stack-size": 8192,
            "target.printf_lib": "std"
        },
        "DISCO_L475VG_IOT01A": {
            "target.components_add": ["ism43362"],
            "ism43362.provide-default": true,
            "target.network-default-interface-type": "WIFI"
        }
    }
}
```

Program Example 12.5 The Mbed_app.json file, with placeholders for Wi-Fi connection details

Finally, you can compile and upload this program to your development board. Run the program and check the output on a serial terminal (setting a data rate of 115,200 baud) to make sure that your program is correctly displaying sensor values every second. So far, the program is behaving very much like Program Example 7.12.

12.4.6 Sending Movement Data

At this stage, we have checked that the on-board IMU is working, and we have the basics of our fitness tracker implemented. It is now just a case of sending the data to the cloud by triggering the **processSensorData( )** function we created earlier. This function expects a batch of sensor data, so we are going to modify our code to send blocks of five samples. This means that we will continue to collect sensor readings every second, and then every fifth collection we will send the data we have collected in that period.

To do this, we will modify the **tick( )** function to detect when there are five samples in the buffer and call a new function **sendSensorReadings( )**. See Program Example 12.6 for the code—note that this new code replaces the existing tick function, which sits in the program listing as a single line just before **main()**. You will need to replace **<YOUR-TRIGGER-URL>** in the program example with the trigger URL for your own **processSensorData( )** cloud function. You can find this URL in the "Trigger" tab of the cloud function configuration section.

```
static void sendSensorReadings(NetworkInterface *net) {
  printf("Sending %d samples...\n", sensorReadings.size());

  auto *req = new HttpRequest(
      net, HTTP_POST,
      "<YOUR-TRIGGER-URL>");

  req->set_header("Content-Type", "application/json");

  stringstream body;
  body << "{ \"readings\": [";

  bool first = true;
  for (const auto &reading : sensorReadings) {
    if (first) {
      first = false;
    } else {
      body << ",";
    }
    body << "{";
    body << "\"accel\": { \"x\": " << reading.accel[0]
         << ", \"y\": " << reading.accel[1] << ", \"z\": " << reading.accel[2]
         << " },";
```

```
    body << "\"gyro\": { \"x\": " << reading.gyro[0] << ", \"y\": " << reading.gyro
[1] << ", \"z\": " << reading.gyro[2] << " }";
    body << "}";
  }
  body << "] }";

  auto bodyString = body.str();
  printf("JSON: %s\n", bodyString.c_str());

  HttpResponse *post_res = req->send(bodyString.c_str(), bodyString.length());
  if (!post_res) {
    printf("Http request failed (error code %d)\n", req->get_error());
  }

  delete req;
  sensorReadings.clear();
}
static void tick(NetworkInterface *net) {
  recordSensors();
  if (sensorReadings.size() >= 5) {
    sendSensorReadings(net);
  }
}
```

Program Example 12.6 The routines to send sensor data to the cloud

You can now build and run this code on your device, and after five sensor readings are taken (i.e., every 5 s), you should be able to see activity data appearing in the Firestore database. To observe this, navigate to the "Firestore" section in the Google Cloud console, and click on the "activity-data" collection. New activity data will appear in the list as it is generated.

12.4.7 Building the Mobile App: Designing the User Interface

The final part of this full-stack IoT project is to build the Android app that will retrieve the predicted activity data and display it in a simple user interface. Building an entire mobile application can be a complex, daunting, and challenging task, so we are only going to touch on the basics of Android development. If you feel comfortable—or adventurous—however, feel free to explore how you can turn this basic app into something better! You may at this stage like to make use of some of the introductory training material on Android Studio that is available on line, starting for example with Reference 12.6. In this project, we will be developing in the Java programming language, although the Kotlin language is also available to use.

To get started with Android development, you will need an Android phone that is in developer mode (details available online on how to do this for your specific model) and an installation of the Android Studio on your PC. Android Studio is available here: https://developer.android.com/studio. Here we are using Version 2022.2.1. USB debugging should be enabled on your phone.

Once you have installed Android Studio, open it and create a new project, selecting **File > New > New Project**. Then, choose the Phone and Tablet "Empty Views Activity" template (be careful not to select "Empty Activity"), click Next, and give the project a suitable name. We use here "My Activity Tracker." Make sure that the programming language is set to "Java" (Kotlin may be the default), accept the further defaults shown, and then click "Finish."

With the project created, plug your development mobile device into your computer with USB. Your device should appear in the "Device Manager" window on the Android Studio toolbar. Select it, click the Run (arrowhead) button to its right, and after a few moments the empty App will appear on your mobile device, as in Fig. 12.14. If you have trouble linking your device to the computer, you will probably see a box saying "No Devices" on

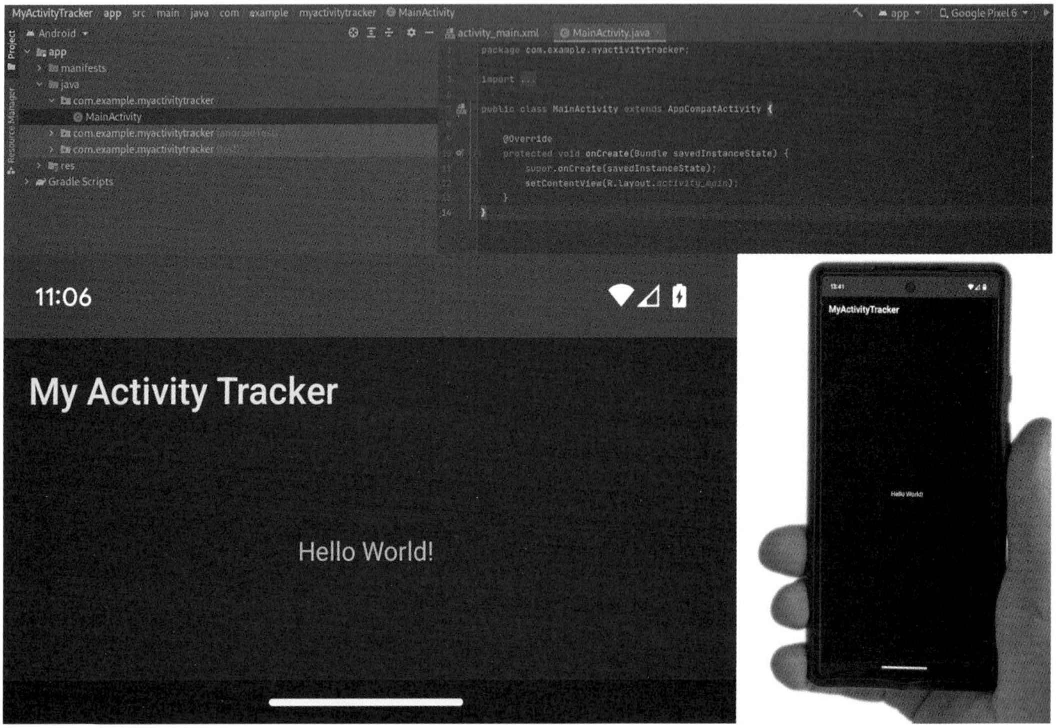

Figure 12.14
Android Studio and the Hello World application running on a mobile device.

the toolbar. You can invoke "Troubleshoot Device Connections" via its pull-down menu, which can be helpful. Note that you may need to allow access every time you reconnect your phone to Android Studio.

When this is working, we are ready to start developing!

This app will provide a switch which, when turned on, will interrogate the previously built cloud function every 5 s and request the data. It will then build a list of activity periods, which are the durations of particular activities, such as walking or running. These activity periods will be shown in a table on the main screen.

The first thing we will do is create the user interface (UI). Android UIs are hierarchical, in that you have components within containers, which can also be in containers. This hierarchy can be seen in the UI designer. Our UI will comprise the following hierarchy:

- Constraint Layout
  - Progress Indicator
  - Scrollable Layout
    - Results Table
  - Read Activities Timer Switch

Start by opening the **activity_main.xml** file in Android Studio by selecting the file from the source-code list on the left. You will find it in **app > res > layout**. You should be presented with a UI designer similar to Fig. 12.15. This file allows us to create the visual layout of the application.

At the top left of the Android Studio screen you can see a "Palette" of components that can be used in the designer. The bottom left shows the "Component Tree" and is very useful for making sure everything is structured correctly. The central main pane is the designer interface itself and is where we will be adding components. On the right is the "Attributes" pane, which presents the various options for the components. Notice you can expand or contract the phone screen image by clicking on the + and − buttons.

Start by deleting the existing "Hello World!" text label (click it and press Delete), and insert a progress indicator by navigating to the "Widgets" section in the palette and dragging the "Progress Bar" component onto the design surface. Note that you can also drag components and drop them directly into the Component Tree, with the same effect.

All components need a unique identifier and will be given one by default. To make it clearer what this is referring to, change the component's ID by looking in the "Attributes" pane on the right and changing the "id" property to **loadingProgress**. A dialog box will ask you to confirm this by clicking "Refactor."

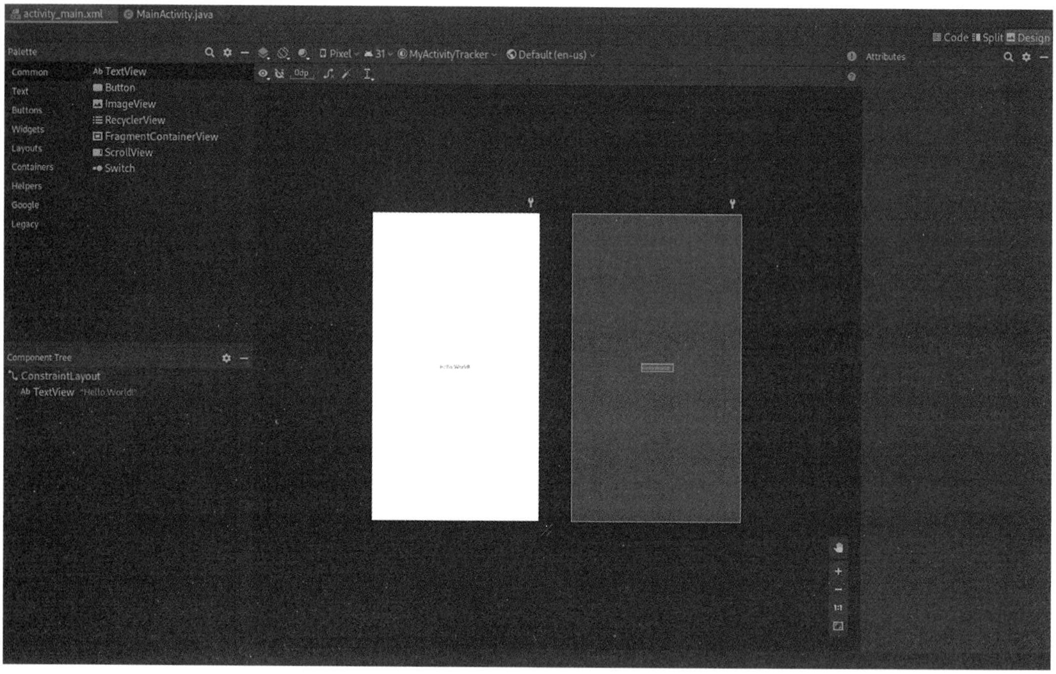

Figure 12.15
The Android Studio User interface designer.

Next, navigate to the "Containers" section of the Palette and drag in a **ScrollView** component. Once added, it should fill the remaining available area. Inside the **ScrollView** is a **LinearLayout** component that was automatically added. Delete this by expanding the **ScrollView** component in the component tree, selecting **LinearLayout**, and pressing Delete.

Navigate to the "Layouts" section of the Palette and drag a **TableLayout** component into the middle of the **ScrollView** component. This should appear as a child of the **ScrollView** component in the Component Tree. Give this an ID: **resultsTable**. Expand the **TableLayout** in the Component Tree and delete all of the automatically created **TableRow** components.

Finally, navigate to the "Buttons" section of the Palette and add a **Switch** component. Drag this directly into the **ConstraintLayout** component in the Component Tree so that it does not become a child of the **ScrollView**. Give the switch an ID, **activateLoaderSwitch**, and give it some descriptive text by modifying the "text" attribute, such as "Activate Loader."

The **ConstraintLayout** is a way of laying out components so that they sit relative to each other, and we will be using this concept to position everything nicely.

Figure 12.16

The main layout of the application, with a loading progress indicator, a display table, and a toggle switch.

Now position the switch, by dragging it into a central position and then constraining the bottom of it to the parent. Do this by dragging the bottom anchor point downwards. Next, position the **ScrollView** by dragging its bottom anchor point, to the top anchor point of the switch, so that it sits in between the **ProgressBar** and the **Switch**. If you have done this, your layout and component tree should look something like Fig. 12.16.

At this point, you can give your app a try by clicking the "Run" button at the top right of the toolbar and seeing it appear on your mobile device.

If everything worked, you should see the progress bar spinning at the top, a blank area in the middle, and a toggle switch (which changes state when you tap it) at the bottom—this is illustrated in Fig. 12.17.

The design of the user interface is now complete, and we can move to the coding.

12.4.8 Building the Mobile App: Reading and Processing Activity Data

We can now start writing code to read and process activity data. There is quite a bit of code to get this working, so we will build it up and test it as we go. In the following pages, code sections are fitted together, piece by piece, with accompanying descriptions. This is informative, but it is quite possible to make a slip as you construct the code. Therefore, you may wish to download the full code from the book website, and use that as a point of reference as you progress from here.

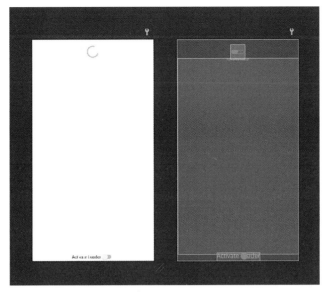

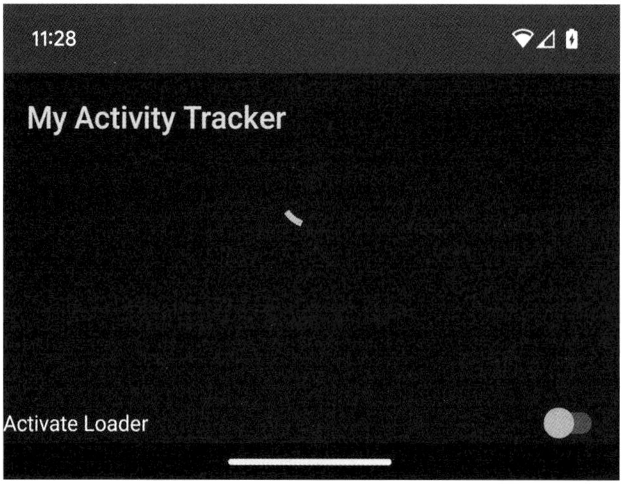

Figure 12.17
The skeleton app running on a mobile device.

Open **MainActivity.java** to see where we will be writing the Java code. In this file there already is a class that represents the "activity" (Android parlance for "view" or "screen") of the app we have just designed. It contains a method, **onCreate( )** (a Java method is similar to a C function), which is invoked as the view is being created, and before it is shown to the user.

The first thing to tackle is getting hold of references to the user interface elements that we want to alter. Program Example 12.7 shows how this process is started. Enter this into **MainActivity.java**, replacing the skeleton code sections which are there.

```
public class MainActivity extends AppCompatActivity {
    private Switch activateLoaderSwitch;
    private View loadingIndicator;
    private TableLayout resultsTable;

    @Override
    protected void onCreate(Bundle savedInstanceState) {
        super.onCreate(savedInstanceState);
        setContentView(R.layout.activity_main);

        loadingIndicator = findViewById(R.id.loadingProgress);
        loadingIndicator.setVisibility(View.INVISIBLE);

        activateLoaderSwitch = findViewById(R.id.activateLoaderSwitch);
        resultsTable = findViewById(R.id.resultsTable);
    }
}
```

Program Example 12.7 Adding fields and modifying onCreate( ) to obtain references to UI elements

As you enter code into the IDE, it may start complaining about missing classes—but it will help you! If you see an error (such as "cannot resolve"), navigate to the error, and it will suggest a fix in a popup. The fix is almost always "import class"—and the IDE will automatically insert import statements at the top of the source code for you. In Program Example 12.7, we've created three fields:

- **activateLoaderSwitch:** for referencing the activateLoaderSwitch UI element
- **loadingIndicator:** for referencing the progress bar
- **resultsTable:** for referencing the table that will show our activities

In the **onCreate( )** method, we have acquired references to those elements by their ID, which we specified in the UI designer. We have also hidden the progress bar, by setting its visibility property to **INVISIBLE**, as we only want this to appear when the app is loading data.

Next, we need to detect when **readActivitiesSwitch** has changed, that is, the user has turned it on or off. To do this, we need to register a checked change event listener, which involves adding an interface implementation to the class, and then adding a new method after **onCreate( )**. See Program Example 12.8 for the modifications.

```
public class MainActivity extends AppCompatActivity implements
CompoundButton.OnCheckedChangeListener {
  @Override
  protected void onCreate(Bundle savedInstanceState) {
    super.onCreate(savedInstanceState);
    setContentView(R.layout.activity_main);

    loadingIndicator = findViewById(R.id.loadingProgress);
    loadingIndicator.setVisibility(View.INVISIBLE);

    activateLoaderSwitch = findViewById(R.id.activateLoaderSwitch);
    activateLoaderSwitch.setOnCheckedChangeListener(this);
    resultsTable = findViewById(R.id.resultsTable);
  }

  @Override
  public void onCheckedChanged(CompoundButton buttonView, boolean isChecked)
  {
    if (buttonView == activateLoaderSwitch) {
      if (isChecked) {
        // When the switch is turned ON
      } else {
        // When the switch is turned OFF
      }
    }
  }
}
```

Program Example 12.8 The modified MainActivity class, which detects changes to the switch

When the read activities switch is turned on, we want to schedule an action to occur every 5 s: namely, make a web request to the cloud function and process the results. To do this, we need to use a **Timer** object. Add a new field in the class definition:

```
private Timer readerTimer;
```

And update the **onCheckedChanged( )** method:

```
if (isChecked) {
  readerTimer = new Timer();
  readerTimer.scheduleAtFixedRate(new TimerTask() {
    @Override
    public void run() {
      // This is invoked every FIVE seconds
    }
  }, 0, 5000);
} else {
  readerTimer.cancel();
  readerTimer = null;
}
```

What is happening here is when the switch is turned on, a new **Timer** object is created, and a task is scheduled to run immediately, and then every 5000 ms after that. If the switch is turned off, the timer is cancelled and the object reference disposed.

We will now turn our attention to the action we are going to perform when the timer elapses. To do this, we need to use something called a **Handler**, because we need to make sure the timer action is performed in the same thread as the user interface. It is not possible to change UI elements from other threads, as this would make thread synchronization very tricky, so instead we use Handlers to perform actions on the UI thread. Add two new fields to the **MainActivity** class definition:

```
private Handler doLoadDataHandler;
private boolean loading;
```

Then, inside the **onCreate( )** method, add the following code at the end:

```
MainActivity outer = this;
doLoadDataHandler = new Handler(this.getMainLooper()) {
  @Override
  public void handleMessage(Message msg) {
    outer.doLoadData();
  }
};
```

What is happening here is that we are creating a new **Handler** object and storing it in the **doLoadDataHandler** field. This handler calls another method, **doLoadData( )**, on the main class. This is for convenience, so that we can keep our processing logic together and out of the way of everything else. We also need to add the **doLoadData( )** method to the class:

```
private void doLoadData() {
  if (this.loading) {
      return;
  }

  this.loading = true;
  loadingIndicator.setVisibility(View.VISIBLE);
}
```

Finally, we need to invoke the handler in the timer run method:

```
readerTimer.scheduleAtFixedRate(new TimerTask() {
  @Override
  public void run() {
    doLoadDataHandler.obtainMessage().sendToTarget();
  }
}, 0, 5000);
```

What's exciting now is that we are at a stage where we can perform a quick test of our app, to ensure that things are working up to this point. When you start the app (click Run in the IDE), you should be able to click on the switch at the bottom, and the loading indicator should be displayed. We haven't written any other logic at the moment, so the indicator will not disappear—but it should be encouraging that we have gotten this far.

To actually read the data from the Cloud, we need to add another field to the **MainActivity** class definition:

```
private RequestQueue requestQueue;
```

A request queue is a way to dispatch an HTTP request and wait for the response without blocking any threads. Add the following line to the end of the **onCreate( )** method:

```
requestQueue = Volley.newRequestQueue(this);
```

The request queue is ready to use. The IDE will at first complain about the "Volley" class being missing, but as before, it will suggest including a reference to the library automatically for you. We also need to add two more interface implementations to the class definition:

```
public class MainActivity extends AppCompatActivity implements
CompoundButton.OnCheckedChangeListener, Response.Listener<String>,
Response.ErrorListener {
```

And then add corresponding methods:

```
@Override
public void onResponse(String response) {
  System.out.println(response);

  loadingIndicator.setVisibility(View.INVISIBLE);
  this.loading = false;
}

@Override
public void onErrorResponse(VolleyError error) {
  System.out.println(error.getMessage());

  loadingIndicator.setVisibility(View.INVISIBLE);
  this.loading = false;
}
```

We can also add the following lines to the end of **doLoadData( )**, which will trigger the request:

```
String url = "URL_TO_CLOUD_FUNCTION";
this.requestQueue.add(new StringRequest(Request.Method.GET, url, this, this));
```

You will need to fill in the URL string with the correct path to the cloud function we created previously, that is, the **getActivityData( )** function. As before, you can get this from the Google Cloud console when you go into the function settings and navigate to the "Trigger" tab.

There is now enough code to debug calls to the cloud function, but we need to give the app permission to access the internet before we can test it. To do this, open the **manifests/AndroidManifest.xml** file and add the following element before **</manifest>** at the bottom:

```
<uses-permission android:name="android.permission.INTERNET" />
```

Now, click the Play button and toggle the switch. If you click on the "Run" tab at the bottom of the IDE, you will see debug messages appear, which should contain the result of calling the cloud function. The last piece of the puzzle is to parse the resulting data and display it in the table.

We'll need a helper function to add rows to the table, so add the **appendActivityPeriod( )** method in Program Example 12.9 into the **MainActivity** class.

```
private void appendActivityPeriod(Date periodStart, Date periodEnd, String
activity) {
  TableRow row = new TableRow(this.getBaseContext());

  DateFormat fmt = DateFormat.getDateTimeInstance();

  TextView startTimestamp = new TextView(this.getBaseContext());
  startTimestamp.setText(fmt.format(periodStart));
  row.addView(startTimestamp);
```

```
    Space s = new Space(this.getBaseContext());
    s.setMinimumWidth(32);
    row.addView(s);

    TextView endTimestamp = new TextView(this.getBaseContext());
    endTimestamp.setText(fmt.format(periodEnd));
    row.addView(endTimestamp);

    Space s2 = new Space(this.getBaseContext());
    s2.setMinimumWidth(32);
    row.addView(s2);

    TextView activityText = new TextView(this.getBaseContext());
    activityText.setText(activity);
    row.addView(activityText);

    resultsTable.addView(row);
  }
```

Program Example 12.9 The appendActivityPeriod( ) helper method, which adds a row to the results table

This method might look complicated, so let's break it down. The method takes three parameters (**periodStart**, **periodEnd**, and **activity**), which are the three cells that will be added to the table as one row. A new row object is created, and a **DateFormat** object is created for displaying date strings. Then each cell is created (via a **TextView** object) and added to the new row, with a spacer between them.

Replace the **onResponse( )** method with the code from Program Example 12.10.

```
  @Override
  public void onResponse(String response) {
    try {
      resultsTable.removeAllViews();

      JSONArray activities = new JSONArray(response);

      JSONObject lastActivity = null;
      Date activityPeriodStart = null;
      for (int i = 0; i < activities.length(); i++) {
        JSONObject activity = activities.getJSONObject(i);
        System.out.println(activity);

        // Check to see if there was a previous activity
        if (lastActivity != null) {
          // Now, compare the activity types
          if (!lastActivity.getString("activity")
         .equalsIgnoreCase(activity.getString("activity"))) {
            // If the activity types are different, then terminate
            // the current activity period, and start a new one.
            appendActivityPeriod(
```

```
            activityPeriodStart,
          new Date(activity.getLong("timestamp")),
          lastActivity.getString("activity"));
        activityPeriodStart = new Date(activity.getLong("timestamp"));
      }
    } else {
      // There was no previous activity, so start a new activity period.
      activityPeriodStart = new Date(activity.getLong("timestamp"));
    }

    lastActivity = activity;
  }
} catch (JSONException e) {
    e.printStackTrace();
} finally {
  loadingIndicator.setVisibility(View.INVISIBLE);
  this.loading = false;
}
}
```

Program Example 12.10 The response handler that runs when the cloud function returns a result

Let's take a look at the high-level structure of this function, which comprises a **try**...**catch**...**finally** block. This is a code construct related to "exception handling," which allows us to detect errors that might occur in the functions we use; look for these three keywords in the code. Inside the **try** block, we can run code that might produce an error—or more precisely, raise an exception. If the code does raise an exception, it immediately stops and jumps to the **catch** block, provided that the **catch** block is defined for the right type of exception. The **finally** block is run regardless of whether or not the code raised an exception, and is always executed last (i.e., after the **try** block finished successfully, or the **catch** block completed).

In this example, we encapsulate our main functionality in the **try** block, to detect errors. If we detect an error, then we print the error out to the console for debugging. In either case, in our **finally** block, we hide the loading progress bar and clear the **loading** flag.

Inside the **try** block comes the main part of the implementation. First, any existing elements in the table are removed. Next, we parse the returned data into a **JSONArray** structure, so that we can iterate over it in the **for** loop. For each activity that is returned to us by the cloud, we check to see if it is the same type as the one we have just seen (in which case we merge it), or if it is different, then we start a new activity period. When we have detected a complete activity period, we use our helper method, **appendActivityPeriod( )**, to add it to the table.

Try this out now, by clicking the Run button. Once the app has loaded, flick the switch, and wait for the data to come in.

Figure 12.18
The finished mobile application running on a mobile device.

Now turn on your IoT Discovery board, with Program Example 12.6 running, and ensure the Google Cloud interface is active. Once the board has connected to the cloud, every 5 s you should see new data coming into the phone and being added to the end of the table, something like Fig. 12.18. Make sure you simulate walking and running on your device by giving it a shake—remember that data is merged into time periods, so you will need to simulate a few different activities. Two or more successive time periods of the same activity type are merged into one line, as the figure shows.

12.5 Summary

The Internet of Things (IoT) is a rapidly growing technology that is transforming the way we live, work, and interact with the world around us. It involves the use of sensors, connectivity, and processing to enable the seamless exchange of information between devices and systems.

The project we have just undertaken combines and demonstrates a number of the key features of the IoT. We have implemented Fig. 12.7, and can reasonably view that as a subset of Fig. 12.1, the IoT overview. Therefore, with the expertise gained from this project and from other parts of the book, you should be able—with care—to build up systems of increasing complexity and sophistication. There's certainly plenty of complexity in what we have done, yet it represents only a tiny fraction of what is possible.

The IoT has the potential to revolutionize a wide range of industries, from healthcare and transportation to agriculture and manufacturing. It is already being used in a variety of applications, such as smart cities, connected vehicles, and smart homes.

However, the rapid growth of the IoT has also raised concerns about privacy, security, and the impact on society and the environment. Researchers and policymakers are working to address these challenges and ensure that the technology is used in a responsible and ethical manner.

Despite these challenges, the IoT is expected to play a major role in shaping the future of our world. It will continue to evolve and become an increasingly integral part of our lives, bringing with it both opportunities and challenges that we will need to navigate as a society.

12.6 Mini Projects

12.6.1 Smart Lightbulb Simulation

Simulate a smart lightbulb using the IoT Discovery Board, by using the on-board LED as the light source. Build a mobile app that allows you to turn this LED on and off.

12.6.2 Reconfiguring the Activity Tracker Example

Reconfigure the example in this chapter to use the mobile app as a gateway, by communicating with the IoT Discovery Board over BLE and forwarding the information to the cloud server. You can tackle this by taking inspiration from the example in Chapter 11 and exploring the Android BLE API (Reference 12.7). Effectively, Fig. 12.7 is reconfigured, with the IoT Board linking to the phone, which then links to the cloud. See Fig. 12.19 for how this might look.

Chapter Review

- The Internet of Things (IoT) is a global network of connected devices, sensors, and other physical objects.
- The four main components of an IoT system are sensors and actuators, connectivity, processing, and applications.
- Sensors are devices that collect data about the environment, such as temperature, light, and motion. Actuators are devices that perform actions in response to the data collected by the sensors.
- Processing in IoT systems includes both edge computing and cloud computing. Edge computing refers to data processing at or near the source of data, whereas cloud computing refers to data processing in a centralized location.

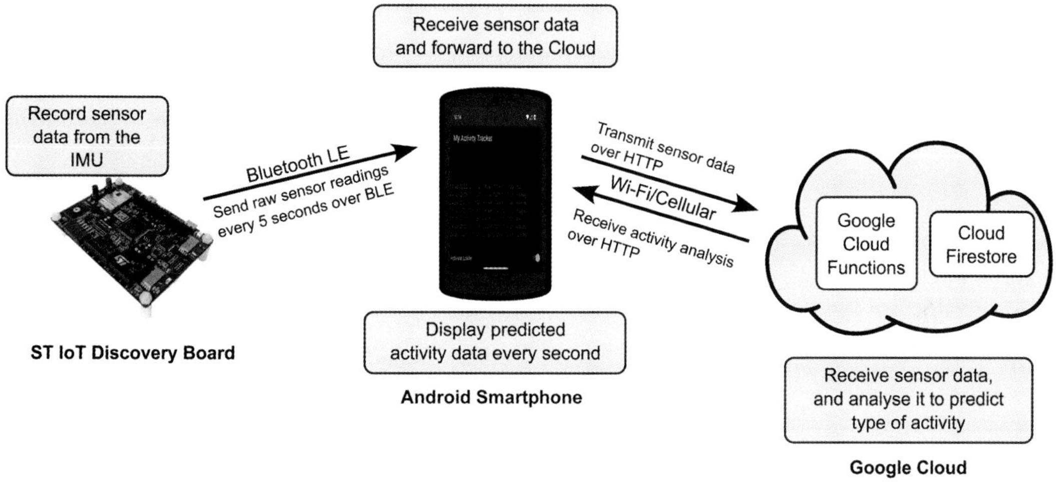

Figure 12.19
A reconfiguring of the Activity Tracker example, to use the smartphone as a gateway.

- Connectivity in IoT systems enables communication between devices and servers. It can be device-to-device or device-to-server communication.
- Apps are the mobile or web applications that let users interact with an IoT system, and offer a way to configure, control, and visualize the data.
- Public cloud services, such as Google Cloud, Microsoft Azure, and Amazon AWS, make building scalable IoT systems easier and more cost-effective for developers. They provide the various resources on demand and offer a wide range of services that can be used for IoT deployments.
- Android Studio and the Android development ecosystem make it easy to get started with mobile app development, much as the Mbed Studio IDE makes it easy to get started with embedded development!

Quiz

1. List all the sensors that have appeared in the book so far. For each, indicate a role it could play in an IoT system.
2. In reference to connectivity, what is the difference between a "direct" connection and an "indirect" connection to the internet?
3. What is edge computing?
4. Why is BLE especially useful for the Internet of Things?
5. Google cloud functions are an implementation of a so-called "serverless" execution environment. What does "serverless" mean, and why is it a misnomer?
6. What are the advantages of using HTTP in the Internet of Things?

7. How does HTTPS (HTTP Secure) extend HTTP?
8. What wireless technologies might be suitable for an IoT system that comprises a large number of sensors that are placed over a wide geographical area, such as a field or forest? List the pros and cons of each.
9. You have been tasked with designing an IoT system to manage an automated production line, that is capable of tracking stock levels at various points in the manufacturing process. What strategies might you employ to ensure that the production line continues to work, even if upstream connectivity to the cloud is lost? Sketch a diagram of this architecture, specifying each component and the role it plays.
10. Your company is developing a "smart" heat-resistant protective suit for fire-fighters, as illustrated in Fig. 12.20. Sensors are embedded in the suit as shown. Overall, they give an indication of the wellbeing of the fire-fighter. Any of them could, however, indicate a life-threatening situation. The sensors are all wired back to a single microcontroller, which transmits data every 10 s, by wireless link. Data from the protective suit is needed in the fire engine cab, in the fire station, by the company developing the suit, and for national fire service records. Sketch and explain an IoT system appropriate for this, indicating what IoT-related equipment should be available, and where.

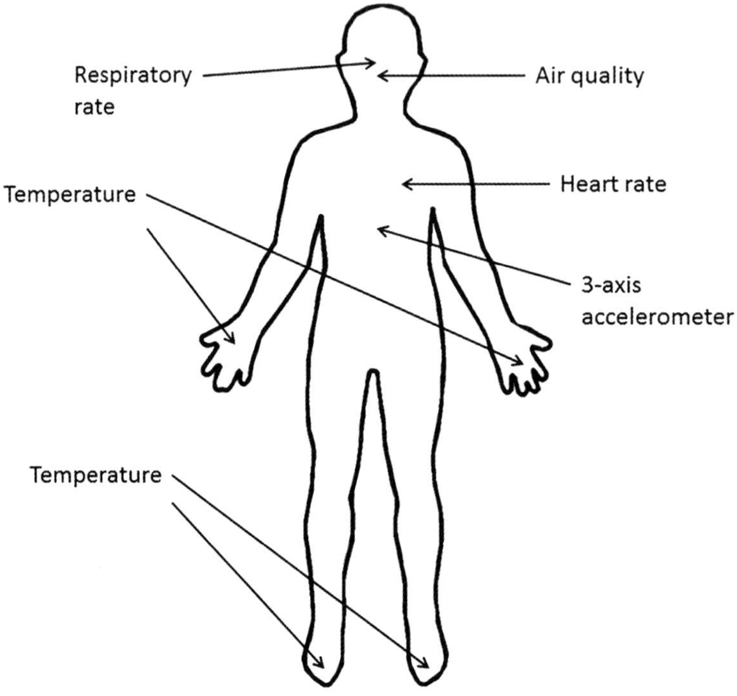

Figure 12.20
A "smart" heat-resistant protective suit.

References

1. H. Fernandez-Lopez, J. A. Afonso, J. H. Correia, and R. Simoes, Remote patient monitoring based on ZigBee: Lessons from a real-world deployment. https://www.ncbi.nlm.nih.gov/pmc/articles/PMC3880127/.
2. The Little Known Story of the first IoT Device. https://www.ibm.com/blogs/industries/little-known-story-first-iot-device/.
3. The Internet of Things (IoT)—Market Forecast. https://nishithsblog.wordpress.com/2014/04/27/the-internet-of-things-iot-market-forecast.
4. The Philips Hue Smart Lightbulb. https://www.philips-hue.com/en-gb.
5. Garmin Fitness and Health Tracking Products. https://www.garmin.com/en-GB/c/sports-fitness/activity-fitness-trackers/.
6. Build Your First Android App. https://developer.android.com/training/basics/firstapp.
7. Android Bluetooth LE. https://developer.android.com/guide/topics/connectivity/bluetooth/ble-overview.

CHAPTER 13

Further Aspects of the IoT

The previous chapter gave a high-level overview of the IoT and how the various components fit together. We saw that the actual "things" are embedded devices which contain sensors and/or actuators and a way to communicate, and that servers and applications receive and process data to run the IoT systems, and provide feedback and control to the user.

In this chapter, we will explore in greater detail how some of this enabling technology works, and what is actually going on "under the hood". Unlike other chapters, this one comes without practical exercises, and is intended simply to increase your theoretical background understanding. We do, however, still link the chapter content to the Mbed OS, which has features appropriate for all chapter sections. Most Mbed OS APIs relevant to the chapter are grouped under the heading "Connectivity," as seen in Table 2.2 and Reference 13.1.

As we have seen, the IoT relies on a variety of communication protocols and strategies to communicate with other devices or servers in the cloud, or equivalent processing setups. Chapter 11 introduced wireless communication technology, and Chapter 12 gave examples of communication in practice. In the following sections, we are going to take a look at network communication in more detail, and explore how these technologies that power our world today fit together.

13.1 Ethernet

We will start by taking a look at the Ethernet protocol, as it is the technology underpinning all modern-day wired communication networks. It may seem a bit strange to focus on a wired protocol, but this is a technology that is heavily involved in communication over the Internet—even if the initial message comes from a wireless device. At some point, Ethernet is used to get data from A to B, and it is important to understand how it works, and its role in the wider communication space.

Ethernet is a serial protocol which is designed to facilitate network communications, particularly on LANs and WANs. Any device connected to an Ethernet network can communicate with any other device connected to the same network. As seen in Table 11.1, Ethernet communications are defined by the IEEE 802.3 standard (see Reference 13.2) and support fast data rates up to 800 Gbps (Gigabits per second). On the physical layer,

Fast and Effective Embedded Systems Design. https://doi.org/10.1016/B978-0-323-95197-5.00013-9

Ethernet uses differential send (TX) and receive (RX) signals, resulting in four wires labeled RX+, RX-, TX+, and TX-.

Ethernet messages are communicated as serial data packets referred to as *frames*. Using frames allows a single message to hold several data values, including a value defining the length of the data packet as well as the data itself. The Ethernet frame therefore defines its own size. Ethernet communications need to pass a large quantity of data at high rates, so data efficiency is a very important aspect; by defining the data size of each packet, rather than relying on a fixed size for all messages, the number of empty data bytes are minimized. Each frame includes a unique source and destination Media Access Control (MAC) address. A MAC address is a 6-byte identifier (usually specified as six hexadecimal numbers separated by colons) that has been allocated partly by the IEEE, and partly by the device manufacturer. This allocation strategy enables MAC addresses to be globally unique, as there is a single registry that allocates numbers from this address space. The frame is wrapped within a set of *preamble* and *start of frame* (SOF) bytes and a *frame check sequence* (FCS), which enables devices on the network to understand the function of each communicated data element. The standard 802.3 Ethernet frame is constructed as shown in Table 13.1.

The minimum Ethernet frame is 72 bytes. However, the preamble, SOF, and FCS are often discarded once a frame has been successfully received. So a 72 byte message "on the wire" is often reported as having just 60 bytes by some Ethernet vendors.

The original Ethernet protocol utilizes the *Manchester encoding* method, which relies on the direction of the edge transition within the timing window, as shown in Fig. 13.1. This shows first a string of zeros being transmitted, then a string of ones, and then alternating ones and zeros. If the edge transition within the timing frame is high-to-low, the coded bit is a 0; if the transition is low-to-high, then the bit is a 1. The Manchester protocol is very simple to implement in integrated circuit hardware and, as there is always a switch from 0 to 1 or 1 to 0 for every data value, the clock signal is effectively embedded within the data. As also shown in Fig. 13.1, even when a stream of zeros (or ones for that matter) is being transmitted, the digital signal still shows transitions between high and low states.

Table 13.1: Ethernet frame structure

Preamble	Start of Frame Delimiter	Destination MAC address	Source MAC Address	Length	Data	Frame Check Sequence	Interpacket Gap
7 bytes of 10101010	1 byte of 10101011	6 bytes	6 bytes	2 bytes	46 – 1500 bytes	4 bytes	

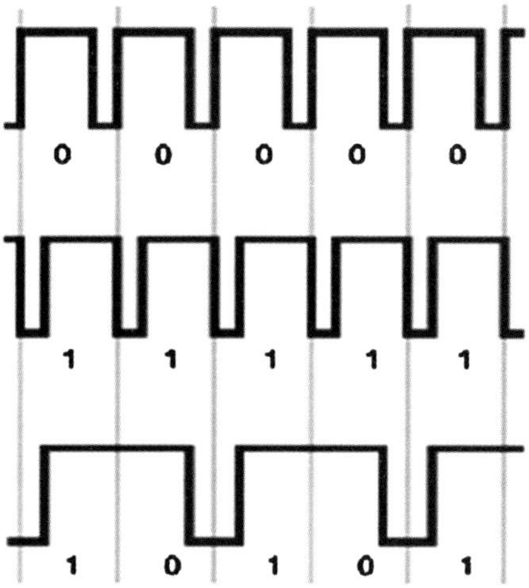

Figure 13.1
Manchester encoding for Ethernet data.

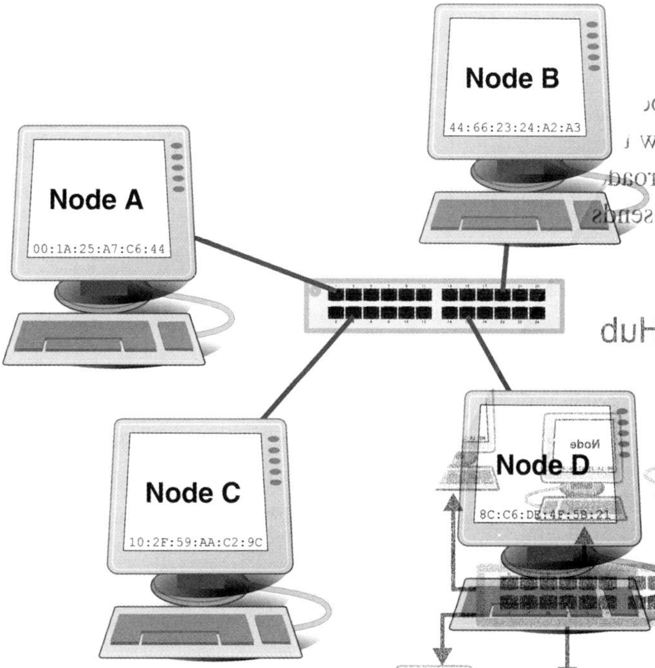

Figure 13.2
Ethernet network star topology.

To build an Ethernet network, devices or "nodes" are connected with each other through either a network switch or a hub. This creates a star topology, as shown in Fig. 13.2. Every node in the network has a unique MAC address, which allows it to be individually addressed and therefore communicated with.

When a data packet is being sent between nodes, the sending node encodes its own MAC address in the "Source MAC Address" field of the frame, and the intended recipient of the packet in the "Destination MAC Address." If the packet is intended for a group of nodes, a special broadcast MAC address is used.

A network switch and a hub are both devices that connect multiple nodes together on a network, but they function quite differently.

A hub is a basic networking device, which effectively connects every device to every other device. When a device sends a data packet to a hub, the hub broadcasts the packet to all connected devices, regardless of their intended destination. This means that all devices connected to the hub receive the packet, but only the device with the matching destination address should process it. This can lead to a lot of unnecessary network traffic and slow down the network's performance.

On the other hand, a switch is a more advanced networking device, which uses the node's MAC addresses to forward data packets to their intended destinations. When a node sends a data packet to a switch, the switch examines the destination MAC address in the packet header and forwards the packet only to the node with the matching address. The switch monitors network traffic to understand which MAC addresses correspond to which port, and this results in better use of the network bandwidth and therefore less network traffic. Fig. 13.3 shows how this arrangement looks. On the left is a hub, and when a device sends a packet, the hub broadcasts that packet to all connected devices. On the right is a switch, and when a device sends a packet, the switch forwards that packet only on the port for the intended device.

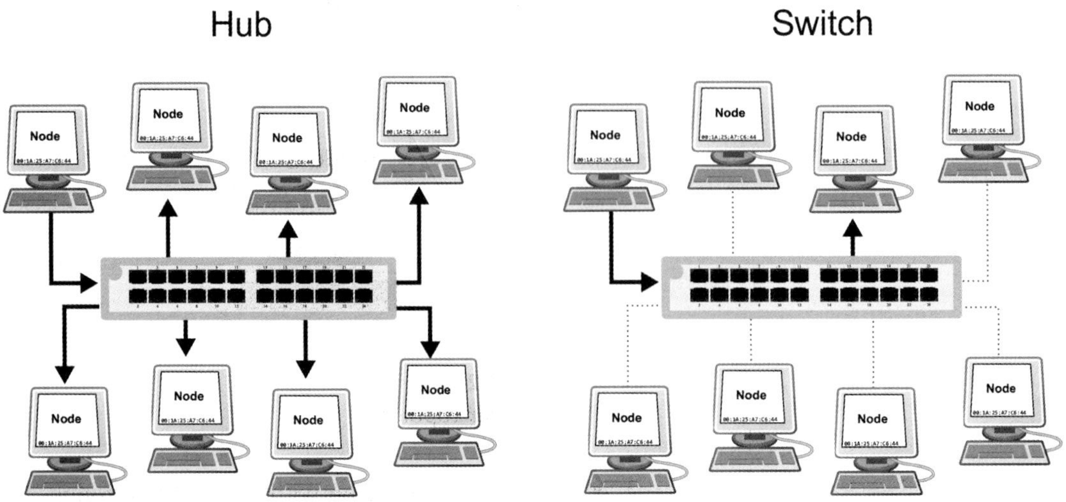

Figure 13.3
Network hubs vs. switches.

A link to the Mbed OS Ethernet API is given in Reference 13.3, along with a simple example program. This could be tried on the Mbed LPC1768, which has an ethernet interface. It is not available for the Nucleo F401RE or IoT Discovery board, as they do not have ethernet connectivity.

13.2 Wi-Fi

In the IoT example in Chapter 12, we used a Wi-Fi connection to link our embedded device to the wider internet. Thankfully, the Mbed OS provides us with an easy-to-use interface to access wireless hardware, and we need not concern ourselves with the details. But it is important to understand what's actually happening, how Wi-Fi networks work, and how the data can get from A to B.

Wi-Fi is a wireless networking technology that allows devices to connect to a network and communicate with each other without the need for physical cables. It operates on the 2.4-GHz and 5-GHz radio frequency bands and uses the IEEE 802.11 standards for communication (Table 11.1). Wi-Fi is based on a set of standards and protocols that define how wireless devices can access a network, authenticate themselves, and exchange data.

The history of Wi-Fi can be traced back to the late 1960s, when the U.S. government began funding research into wireless networking technology. In the 1980s, the IEEE began working on a wireless networking standard, which eventually led to the creation of the 802.11 standard in 1997. This standard, known as 802.11b, provided speeds of up to 11 Mbps and had a range of about 300 feet. In the following years, the IEEE developed faster and more advanced versions of the 802.11 standard, called 802.11a, 802.11g, and 802.11n, which increased speeds and range. The most recent version is 802.11ax (Wi-Fi 6), which was ratified in 2019, and promises to increase speed and capacity by several times compared with the previous version.

Wi-Fi networks can be set up and configured in different ways, depending on the requirements of the user. A basic setup can be created with a single access point that connects to a wired network and broadcasts a wireless signal. More complex setups, such as those in large buildings or public spaces, can use multiple wireless access points to provide coverage over a wider area.

Just as in Ethernet, data in a Wi-Fi network is transmitted and received using a series of frames. These frames are called 802.11 frames, as they are defined by the various 802.11 standards. The two main types of frames are management frames and data frames.

Management frames are, as they sound, used for network management and control tasks such as connecting a device to a network, performing authentication, and managing network connections. These include the following:

- **Beacon frames:** transmitted periodically by an access point to advertise its presence and provide information about the network.
- **Probe request frames:** sent by a device looking for available networks to connect to.
- **Probe response frames:** sent by an access point in response to a probe request to provide information about the network.
- **Association request frames:** sent by a device to request access to a network.
- **Association response frames:** sent by an access point to grant or deny access to a network.

Data frames are used to transmit data between devices on a network. Some examples are the following:

- **Data frames:** contain the payload of the data being transmitted.
- **Control frames:** used for flow control and error correction.
- **Acknowledgment frames:** sent by a device to confirm receipt of a data frame.
- **Request-to-send/clear-to-send frames:** used for managing access to the wireless medium and avoiding collisions.

Every frame has a header that contains information such as the source and destination MAC addresses, the frame type, and other control information. The header is followed by the payload, which contains the actual data being transmitted.

There are two main types of devices that take part in a Wi-Fi network: access points and stations.

An access point (AP) is a device that connects wireless devices to a wired network. An access point acts as a central hub for wireless devices and allows them to communicate with each other and with wired devices on the network—just like a network switch/hub. Access points typically have a wired connection to a switch, which provides connectivity to the rest of the wired network and beyond.

A station (STA) is a device that connects to a wireless network through an access point. A station could be a laptop, smartphone, tablet, or any other device that has wireless capabilities—like the ST IoT Discovery board! Stations use the wireless network to communicate with other devices on the network, as well as with devices on the Internet or other networks.

When a station wants to connect to a wireless network, it sends a probe request to the access point. The access point responds with a probe response, providing information about the network such as the network name (also known as the service set identifier, or SSID), the type of security used, and the supported data rates. Once the station receives this information, it can decide whether to connect to the network.

If the station decides to connect to the network, it sends an association request to the access point. The access point responds with an association response, either accepting or rejecting the connection, based on the network's security settings, the number of connected devices, and other factors. This flow is demonstrated in Fig. 13.4.

Once the station is associated with the access point, it can communicate with other devices on the network. The access point is responsible for managing the wireless network, including controlling the wireless medium, authenticating and managing connections, and routing data between the wired and wireless network.

All 802.11 frames are structured in the same way, and contain several fields in their headers. Some fields are unused for some frame types. The following fields are present in the header:

- **Frame control:** This field contains information about the type of frame, such as various control flags, including whether the frame is a data frame, the type of data, and whether the frame is protected by encryption.
- **Duration/ID:** This field contains the duration of the transmission in microseconds or an identification number for the transmission.
- **Address fields:** These fields contain the source and destination MAC addresses, as well as the address of the device that the frame will be sent to next.
- **Sequence control:** This field contains a sequence number for the frame and a fragment number, used for fragmentation and reassembly of large data packets.

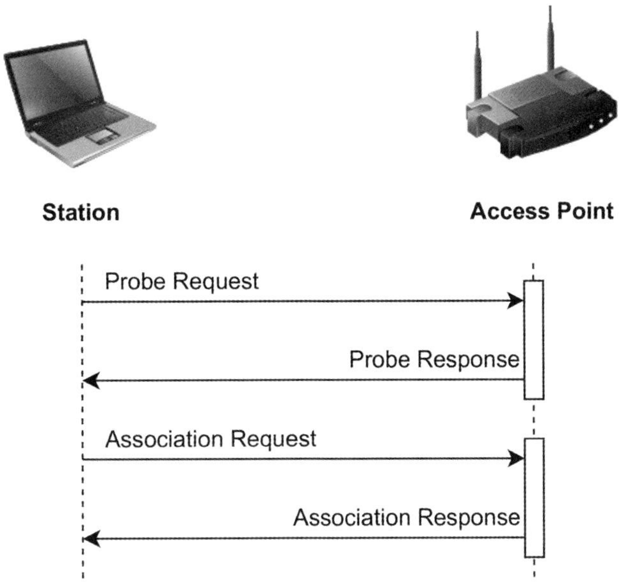

Figure 13.4
A wireless station connecting to an access point.

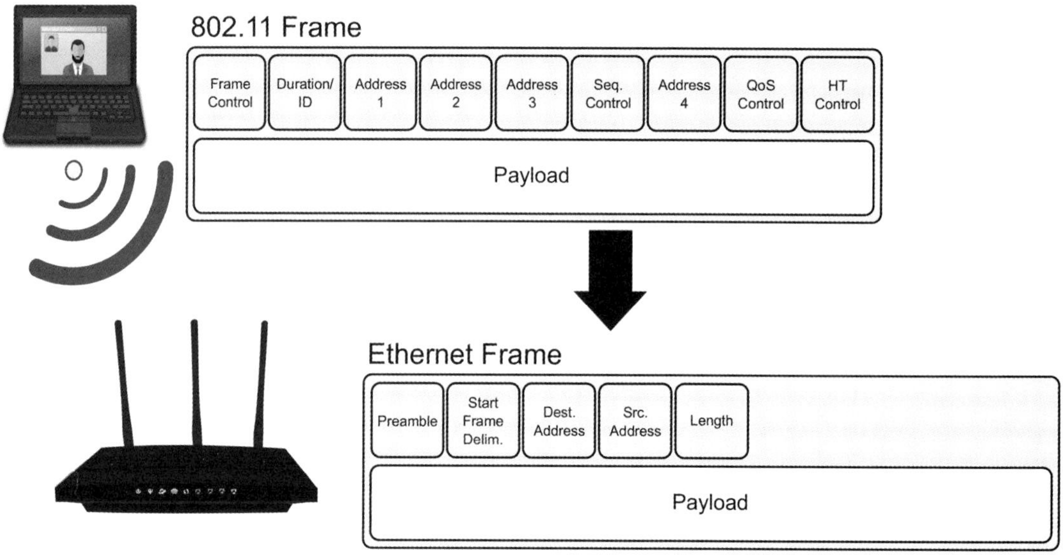

Figure 13.5
An access point translating an 802.11 frame into an Ethernet frame.

- **Quality of service (QoS):** This field contains information about the priority and other characteristics of the data being transmitted.
- **High-throughput (HT) control:** This field contains control data for facilitating high data rates, and is specified by the 802.11n variant of the protocol.

Following the header is the frame body, which contains the payload of the data being transmitted.

A data frame may also contain a "trailer" that contains the integrity check value or the cyclic redundancy check. It is the job of the access point to translate an 802.11 data frame into an Ethernet frame, when forwarding traffic onto the wired network. This translation is shown in Fig. 13.5.

A link to the Mbed OS Wi-Fi API is given in Reference 13.4, along with a simple example program. This is available to the IoT Discovery board, which has Wi-Fi connectivity. It is not available for the Mbed LPC1768 or Nucleo F401RE.

13.3 Getting Data from A to B

We have looked at some of the physical technology involved in transmitting data, and we've described how local devices can exchange data. Now it's time to consider how data can travel from one location to another, when those locations could be on opposite sides of the world.

13.3.1 Routing and IP Addresses

An Ethernet or Wi-Fi network can be thought of as an isolated group of devices communicating with each other purely within the confines of the network infrastructure, that is, the physical topology defined by switches, hubs, and wireless access points. Therefore, for a device in one network to communicate with a device in another, there needs to be a means of routing this data between the two networks.

Because MAC addresses are unique, it can be tempting to think that routing data is not actually a problem, as you simply address the device you want to speak to. But when you consider the wider internet as a whole, MAC addresses are not suitable at this scale, as they are effectively random—they have no structure, which makes it very hard to find a path from A to B. If you want to communicate with a machine that is on the other side of the world, which can only be reached through a number of different networks, how can a MAC address help? Additionally, MAC addresses are tied to the physical hardware device, and cannot (easily) be changed, making them a very inflexible addressing method. If you had to replace the network card in your server, suddenly it has a new MAC address.

Instead, Internet Protocol (IP) technology is the answer. It is layered on top of Ethernet/Wi-Fi to enable "inter-network" (Internet!) communications. This works by assigning unique IP addresses to nodes, and having those addresses structured in a hierarchy, such that network routers can decide which path to send a particular packet down. Routing is solved by being able to split the IP address up and figure out where to send it next.

IP addresses are assigned at the network layer (Layer 3) of the OSI model (see Fig. 11.4). One address actually contains two key pieces of information: the network and the host. The network mask identifies which portion of the address represents the network, and which represents the host. Routers use the network portion of the address to determine which individual network a packet is destined for. When a packet is sent to an IP address that does not exist on the local network, a router picks up the packet and sends it on. This process happens multiple times (each transmission between a router is called a "hop") until the packet eventually arrives at the destination network. Internally, routers use routing tables, which are essentially lists of networks and the IP addresses of either the routers responsible for those networks, or other routers that know about the ultimate destination. Routing tables are normally automatically populated using routing algorithms, but sometimes they are manually programmed.

The IP addressing system in use today has two main types: IPv4 and IPv6. IPv4 addresses are 32-bit addresses that are typically represented in dotted decimal notation (e.g., 192.168.1.1). IPv6 addresses are 128-bit addresses represented in hexadecimal notation (e.g., 2001:ae2:7765::290:1337). IPv4 addresses are the most common form of IP addresses in use, but are suffering from a severe shortage. This is because a 32-bit IP address can only address 2^{32} (4.2 billion) different nodes, and there are more than 4.2 billion devices connected to the internet! Although IPv6 solves this problem by providing over 3.4×10^{38} IP addresses, its adoption has been slow. Like MAC addresses, so-called "public" IP addresses are globally unique. They are assigned by the Internet Assigned Numbers Authority (IANA) and then delegated to regional Internet registries.

Due to the exhaustion of IPv4 addresses, techniques such as network address translation (NAT) are used to "virtually" inflate the IPv4 address space. This works by keeping a range of IP addresses private to a network, and exposing a single public IP address on the router. Any packets coming from any host on the private network will look like they have come from the single public IP address, and any packets coming back into the network will be translated back into the corresponding private IP address. This technique allows multiple networks to use the same IP address space, but still allows traffic to be routed between them.

Fig. 13.6 demonstrates how NAT works. In (1), a packet is sent from a device with the private IP address 192.168.0.5 to the server with the public IP address 212.24.55.60. At (2) the packet has been modified by the router to replace the source IP address with the router's public IP address (84.23.34.77). This is because the private IP address has no meaning outside the network. At (3), the server responds to the original packet with a new packet, addressed to the public IP address of the router (84.23.34.77). Finally, at (4), the router has re-written the packet and replaced the destination IP address with the private IP address of the original device (192.168.0.5)—the one for which the response was intended. NAT relies on a technique called connection tracking to remember which internal IP address corresponds to which active connection.

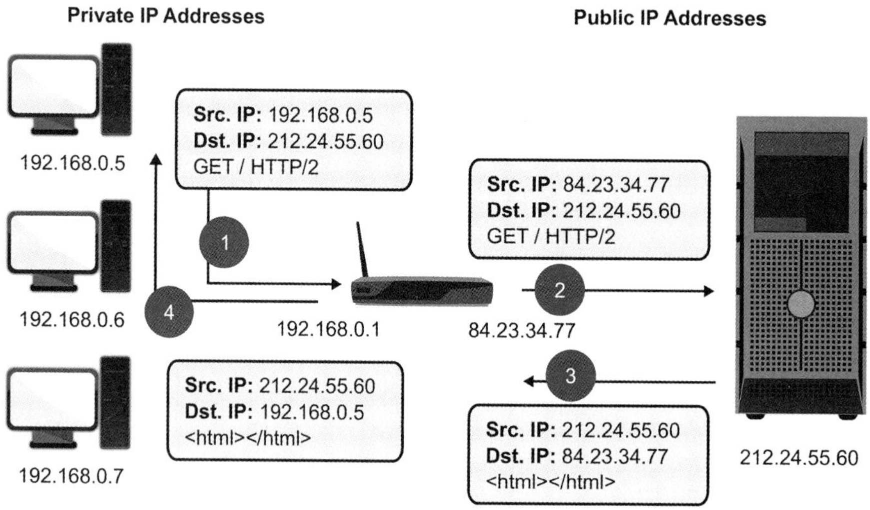

Figure 13.6
Network address translation.

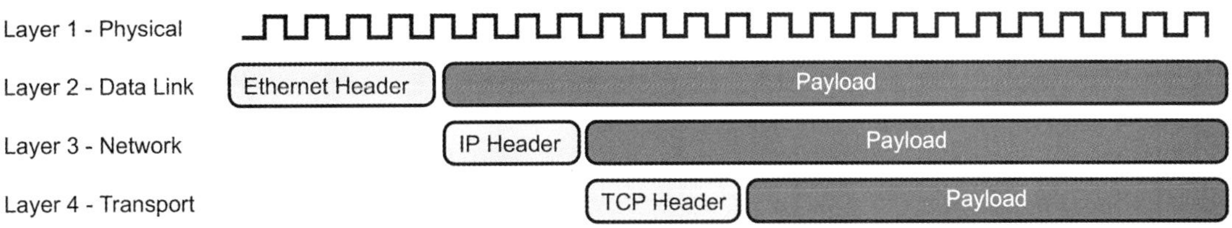

Figure 13.7
Packet encapsulation: A data packet from a higher-level protocol (such as HTTP) is contained within a TCP packet, which is inside an IP packet, which is inside an Ethernet frame.

IP packets are encapsulated within data-link layer (Layer 2) protocols, such as Ethernet or Wi-Fi. An IP packet, like the outer Ethernet packet, contains a source and destination IP address, along with additional information describing the inner protocol being used. Transmission control protocol (TCP) and user datagram protocol (UDP) are the most prevalent protocols that run on top of the IP protocol; often TCP will be qualified as TCP/IP to show that it is the TCP protocol running on top of the IP protocol. The internet control message protocol (ICMP) exists to facilitate network diagnostics and management. An example of this encapsulation is given in Fig. 13.7, and it is this which reflects the layering seen in the OSI model (Fig. 11.4).

Every node taking part in an IP network needs to have its own unique IP address, and these addresses can be assigned either manually or automatically. Manual configuration requires the network administrator to input the IP address and other network details on the node itself, but this process can be error prone, and is not scalable to a large number of devices. Automatic configuration allows devices to receive IP addresses automatically each time they connect to a network. An IP address assigned to a node can either be static or dynamic. A static address does not change for that node over its entire lifetime (unless the network administrators want to change it!), whereas a dynamic address means that a node might receive a different IP address each time it connects. To facilitate automatic address assignment, a service called the dynamic host configuration protocol (DHCP) is used. DHCP provides IP addresses (either static or dynamic) and other useful network information, such as the router's IP address, to nodes each time they connect.

An interesting introduction to the Mbed OS approach to IP networking and support for it is given in Reference 13.5, with links to a range of APIs. This will become essential reading if your IoT designing takes you in this direction.

13.3.2 The Role of the Gateway

Many IoT devices do not have direct connections to the internet; that is, they do not have an IP address with which they can communicate. An example might be a fitness tracker, which only has a Bluetooth LE connection to a mobile phone. In principle, it might be possible to establish an IP link over Bluetooth, but it's unlikely to be worth it, as the device is constrained and may not have the capability of processing IP packets.

Therefore, there needs to be another way for the device to communicate with a server, and this is the job of a network gateway. A network gateway can simply be a forwarding device, which takes data from a device and sends it on to a server, relaying any responses back to the device. It could, however, be more complex and perform some local data processing before sending the information on.

Fig. 13.8 shows how a fitness tracker that only has a Bluetooth LE connection might upload its sensor data to a remote server. In this example, the fitness tracker communicates with a mobile device and sends its sensor data to an app. The app then connects to the data center via a Wi-Fi connection to transmit the collected sensor data from the tracker. Here, the mobile device is the gateway for the fitness tracker.

As described in Chapter 11, LoRaWAN is an excellent example of an architecture that uses network gateways. Here, the LoRaWAN gateways provide the RF front end (the actual physical radio hardware), and are responsible for routing data from the IoT devices to the network server. Multiple LoRaWAN gateways extend the coverage of the network, and can implement QoS control through methods such as load balancing, failover (i.e., switching to a reserve system if the first fails), and redundancy. This helps to ensure that the network continues to operate even if one or more devices or network components fail.

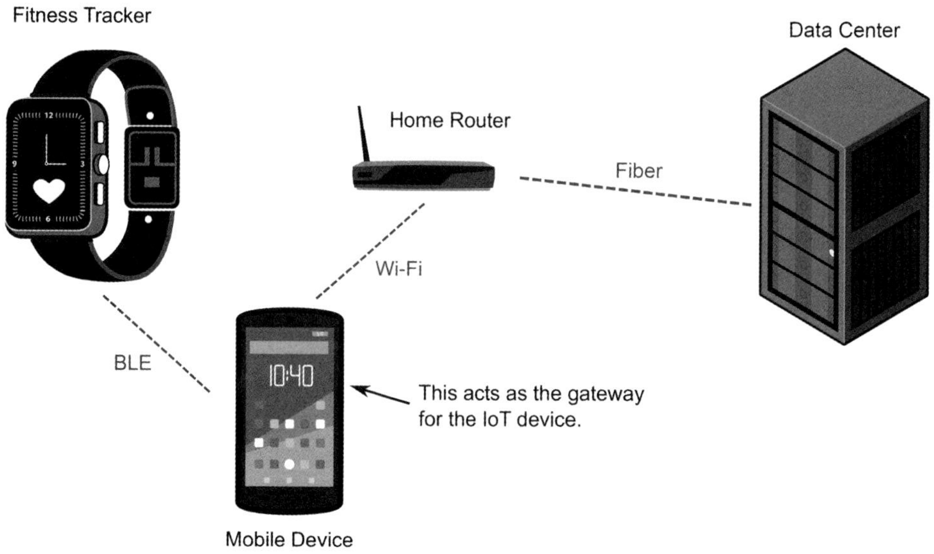

Figure 13.8
A fitness tracker using its BLE connection to transmit data to a gateway (the mobile device).

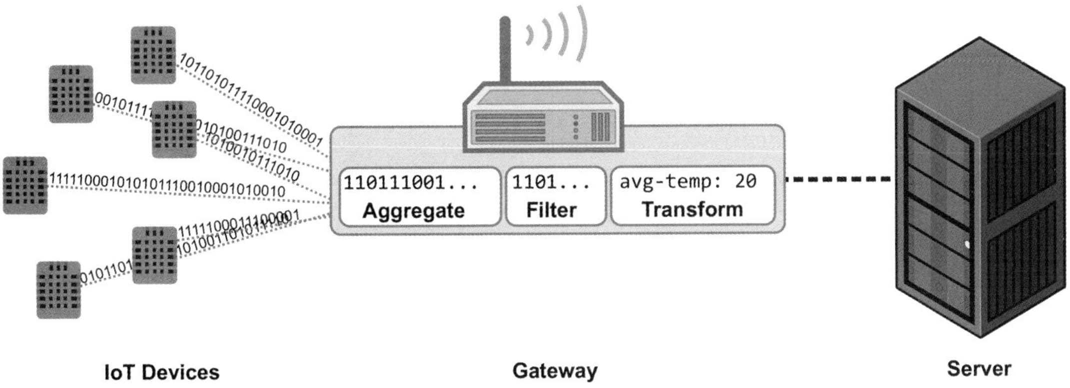

Figure 13.9
A network gateway performs aggregation, filtering, and transformation on the data received from several IoT devices.

Network gateways often include advanced functionality that operates on the data that they are responsible for, such as aggregation, filtering, and transformation:

- **Data aggregation:** Network gateways collect data from multiple devices and aggregate it into a single, centralized data stream. This allows easier analysis and management of the data.
- **Data filtering:** Network gateways can be used to filter data to reduce the volume transmitted over the network, and to prevent sensitive or unnecessary data from being transmitted.
- **Data transformation:** Network gateways can perform data transformations, such as converting data from one format to another, to make it compatible with the backend infrastructure.

In Fig. 13.9, a network gateway aggregates data from multiple IoT devices and turns it into a single data stream. It also strips out unnecessary information that the devices send across, and then transforms it into a format that the server can understand.

13.4 Data Exchange

After considering how data is moved around, we now turn to the actual data itself, and the protocols used to exchange data between device and server, and the forms in which data is encoded.

13.4.1 TCP and UDP

Underpinning most of the application-level network protocols (Layer 7 in the OSI model, see Fig. 11.4) in use today are the transmission control protocol (TCP) and user datagram protocol (UDP) Layer 4 transport protocols. Both of these protocols are used for transferring data between applications over an IP network, but they differ in their approach and characteristics.

TCP is a connection-oriented, stream-based protocol, which means that it establishes a reliable two-way communication stream between two devices before any data transfer begins. TCP ensures that data is received in the correct sequence, and that any lost packets are retransmitted, thus providing the illusion of a continuous flow of bytes between devices. To achieve this, TCP uses a range of techniques to split data up into packets, and to detect when packets go missing, or arrive at the destination out of order. TCP will insert missing packets and reconstruct the original ordering, ensuring that the application does not see anything out of order. However, this reliability doesn't come for free, as it introduces overhead into the processing pipeline. If an earlier packet goes missing, the entire connection is blocked until it is retransmitted and reinserted in the correct position. TCP is widely used in applications that require a high level of reliability, such as file transfers, email, and web browsing.

UDP is a connectionless datagram-based protocol, which means that it does not establish a connection before data transfer, and individual messages are transferred one at a time, rather than being a continuous stream. Conceptually, it is "faster" than TCP because it does not have the overhead of connection establishment, sequence checking, and retransmission. However, UDP does not provide reliable data transmission or congestion control, which means that lost packets may not be retransmitted, and data may go missing in the case of network congestion. UDP is commonly used in applications that require real-time data transfer, but are tolerant of dropped packets, such as online gaming, video streaming, and voice-over-internet protocol (VoIP).

13.4.2 HTTP and HTTPS

HTTP (hypertext transfer protocol) is a request—response protocol that allows clients to communicate with servers over the internet. It is based on the client—server model, where the client initiates a request, and the server responds to the request. An HTTP connection is usually established over a TCP/IP link, although some recent developments (e.g., QUIC, a protocol developed by Google) have started using UDP communication to improve performance.

When a client makes an HTTP request, it sends an HTTP message to the server. The message contains several parts, including a request line, headers, and an optional message body. The request line includes the method (e.g., GET, POST, PUT, DELETE), the uniform resource identifier (URI) of the requested resource, and the protocol version (e.g., HTTP/1.1). The headers provide additional information about the request, such as the type of content being requested and the client's preferred language. The message body may contain data, such as form data or a JavaScript Object Notation (JSON) payload.

The server then processes the request and sends an HTTP response back to the client. Just like the request, the response contains several parts, including a status line, headers, and a message body. The status line includes the protocol version, a numeric status code, and a brief reason phrase (e.g., "200 OK", or "404 Not Found"). The headers provide additional information about the response, such as the type of content being returned and the server's software version. The message body contains the requested resource or an error message.

HTTP is a stateless protocol, meaning that the server does not retain any information about the client between requests. However, cookies and sessions can be used to maintain states across multiple requests, should the application require this.

There are a few different versions of HTTP, the most widely used being HTTP/1.1, although recently HTTP/2 and HTTP/3 have emerged, which are designed to improve the performance, security, and functionality of HTTP.

HTTPS (HTTP Secure) is a variant of HTTP that uses SSL (secure sockets layer) or TLS (transport layer security) to encrypt data being transmitted over the connection. This added security helps to protect sensitive information, such as login credentials and financial data, from being intercepted by malicious actors.

HTTP is well suited to IoT applications for several reasons:

- **Widespread adoption:** HTTP is the foundation of the World Wide Web and is widely supported by web browsers, servers, and other devices. This means that most IoT devices will be able to communicate using HTTP without the need for additional software or libraries.
- **Simple and well documented:** HTTP is a simple protocol that is easy to implement and understand. It is well documented, with many resources and tutorials available to help developers get started.
- **Efficient:** HTTP is designed to be efficient, with minimal overhead and a small packet size. This makes it well suited for use in low-power, resource-constrained IoT devices.
- **Proven:** HTTP has been in use for decades and has been proven to be a robust and reliable protocol for communication over the internet.
- **RESTful Web Services:** HTTP is the foundation of RESTful (representational state transfer) web services, which is an architectural style for building web services. These services are lightweight, easy to understand, and easy to implement, which makes them an ideal choice for IoT applications.
- **Security:** HTTPS can be used to encrypt data being transmitted over the internet, which helps to protect sensitive information from being intercepted by malicious actors. This added security is important for many IoT applications that deal with sensitive data.

13.4.3 Data Interchange Formats

As we have seen, the IoT necessarily requires transmission of data between devices, and that data needs to be encoded in some form. Rather than bespoke, proprietary data encodings having to be invented, there are a number of popular formats that can be used to transmit data, which ultimately maximize compatibility with other systems. These formats are usually used in conjunction with an HTTP connection, as a way to send and receive data to/from a server.

JSON, which we used in the example in Chapter 12 and first mentioned in Chapter 5, is a lightweight data-interchange format that is easy for humans to read and write, and easy for machines to parse and generate. It is based on a subset of the JavaScript programming language, and is often used as the format for data exchange in web APIs. This is because it is lightweight, and supported by most programming languages. Program Example 13.1 gives an example of a JSON data structure, which describes the values of some IoT sensors.

```
{
  "sensors": [
    { "id": 1000, "temperature": 18.5 },
    { "id": 1001, "temperature": 19.0 },
    { "id": 1002, "temperature": 17.5 }
  ],
  "avg_temp": 18.3
}
```

Program Example 13.1 An example JSON data structure

SOAP (simple object access protocol) is an alternative (and older) protocol for exchanging structured information in the implementation of web services. SOAP uses XML (extensible markup language) as its message format, and can be carried over a variety of lower-level protocols. SOAP is more verbose than JSON and requires more bandwidth to transmit, which can make it less efficient for some use cases. However, SOAP provides more built-in functionality for things like data validation, error handling, and support for complex data types. Program Example 13.2 gives an example of the equivalent SOAP representation of the data from Program Example 13.1.

```
<?xml version = "1.0"?>
<soapenv:Envelope
    xmlns:soapenv="http://schemas.xmlsoap.org/soap/envelope/"
    soapenv:encodingStyle="http://www.w3.org/2001/12/soap-encoding">

    <soapenv:Body xmlns:iot="http://iot/sensor">
    <iot:SensorData>
      <iot:Sensor>
        <iot:SensorId>1000</iot:SensorId>
        <iot:Temperature>18.5</iot:Temperature>
      </iot:Sensor>
      <iot:Sensor>
        <iot:SensorId>1001</iot:SensorId>
        <iot:Temperature>19.0</iot:Temperature>
      </iot:Sensor>
      <iot:Sensor>
        <iot:SensorId>1002</iot:SensorId>
        <iot:Temperature>17.5</iot:Temperature>
      </iot:Sensor>
      <iot:AvgTemp>18.3</iot:AvgTemp>
    </iot:SensorData>
  </soapenv:Body>
</soapenv:Envelope>
```


Program Example 13.2 An example SOAP representation of the same data from Program Example 13.1

Protocol buffers (often called Protobuf) is a binary data serialization format developed by Google. It is similar to JSON and XML, although it is not "self-describing" in the sense that the data being transmitted does not define its own structure. This means that both sides of a Protobuf link must agree on the structure beforehand, which in turn makes it more efficient and smaller in size, especially when it comes to larger data sets. It uses a specific interface definition language to define the data structure and then generates code in several programming languages that can be used to read and write data in the format. It is particularly useful in situations where the data is being transmitted over a network and bandwidth is at a premium—thus well suited for the IoT! However, Protobuf requires additional setup and configuration compared with JSON and SOAP, and it is not as widely supported. That being said, there are many different language implementations for Protobuf, so it can be a good choice for some applications. Fig. 13.10 shows how a Protobuf specification translates to raw data bytes. Unlike JSON and SOAP, this encoding is not human-readable.

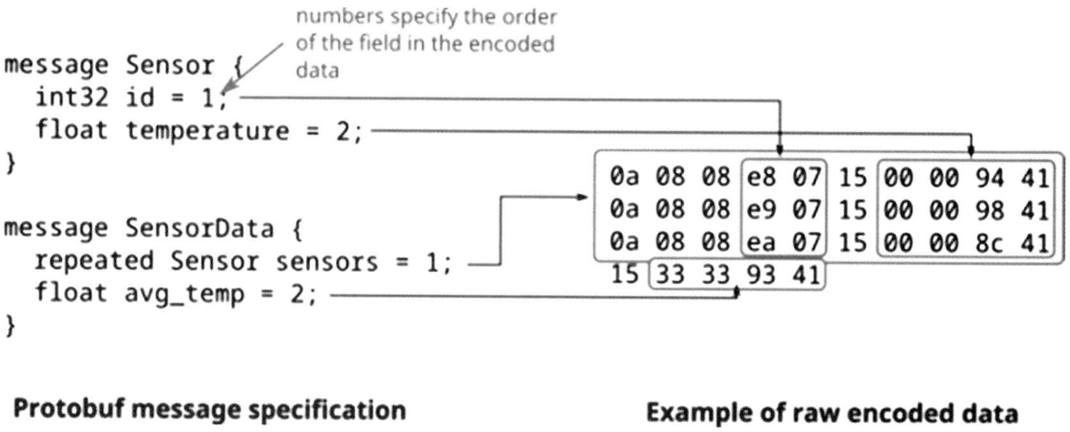
numbers specify the order of the field in the encoded data
message Sensor {
int32 id = 1;
float temperature = 2;
}
message SensorData {
repeated Sensor sensors = 1;
float avg_temp = 2;
}
0a 08 08 e8 07 15 00 00 94 41
0a 08 08 e9 07 15 00 00 98 41
0a 08 08 ea 07 15 00 00 8c 41
15 33 33 93 41
Protobuf message specification
Example of raw encoded data

Figure 13.10
How a Protobuf specification maps data to an efficient byte format; here, the encoded data representing the same information as the JSON and SOAP example is only 35 bytes.

13.5 Challenges and Risks in Building IoT Systems

Building IoT systems can be challenging for a variety of reasons. This is a big topic, and we just give an overview here. Some of the key challenges include the following:

Complexity: IoT systems are inevitably complex, involving many different types of devices and technologies that need to work together seamlessly. This can be difficult to achieve, especially when dealing with many devices or a wide range of environments. The example in Section 12.4 is a very basic introduction to a full-stack IoT system, but it shows how many different technologies must work together, just to get it started!

Scalability: IoT systems often need to be able to scale to millions or even billions of devices. This can be a challenge because each device needs to be able to communicate either with others, or with the cloud. The system needs to be designed to handle the data generated by all these devices.

Security: IoT systems can be vulnerable to security threats, such as hacking or malware. This is because many IoT devices have limited processing power and are connected to the internet, making them easier to attack. This topic is developed further in the Section which follows.

Interoperability: Different IoT devices and systems often need to be able to work together, regardless of their manufacturer or technology. This can be a challenge because there are many differing ways for IoT devices to communicate, and different devices may use different protocols or technologies.

Privacy: IoT systems can generate a large amount of data about people's habits and behavior. Ensuring the privacy of this data is a challenge, especially as IoT systems become more widespread and the data they generate becomes more valuable.

Cost: Building and maintaining an IoT system can be costly, especially when dealing with many devices or complex environments. This can be a challenge for businesses and organizations that want to adopt the IoT, but may not have the resources or budget to do so.

Although there are clear benefits to adopting an IoT system, there are also several risks that businesses and organizations need to consider. Some of the key risks include the following:

Legal risks: There are several legal risks associated with the IoT, such as liability for data breaches or privacy violations. There are also questions about who is responsible for the actions of IoT devices, such as self-driving cars or industrial robots.
Operational risks: IoT systems can be complex and can involve many different devices and technologies that need to work together seamlessly. This can be a risk to organizations that adopt IoT systems, as problems with the system could result in disruptions or downtime, leading to loss of revenue.
Strategic risks: Adopting IoT systems can involve significant investments of time and resources, and there is always the risk that the technology will not deliver the expected benefits. This could be a risk to organizations that adopt IoT systems, as they could end up with a system that does not meet their needs or that becomes obsolete quickly.

Overall, adopting IoT systems involves balancing the potential benefits against the risks. It is important for organizations to carefully assess these risks and put measures in place to mitigate them.

13.5.1 Security and Privacy

Security and privacy issues in IoT systems include a wide range of potential vulnerabilities and threats, such as unauthorized access to device data, data breaches, denial of service attacks, and malware infections. The Open Web Application Security Project (OWASP, Reference 13.6) is a community-led organization which aims to improve software security. It maintains an interesting list of the top 10 web security risks, which can be seen on its website.

Numerous examples have appeared in the press and on the internet reporting IoT security failings. A well-known one, which links to an example given in Chapter 12, relates to a reported vulnerability of the Philips Hue lightbulb, Reference 13.7. In a scenario such as this, the attacker (in range of the bulb) plays around with the color or brightness settings to convince the controller that the bulb is faulty. The controller tries to reset it, deletes the bulb, and attempts to reconnect. The bulb (now hacker-controlled, with updated firmware) re-joins the IoT network. The bulb bombards the controller with data requests, triggering a buffer overflow. The hacker installs malware on the controller, which is connected to the target IP network. The attacker can now infiltrate the target network to spread ransomware or spyware.

These issues can be mitigated through a variety of technologies and best practices, including the following:

Device authentication and access control: Ensuring that only authorized devices and users can access the network and control devices. This can be accomplished through various authentication methods, such as password protection, biometric authentication (fingerprints), and public key infrastructure (PKI).

Network security: Protecting the communication channels and infrastructure used by IoT devices. This can be done through encryption and the use of secure protocols such as HTTPS.

Firewalls: A firewall is a network security system that monitors and controls incoming and outgoing network traffic based on predetermined security rules and policies. Firewalls are the first line of defense in a network, and should be configured to be as restrictive as possible.

Intrusion detection and prevention systems (IDPS): These systems monitor network traffic for suspicious activity, and can take action to block or alert on potential threats.

Software updates and patch management: Keeping devices up to date with the latest security patches and software updates can help prevent known vulnerabilities from being exploited.

Data privacy: ensuring that personal information is protected, and only shared with authorized parties, through techniques such as anonymization or using pseudonyms to protect personal identity.

Compliance and regulations: Adhering to industry-specific regulations, such as the general data protection regulation (GDPR), can help ensure that IoT systems are secure and compliant with relevant laws and regulations.

It is important to note that there is no one-size-fits-all solution for securing IoT systems, and organizations should take a holistic approach to identifying and addressing potential vulnerabilities and threats. Additionally, security should be built in from the design stage and should be continuously monitored and updated over time; otherwise vulnerabilities may creep in undetected.

Implementing security and privacy on constrained IoT devices can be challenging due to their limited resources, such as processing power, storage, and memory. However, there are several techniques and technologies that can be used to effectively secure these devices, such as the following:

Lightweight security protocols: Constrained devices may not have the resources to support complex security protocols, so it's important to use lightweight and efficient protocols such as datagram transport layer security (DTLS) or secure MQTT (a lightweight machine-to-machine network protocol; MQTT is now the name, and not an initialism).

Secure boot: Secure boot is a process that ensures that only authenticated and trusted firmware can be executed on the device. This can help prevent unauthorized access to the device and protect against malware.

Secure storage: Constrained devices may not have the resources for full-disk encryption, so it is important to use secure storage methods such as trusted platform modules (TPMs) to protect sensitive data.

Over-the-air (OTA) updates: Constrained devices may not have the ability to physically connect to a computer for software updates, so it is important to implement OTA update mechanisms to ensure that devices are kept up to date with the latest security patches.

Minimizing the attack surface: The "attack surface" defines the range of possible entry points that an attacker can use to attempt to gain access to a device. Keeping this minimal reduces the number of opportunities for attackers to infiltrate a system. This can be achieved by disabling unnecessary features and services, and by using minimalistic firmware.

Code signing: Ensuring that the firmware running on devices is authentic and has not been tampered with by signing the code digitally before it is flashed on the device.

The Mbed OS makes a particular claim to offer secure software, and security is a distinct category in its API listing, as seen in Table 2.2 and linked through Reference 13.8. It is worth checking what is available, although these features are mainly for the advanced player in the field. Finally, there are a number of useful and interesting publications on this topic, available for further reading. Try Reference 13.9, for example.

13.5.2 Dealing with IoT Scale and Complexity

There are several strategies that can be employed to manage the scale and complexity of IoT systems effectively. They start with a well-designed architecture that separates out the different components of the IoT system. For example, an architecture that employs so-called "microservices" can help to break down the system into smaller, more manageable components that can be developed, deployed, and maintained independently.

Considering the actual IoT devices themselves, implementing an effective device management solution can help to manage the scale of such systems. Imagine an IoT system with millions of devices—it is practically impossible to manually configure, provision, and update each of those devices individually, and so a solution that largely allows management and maintenance tasks to be automated is critical to supporting an IoT system on this scale.

Along with managing the physical devices is managing the flow and storage of data that they generate. This can include implementing data storage solutions that are designed for the high volume, high data rate and variety of IoT data, and using data analytics tools to gain insights into the data and make informed decisions.

As we have seen, the cloud often underpins IoT systems, as it provides resources (such as computation and storage) that can scale on demand. Choosing scalable infrastructure solutions, such as these cloud-based platforms, can help to manage the growth of IoT systems over time. This can help to ensure that the system remains performant and reliable as the number of devices and amount of data grows. Public cloud providers offer a wide range of services (sometimes too many to choose from!) spanning from data storage solutions through to complex artificial intelligence frameworks. These enable scalability, as the providers have the power of large data center resources available on demand, and they give developers access to advanced technologies, which would require specialist expertise to implement internally. Not only that, it is usually more cost-effective to use these services than it is to build and maintain your own hardware.

13.6 Powering the IoT "Things"

It's clear that many IoT "things" are battery-powered, and hence need to optimize their power consumption. We consider the theoretical underpinning to this topic in Sections 15.4 to 15.7 but use this section to give an introduction and overview.

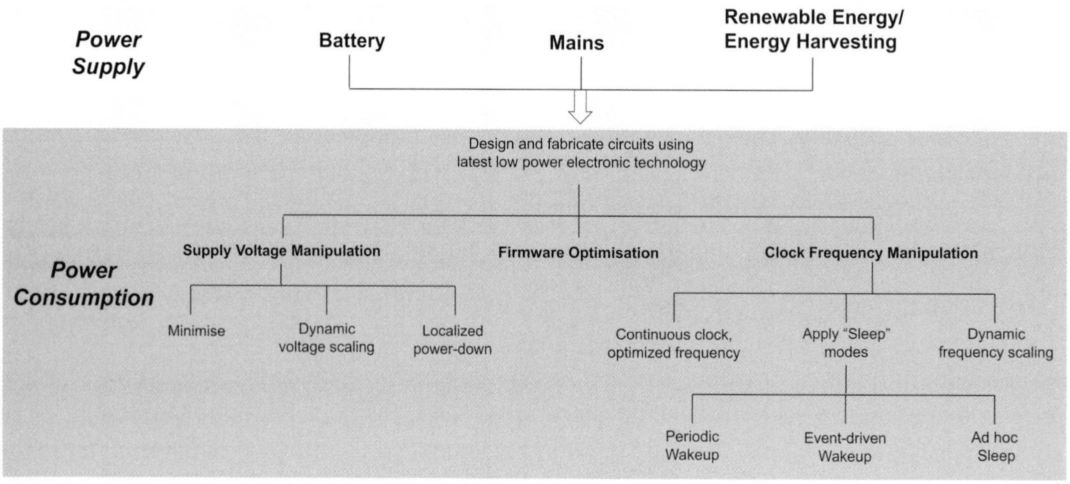

Figure 13.11
A low power "street map."

Fig. 13.11 presents a low-power "street map," which summarizes the options available to a designer wishing to optimize power consumption. Starting at the top of the diagram, for *any* electronic product, we need to consider its power source, or sources. As options, we have conventional mains power, battery, or renewable. For our remote IoT device, let's discard mains power, as being inappropriate for the sort of device that we are likely to be thinking of. Then we note that battery power is considered further in Chapter 15, so we will not view it further here.

In terms of renewables, it is easy to propose the familiar ones, solar or wind. However, as we are dealing with tiny demands for power, a number of less-familiar sources present themselves. These include thermal gradients, electromagnetic radiation, vibration, and movement. In the last case, a device carried in the pocket could be recharged simply through the movements of the person carrying it. There are specialist companies that offer techniques and technologies to extract energy from such sources. Reference 13.10, for example, is well worth consulting.

Still with Fig. 13.11, let's consider the design and fabrication of low-power circuits, and especially microcontrollers and their ancillaries. At the beginning of this book, in Table 1.1, we made reference to low-power CPU design and recognized that ARM is a leading exponent of this. For minimum consumption, we saw that microcontrollers based on the Cortex-M0 or M0+ cores give the best performance. Section 15.7 describes the performance of the *Zero Gecko* Mbed-enabled card, with its Cortex-M0+ core, reporting the astonishingly low currents that it consumes. Regarding other circuit elements, LCD technology should of course be favored over LED for displays, unless extreme temperatures are anticipated. The opportunity also exists here to design localized shutdown into the circuit, so that parts of the circuit can be switched off when not in use.

In Chapter 15 we will demonstrate the importance of low-power electronic technology, in the form of CMOS (complementary metal oxide semiconductor), and for digital electronics, the inextricable link between power consumption, clock frequency, and supply voltage. Higher clock frequencies or higher supply voltages always lead to higher power consumption. Glance forward for example to Section 15.4 and Eq. 15.1. The other side to this, however, is that programs which need to execute fast need to run at a high clock frequency, with a possible implication of higher-voltage supply.

So, assuming now that we have a more or less given circuit, we have three broad headings under which we can reduce power consumption: firmware, supply voltage, and clock frequency. Let's start with firmware, the code we write. Complex, "busy," inefficient code requires more clock cycles to get through, implying a higher clock frequency and higher power consumption. Hence, code for a low-power device should be as simple and efficient as possible. Simple choices, like choosing integer arithmetic over floating point, may allow you to run at a reduced clock frequency and save power. Other options may center on reduction of sampling rates, or reducing the amount of data transmitted or its frequency of transmission.

Regarding clock frequency manipulation, in Chapter 15 we consider in detail how the clock frequency can be adjusted, and we note the effect of this. One possibility is simply to choose the lowest continuous clock frequency that will allow the program to run satisfactorily. An alternative is to vary clock frequency according to need, a technique sometimes known as *dynamic frequency scaling*. Where intensive calculations are required, for example, the CPU clock frequency is increased. Where only slow background tasks are needed, the clock frequency and power consumption can be reduced. Quiz question 12 of Chapter 15 performs some calculations around this topic.

A more extreme form of clock frequency manipulation is to switch the clock off entirely to unneeded parts of the microcontroller circuit, including possibly the CPU, resulting in a range of *idle, sleep,* or *power-down* modes. We have been applying this capability from our very first program, with statements like

```
ThisThread::sleep_for(500ms);
```

Sleep functions such as these allow "ad hoc" moments of power consumption reduction, taken as program design allows. Depending on the structure of the program, sleep may also be taken periodically, or when waiting for an event to occur.

Finally, considering supply voltage manipulation, it is always best to operate at the lowest possible supply voltage, as long as circuit performance is not compromised. For example, Table C.1 in Appendix C gives operating conditions for the LPC1768 microcontroller. Although the "typical" supply voltage is 3.3 V, the minimum is quoted as 2.4 V. Possibly there is power saving to be done by the supply voltage reduction that this appears to afford, though in practice a very careful evaluation of other operating conditions would need to be made before this move is taken. There are also sometimes opportunities to power down unused parts of a circuit temporarily, or even subsystems within a microcontroller. Finally, let's mention a technique called *dynamic voltage scaling.* Like dynamic frequency scaling, this allows the operating voltage to be adjusted according to need, noting that a higher supply voltage tends to allow faster operation. Though this technique is applied elsewhere, there is only modest scope for it to be applied on the range of microcontrollers we are using, as they have a comparatively narrow range of operating voltages.

Chapter Review

- There are several different networking technologies involved in making a communication systems work.

- The Ethernet protocol underpins many of the wired communication networks today, and allows data to be exchanged within a local network.
- The Wi-Fi protocol performs a function similar to that of the Ethernet protocol, but extends this to work wirelessly.
- Wi-Fi is the most widely used wireless networking protocol in the world, and is designed to integrate seamlessly with ethernet networks.
- The IP allows data to be exchanged between networks by associating IP addresses with devices, and specifying how routes to other networks can be discovered and traversed.
- Web services are a popular way of implementing internet-based services, and typically use the HTTP protocol, or its secure HTTPS variant, to operate.
- Data transfer between services requires that data to be encoded in ways that the communicating parties can understand. JSON, SOAP, and Protobuf are examples of popular data exchange formats that are well supported by languages and frameworks.
- There are a wide range of challenges and risks, such as complexity, scalability, interoperability, and cost, that need to be considered when designing and implementing an IoT system.
- Security and privacy are challenges that exist in every part of an IoT system, from the embedded device to the processing applications, and to the mobile or web applications used for control.
- There is a range of very effective techniques available to minimize power consumption in an IoT "thing," including best component choice, and optimization of firmware, clock frequency, and supply voltage.

Quiz

1. Describe the "Manchester" digital communication format for Ethernet signals.
2. Sketch the following Ethernet data streams, as they would appear on an analog oscilloscope, labeling all points of interest:
    a. 0000
    b. 0101
    c. 1110
3. What are the minimum and maximum Ethernet data packet sizes, in bytes?
4. Describe the procedure by which a Wi-Fi station initiates and connects to a wireless access point.
5. What are the differences between a public IP address and a private IP address?
6. Why might static IP addresses be useful?
7. What role does the Uniform Resource Identifier (URI) play in making an HTTP request?
8. Describe how the Protobuf data interchange format might be more efficient than JSON.

9. How do device authentication and access control mitigate some security issues, in the context of an IoT system?
10. What makes the Cortex-M0 and M0+ cores so power-efficient? (To answer, revisit Chapter 1, and do a little background reading.)
11. Identify the correct statements below:
    a. It's usually best to power IoT "things" from the mains supply, to ensure that they have a continuous and reliable power source.
    b. It's good to aim for the highest clock frequency possible on a remote IoT device, so that it can execute as much code as possible, without worrying about code efficiency.
    c. Reducing the supply voltage to an IoT device reduces power consumption; one must just ensure that circuit performance is not compromised.
    d. Power consumption does depend on supply voltage and clock frequency, but there isn't any dependence on the firmware, as that is just lines of code.
    e. The best electronic technology for low power consumption is called TTL, or transistor transistor logic.
12. You have recently joined a company which makes electrical kitchen items. The products are successful but conventional, and sales have been falling in the past few years. The managing director, who is not a technical person, has become very excited about adopting IoT technology, and sees it as an easy and quick way to rebuild product sales. In discussion, you become increasingly concerned that he does not recognize some of the pitfalls associated with IoT adoption. Outline what concerns there may be with the adoption of IoT technology, explaining why this is so.

References

13.1. Connectivity APIs (Mbed OS). https://os.Mbed.com/docs/Mbed-os/v6.15/apis/connectivity.html.
13.2. IEEE 802.3 Ethernet Working Group. https://standards.ieee.org/develop/wg/WG802.3.html.
13.3. Ethernet (Mbed OS). https://os.Mbed.com/docs/Mbed-os/v6.15/apis/ethernet.html.
13.4. Wi-Fi (Mbed OS). https://os.Mbed.com/docs/Mbed-os/v6.15/apis/wi-fi.html.
13.5. IP Networking APIs (Mbed OS). https://os.Mbed.com/docs/Mbed-os/v6.15/apis/connectivity-architecture.html.
13.6. Open Web Application Security Project (OWASP) home page. https://owasp.org/.
13.7. IoT Network Flaw Left Philips Hue Bulbs Open to Attack. https://www.computerweekly.com/news/252477918/IoT-network-flaw-left-Philips-Hue-bulbs-open-to-attack.
13.8. Security APIs (Mbed OS). https://os.Mbed.com/docs/Mbed-os/v6.15/apis/security.html.
13.9. Chantzis F, Deirme E, Stais I, Calderon P, Woods B. *Practical IoT Hacking: The Definitive Guide to Attacking the Internet of Things*. San Francisco: No Starch Press; 2021.
13.10. Energy Harvesting and Storage, Cymbet Corporation. https://www.cymbet.com/industry-solutions/energy-harvesting-storage/.

PART 3

Deeper Details

CHAPTER 14

Working Directly with the Control Registers

14.1 Introduction: Given the Mbed OS, Why Learn about Control Registers?

The Mbed OS contains countless useful functions, which allow the programmer to write simple and effective code. This seems like a good thing, but it is also sometimes limiting. What if we want to use a peripheral in a way not allowed by any of the functions, or to write code that is more efficient than the OS allows? It is useful to understand how peripherals can be configured by direct access to the microcontroller's registers. In turn, this leads to deeper insight into some aspects of how a microcontroller works. As a further by-product, and because we will be working at the bit and byte level, this study develops further skills in C/C++ programming.

It is worth issuing a very clear health warning at this early stage – this chapter is technically demanding. It introduces some of the complexity of the LPC1768 microcontroller, which lies at the heart of the Mbed LPC1768, a complexity which developers of the Mbed Ecosystem rightly wish to keep from you! Your own curiosity, ambition, or professional needs, may however lead you to want to work at this deeper level.

In exploring the details of the LPC1768 microcontroller, we will refer to Reference 2.3, the LPC1768 datasheet, and even more to Reference 2.4, its user manual. Because we are now working at the microcontroller level, rather than the Mbed target level, we will have to take more care about how the microcontroller pins connect with the Mbed pins. Therefore, you may wish to have the Mbed LPC1768 schematics, Reference 2.2, ready. As a start from these schematics, Fig. 14.1 shows which microcontroller pin connects to which I/O pin of the target board. For example, pin 5 of the Mbed board connects to pin 76 of its microcontroller.

Working with the complexity of this chapter may result in two opposing feelings. One is a sense of gratitude to the writers of the Mbed APIs, that they have saved you the difficulty of controlling the peripherals directly. At the opposite extreme, getting to grips with the chapter could also be a liberating experience—like throwing away the water wings after you have learned to swim. At the moment, you probably think that you cannot write any

Fast and Effective Embedded Systems Design. https://doi.org/10.1016/B978-0-323-95197-5.00014-0

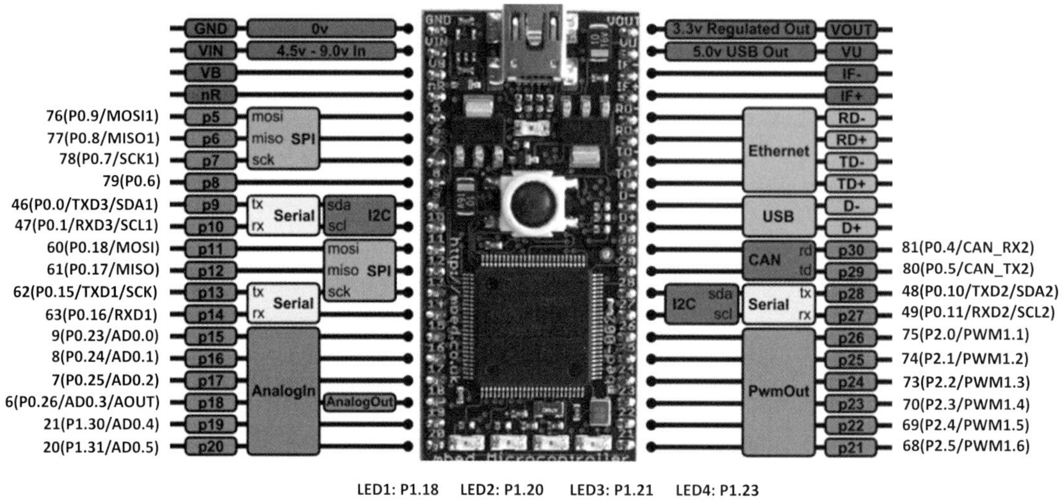

Figure 14.1

Connections between the LPC1768 microcontroller and the target board pins.

program unless you have the API functions ready. When you are through with this chapter you will realize that you are no longer dependent on the APIs; you use them when you want to (probably most of the time), but you can also write your own routines—the choice becomes yours!

This chapter may be read in sequence with all of the other chapters. Alternatively, it can be read in different sections as extensions of earlier chapters. Note that all data and examples in the chapter relate only to the Mbed LPC1768, it would be just too complex to attempt to describe two microcontrollers in parallel. Similar lines of exploration can, however, be followed with the Nucleo F401RE, or indeed any other target board.

14.2 Control Register Concepts

It is useful here to understand a little more about how the microcontroller CPU interacts with its peripherals. Each of these has one or more *system control registers* which act as the doorway between the CPU and the peripheral. To the CPU, these registers look just like memory locations. They can usually be written to and read from, and each has its own address within the memory map. The clever part is that each of the bits in the register is wired across to the peripheral. Each bit might carry control information, sent by the CPU to the peripheral; or it might return status information from the peripheral. It might also provide a path for data in either direction between the CPU and peripheral. The general idea of this is illustrated in Fig. 14.2. The microcontroller peripherals also usually generate

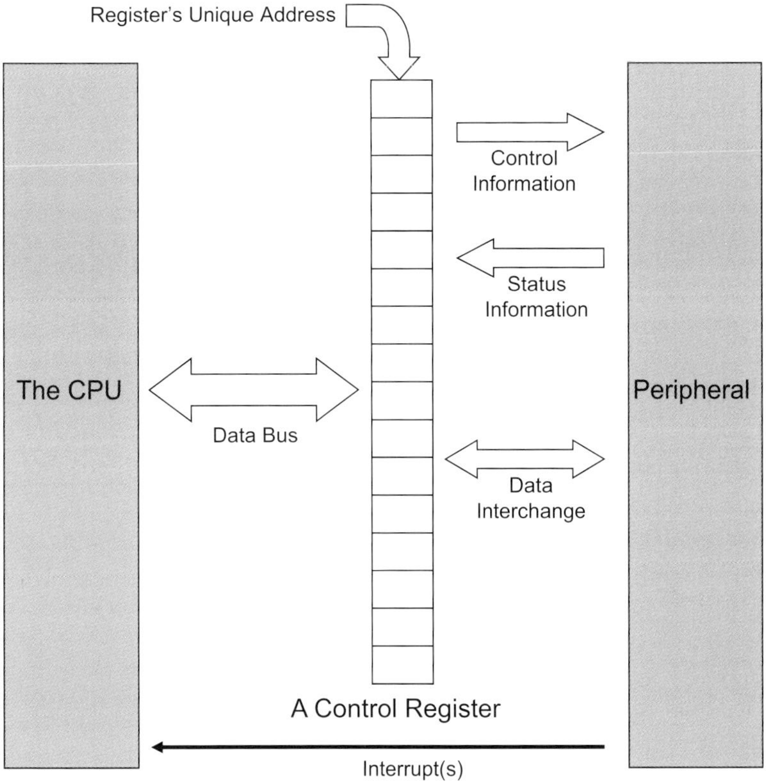

Figure 14.2
The principle of a control register.

interrupts, for example, to flag when an ADC conversion is complete, or a new word has been received by a serial port.

Early in any program, the programmer must write to the control registers, to set the peripherals to the configuration needed for that application. This is called *initialization*. Using the Mbed API, this task is undertaken in the Mbed utilities; in this chapter, we move to doing this work ourselves. In all of this, an important question arises: what happens in that short period of time *after* power has been applied, but *before* the peripherals have been set up by the program? At this time an embedded system is powered, but not under complete control. Fortunately, all makers of microcontrollers design in a *reset condition* of each control register. These reset values appear in many of the tables later in this chapter. This generally puts a peripheral into a predictable, and often inactive, state. We can look this up, and write the initialization code accordingly. On some occasions, the reset state may be the state that we need.

A central theme of this chapter is the exploration and use of some of the LPC1768 control registers. Although we draw information from the LPC1768 data, the chapter is self-contained, with all necessary data included. However, we show which table in the manual the data is taken from, so it is easy to cross-refer. We hope this gives you the confidence and ability to move on to explore and apply the other registers.

14.3 Digital I/O

14.3.1 LPC1768 Digital I/O Control Registers

The digital I/O is a useful place to start our study of control registers, as this is simpler than any other peripheral on the LPC1768. This has its digital I/O arranged nominally as five 32-bit ports; yes, that implies a stunning 160 bits. Only parts of these are used, however; for example, Port 0 has 28 bits implemented, Port 2 has 14, and Ports 3 and 4 have only 2 each. In the end, among its 100 pins, the LPC1768 has around 70 general purpose I/O pins available. However, the Mbed has only 40 accessible pins, so only a subset of the microcontroller pins actually appear on the Mbed interconnect, and many of these are shared with other features.

As we saw in Chapter 3, it is possible to set each port pin as an input or as an output. Each port has a 32-bit register which controls the direction of each of its pins. These are called the **FIODIR** registers. To specify which port the register relates to, the port number is embedded within the register name, as shown in Table 14.1. For example, **FIO0DIR** is the direction register for port 0. Each bit in this register then controls the corresponding bit in the I/O port; for example, bit 0 in the direction register controls bit 0 in the port. If the bit in the direction register is set to 1, then that port pin is configured as an output; if the bit is set to 0, the pin is configured as an input.

It is sometimes more convenient not to work with the full 32-bit direction register, especially when we might just be thinking of one or two bits within the register. For this reason, it is also possible to access any of the bytes within the larger register, as a single-byte register. These registers have number codes at their ends. For example, **FIO2DIR0** is byte 0 of the Port 2 direction register, also seen in Table 14.1. From the table, you can see that the address of the whole word is shared by the address of the lowest byte.

A second set of registers, called **FIOPIN**, hold the data value of the microcontroller's pins, whether they have been set as input or output. Again, two are seen in Table 14.1, with the same naming pattern as the **FIODIR** registers. If a port bit has been set as an output, then writing to its corresponding bit in its **FIOPIN** register will control the logic value placed on that pin. If the pin has been set as input, then reading from that bit will tell you the logic value asserted at the pin. The **FIODIR** and **FIOPIN** registers are the only two register sets we need to worry about for our first simple I/O programs.

Table 14.1: Example digital I/O control registers

Register Name	Register Function	Register Address
FIO*n*DIR	Sets the data direction of each pin in Port *n*, where *n* takes the value 0 to 4. A port pin is set to output when its bit is set to 1, and to input when it is set to 0. Accessible as a 32-bit word. Reset value = 0; i.e., all bits are set to input on reset.	-
FIO0DIR FIO1DIR FIO2DIR	Example of above for Port 0. Example of above for Port 1. Example of above for Port 2.	0x2009C000 0x2009C020 0x2009C040
FIO*n*DIR*p*	Sets the data direction of each pin in byte *p* of Port *n*, where *p* takes value 0 to 3. A port pin is set to output when its bit is set to 1. Accessible as a byte.	-
FIO0DIR0 FIO0DIR1 FIO2DIR0	Example of above, Port 0 byte 0. Example of above, Port 0 byte 1. Example of above, Port 2 byte 0.	0x2009C000 0x2009C001 0x2009C040
FIO0PIN FIO1PIN FIO2PIN	Sets the data value of each bit in Port 0 or 2. Accessible as a 32-bit word. Reset value = 0.	0x2009C014 0x2009C034 0x2009C054
FIO0PIN0 FIO0PIN1 FIO2PIN0	Sets the data value of each bit in selected byte of Port 0 or 2. Accessible as a byte. Reset value = 0.	0x2009C014 0x2009C015 0x2009C054

14.3.2 A Digital Output Application

As we know well, a digital output can be configured and then tested simply by flashing an LED. We did this back at the beginning of the book, in Program Example 2.1. We will look at the complete method for making this happen, working directly with the microcontroller registers. In this first program, Program Example 14.1, we will use Port 2 Pin 0 as our digital output. You can see from Fig. 14.1, or the Mbed schematic (Reference 2.2), that this pin is routed to the Mbed pin 26.

C/C++ code feature

Follow through Program Example 14.1 with care. Notice that for once we are *not* writing **#include "Mbed.h"**! At the program's start, the names of two registers from Table 14.1 are defined to equal the contents of their addresses. This is done by defining the addresses as pointers (see Section B.8.2), using the * operator. The format of this is not entirely simple; for our purposes we can use it as shown. Having defined these pointers, we can now refer to the register names in the code rather than worrying about the addresses. The C keyword **volatile** is also applied. This is used to define a data type whose value may change outside program control. This is a common occurrence in embedded systems, and is particularly applicable to memory-mapped registers, such as we are using here, which can be changed by the external hardware.

C/C++ code feature

Within the **main( )** function, the data direction register of Port 2, byte 0 is set to output by setting all bits to Logic 1. A **while** loop is then established, just as in Program Example 2.1. Within this loop, pin 0 of Port 2 is set high and low in turn. Check the method of doing this, if you are not familiar with it, as this shows one way of manipulating a single bit within a larger word. To set the bit high, it is ORed with logic 1. All other bits are ORed with logic 0, so they will not change. To set it low, it is ANDed with binary 1111 1110. This has the effect of returning the LSB to 0, while leaving all other bits unchanged. The 1111 1110 value is derived by taking the logic inversion of 0x01, using the C ~ operator.

The delay function should be self-explanatory and is explored more in Exercise 14.1. Its function prototype appears early in the program. The compiler has a tendency to "optimize out" sections of code which it identifies as having no use; there is a probability that it will do this to part or all of the **delay()** function. The **volatile** keyword is used here to "trick" the compiler into leaving the **delay()** function unchanged. There are ways of controlling the compiler optimization settings, but we don't address those here.

```
/*Program Example 14.1: Sets up a digital output pin using control registers, and
flashes an led. For LPC1768 only.
                                                                              */
// function prototype
void delay(void);

//Define addresses of digital i/o control registers,
    //as pointers to volatile data
#define FIO2DIR0       (*(volatile unsigned char *)(0x2009C040))
#define FIO2PIN0       (*(volatile unsigned char *)(0x2009C054))

int main() {
  FIO2DIR0 = 0xFF;   // set port 2, lowest byte to output
  while(true) {
    FIO2PIN0 |= 0x01;    // Mbed pin 26. OR bit 0 with 1 to set pin high
    delay();
    FIO2PIN0 &= ~0x01;   // AND bit 0 with 0 to set pin low
    delay();
  }
}
//delay function
void delay(void){
  //loop variable j, volatile inhibits optimisation
  volatile int j;
  for (j = 0;j < 1000000;j++) {
    j++;
    j--;                         //waste time
  }
}
```

Program Example 14.1 Manipulating control registers to flash an LED

Connect an LED between an Mbed LPC1768 pin 26 and 0 V, and compile, download, and run the code. Press reset and the LED should flash.

Exercise 14.1

With an oscilloscope, carefully measure the duration of the delay function in Program Example 14.1. Adjust the delay function experimentally so that it is precisely 100 ms. Then create a new 1 s delay function, which works by calling the 100 ms function 10 times. Now write a library delay routine which gives a delay of *n* ms, where *n* is the parameter sent.

14.3.3 Adding Further Digital Outputs

Following the principles outlined above, we can add further digital outputs, choosing pins on either side of the Mbed pin used so far. We add two further LEDs to make a flashing pattern. Mbed pin 25 connects to Port 2 Pin 1; this is straightforward, as we have just been using this byte of this port. However, in adding Mbed pin 27, we see from Fig. 14.1 that it connects to bit 11 (that is "eleven," decimal) of Port 0. So we will need to make further declarations and puzzle over how to write to a bit deep within the 32-bit port.

Create a new project called Program Example 14.2, but copy Program Example 14.1 into it. Now take a look at the code fragments in Program Example 14.2. This is effectively a kit of parts, which you need to insert with care into appropriate locations in the new project. For simplicity, you can start by copying every section across. You will then have a program which will compile successfully, but which you can adapt and reduce as you try different options, explained below or in the comments.

We need to address byte 1 of Port 0, and Table 14.1 shows that we can address either the individual byte, or the full 32-bit word. Identify those lines of code which use the former, and those which use the latter. You can choose to use one or the other, or mix them pretty much as you wish (probably then leading to some inefficient code). In the opening definitions, notice that if the full 32-bit word is used, it must be defined as **int**, not **char.**

The Port 2 bits are switched using the method already seen, with bit 1 being ORed with 0x02, and then ANDed with its inverse. When switching the Port 0 bit, notice carefully the four options given. We can again AND and OR with a bit pattern, expressed in hexadecimal, in either 8 or 32 bits. These bit patterns get quite tricky to deal with in larger words, so a useful alternative is left-shifting in a single bit. The convenient thing here is that the number of times the bit is shifted (in this case 11) is the bit number in the word, which gives considerable clarity when writing or reading the program.

```
...
//choose either pair, or both.
#define FIO0DIR       (*(volatile unsigned int *)(0x2009C000))
#define FIO0PIN       (*(volatile unsigned int *)(0x2009C014))
#define FIO0DIR1      (*(volatile unsigned char *)(0x2009C001))
#define FIO0PIN1      (*(volatile unsigned char *)(0x2009C015))
...
char i;
...
//Insert at start of main(), either will work
  FIO0DIR1 = 0xFF;          // set port 0, byte 1 to output
  FIO0DIR = 0x0000FF00;     // set port 0, byte 1 to output
...
//insert the following at the very end of, but within, the while loop
  for (i = 1;i <= 3;i++){     //Blink Mbed pin 25 three times
    FIO2PIN0 |= 0x02;        // set port 2 bit 1 high
    delay();
    FIO2PIN0 &= ~0x02;       // set port 2 bit 1 low
    delay();
  }
  for (i = 1;i <= 4;i++){     //Blink Mbed pin 27 four times
    FIO0PIN1 |= 0x08;        // set port 0 bit 11 high
    //Or use: FIO0PIN |= 0x0800;
    //Or use: FIO0PIN = FIO0PIN | (1<<11);
    //Or use: FIO0PIN |= (1 << 11);
    delay();
    FIO0PIN1 &= ~0x08;       // set port 0 bit 11 low
    //Or use: FIO0PIN &= ~0x0800;
    //Or use: FIO0PIN = FIO0PIN & ~(1<<11);
    //Or use: FIO0PIN &= ~(1 << 11);
    delay();
  }
```

Program Example 14.2 (code fragments): Controlling further LED outputs

Add LEDs to pins 25 and 27 of the Mbed, and run the program. It should flash the first LED (on pin 26) once, followed by three flashes on pin 25, and four on pin 27.

Exercise 14.2

From Fig. 14.1, identify the LPC1768 pins that drive the on-board LEDs. Rewrite Program Example 14.1/14.2 so that the on-board LEDs are activated, instead of external ones.

14.3.4 Digital Inputs

We can create digital inputs simply by setting a port bit to input, using the correct bit in an **FIODIR** register. Program Example 14.3 now develops the previous example by including a digital input from a switch. The state of the switch changes the loop pattern, determining which LED flashes three times and which flashes just once in a cycle. This then gives a control system which has outputs that are dependent on particular input characteristics. It uses two of the outputs already used, and adds bit 7 of Port 0 as a digital input. Looking at Fig. 14.1, or the schematic, we can see that this is pin 7 of the Mbed.

It should not be difficult to follow Program Example 14.3 through. As before, we see the necessary register addresses defined. Directly inside the **main( )** function, Port 0 byte 0 is set as a digital input, noting that a Logic 0 sets the corresponding pin to input. We have set all of byte 0 to input by sending 0x00; we could of course set each pin within the byte individually if needed. Moreover, this setting is not fixed; we can change a pin from input to output as a program executes. After all this, a reading of Table 14.1 reminds us that the reset value of all ports is as input. Therefore, this little bit of code is not actually necessary—try removing it when you run the program. However, it is good practice to reassert values which are said to be in place due to reset; it gives you the confidence that the value is in place, and it is a definite statement in the code of a setting that you want.

The **while** loop then starts, and at the beginning of this we see the **if** statement testing the digital input value. The **if** condition uses a bit mask to discard the value of all the other pins on Port 0 byte 0 except for bit 7, which represents the switch input value. To test if bit 7 is a 1, the byte needs to be compared to 0x80, that is, the value of the bit mask. Notice from Table B.5 that bitwise AND and OR operators have a precedence *below* the equals operator, so the extra parentheses enclosing the **&** evaluation are essential.

The variables **a** and **b** hold values which change depending on the switch position. A little later, we see the **a** and **b** values being output to the port, driving the green and red LEDs. If **b** has been set to 0x01 before the **for** loop, then the red LED will flash three times; if it has been set to 0x02, then the green LED flashes three times.

```
/* Program Example 14.3: Uses digital input and output using control
registers, and flashes an LED. LEDS connect to Mbed pins 25 and 26. Switch
input to pin 7. For LPC1768 only.
```

```
                                        */
//function prototypes
void delay(void);

//Define Digital I/O registers
#define FIO0DIR0 (*(volatile unsigned char *)(0x2009C000))
#define FIO0PIN0 (*(volatile unsigned char *)(0x2009C014))
#define FIO2DIR0 (*(volatile unsigned char *)(0x2009C040))
#define FIO2PIN0 (*(volatile unsigned char *)(0x2009C054))
//some variables
char a;
char b;
char i;

int main() {
  FIO0DIR0 = 0x00;             // set all bits of port 0 byte 0 to input
  FIO2DIR0 = 0xFF;             // set port 2 byte 0 to output
  while(true){
    if ((FIO0PIN0 & 0x80) == 0x80){    // bit test port 0 bit 7 (Mbed pin 7)
      a = 0x01;                // this reverses the order of LED flashing
      b = 0x02;                // based on the switch position
    }
    else {
      a = 0x02;
      b = 0x01;
    }
    FIO2PIN0 |= a;         //set port bit outputs based on value of a
    delay();
    FIO2PIN0 &= ~a;
    delay();
    for (i = 1;i <= 3;i++){
      FIO2PIN0 |= b;     //set port bit outputs based on value of b
      delay();
      FIO2PIN0 &= ~b;
      delay();
    }
  }                           //end while loop
}

//delay function
void delay(void){
  volatile int j;              //loop variable j
  for (j = 0;j < 1000000;j++) {
    j++;
    j--;                      //waste time
  }
}
```

Program Example 14.3 Combined digital input and output

To run this program, set up a circuit similar to Fig. 3.7, except that the green and red LEDs should connect to Mbed pins 25 and 26, respectively, while the switch input remains on pin 7. Compile and run. You should see that the position of the switch toggles the flashing LEDs between the patterns:

green – red – red – red – green – red – red – red – green - ...

and

green – green – green – red – green – green - green – red – green - ...

Exercise 14.3

Rewrite Program Example 14.3 so that it runs on the exact circuit of Fig. 3.7.

14.4 Getting Deeper into the Control Registers

We now go further into the use of the LPC1768 control registers. In this section we look at some of the registers which control features across the microcontroller—we could call them informally "global" registers—which we will need to use in later sections. These relate to the allocation of pins, setting clock frequency, and controlling power. This is by no means a complete survey, and we will not even see or make use of many of the features of this microcontroller.

14.4.1 Pin Select and Pin Mode Registers

One of the reasons that modern microcontrollers are so versatile is that most pins are multifunctional. Each pin can be allocated to more than one peripheral, and used in different ways. With the Mbed OS, this flexibility is (quite reasonably) more or less hidden from you; the APIs tidily make the allocations for you, without you even knowing. If they did not, you would be faced with a bewildering choice of possibilities every time you tried to develop an application. As our expertise grows, however, it is good to know that some of these possibilities are available.

Two important sets of registers used in the LPC1768 are called **PINSEL** and **PINMODE**. The **PINSEL** register can allocate each pin to one of four possibilities. An example of part of one register which we will be using soon, **PINSEL1**, is shown in Table 14.2. This controls the upper 16 bits of Port 0. The first column shows the bit number within the register; each line details two bits. The second column shows the microcontroller pin which is being controlled. The two bits under consideration can have four possible combinations; each of these connects the pin in a different way. These are shown in the

Table 14.2: PINSEL1 register (address 0x4002 C004).

PINSEL1	Pin Name	Function when 00	Function when 01	Function when 10	Function when 11	Reset Value
1:0	P0.16	GPIO Port 0.16	RXD1	SSEL0	SSEL	00
3:2	P0.17	GPIO Port 0.17	CTS1	MISO0	MISO	00
5:4	P0.18	GPIO Port 0.18	DCD1	MOSI0	MOSI	00
7:6	P0.19[1]	GPIO Port 0.19	DSR1	Reserved	SDA1	00
9:8	P0.20[1]	GPIO Port 0.20	DTR1	Reserved	SCL1	00
11:10	P0.21[1]	GPIO Port 0.21	RI1	Reserved	RD1	00
13:12	P0.22	GPIO Port 0.22	RTS1	Reserved	TD1	00
15:14	P0.23[1]	GPIO Port 0.23	AD0.0	I2SRX_CLK	CAP3.0	00
17:16	P0.24[1]	GPIO Port 0.24	AD0.1	I2SRX_WS	CAP3.1	00
19:18	P0.25	GPIO Port 0.25	AD0.2	I2SRX_SDA	TXD3	00
21:20	P0.26	GPIO Port 0.26	AD0.3	AOUT	RXD3	00
23:22	P0.27[1][2]	GPIO Port 0.27	SDA0	USB_SDA	Reserved	00
25:24	P0.28[1][2]	GPIO Port 0.28	SCL0	USB_SCL	Reserved	00
27:26	P0.29	GPIO Port 0.29	USB_D+	Reserved	Reserved	00
29:28	P0.30	GPIO Port 0.30	USB_D−	Reserved	Reserved	00
31:30	-	Reserved	Reserved	Reserved	Reserved	00

[1]Not available on 80-pin package.
[2]Pins P0[27] and P0[28] are open-drain for I^2C-bus compliance.
From Table 81 of the LPC1768 User Manual.

next four columns. Don't worry if some of the abbreviations shown have little meaning to you; we will pick out the ones we need, when we need them.

Let's take as an example the line showing the effect of bits 21:20; that is, line 11 of the table: these control bit 26 of Port 0. Column 3 shows that if the bits are 00, the pin is allocated to Port 0 bit 26; that is, the pin is connected as general purpose I/O. Importantly, the final column shows that this is also the value when the chip is reset. In other words, as long as we only want to use digital I/O, we don't need to worry about this register at all, as the reset value is the value we want. If the bits are set to 01, the pin is allocated to input 3 of the ADC. If they are set to 10, the pin is used for analog output (i.e., the DAC output). If they are set to 11, the pin is allocated as receiver input for UART 3.

Table 14.3: PINMODE0 register (address 0x4002 C040).

PINMODE0	Symbol	Value	Description	Reset Value
1:0	P0.00MODE		Port 0 pin 0 on-chip pull-up/down resistor control.	00
		00	P0.0 pin has a pull-up resistor enabled.	
		01	P0.0 pin has repeater mode enabled.	
		10	P0.0 pin has neither pull-up nor pull-down.	
		11	P0.0 has a pull-down resistor enabled.	
3:2	P0.01MODE		Port 0 pin 1 control, see P0.00MODE.	00

continued to P0.15MODE
From Table 88 of the LPC1768 User Manual.

Turning to the **PINMODE** registers, partial details of one of these are shown in Table 14.3. This is **PINMODE0,** which controls the input characteristics of the lower half of Port 0. The pattern is the same for every pin, so there is no need for repetition. It is easy to see that pull-up and pull-down resistors (as seen in Fig. 3.6) are available; we explore this in Exercise 14.4. The repeater mode is a neat little facility which enables a pull-up resistor when the input is a Logic 1, and a pull-down resistor when it is low. If the external circuit changes so that the input is no longer driven, then the input will hold its most recent value.

■ Exercise 14.4

Change the hardware for Program Example 14.3 so that you use an SPSTswitch, for example, a push button, instead of the toggle (SPDT) switch. Connect it first between pin 9 and ground, and run the program with no change. This should run as before, because you are depending on the pull-up resistor being in place due to the reset value of the **PINMODE** register. In diagrammatic terms, you have moved from the switch circuit of Fig. 3.6(a) to that of Fig. 3.6(b). Now change the setting of **PINMODE0** so that the pull-down resistor is enabled instead, and connect the switch between pin 9 and 3.3 V, applying Fig. 3.6(c). The program should again work, but with the changed input mode selection.

■

14.4.2 Power Control and Clock Select Registers

As Chapter 15 will describe in detail, power control and clock frequency are very closely linked. Every clock transition causes the circuit of the microcontroller to take a tiny pulse of current; the more transitions, the more current taken. Hence, a processor or peripheral running at a high clock speed will cause high-power consumption; one running at a low clock frequency will consume less power. One with its clock switched off, even if it is powered up, will (if a purely digital circuit using CMOS technology) take negligible power. To conserve power, it is possible to turn off the clock source to many of the LPC1768 peripherals. This power management is controlled by the **PCONP** register, seen in part in Table 14.4. Where a bit is set to 1, the peripheral is enabled; where set to 0, it is disabled. It is interesting to note that some peripherals (like the SPI) are reset in the enabled mode, and others (like the ADC) are reset disabled.

Aside from being able to switch the clock to a peripheral on or off, there is some control over the peripheral's clock frequency itself. This controls the peripheral's speed of operation as well as its power consumption. This clock frequency is controlled by the **PCLKSEL** registers. Peripheral clocks are derived from the clock that drives the CPU, which is called **CCLK** (or **cclk**). For the Mbed, **CCLK** normally runs at 96 MHz. Partial details of **PCLKSEL0** are shown in Table 14.5. We can see that two bits are used per peripheral to control the clock frequency to each. The four possible combinations are shown in Table 14.6. This shows that the **CCLK** frequency itself can be used to drive the

Table 14.4: Power control register PCONP (address 0x400F C0C4, part of).

Bit	Symbol	Description	Reset Value
0	-	Reserved.	NA
1	PCTIM0	Timer/Counter 0 power/clock control bit.	1
2	PCTIM1	Timer/Counter 1 power/clock control bit.	1
3	PCUART0	UART0 power/clock control bit	1
4	PCUART1	UART1 power/clock control bit.	1
5	-	Reserved.	NA
6	PCPWM1	PWM1 power/clock control bit.	1
7	PCI2C0	The I^2C0 interface power/clock control bit.	1
8	PCSPI	The SPI interface power/clock control bit.	1
9	PCRTC	The RTC power/clock control bit.	1
10	PCSSP1	The SSP 1 interface power/clock control bit.	1
11	-	Reserved.	NA
12	PCADC	A/D converter (ADC) power/clock control bit.	0

continued to bit 28

From Table 46 of the LPC1768 User Manual.

Table 14.5: Peripheral clock selection register PCLKSEL0 (address 0x400F C1A8).

Bit	Symbol	Description	Reset Value
1:0	PCLK_WDT	Peripheral clock selection for WDT.	00
3:2	PCLK_TIMER0	Peripheral clock selection for TIMER0.	00
5:4	PCLK_TIMER1	Peripheral clock selection for TIMER1.	00
7:6	PCLK_UART0	Peripheral clock selection for UART0.	00
9:8	PCLK_UART1	Peripheral clock selection for UART1.	00
11:10	-	Reserved.	NA
13:12	PCLK_PWM1	Peripheral clock selection for PWM1.	00
15:14	PCLK_I2C0	Peripheral clock selection for I^2C0.	00
17:16	PCLK_SPI	Peripheral clock selection for SPI.	00
19:18	-	Reserved.	NA
21:20	PCLK_SSP1	Peripheral clock selection for SSP1.	00
23:22	PCLK_DAC	Peripheral clock selection for DAC.	00
25:24	PCLK_ADC	Peripheral clock selection for ADC.	00
27:26	PCLK_CAN1	Peripheral clock selection for CAN1.[1]	00
29:28	PCLK_CAN2	Peripheral clock selection for CAN2.[1]	00
31:30	PCLK_ACF	Peripheral clock selection for CAN acceptance filtering.[1]	00

[1]PCLK_CAN1 and PCLK_CAN2 must have the same PCLK divide value when the CAN function is used.
From Table 40 of the LPC1768 User Manual.

peripheral. Alternatively, it can be divided by 2, 4, or 8. **CCLK** is derived from the main oscillator circuit and can be manipulated in a number of interesting and useful ways, as described in Chapter 15.

14.5 Using the DAC

We now turn to controlling the DAC through its registers, trying to replicate and develop the work we did in Chapter 4. Remind yourself of the general block diagram of the DAC, as seen in Fig. 4.1. It is worth mentioning here that the positive reference voltage input to the LPC1768, shared by both ADC and DAC, is called V_{REFP}, and is connected to the power supply 3.3 V. The negative reference voltage input, called V_{REFN}, is connected directly to ground. We explore this further in Chapter 15.

Table 14.6: Peripheral clock selection register bit values.

PCLKSEL0 and PCLKSEL1 individual peripheral's clock select options	Function	Reset Value
00	PCLK_peripheral = CCLK/4	00
01	PCLK_peripheral = CCLK	
10	PCLK_peripheral = CCLK/2	
11	PCLK_peripheral = CCLK/8, except for CAN1, CAN2, and CAN filtering when "11" selects = CCLK/6	

From Table 42 of the LPC1768 User Manual.

14.5.1 LPC1768 DAC Control Registers

As with all peripherals, the DAC has a set of registers which control its activity. In terms of the "global" registers which we have just seen, the DAC power is always enabled, so there is no need to consider the **PCONP** register. The *only* pin that the DAC output is available on is Port 0 pin 26, so we must allocate this pin appropriately through the **PINSEL1** register, as seen in Table 14.2. The DAC output is labeled AOUT here. It is no surprise to see in the Mbed schematics that this pin is connected to pin 18, the Mbed's only analog output.

The only register specific to the DAC that we will use is the **DACR** register. This is comparatively simple to grasp, and is shown in Table 14.7. We can see that the digital input to the DAC must be placed in here, in bits 6 to 15. Most of the rest of the bits are unused, apart from the bias bit, explained in the table.

Applying Eq. 4.1 to this 10-bit DAC, its output is given by:

$$V_0 = (V_{REFP} \times D) / 1024 = (3.3 \times D) / 1024 \quad (14.1)$$

where D is the value of the 10-bit number placed in bits 15 to 6 of the **DACR** register.

14.5.2 A DAC Application

Let's try a simple program to drive the DAC, accessing it through the microcontroller control registers. Program Example 14.4 replicates the simple sawtooth output, which we first achieved in Program Example 4.2. The program follows the familiar pattern of defining register addresses, and then setting these appropriately early in the **main( )** function. The **PINSEL1** register is set to select DAC output on port bit 0.26. An integer variable, called **dac_value**, is then repeatedly incremented by four and transferred to the DAC input, in register **DACR**. It has to be shifted left six times, to place it in the correct

Table 14.7: The DACR register (address 0x4008 C000).

Bit	Symbol	Value	Description	Reset Value
5:0	-		Reserved, user software should not write ones to reserved bits. The value read from a reserved bit is not defined.	NA
15:6	VALUE		After the selected settling time after this field is written with a new VALUE, the voltage on the AOUT pin (with respect to V_{SSA}) is VALUE x $((V_{REFP} - V_{REFN})/1024) + V_{REFN}$.	0
16	BIAS	0	The settling time of the DAC is 1 μs max, and the maximum current is 700 μA. This allows a maximum update rate of 1 MHz.	0
		1	The settling time of the DAC is 2.5 μs and the maximum current is 350 μA. This allows a maximum update rate of 400 kHz.	
31:17	-		Reserved, user software should not write ones to reserved bits. The value read from a reserved bit is not defined.	NA

From Table 540 of the LPC1768 User Manual.

bits of the **DACR** register. After each new value of DAC input, a delay (a shorter version of the one used previously) is introduced. We explore the effect of this in Exercise 14.5.

```
/* Program Example 14.4: Sawtooth waveform on DAC output. View on oscilloscope.
Port 0.26 is used for DAC output, i.e. Mbed Pin 18
For LPC1768 only.                                                          */

// function prototype
void delay(void);
// variable declarations
int dac_value;             //the value to be output
//define addresses of control registers, as pointers to volatile data
#define DACR (*(volatile unsigned long *)(0x4008C000))
#define PINSEL1 (*(volatile unsigned long *)(0x4002C004))

int main(){
  PINSEL1 = 0x00200000; //set bits 21-20 to 10 to enable analog out on P0.26
  //or use: PINSEL1 = PINSEL1 | (1<<21); //(all bits in register reset to 0)
  while(true){
    for (dac_value = 0;dac_value < 1024;dac_value = dac_value + 4){
      DACR = (dac_value << 6);
      delay();
    }
  }
}
```

```
//delay function
void delay(void){
  volatile int j;            //loop variable j
  for (j = 0;j < 1000;j++) {
    j++;
    j--;                     //waste time
  }
}
```

Program Example 14.4 Sawtooth output on the DAC

Compile the program and run it on an Mbed LPC1768; no external connections are needed. View the output from pin 18 on an oscilloscope.

Exercise 14.5

Measure the period of the sawtooth waveform. How does it relate to the delay value you measured in Exercise 14.1 (noting that this one has been reduced by a factor of 1000)? Try varying the period by varying the delay value, or removing it altogether.

14.6 Using the ADC

We now turn to controlling the ADC through its registers, trying to replicate and develop the work we did in Chapter 5. It is worth glancing back at Fig. 5.1, as this represents many of the features that we need to control. This includes selecting (or at least knowing the value of) the voltage reference, clock speed, and input channel, starting a conversion, detecting a completion, and reading the output data. The LPC1768 microcontroller has eight inputs to its ADC, which appear—in order from input 0 to input 7—on pins 9–6, 21, 20, 99, and 98 of the microcontroller. A study of the Mbed LPC1768 circuit shows that the lower six of these are used, connected to pins 15 to 20 of the Mbed, inclusive.

14.6.1 LPC1768 ADC Control Registers

The LPC1768 has a number of registers which control its ADC, particularly in its more sophisticated operation. However, we will apply only two of these, the ADC control register, **ADCR,** and the global data register, **ADGDR**. These are detailed in Tables 14.8 and 14.9. As we have seen, the ADC can also be powered down; indeed, on microcontroller reset, it is

Table 14.8: The AD0CR register (address 0x4003 4000).

Bit	Symbol	Value	Description	Reset Value
7:0	SEL		Selects which of the AD0.7:0 pins is (are) to be sampled and converted. For AD0, bit 0 selects Pin AD0.0, and bit 7 selects pin AD0.7. In software-controlled mode, only one of these bits should be 1. In hardware scan mode, any value containing 1 to 8 ones is allowed. All zeroes is equivalent to 0x01.	0x01
15:8	CLKDIV		The APB clock (PCLK_ADC0) is divided by (this value plus one) to produce the clock for the A/D converter, which should be less than or equal to 13 MHz. Typically, software should program the smallest value in this field that yields a clock of 13 MHz or slightly less, but in certain cases (such as a high-impedance analog source) a slower clock may be desirable.	0
16	BURST	1	The AD converter does repeated conversions at up to 200 kHz, scanning (if necessary) through the pins selected by bits set to ones in the SEL field. The first conversion after the start corresponds to the least-significant 1 in the SEL field, then higher numbered 1-bits (pins) if applicable. Repeated conversions can be terminated by clearing this bit, but the conversion that's in progress when this bit is cleared will be completed. **Remark:** START bits must be 000 when BURST = 1 or conversions will not start.	0
		0	Conversions are software controlled and require 65 clocks.	
20:17	-		Reserved, user software should not write ones to reserved bits. The value read from a reserved bit is not defined.	NA
21	PDN	1	The A/D converter is operational.	0
		0	The A/D converter is in power-down mode.	
23:22	-		Reserved, user software should not write ones to reserved bits. The value read from a reserved bit is not defined.	NA
26:24	START		When the BURST bit is 0, these bits control whether and when an A/D conversion is started:	0
		000	No start (this value should be used when clearing PDN to 0).	
		001	Start conversion now.	

Note: Further, more advanced options for START control are available; see full table.
From Table 532 of the LPC1768 User Manual.

Table 14.9: The AD0GDR register (address 0x4003 4004).

Bit	Symbol	Description	Reset Value
3:0	-	Reserved, user software should not write ones to reserved bits. The value read from a reserved bit is not defined.	NA
15:4	RESULT	When DONE is 1, this field contains a binary fraction representing the voltage on the AD0[n] pin selected by the SEL field, as it falls within the range of V_{REFP} to V_{REFN}. Zero in the field indicates that the voltage on the input pin was less than, equal to, or close to that on V_{REFN}, while 0x3FF indicates that the voltage on the input was close to, equal to, or greater than that on V_{REFP}.	NA
23:16	-	Reserved, user software should not write ones to reserved bits. The value read from a reserved bit is not defined.	NA
26:24	CHN	These bits contain the channel from which the RESULT bits were converted (e.g. 000 identifies channel 0, 001 channel 1...).	NA
29:27	-	Reserved, user software should not write ones to reserved bits. The value read from a reserved bit is not defined.	NA
30	OVERRUN	This bit is 1 in burst mode if the results of one or more conversions was (were) lost and overwritten before the conversion that produced the result in the RESULT bits. This bit is cleared by reading this register.	0
31	DONE	This bit is set to 1 when an A/D conversion completes. It is cleared when this register is read and when the ADCR is written. If the ADCR is written while a conversion is still in progress, this bit is set and a new conversion is started.	0

From Table 533 of the LPC1768 User Manual.

switched off. Therefore, to enable it, we will have to set bit 12 in the **PCONP** register, seen in Table 14.4.

14.6.2 An ADC Application

Program Example 14.5 provides a good opportunity to see many of the control registers in action. Channel 1 of the ADC is applied, which connects to pin 16 of the Mbed.

The configuration of the ADC has some complexity, so read the program with care, starting from the comment "initialize the ADC." The ADC channel we want is multiplexed with bit 24 of Port 0, so first we must allocate this pin to the ADC. This is done through bits 17 and 16 of the **PINSEL1** register, seen in Table 14.2. We then enable the ADC, through the relevant bit in the **PCONP** register, Table 14.4. There follows quite a complex process of configuring the ADC, through the **AD0CR** register. We could set this register by transferring a single word. Instead, as can be seen, the required value is constructed by taking the OR combination of a series of bit shifts. This approach can provide better visibility of how the individual bits are being determined. Compare the settings with Table 14.8.

The data conversion then starts in the **while** loop which follows. The comments contained in the program listing should give you a good picture of each stage.

```
/* Program Example 14.5: A bar graph meter for ADC input, using control registers
to set up ADC and digital I/O
For LPC1768 only.                                                          */

// variable declarations
char ADC_channel = 1;        // ADC channel 1
int ADCdata;               // this will hold the result of the conversion
int DigOutData = 0;          //a buffer for the output display pattern

// function prototype
void delay(void);
//define addresses of control registers, as pointers to volatile data
//(i.e. the memory contents)
#define PINSEL1       (*(volatile unsigned long *)(0x4002C004))
#define PCONP         (*(volatile unsigned long *)(0x400FC0C4))
#define AD0CR         (*(volatile unsigned long *)(0x40034000))
#define AD0GDR        (*(volatile unsigned long *)(0x40034004))
#define FIO2DIR0      (*(volatile unsigned char *)( 0x2009C040))
#define FIO2PIN0      (*(volatile unsigned char *)( 0x2009C054))

int main() {
  FIO2DIR0 = 0xFF;// set lower byte of Port 2 to output, this drives bar graph

//initialise the ADC
  PINSEL1 = 0x00010000; //set bits 17-16 to 01 to enable AD0.1 (Mbed pin 16)
  PCONP |= (1 << 12);              // enable ADC clock
  AD0CR = (1 << ADC_channel)       // select channel 1
         | (4 << 8)       // Divide incoming clock by (4+1), giving 4.8MHz
         | (0 << 16)      // BURST = 0, conversions under software control
         | (1 << 21)      // PDN = 1, enables power
         | (1 << 24);     // START = 1, start A/D conversion now
```

```
  while(true){                        // infinite loop
    AD0CR = AD0CR | 0x01000000;       //start conversion by ORing bit 24 to 1.
   //Or use: AD0CR = AD0CR | (1 << 24);
   // wait for it to finish by polling the ADC DONE bit
   while ((AD0GDR & 0x80000000) == 0) { //test DONE bit, wait till it's 1
   }
   ADCdata = AD0GDR;                  // get the data from AD0GDR
   AD0CR &= 0xF8FFFFFF;               //stop ADC by setting START bits to zero
  // Shift data 4 bits to right justify, and 2 more to give 10-bit ADC
  // value - this gives convenient range of just over one thousand.
   ADCdata = (ADCdata >> 6) & 0x03FF;      //and mask
   DigOutData = 0x00;                  //clear the output buffer
   //display the data
   if (ADCdata > 200)
    DigOutData = (DigOutData|0x01);  //set the lsb by ORing with 1
   if (ADCdata > 400)
    DigOutData = (DigOutData|0x02);  //set the next lsb by ORing with 1
   if (ADCdata > 600)
    DigOutData = (DigOutData|0x04);
   if (ADCdata > 800)
    DigOutData = (DigOutData|0x08);
   if (ADCdata > 1000)
    DigOutData = (DigOutData|0x10);

   FIO2PIN0 = DigOutData;          // set port 2 to Digoutdata
   delay();      // pause
  }
}

  //delay function
void delay(void){
  volatile int j;              //loop variable j
  for (j = 0;j < 1000000;j++) {
    j++;
    j--;                       //waste time
  }
}
```

Program Example 14.5 Applying the ADC as a bar graph

Connect an Mbed LPC1768 with a potentiometer between 0 and 3.3 V with the wiper connected to pin 16. Connect five LEDs between pin and ground, from pin 22 to pin 26, inclusive. Compile and download your code, and press reset. Moving the potentiometer should alter the number of LEDs lit, from none at all to all five.

Exercise 14.6

Add a digital input switch as in the previous example to reverse the operation of the analog input. With the digital switch in one position the LEDs will light from right to left; with the switch in the other position the **DigOutData** variable is inverted to light LEDs in the opposite direction, from left to right.

Exercise 14.7

Extend the bar graph so that it has 8 or 10 LEDs. You will of course need to work out how to drive the extra I/O pins that you choose to use.

14.6.3 Changing ADC Conversion Speed

One of the limitations of the original Mbed AnalogIn API is the comparatively low speed of the ADC conversion. We explored this in Exercise 5.7. Let's try now to vary this conversion speed by adjusting the ADC clock speed.

Table 14.8 tells us that the ADC clock frequency should have a maximum value of 13 MHz, and that it takes 65 cycles of the ADC clock to complete a conversion. A quick calculation shows that the minimum conversion time possible is therefore 5 μs. It takes a very careful reading of the LPC1768 user manual, Reference 2.4, to get a full picture of how the ADC clock frequency is controlled. The ADC clock is derived from the main microcontroller clock; there are several stages of division that the user can control in order to set a frequency as close to 13 MHz as possible. The first is through register **PCLKSEL0**, detailed in Tables 14.5 and 14.6. Bits 25 and 24 of **PCLKSEL0** control the ADC clock division. We have seen that for most peripherals, including the ADC, the clock can be divided by one, two, four, or eight. On power-up the selection defaults to divide-by-four. The clock may be further divided through bits 15 to 8 of the **AD0CR** register, seen in Table 14.8.

Program Example 14.6 is similar to Program Example 5.5, with some interesting results. It is also a useful example, as it combines ADC, DAC, and digital I/O, therefore illustrating how these can be used together. It is made up of elements from programs earlier in this chapter, sometimes with adjustments; it should be possible to follow it through without too much difficulty. As sections of the program repeat from earlier examples, only the newer parts are reproduced here. The full program listing can be downloaded from the book website.

```
/* Program Example 14.6: Explore ADC conversion times, programming control
registers directly. ADC value is transferred to DAC, while an output pin is
strobed to indicate conversion duration. Observe on oscilloscope.
For LPC1768 only.                                         */
...
//insert all necessary #define statements here
...
int main() {
  FIO2DIR0 = 0xFF;             // set lower bits port 2 to output
  PINSEL1 = 0x00210000;       //set bits 21-20 to 10 for analog output (Mbed p18)
        //and bits 17-16 to 01 to enable ADC channel 1 (AD0.1, Mbed pin 16)
  ...
//insert ADC initialisation here
...
  while(true){              // infinite loop
    // start A/D conversion by modifying bits in the AD0CR register
    AD0CR = (AD0CR & 0xFFFFFF00);
    FIO2PIN0 |= 0x01;            // set strobe pin (Mbed pin 26) high
    AD0CR |= (1 << ADC_channel) | (1 << 24); //select ADC channel, and START.
    // wait for it to finish by polling the ADC DONE bit
    while((AD0GDR & 0x80000000) == 0) {
    }
    FIO2PIN0 &= ~0x01;          // set strobe pin low
    ADCdata = AD0GDR;           // get the data from AD0GDR
    AD0CR &= 0xF8FFFFFF;       //stop ADC by setting START bits to zero
   // shift data 4 bits to right justify, and 2 more to give 10-bit ADC value
    ADCdata = (ADCdata>>6) & 0x03FF;  //and mask
    DACR = (ADCdata<<6);       //could be merged with previous line,
                               // but separated for clarity
    //delay();                 //insert delay if wished
  }
}
...
//insert delay() function from Program Example 14.5 here
```

Program Example 14.6 Applying ADC, DAC, and digital output to measure conversion duration

You can use the same Mbed circuit for this as you did for Program Example 14.5, although only one LED, on pin 26, is necessary. Compile and run the program. First put an oscilloscope probe on pin 18, the DAC output. The voltage on this pin should change as the potentiometer is adjusted. This confirms that the program is running. Now move the probe to pin 26; you will see the pin pulsing high for the duration of the ADC conversion. If you measure this, you should find that it is close to 14 μs.

To calculate the ADC clock frequency, and hence the conversion time, remember that the Mbed **CCLK** frequency is 96 MHz. We have not touched the **PCLKSEL0** register, so the clock setting for the ADC will be 96 MHz divided by the reset value of 4, that is, 24

MHz. This is further divided by 5 in the **ADC0CR** setting seen in the program example, leading to an ADC clock frequency of 4.8 MHz, or a period of 0.21 μs. Sixty-five (65) cycles of 0.21 μs leads to the measurement duration mentioned in the previous paragraph.

Exercise 14.8

1. Adjust the setting of the **CLKDIV** bits in **AD0CR** in Program Example 14.6 to give the fastest permissible conversion time. Run the program, and check that your measured value agrees with that predicted.
2. Can you now account for the value of ADC conversion time you measured in Exercise 5.7?

14.7 A Conclusion on Using the Control Registers

In this chapter we have explored the use of the LPC1768 control registers, in connection with use of the digital I/O, ADC, and DAC peripherals. We have demonstrated how these registers allow the peripherals to be controlled directly, without using the Mbed APIs. This has allowed greater flexibility of use of the peripherals, at the cost of getting into the tiny details of the registers, and programming at the level of the bits that make them up. Ordinarily we probably wouldn't want to program like this; it is time-consuming, inconvenient, and error-prone. However, if we need a configuration or setting not offered by the APIs, this approach can be a way forward. While we have only worked in this way in connection with three of the peripherals, it is possible to do it with any of them. It is worth mentioning that the three that we have worked with are some of the simpler ones; others require even more attention to detail.

Chapter Review

- In this chapter we have recognized a different way of controlling the Mbed LPC1768 peripherals. It demands a much deeper understanding of the Mbed microcontroller, but allows greater flexibility in using the microcontroller.
- There are registers which relate just to one peripheral, and others which relate to microcontroller performance as a whole.
- We have found that we can exploit features that may not be available in the Mbed APIs, for example, in the change of the ADC conversion speed.

- The chapter only introduces a small range of the control registers which are used by the LPC1768. However, it should have given you the confidence to look up and begin to apply any that you need.

Quiz

1. The I/O bits on any microcontroller are grouped into "ports."
   i. How many I/O ports does the LPC1768 microcontroller have?
   ii. Theoretically how many bits does this lead to?
   iii. How many I/O bits does the LPC1768 actually have?
   iv. How many port bits does the Mbed LPC1768 target board make available?
2. What is the name of the register which sets the data direction of Port 1, byte 2?
3. Explain the function of the **PINSEL** and **PINMODE** register groups.
4. The initialization section of a certain program reads as follows:

   ```
   FIO0DIR0 = 0xF0;
   PINMODE0 = 0x0F;
   PINSEL1 = 0x00204000;
   ```

   Explain the settings that have been made.
5. An LPC1768 microcontroller is connected with a 3.0 V reference voltage. Its DAC output reads 0.375 V. What is its input digital value?
6. A designer is developing a low-power application, where high speed is not essential. Which aspect of the DAC control, excluding clock frequency considerations, allows trade-off between speed of conversion and power consumption?
7. On an LPC1768 microcontroller, the ADC clock is set at 4 MHz. How long does one conversion take?
8. A user wants to sample an incoming signal with an Mbed LPC1768 ADC at 44 kHz or greater. What is the minimum permissible ADC clock frequency?
9. Describe how the ADC clock frequency calculated in Question 8 can be set up.
10. Which feature of the ADC allows a set of inputs to be selected and scanned, under hardware control? Which feature detects if use of this mode leads to an error?

CHAPTER 15

Some Hardware Insights: Clocks, Resets, and Power Supply

15.1 Hardware Essentials—Power Supply

In the days when cars and motorcycles were simple mechanical systems (i.e., before electronics and embedded systems got into them!) they seemed easier to fix. If something went wrong, there were always two things to check first: was there fuel, and was there a spark? No petrol engine could run without these. A microcontroller has similar needs, except that they are for power supply and clock. Any serious designer needs to know some underlying details about these, and how they can be manipulated to best effect. Directly linked with these two basics come several other important features, including:

- start-up, restart, and reset features;
- mechanisms to enhance reliability, like the watchdog timer and brownout detect;
- and the range of low-power modes that are available in most modern microcontrollers and their use.

This chapter starts with power supply, and moves on to cover the other points listed. The principles of these features are broadly shared by all microcontrollers, though the practice may differ. Because the internal details of a microcontroller are so intricate, we develop the central ideas of the chapter mainly through just one device, the Mbed LPC1768. We then turn to the Nucleo-F401RE for comparison, and finally to another Mbed-enabled development board, the EFM32 Zero Gecko.

15.1.1 Powering the LPC1768 Microcontroller

We have already seen an overview of the Mbed LPC1768 power distribution in Fig. 2.2, and how a 3.3 V supply is generated from one of the incoming power sources. Fig. 15.1, taken from the LPC1768 data sheet (Reference 2.3), shows that the microcontroller itself makes many uses of that 3.3 V. The diagram identifies three power supply domains: main, RTC (real-time clock), and ADC. Two power inputs supply the main domain, labeled $V_{DD(3V3)}$ and $V_{DDREG(3V3)}$. It is easy to see that $V_{DD(3V3)}$ simply supplies the I/O ports. However, $V_{DDREG(3V3)}$ goes to a further regulator, which then supplies the inner workings of the microcontroller. Another power input, V_{BAT}, is the main power source for the RTC

Fast and Effective Embedded Systems Design. https://doi.org/10.1016/B978-0-323-95197-5.00015-2

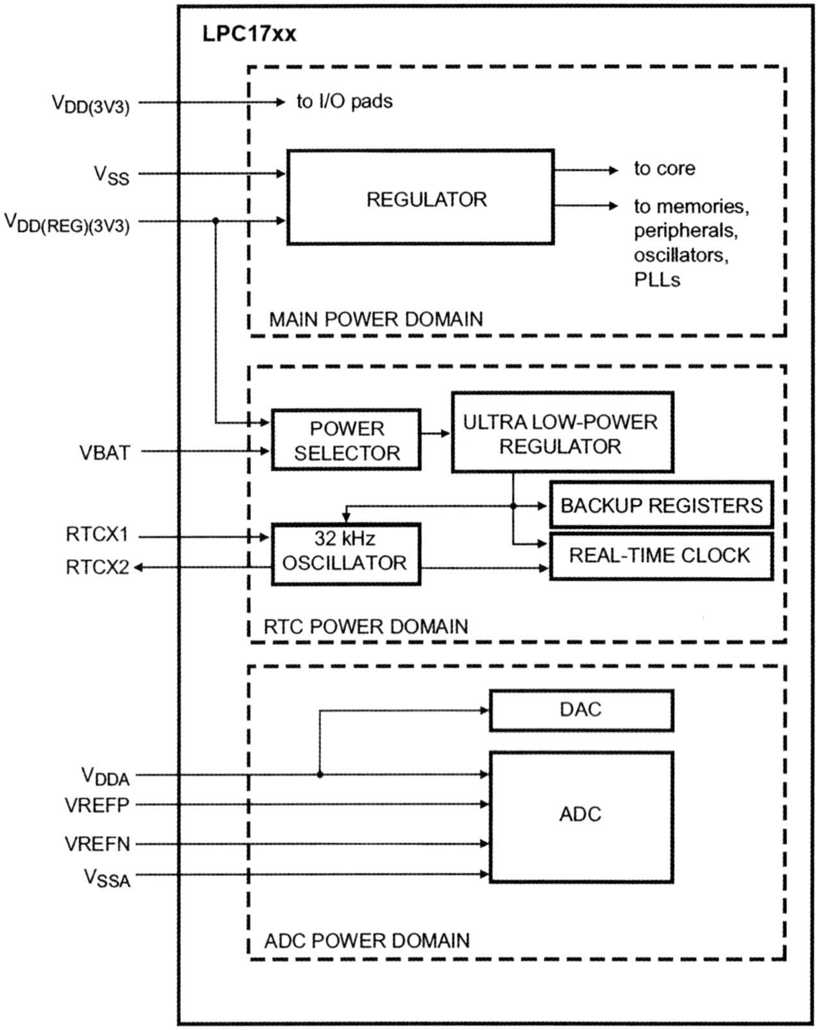

Figure 15.1
Power distribution in the LPC1768. *Image courtesy of NXP.*

domain; this can be connected to an external battery, to sustain RTC and backup registers when all other power is lost. Finally, the ADC and DAC can be powered separately from the rest of the microcontroller, with their own private and pristine power supply. A voltage reference also needs to be provided, through the positive and negative voltage pins, V_{REFP} and V_{REFN}.

The exact supply voltage requirements of the LPC1768, for each of the supply input pins seen in Fig. 15.1, are shown in Table 15.1. All are centered at 3.3 V, but actually there is flexibility around this in each case, useful for example to tolerate a diminishing battery voltage. V_{BAT} has the lowest minimum value, allowing RTC operation to be sustained with

Table 15.1: Supply voltage requirements for the LPC1768.

T_{amb} = −40 °C to +85 °C, unless otherwise specified.					
Symbol	**Parameter**	**Min**	**Typ**	**Max**	**Unit**
Supply pins					
$V_{DD(3V3)}$	supply voltage (3.3 V)	2.4	3.3	3.6	V
$V_{DD(REG)(3V3)}$	regulator supply voltage (3.3 V)	2.4	3.3	3.6	V
V_{DDA}	analog 3.3 V pad supply voltage	2.5	3.3	3.6	V
$V_{i(VBAT)}$	input voltage on pin VBAT	2.1	3.3	3.6	V
$V_{i(VREFP)}$	input voltage on pin VREFP	2.5	3.3	V_{DDA}	V

From Table 8 of LPC1768 data sheet.

a battery voltage down to 2.1 V. All supplies can go up to 3.6 V, except for the ADC positive reference voltage, which must not exceed the ADC supply voltage.

15.1.2 Powering the Mbed LPC1768

Turning back to the Mbed LPC1768 board, the diagram of Fig. 15.2 shows how the designers connect the microcontroller power inputs, and a number of other useful things besides. The main block in the diagram is of course the LPC1768 itself; each connection shows the microcontroller pin number and the name of the signal. It is first worth noting that a complex integrated circuit tends to have more than one ground connection, and similar multiple power supply connections. This is because the interconnecting wires inside the IC are so very thin that they can have significant resistance. These multiple connections are dotted around the IC interconnect pins, to enhance connectivity to the circuit chip itself. We can note that there are no fewer than six ground connections, labeled V_{SS}. There are four $V_{DD(3V3)}$ connections and two for $V_{DDREG(3V3)}$. The designers don't take the opportunity to differentiate between these latter two connections; they just join them together. The V_{BAT} connection is, however, kept separate, and *can* be supplied via pin 3, VB, of the Mbed board. Note how the V_{DD} supply is smoothed by the distributed capacitors C_{15}−C_{17} and C_{20}, each of 100 nF, and C_{21}, of value 10 μF, placed according to the guidance embedded in the diagram. Yes, power supply distribution is a sophisticated art in a complex circuit board.

It is interesting to see how the ADC is powered and referenced. We did not get into much detail on this in Chapter 5. The designers make a happy compromise here. They do not provide a separate supply, but do filter V_{DD} with the severe low-pass filter made up of L1 and C14, which will remove much of the high-frequency interference which V_{DD} may pick up. The ADC uses the same voltage for its supply (V_{DDA}), and for its positive

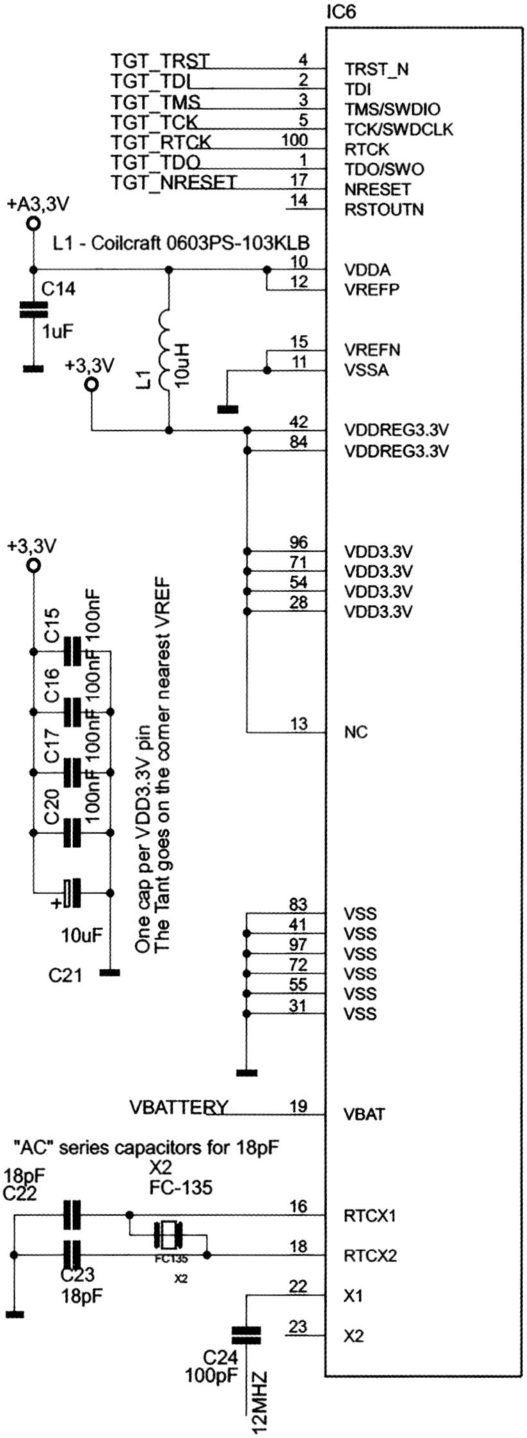

Figure 15.2

LPC1768 microcontroller connections on the Mbed board. *Image courtesy of ARM.*

reference (V_{REFP}). This is common practice, although the use of a voltage regulator as a reference risks introducing inaccuracy into the measurement (try Reference 3.1 to get into the interesting detail of this). Meanwhile, the negative side of the ADC supply and reference are connected straight to system ground.

15.2 Clock Sources and Their Selection

15.2.1 Some Clock Oscillator Preliminaries

An essential part of the microcontroller system is the clock oscillator, providing a continuous square wave which relentlessly drives the microcontroller action forward. Besides this, it is also the basis of any accurate time measurement or generation. Clock oscillators can be based on resistor—capacitor (R—C) networks, or ceramic or crystal resonators. The designer should be aware of the options that are available, and their relative advantages.

A popular R—C oscillator circuit is shown in Fig. 15.3(a). Here the capacitor C is charged from the supply rail through resistor R. The voltage at the R—C junction connects to the input of a logic gate. When the input threshold of the logic gate is passed, its output goes high and switches on a transistor. This rapidly discharges the capacitor. The gate output goes low again, and the capacitor restarts its charging. This action continues indefinitely. Operating frequencies depend on values of R and C and the input thresholds of the gate. This incidentally has a *Schmitt trigger* input stage, shown by that little symbol inside the gate symbol. A Schmitt trigger tidies up a poorly defined logic signal, such as appears at the R—C junction, and its *hysteresis* action ensures good discharge of the capacitor.

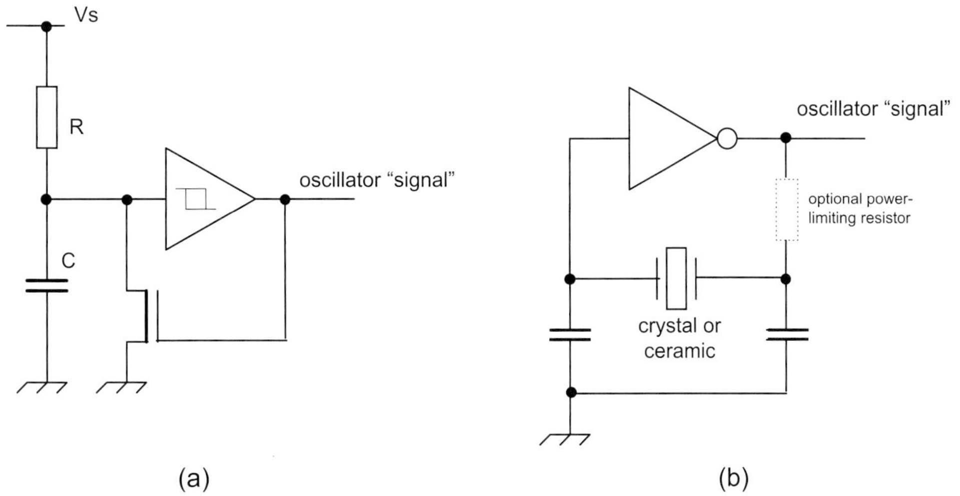

Figure 15.3
Oscillator circuits: (a) the resistor—capacitor (R—C) oscillator; (b) the crystal oscillator.

The R—C circuit is the cheapest form of clock oscillator available and is widely used. Moreover, all components can be integrated on chip, and it is resistant to mechanical shock. Because of this, an on-chip R—C oscillator is highly reliable. With some microcontrollers (though not with the LPC1768), the resistor and capacitor can also be placed externally, so the operating frequency can be approximately selected. In this case, only one IC interconnection pin is required. The main disadvantage of the circuit, which in many situations is decisive, is that the precise frequency of oscillation is unpredictable, as it drifts with temperature and time. Recent technical advances have improved this situation, and many on-chip R—C oscillators are now surprisingly stable, though never as good as a crystal oscillator, described next.

A quartz *crystal oscillator* is based on a very thin slice of crystal. The crystal is piezoelectric, which means that if it is subjected to mechanical strain and then a voltage appears across opposite surfaces, and if a voltage is applied to it, then it experiences mechanical strain; that is, it distorts slightly. The crystal is cut into shape, polished, and mounted so that it can vibrate mechanically, the frequency depending on its size and thickness. The resonant frequency of vibration is very stable and predictable. If electrical terminals are deposited on opposite surfaces of the crystal, then due to its piezoelectric properties, vibration can be induced and then sustained electrically by applying a sinusoidal voltage at the appropriate frequency. A suitable circuit is shown in Fig. 15.3(b). A crystal can never be integrated on an IC, so must always be external, though the logic gate is usually on chip. It is very common to see an external crystal connected to a microcontroller through two terminals, with the two associated low-value capacitors. You can see this applied for the LPC1768 RTC oscillator in Fig. 15.2, connected to pins RTCX1 and RTCX2.

15.2.2 LPC1768 Clock Oscillators and the Mbed Implementation

Fig. 15.4 shows how the clock sources within the LPC1768 are acquired and distributed. Note that this is a generic diagram; the Mbed LPC1768 does not have an external crystal connected directly to the main oscillator terminals. There are three sources available, each of which can be used in many ways. They are the main oscillator (an external crystal—as shown—*or* an external clock signal), the internal R—C oscillator, and the RTC oscillator. This requires another crystal, of low frequency. All three sources are seen to the left of the diagram. The main oscillator, based on the circuit of Fig. 15.3(b), can operate between 1 MHz and 25 MHz. However, this oscillator circuit can also act in *slave mode*, in which case an external clock signal can be connected to one of the crystal pins via a capacitor. This approach is implemented in the Mbed LPC1768, where a single 12 MHz oscillator supplies both microcontrollers on the development board. The Internal R—C oscillator runs at a nominal frequency of 4 MHz. Fig. 15.4 shows that it *can* be used as the main clock source, though it will lack the precision required for certain time-based activities. It is also available to drive the watchdog timer. The RTC oscillator is generally expected to

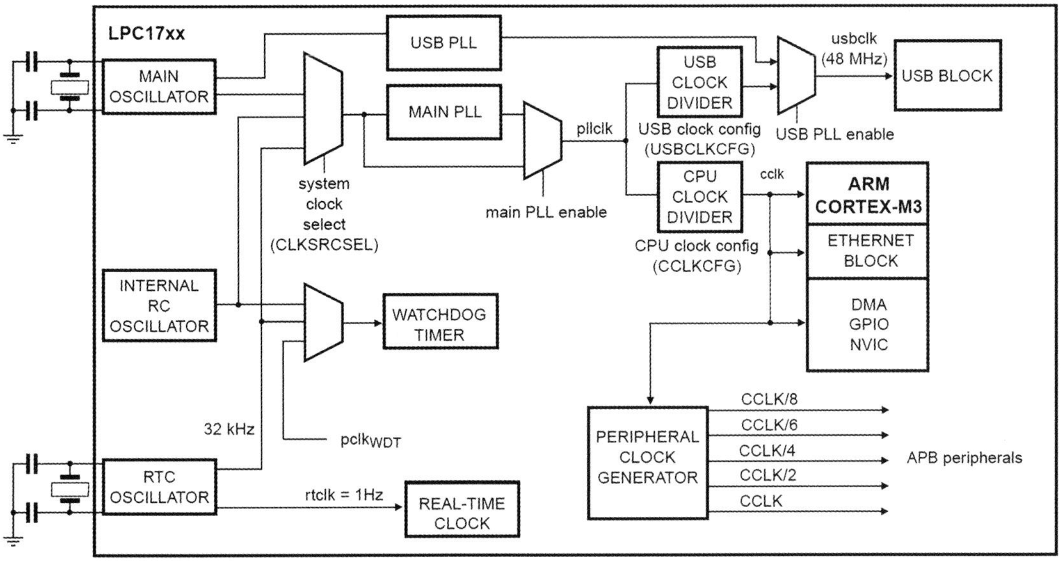

Note: MAIN PLL is also called PLL0

<u>Key</u>
APB: Advanced Peripheral Bus
GPIO: General Purpose Input/output
PLL: Phase Locked Loop
USB: Universal Serial Bus
DMA: Direct Memory Access
NVIC: Nested Vectored Interrupt Controller
RTC: Real Time Clock

Figure 15.4
The LPC1768 clock distribution circuit. *Image courtesy of NXP.*

be 32.768 kHz. This is 2^{15}, which divides tidily down to a 1 Hz, one pulse per second, signal. This is useful as the basis for real-time clock applications, where seconds, minutes, hours, and days are counted.

The three clock sources all enter a multiplexer, one of several "wedge-shaped" symbols in the circuit. This is effectively a selector circuit, controlled by a couple of bits from the clock source select register **(CLKSRCSEL)**; we will see more of this soon. The output of this multiplexer can go through a *phase-locked loop* (PLL), a clever circuit which can *multiply* frequencies. Hence, it could take a 10 MHz oscillator input, multiply by 8, to give an output of 80 MHz. This is useful—indeed essential—because it allows an internal clock frequency *higher* than the frequency a crystal can supply. The main PLL is called PLL0; we return to it in Section 15.2.4. Another multiplexer then selects whether the PLL is used or not. The resulting signal, **pllclk**, is available to both the USB and the CPU. Following the CPU path, it can then be *divided,* producing a signal called **cclk**. Why divide the frequency, when we have just been talking about multiplying it? Different combinations of multiplication and division allow a very wide range of oscillator frequencies to be selected. The **cclk** signal goes to four further system blocks, including the Cortex core itself and the peripherals. Details of registers in the peripheral clock generator have already been covered in Chapter 14, Section 14.4.2. Note that the maximum permissible frequency for the LPC1768 CPU is 100 MHz, so all settings should reflect this.

How does the Mbed board use the three clock sources available to the LPC1768? Fig. 15.2 will help to answer this question. It is easy to see the RTC crystal, connected to pins 16 and 18. A quick check of the data shows that the FC-135 crystal, identified in the diagram, runs at 32.768 kHz, as expected. The internal R–C oscillator is entirely on chip, so it has no visibility here; yet Section 15.3 will show what an important role it has. The main oscillator is connected to pin 22. The Mbed has a 12-MHz crystal oscillator source, already mentioned, which links to the LPC1768 through C24. We know, however, that the LPC1768 itself runs at 96 MHz (Appendix C). We must deduce that the incoming 12 MHz is multiplied by 8 to achieve this value.

Would it be possible to change the operating frequency of the microcontroller? Answers to this are explored in the next few sections. Fig. 15.4 suggests that there are three ways to manipulate the frequency of the main clock: by selecting a different clock source, by changing the PLL setting, or by changing the divider. Control of the PLL carries some complexities, so let's leave that to a later section. The simplest clock frequency change can be made by changing the CPU clock divider; we make that our next topic.

15.2.3 Setting the LPC1768 Clock Configuration Register

The CPU clock divider block, seen in Fig. 15.4, is controlled by the clock configuration register **CCLKCFG,** shown in Table 15.2. It is easy to see that the frequency of **pllclk** is divided by the number held in the lower 8 bits of this register, plus one; this produces **cclk**. The value held in this register is accessed and adjusted in Program Example 15.1.

Table 15.2: LPC1768 clock configuration register CCLKCFG.

Bit	Symbol	Value	Description	Reset Value
7:0	CCLKSEL		Selects the divide value for creating the CPU clock (CCLK) from the PLL0 output.	0x00
		0 to 1	Not allowed, the CPU clock will always be greater than 100 MHz.	
		2	PLL0 output is divided by 3 to produce the CPU clock.	
		3	PLL0 output is divided by 4 to produce the CPU clock.	
		4	PLL0 output is divided by 5 to produce the CPU clock.	
		:	:	
		255	PLL0 output is divided by 256 to produce the CPU clock.	
31:8	-		Reserved, user software should not wnte ones to reserved bits. The value read from a reserved bit is not defined.	NA

From Table 38 of LPC1768 User Manual.

Like Program Example 14.1, Program Example 15.1 approximately replicates the original "blinky" program. However, it starts by resetting the value of **CCLKCFG**, so that the program then runs with a changed clock frequency. The program uses techniques introduced in Chapter 14 to access the microcontroller control registers.

```
/*Program Example 15.1 Adjusts clock divider through register CCLKCFG,
with trial blinky action. For Mbed LPC1768 only.                */

#include "mbed.h"      //keep this, as DigitalOut is used
DigitalOut myled(LED1);
#define CCLKCFG (*(volatile unsigned char *)(0x400FC104))

// function prototypes
void delay(void);

int main() {
  CCLKCFG=0x05; // divider divides by this number plus 1
    while(true){
    myled = 1;
    delay();
    myled = 0;
    delay();
  }
}

void delay(void){            //delay function.
  //loop variable j, volatile to stop compiler optimisation
  volatile int j;
  for (j = 0;j < 5000000;j++) {
    j++;
    j--;                    //waste time
  }
}
```

Program Example 15.1: Changing the CPU clock divider settings.

Exercise 15.1

Compile and download Program Example 15.1 to an Mbed LPC1768, first with the line `CCLKCFG=0x05;` commented out. The clock frequency will not be changed. Carefully record how many times the LED flashes in 30 s. Now enable the divider code line, and run the program several times, with different values entered for CCLKCFG, initially in the range from 2 to 9. For each, record how many times the LED flashes in 30 s.

1. Deduce what is the approximate duration of the delay function.
2. Which setting of CCLKCFG most closely matches your original reading?

You *may* want to look forward and do Exercise 15.7 at the same time; this applies the same program.

■

15.2.4 Adjusting the LPC1768 PLL

The PLL has already been mentioned as a circuit which can multiply frequencies. The main PLL of the LPC1768, PLL0, is actually made up of a divider followed by the PLL. Hence (perhaps strangely), it can divide frequencies as well as multiply them. Different combinations of multiply and divide give a huge range of possible output frequencies, which can be extremely useful in some situations. As its name suggests, a PLL needs to "lock" to an incoming frequency. However, it only locks if conditions are right, and it may take finite time to do this. These conditions include the requirement that the input to PLL0 must be in the range from 32 kHz to 50 MHz. It is interesting to see that the PLL can therefore be used to multiply the RTC frequency if required.

Full use of the PLL0 subsystem is complex, and requires a very careful reading of relevant sections of the LPC1768 user manual (Reference 2.4). However, we can still gain some useful insights by accessing its features in a limited way. The PLL is controlled by four registers, outlined in Table 15.3. We can readily see that the PLL can be enabled and connected through **PLL0CON**, with multiply and divide values set through **PLL0CFG**. Because it sits in the path of the main oscillator, and because PLLs sometimes act in a way which can be described as temperamental, there are two further important registers. The Feed Register, **PLL0FEED**, is a safety feature which blocks accidental changes to **PLL0CON** and **PLL0CFG**. A valid "feed sequence" is required before any update can be configured. Once implemented, changes can be tested in **PLL0STAT**. This further carries the important **PLOCK0** bit, which tests whether successful lock has been achieved.

The setup sequence for PLL0 is defined in the LPC1768 user manual; it must be followed precisely. Program Examples 15.2 and 15.3, both adapted from Reference 15.1, illustrate aspects of this. The **main( )** function immediately sets about disconnecting and turning off the PLL, "feeding" the PLL control as required. The microcontroller then runs with the PLL disabled and bypassed.

Table 15.3: PLL0 control registers.

Name	Description	Access	Address
PLL0CON	Control register. Holding register for updating PLL0 control bits. Values written to this register do not take effect until a valid PLL0 feed sequence has taken place. There are only 2 useful bits: bit 0 to enable; bit 1 to connect. Connection must only take place after the PLL is enabled, configured, and locked.	Read/Write	0x400F C080
PLL0CFG	Configuration register. Holding register for updating PLL0 configuration values. Bits 14:0 hold the value for the frequency multiplication, less one; Bits 23:16 hold the value for the pre-divider, less one. Values written to this register do not take effect until a valid PLL0 feed sequence has taken place.	Read/Write	0x400F C084
PLL0STAT	Status register. Read-back register for PLL0 control and configuration information. If **PLL0CON** or **PLL0CFG** have been written to, but a PLL0 feed sequence has not yet occurred, they will not reflect the current PLL0 state. Reading this register provides the actual values controlling PLL0, as well as the PLL0 status. Bits 14:0 and bits 23:16 reflect the same multiply and divide bits as in **PLL0CFG**. Bits 24 and 25 give enable and connect status of PLL. When either is zero, PLL0 is bypassed. When both are 1, PLL0 is selected. Bit 26, **PLOCK0**, gives the lock status of the PLL.	Read Only	0x400F C088
PLL0FEED	Feed register. Correct use of this register enables loading of the PLL0 control and configuration information from the **PLL0CON** and **PLL0CFG** registers into the shadow registers that actually affect PLL0 operation. The required feed sequence is 0xAA followed by 0x55.	Write Only	0x400F C08C

Based on Table 18 and following, of LPC1768 User Manual.

```
/*Program Example 15.2 Switches off PLL0, with blinky action
For Mbed LPC1768 only.                                        */

#include "mbed.h"
DigitalOut myled(LED1);
#define CCLKCFG (*(volatile unsigned char *)(0x400FC104))
#define PLL0CON (*(volatile unsigned char *)(0x400FC080))
#define PLL0FEED (*(volatile unsigned char *)(0x400FC08C))
#define PLL0STAT (*(volatile unsigned int *)(0x400FC088))

// function prototypes
void delay(void);

int main() {
  // Disconnect PLL0
  PLL0CON &= ~(1<<1);  // Clears bit 1 of PLL0CON, the Connect bit
  PLL0FEED = 0xAA;     // Feed the PLL. Enables action of above line
  PLL0FEED = 0x55;     //
```

```
  // Wait for PLL0 to disconnect, ie bit 25 to become 0.
  while ((PLL0STAT & (1<<25)) != 0x00);//Bit 25 shows connection status
  // Turn off PLL0; on completion, PLL0 is bypassed.
  PLL0CON &= ~(1<<0); //Bit 0 of PLL0CON disables PLL
  PLL0FEED = 0xAA;     // Feed the PLL. Enables action of above line
  PLL0FEED = 0x55;
  // Wait for PLL0 to shut down
  while ((PLL0STAT & (1<<24)) != 0x00);//Bit 24 shows enable status

  /****Insert Optional Extra Code Here****
         to change PLL0 settings or clock source.
  **OR** just continue with PLL0 disabled and bypassed*/

  //blink at the new clock frequency
  while(true) {
    myled = 1;
    delay();
    myled = 0;
    delay();
  }
}

void delay(void){          //delay function.
  volatile int j;          //loop variable j
  for (j = 0;j < 5000000;j++) {
    j++;
    j--;                   //waste time
  }
}
```

Program Example 15.2: Switching off the main PLL.

Exercise 15.2

Compile and download Program Example 15.2 to an Mbed LPC1768. Carefully measure how many times the LED flashes in 1 minute. It will be very slow. Can you explain the flashes per minute that you measure for this, comparing with the first value recorded in Exercise 15.1?

You *may* want to look forward and do Exercise 15.7 at the same time; this applies the same program.

■ Exercise 15.3

Adjust Program Example 15.2 to set a multiply value for the PLL. To do this, apply the code fragment of Program Example 15.3, inserting it before the **while** loop. Basic information on setting **PLL0CFG** is given in Table 15.3. You can try your own experimental values, using information supplied in this chapter. In each case, measure the LED blink rate, and try to correlate it with the setting you have made. To work with a deeper understanding of the limits and possibilities, refer to Section 4.5.10 of the User Manual, Reference 2.4.

■

```
#define PLL0CFG (*(volatile unsigned int *)(0x400FC084))
...
      // Set PLL0 multiplier
      PLL0CFG = 07; //arbitrary multiply by 8, divide value left at 1
      PLL0FEED = 0xAA; // Feed the PLL
      PLL0FEED = 0x55;
      // Turn on PLL0
      PLL0CON |= 1<<0;
      PLL0FEED = 0xAA; // Feed the PLL
      PLL0FEED = 0x55;
      // Wait for main PLL (PLL0) to come up
      while ((PLL0STAT & (1<<24)) == 0x00);
      // Wait for PLOCK0 to become 1
      while ((PLL0STAT & (1<<26)) == 0x00);
      // Connect to the PLL0
      PLL0CON |= 1<<1;
      PLL0FEED = 0xAA; // Feed the PLL
      PLL0FEED = 0x55;
      while ((PLL0STAT & (1<<25)) == 0x00); //Wait for PLL0 to connect
```

Program Example 15.3: Code fragment to set PLL0 multiplier.

15.2.5 Selecting the LPC1768 Clock Source

If the clock source is to be changed, through the input multiplexer top left of Fig. 15.4, it must be done with PLL0 shut down. This change is controlled by the clock source select register, **CLKSRCSEL,** with details shown in Table 15.4.

Table 15.4: Clock source select register, CLKSRCSEL.

Bit	Symbol	Value	Description	Reset Value
1:0	CLKSRC		Selects the clock source for PLL0 as follows:	0
		00	Selects the Internal RC oscillator as the PLL0 clock source (default).	
		01	Selects the main oscillator as the PLL0 clock source. **Remark**: Select the main oscillator as PLL0 clock source if the PLL0 clock output is used for USB or for CAN with baudrates > 100 kBit/s.	
		10	Selects the RTC oscillator as the PLL0 clock source.	
		11	Reserved, do not use this setting.	
			Warning: Improper setting of this value, or an incorrect sequence of changing this value may result in incorrect operation of the device.	
31:2	-	0	Reserved, user software should not write ones to reserved bits. The value read from a reserved bit is not defined.	NA

From Table 17 of LPC1768 User Manual.

Exercise 15.4

Write a program which turns off PLL0, changes the clock source, and starts up PLL0 again. To do this, combine Program Examples 15.2 and 15.3 and apply the information in Table 15.4. Try this for both internal R–C and RTC clock sources.

■

15.3 Reset

At certain times in its use, any microcontroller is required to start its program from the beginning, the most obvious being when power is applied. At this moment, it also needs to put all its control registers into a known state, so that peripherals are safe and initially disabled. This "ready-to-start" condition is called *reset*. Apart from power-up, there are other times when this is needed, including the possibility that the user may want to force a reset if a system locks or crashes. The CPU starts running its program when it leaves the reset condition. In an advanced processor like the LPC1768, the user code is preceded by some "boot code," hard-wired into the processor, which undertakes preliminary configuration. How reset is implemented in any microcontroller, and what it actually does, is an important part of any microcontroller-based design.

15.3.1 Power-On Reset

The moment that power is applied is a critical one for any embedded system. Both the power supply and the clock oscillator take finite time to stabilize, and in a complex system power to different parts of the circuit may stabilize at different times. Clearly, this takes some careful handling. How can the start of program execution be delayed until power has settled? How can a complex system be kept in a safe state while all its subsystems initialize? This will only happen if explicit circuitry is built in to detect power-up, to preset control registers, and to force a program restart.

The LPC1768 has an interesting but complex circuit to manage its reset routine, which you can see in Fig. 4 of the User Manual (Reference 2.4). The effect of that circuit, when power is applied, is summarized in Fig. 15.5. Here we see the power supply voltage rising from 0 V to 3.3 V; as a consequence, the internal R–C oscillator starts oscillating. An internal timer starts measuring a fixed delay of 60 μs from when the supply voltage

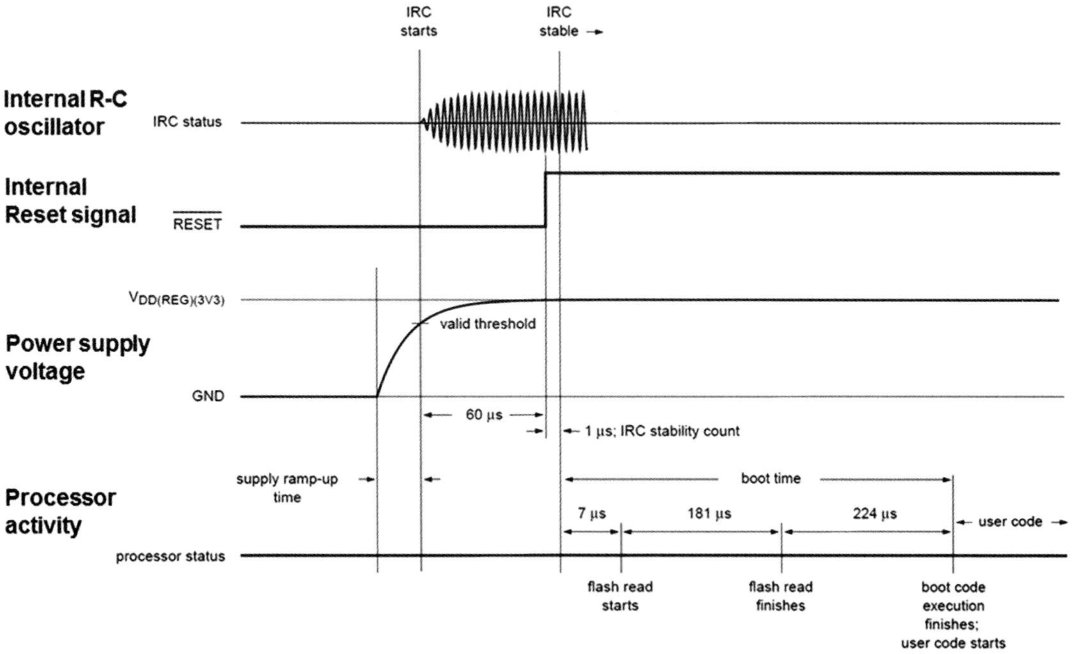

Figure 15.5
LPC1768 internal signals during start-up. *Image courtesy of NXP.*

reaches a valid threshold; this gives time for the oscillator to stabilize. The internal reset signal initially stays low, disabling all activity, and forcing all registers to their reset conditions. On completion of the 60-μs delay, however, this reset signal is set high, and processor activity can start. Summing all the delays, one can see that the user code starts to execute around 0.5 ms after power is switched on.

15.3.2 Other Sources of Reset

Three further sources of reset are described in overview here. The first, external reset, is implemented in just about every development board, including the Mbed LPC1768.

External Reset

The LPC1768 has an external reset input. As long as this is held low, the microcontroller is held in reset. When it is taken high, a sequence is followed very similar to the power-on reset described above. If the pin is taken low while the program is running, then program execution stops immediately and the microcontroller is forced back into reset mode. This allows an external push-switch to be connected, so that a user can force a reset if needed. For a product on the market it is unlikely that this facility would be used—it is almost an expression of mistrust in the design if a reset is made easily available to the user. In a prototype environment, however, it can be useful, as program crashes are more likely to occur. As we know, the Mbed LPC1768 has a reset button. However, this is connected to the interface microcontroller (Fig. 2.2), which can then force a reset to the LPC1768 itself (via pin 17, Fig. 15.2).

Watchdog Timer

A common failure mode of any computer-based system is for the computer to lock up and cease all interaction with the outside world. Maybe it is trapped in an endless loop, or is waiting for an external event that never happens, or is executing corrupted code. For almost any computer, and especially an embedded system, this is unacceptable. An uncompromising solution to the problem is the watchdog timer (WDT), which *resets the processor if the WDT is ever allowed to overflow.* The WDT runs continuously, counting toward its overflow value. It is up to the programmer to ensure that this overflow never happens in normal program operation. This is done by including periodic WDT resets throughout the program. *If* the program crashes, then the WDT overflows, the controller resets, and the program starts from its very beginning, with the program counter set to its reset value. A WDT overflow causes a reset very similar to the power-on reset described above.

The WDT for the LPC1768 appears in Fig. 15.4, with its three possible clock sources. Two of these are direct from R–C and RTC oscillators. The third, $\mathbf{pclk_{WDT}}$, is derived from the peripheral clock generator and can be seen at the top of Table 14.5.

Table 15.5: Watchdog API summary

Functions	Usage
`Watchdog::get_instance`	Get a reference for the WDT (establishes link to the hardware WDT).
`bool start()`	Start the WDT with the maximum timeout value available for the target. Returns true if WDT started successfully.
`bool start(uint32_t timeout)`	Start the WDT with value *timeout*, in ms. Returns true if WDT started successfully.
`void kick()`	Reset the WDT.
`uint32_t get_timeout() const`	Returns refresh value of the WDT.
`bool stop()`	Stops the WDT. Returns true if WDT stopped successfully.

The Mbed OS provides a useful API for the WDT, summarized in Table 15.5. This allows the programmer to start, load, check, and refresh ("kick") the WDT. We use some of these features in Program Example 15.4.

Brownout Detect

An awkward failure condition for an embedded system is when the power just dips, and then returns to normal. This is called a *brownout*, and is illustrated in Fig. 15.6. A brownout will not be detected as a full loss of power, and may not be noticed at all. However, that momentary loss of full power could cause partial system failure—for example, some settings becoming corrupted, resulting in possible malfunction later. Brownouts can be due to many things, including the switching on of a motor or other high-current actuator, a supply dip due to poor power supply design, or severe interference picked up on the supply lines. Like many microcontrollers, the LPC1768 has a brownout detect capability, which must be enabled by the user. It is not normally enabled in the Mbed LPC1768. Detection of brownout, when enabled, causes a reset very similar to the power-on reset described above.

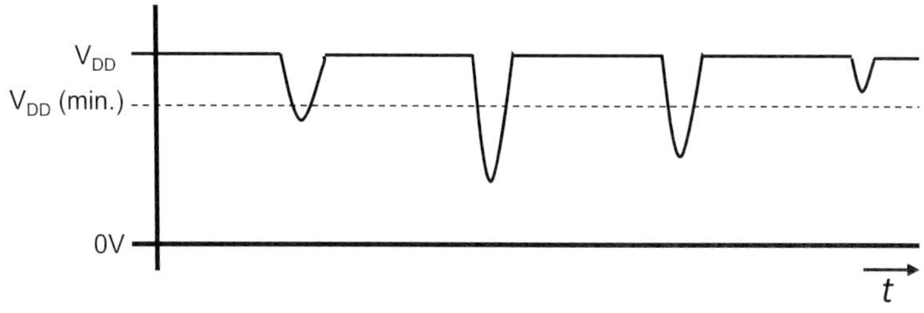

Figure 15.6
Voltage brownouts.

15.3.3 Reason for Reset

Sometimes it is very useful, indeed essential, to know why a system has reset. Did the power dip, did the watchdog time out, was a physical reset pressed? There is a useful OS API, called **ResetReason**, which allows the reason for the most recent reset to be acquired for evaluation. The API accesses those internal registers which hold the reason. Program Example 15.4, which adapts and combines two examples from the Mbed website, demonstrates this. It also shows use of the WDT.

Checking details of the program, **main( )** starts immediately with an access to **ResetReason,** seeking to report on what reset event has caused the fresh start of program execution. It then launches the WDT, preloading with an equivalent value of 5 s. It establishes an interrupt input linked to the pushbutton with ISR, which resets the WDT. The program loops, printing a countdown value to screen. If the button is not pressed, the WDT times out, the processor resets, the program relaunches, and the reset reason (WDT timeout in this case) is extracted and displayed. Other forms of reset can also be tested.

```
/* Program Example 15.4. Implements Watchdog Timer, and tests for reset reason.
Works on Mbed LPC1768 or F401RE with no additional components.
*/

#include "mbed.h"
#include "ResetReason.h"

#include <string>                    //will be manipulating strings

const uint32_t TIMEOUT_MS = 5000;   //Timeout value loaded to WDT
InterruptIn button(p5);        //Replace p5 with BUTTON1 for F401RE
volatile int countdown = 9;

std::string reset_reason_to_string(const reset_reason_t reason){
  switch (reason) {
    case RESET_REASON_POWER_ON:
      return "Power On";
    case RESET_REASON_PIN_RESET:
      return "Hardware Pin";
    case RESET_REASON_SOFTWARE:
      return "Software Reset";
    case RESET_REASON_WATCHDOG:
      return "Watchdog";
    default:
      return "Other Reason";
  }
}

//ISR response to button push, resets Watchdog
void trigger(){
```

```
    Watchdog::get_instance().kick();  //Reset WDT
    countdown = 9;
  }

  int main(){
    //Acquire then print reset reason
    const reset_reason_t reason = ResetReason::get();
    printf("Last system reset reason: %s\r\n", reset_reason_to_string(reason).c_str());
    Watchdog &watchdog = Watchdog::get_instance();
    watchdog.start(TIMEOUT_MS);  //Start WDT with value of TIMEOUT_MS
    button.rise(&trigger);       //Link rise on Interrupt input to ISR
    uint32_t watchdog_timeout = watchdog.get_timeout();
    printf("Watchdog initialized to %lu ms.\r\n", watchdog_timeout);
    printf("Press BUTTON1 at least once every %lu ms to kick the "
           "watchdog and prevent system reset.\r\n", watchdog_timeout);

    while (true) {
      printf("\r%3i", countdown--);
      fflush(stdout);            //Flush the buffer
      ThisThread::sleep_for(TIMEOUT_MS/10); //Sleep for 500ms
    }
  }
```

Program Example 15.4: Implementing the WDT and reset reason

The program can run on either the Mbed LPC1768 or the Nucleo F401RE. It is more effective on the latter, as it is easy to interrupt the power supply to the microcontroller, while leaving the USB link intact (necessary for printing back to the screen). For the Mbed LPC1768, use the circuit of Fig. 6.3. The F401RE uses the onboard pushbutton, so no extra components are needed. Power supply to the microcontroller on the F401RE is easily interrupted by removing the jumper labeled IDD, which lies between the microcontroller itself and the two pushbutton switches, shown as JP6 in Fig. 2.5.

Compile, download, and run the program, and you should see a screen similar to Fig. 15.7. This shows three different reset reasons occurring in turn. Trial the different reset reasons that are available to you. Hardware reset and WDT time out reset are available on both boards. A momentary removal of the IDD jumper on the F401RE board will give a power on reset.

Exercise 15.5

The Mbed OS provides a reset function, **system_reset()**, which can also be used to reset a microcontroller. Insert this function at an appropriate place in Program Example 15.4. Run the program again, and see what reset source is reported.

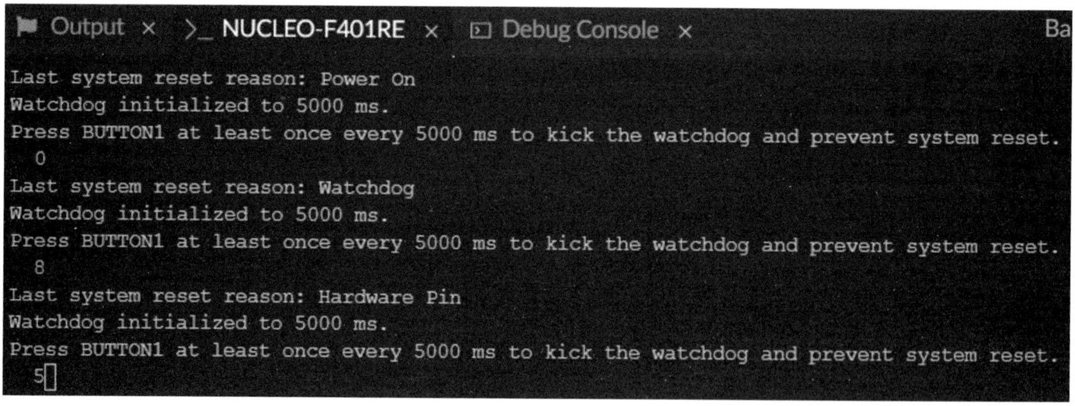

Figure 15.7
Example "reason for reset" readout.

15.4 Toward Low Power

In Section 13.6 we overviewed the challenge of powering IoT "things," in situations where power is at an absolute premium. In the sections which follow we now get into some of the underlying technical detail of how power consumption can be managed and minimized.

15.4.1 How Power Is Consumed in a Digital Circuit

Since the large-scale adoption of digital electronics, a number of logic families have competed for the designer's attention. Each provides modular electronic circuits which implement the essential logic functions of AND, NAND, OR, NOR, and so on; and each has its own advantages, for example, in terms of speed or power consumption. The only logic family that is suitable for low power circuits is called *complementary metal oxide semiconductor*, or CMOS for short. CMOS technology has transformed our lives, as it is the basis for the mobile phone, the laptop, and any other portable electronic device.

To understand how to minimize the power consumption of CMOS, it is useful to have some understanding as to how it consumes that power. Let's look at a simple circuit. An inverting CMOS buffer is shown in different guises in Fig. 15.8. All other CMOS logic gates are extensions of this simple circuit, and a microcontroller is essentially made up of millions of these circuits, grouped together to form all of the essential subsystems. So if we can understand power consumption in one, we have a key to what goes on in much larger systems.

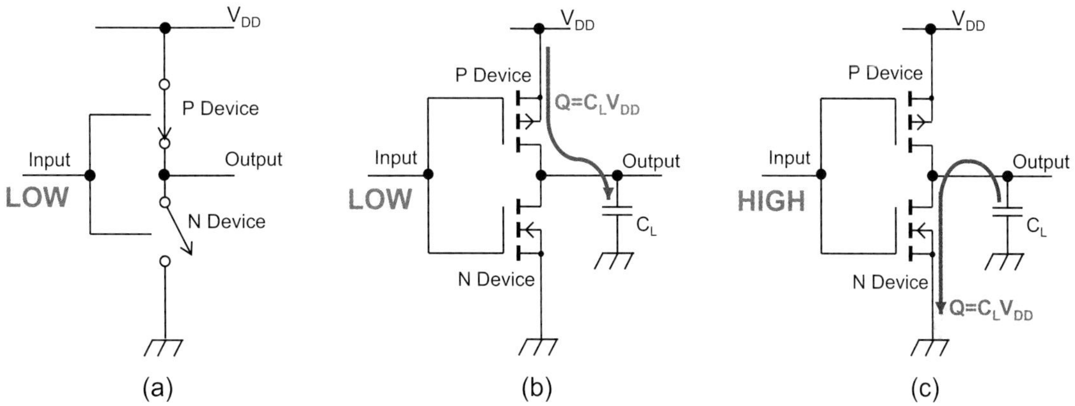

Figure 15.8
Power consumption in a CMOS inverter: (a) Idealized circuit; (b) actual circuit, input going low; (c) actual circuit, input going high.

The CMOS buffer is simply made up of two transistors, one fabricated on n-type and the other on p-type semiconductor substrates (the "complementary" bit). Most of the time, each transistor acts as a switch; to aid understanding, the circuit can first be simplified to Fig. 15.8(a). When the input is low (as is shown at Logic 0), then the lower transistor is switched *off*; that is, the switch is open. The upper switch is meanwhile closed, or switched *on*. The output is therefore connected to the supply voltage; that is, it is high, or at Logic 1. If the input flips to Logic 1, then the transistors change accordingly. What is important is that in either of these states there is no current path from the supply to ground. This, in a nutshell, is why CMOS consumes so little power. The actual circuit is shown in Figs. 15.8(b) and 15.8(c); don't worry about the curvy arrows just yet.

It is when the input is changing state that awkward things happen, as far as power consumption is concerned. First of all, there is always "stray" capacitance in the circuit. This is symbolized as C_L in Figs. 15.8(b) and 15.8(c). This capacitance is due to the interconnecting wires, and also the very structure of the CMOS transistors themselves (the parallel lines in the symbol give the clue). So on every change of input, the stray capacitance is charged or discharged, hence taking momentary current from the supply. This is symbolized by the curvy arrows, which show C_L being charged as the input goes low (output goes high), and discharged as the output goes low. This may not be too bad for one logic gate going through one transition, but if millions of them are doing it millions of times a second, a lot of charge can get dumped to ground.

The process just described is called *capacitive* power consumption. To add to that, as it changes state, the input voltage passes a middle region where both devices are momentarily partially turned on. A tiny pulse of current can flow straight through them at this instant; this is called *shoot-through* power consumption. Finally, there is a tiny bit of leakage current which flows continuously, often so small as to be negligible; this is called *quiescent* power consumption.

It is beyond the scope of this book to fully analyze this power consumption behavior, but it is useful to look briefly at Eq. 15.1, which describes it. Here the total supply current I_T is made up of the small and moderately constant quiescent current I_Q, plus a term which depends on supply voltage V_{DD}, clock frequency f, and a capacitive term, C_{eq}. This "equivalent" capacitance is a complex thing, which in brief lumps together interconnection capacitance in the IC and capacitance due to external interconnections, plus an equivalent capacitance which represents the shoot-through behavior. To convert this current consumption to power consumption, simply multiply each side by V_{DD}:

$$I_T = I_Q + \{V_{DD} \times fC_{eq}\} \tag{15.1}$$

Neglecting for a moment the very small quiescent current, this equation shows that supply current is effectively proportional to supply voltage, switching frequency, and equivalent capacitance. If we can tame these things, we are on our way to taming power consumption! Finally, a tiny health warning—quiescent current is very small at room temperature, but increases rapidly with temperature. If you are implementing systems which run in hot places, you may need to take some notice of it.

15.4.2 A Word on Cells and Batteries

If we are considering power optimization, then it is almost certainly because the plan is to run from battery power. Battery technology and behavior are in themselves a huge topic, but we introduce some key aspects here. This helps to evaluate current consumption in a microcontroller. Correctly speaking, a battery is made up of a collection of cells.

Cells are classified either as primary (non-rechargeable), or as secondary (rechargeable). They are based on a variety of metal/chemical combinations, each of which has special characteristics in terms of energy density, whether rechargeable or not, and other electrical properties. Primary cells like alkaline cells tend to have the highest energy density, so they are widely used for applications of greatest power demand. They are also used for low-power or occasional applications, where replacement is infrequent. Of course, in applications like the mobile phone or laptop, it is unthinkable to consider primary cells, and a range of sophisticated and specialist rechargeable batteries exist. Cells are available in a wide variety of packages. These include the more traditional cylinder formats, now most seen in AA or AAA versions, a range of button cells, and numerous specialist batteries for laptop and mobile phones. Two familiar types are shown in Fig. 15.9, with some basic characteristics given in Table 15.6.

Figure 15.9
Example cells.

Table 15.6: Example cell/battery data.

Manufacturer	Technology	Shape/ Package	Nominal Terminal Voltage (V)	Capacity (mAh)	Capacity (J)
Varta	Silver Oxide	V301, button	1.55	96	536
Varta	Silver Oxide	V303, button	1.55	160	893
Procell	Alkaline	AAA cylinder	1.5	1175	6,345
Procell	Alkaline	AA cylinder	1.5	2700	14,580
Procell	Alkaline	PP3	9.0	550	17,820

The two most important electrical characteristics of a cell are terminal voltage and capacity, the latter generally measured informally in ampere-hours (Ah), or milliampere-hours (mAh). The inference is that a battery of 500 mAh can sustain a 500 mA current for 1 h, or a 1 mA current for 500 h, and so on. In reality, the situation is not so simple; batteries do tend to recover between periods of use, and display somewhat different capacities depending on load current, temperature, battery age, and a number of other things. The mAh rating of a cell or battery can be converted to the fundamental unit of energy, Joules, by multiplying by terminal voltage and $(60 \times 60)/1000$, or 3.6. Thus, the PP3 battery in Table 15.6, which appears to have a modest capacity of 550 mAh, has the highest energy capacity of all the list, though not so far above the AA cell. Notice also the similarity in energy capacity between three AAA cells and the PP3.

As a simple example, suppose a battery-powered product is drawing an average current of 160 mA, taken from three AA alkaline cells in series. Each cell is delivering the same current, so an approximate value for the cell life would be given by the mAh capacity, divided by the load current, also in mA; that is, 2700/160, or just under 17 hours.

Suppose instead that a device was to be powered from a V303 button cell, and a year's continuous operation is required. This cell has a much smaller capacity, 160 mAh. There are 365 $\times$ 24 hours in a year (neglecting leap years), so the maximum allowable average current is 160/(365 $\times$ 24) mA, or 18.3 μA.

Simple calculations like these allow useful estimates to be made relating battery life to circuit current consumption, and give a feel of magnitudes that can be expected. With care, the calculations can be adapted to more realistic situations, where the battery use is intermittent and/or varying.

15.4.3 Microcontroller Low-Power Modes and OS Support

Manipulating the clock frequency is a simple way to influence microcontroller power consumption. Yet there are other, more effective techniques which can be applied as a program runs. These involve either switching off the clock altogether or switching it off to certain parts of a microcontroller, at times when nothing needs doing. For example, the CPU could be switched off, while certain peripherals were left running, if active computation was not needed at the time. These low-power modes are often called sleep or idle, though the terminology varies widely between different microcontrollers. The modes are entered by special instructions in the processor instruction set, and generally exited through an interrupt occurring, or a system reset. We review the low-power modes of both LPC1768 and F401RE microcontrollers in the coming sections.

The Mbed OS provides excellent support for low-power operation. This is described in Reference 15.2 and summarized here. We have frequently met this capability from the beginning of the book, in functions like **thread_sleep_for( ).**

While different microcontrollers may have three or four low-power modes, the OS offers only two sleep modes. They are summarized here:

Sleep: The clock to the CPU is stopped, memory states are retained, and peripherals continue to work and can generate interrupts. Exit from the state occurs through interrupt or reset.

Deep sleep: The clock to the CPU is stopped and all high-speed clocks are turned off. Due to this clock limitation, many peripherals cannot run, and so cannot generate interrupts. All memory states are retained. Exit occurs through interrupt or reset, though fewer interrupts are available. Exit may be slower than for sleep.

For each Mbed-enabled development board, the OS maps these modes to the most appropriate of the microcontroller low-power modes.

Table 15.7: Example sleep-oriented OS functions.

Functions	Usage
`static void sleep(void)`	Send the microcontroller to sleep.
`bool sleep_manager_can_deep_sleep(void)`	Test if target can enter deep sleep, in which case return true.
`void sleep_manager_lock_deep_sleep_internal(void)`	Lock the deep sleep (lock is a counter).
`void sleep_manager_sleep_auto(void)`	Enter auto-selected sleep mode; choice depends on deep sleep lock.
`ThisThread::sleep_for(chrono::milliseconds(int ms));`	Sleep for the number of milliseconds specified .

The use of sleep modes in program execution is managed by the OS *sleep manager.* At any time a system is idle, program execution automatically enters a sleep mode. While regular sleep mode is generally available, deep sleep is not. In the case where a high-speed clock is running, the sleep manager blocks deep sleep access through use of a *deep sleep lock*. This action may be invisible to the programmer. However, it is possible for the programmer to access the deep sleep lock, to see if it is set.

Example sleep-oriented OS functions are shown in Table 15.7. We have of course already met the last many times; we see examples of the use of others a little later, in Program Example 15.6.

15.5 Exploring Mbed LPC1768 Power Consumption

The designers of the Mbed LPC1768 would never claim that the device was designed for low power, yet it does provide a good opportunity to study power consumption in an embedded system, and to explore ways of reducing that power. It is interesting and revealing to set up a simple current measurement circuit, and to do the exercises which follow.

■ Exercise 15.6

Download Program Example 2.1 (or indeed any program which does not require external connections) to an Mbed LPC1768. Disconnect the USB cable and power the circuit as shown in Fig. 15.10. Use a battery pack or bench DC supply, set at around 6 V. The supply should link to the VIN Mbed pin (pin 2), with an ammeter—for example, a digital multimeter in its 200-mA range—inserted between supply positive and the VIN pin. The Mbed will draw an approximately constant current, so the precise supply voltage does not matter, only that it lies in the specified range of 4.5 V to 9 V.

Now measure and record the current supplied to the Mbed. You should find this somewhere in the region of 140 mA.

Complete the calculation of Quiz Question 5.

■

15.5.1 LPC1768 Current Consumption Characteristics

The current consumption characteristics of the LPC1768 itself are shown in Table 15.8. Also shown is the LPC1769, which has the ability to run at a slightly higher clock frequency. We shall return to this table several times. The entries are self-explanatory, and reinforce the message that clock frequency dominates current consumption. The table also refers to the microcontroller sleep and power-down modes that we shall meet soon. Reference 15.3 is useful in applying this and associated data.

Applying Table 15.8 to the Mbed board, we know that it runs with a clock frequency of 96 MHz, with PLL enabled. Hence, we could estimate that the LPC1768 on the Mbed is

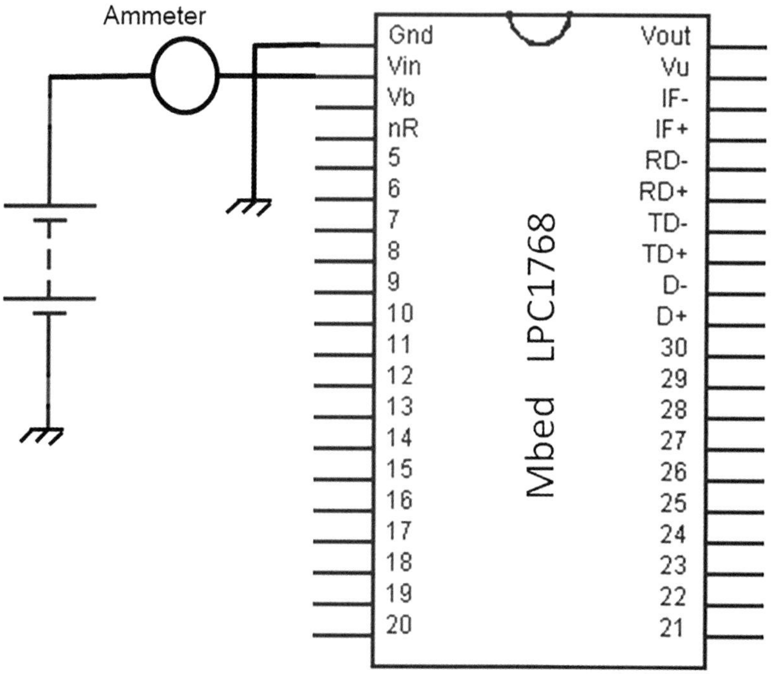

Figure 15.10
Measuring Mbed LPC1768 current consumption.

Table 15.8: LPC1768 current consumption characteristics.

$I_{DD(REG)(3V3)}$	regulator supply current (3.3 V)	active mode; code `while (1) { }` executed from flash; all peripherals disabled; PCLK = CCLK/8					
		CCLK = 12 MHz; PLL disabled	[6][7]	-	7	-	mA
		CCLK = 100 MHz; PLL enabled	[6][7]	-	42	-	mA
		CCLK = 100 MHz; PLL enabled (LPC1769)	[6][8]	-	50	-	mA
		CCLK = 120 MHz; PLL enabled (LPC1769)	[6][8]	-	67	-	mA
		sleep mode	[6][9]	-	2	-	mA
		deep sleep mode	[6][10]	-	240	-	µA
		power-down mode	[6][10]	-	31	-	µA
		deep power-down mode; RTC running	[11]	-	630	-	nA
I_{BAT}	battery supply current	deep power-down mode; RTC running					
		$V_{DD(REG)(3V3)}$ present	[12]	-	530	-	nA
		$V_{DD(REG)(3V3)}$ not present	[13]	-	1.1	-	µA

The footnotes referenced provide the fine detail of how the measurements apply or are made, and may be accessed from Reference 2.3.
From Table 8 of LPC1768 Data Sheet.

taking a supply current just under 42 mA, let's say 40 mA. However, the values shown are with all peripherals disabled, so in practice the consumption will be somewhat higher. This information, and the result of Exercise 15.5, are sobering. If our estimate of 40 mA for the LPC1768 is true, then the circuitry external to the microcontroller is taking around 100 mA! These are big figures, and far removed from what would be required for realistic battery supply.

In reality, this early Mbed board was never designed for low power. A brief glance at its circuit diagram, or the block diagram of Fig. 2.2, reminds you of how much there is in the overall circuit. And much of this is continuously powered, whether you use it or not! In the early days of its use, much was written about how its consumption could be reduced, including in earlier editions of this book. A low-power sister board, the LPC11U24, was also produced. In the next sections we will explore some ways in which the LPC1768 power consumption can be reduced. However, for true low-power applications, it is better to recognize that this is unlikely to be an appropriate development board to work with.

15.5.2 Controlling Power to Microcontroller Peripherals

Just focusing on the microcontroller itself, Program Example 15.5 provides a simple demonstration of power consumed by the peripherals. We saw in Table 14.4 that register **PCONP** controls the power to many peripherals. The program example starts by switching *on* all peripherals that are controlled through **PCONP**. It then blinks an LED 10 times, during which time an initial current measurement can be made. Then all peripherals are switched off, and the program then sits indefinitely in a **while(true)** loop. A further current measurement can be made. A practical application is of course likely to sit somewhere between the "all or nothing" extremes represented in this example.

```
/*Program Example 15.5
Powers down peripherals through PCONP register. Mbed LPC1768 only.
                                                        */

#include "mbed.h"

DigitalOut led1(LED1);

#define PCONP (*(volatile unsigned long *)(0x400FC0C4))

int main() {
  PCONP = 0xFFFFFFFF;  //switch all peripherals ON. For simplicity, write 1s
                      //even to reserved bits, tho normally not good practice.
    for(int i=10; i>0; i-){
      led1=1;
      wait_us(100);  //tiny LED flashes to limit ave current
      led1=0;
      wait_us(499900);
    }
  //Turn all peripherals controlled through PCONP OFF
  PCONP = 0x00000000;
  while (true) {
  }
}
```

Program Example 15.5: Switching off unused peripherals

Create a new program using Program Example 15.5. Download to an Mbed LPC1768 and disconnect the USB cable. Then, applying the circuit of Fig. 15.10, measure and record current consumption in the two phases of program operation. We can expect the first to be greater than that found in Exercise 15.5, the second less. We are beginning to see that we can influence power consumption.

15.5.3 Manipulating the Clock Frequency

We saw earlier in this chapter that the LPC1768 has extensive capabilities to vary the clock frequency. Now we understand one of the reasons—we can trade off speed of execution with power consumption. A program with significant computational demands will need to run fast, and will consume more power; one with low computational demands can run slower, and consume less power.

Exercise 15.7

Rerun both Program Examples 15.1 and 15.2 in turn. Measure current supply to the Mbed for a range of different clock frequencies. Record your results.

It is attractive to imagine that clock speeds can be reduced at will to reduce current consumption. While this is true in principle, take care! Many of the peripherals depend on that clock frequency as well, along with their Mbed API libraries, for example, for the setting of a serial port bit rate. If you are using any such peripheral, say the SPI port, you may need to adjust its clock frequency setting to compensate.

15.5.4 LPC1768 Low-power Modes

The LPC1768 microcontroller has available the low-power modes summarized below, with power consumption given in Table 15.6.

Sleep mode: The clock to the core, including the CPU, is stopped, so program execution stops. Peripherals can continue to function. Exit from this mode is achieved by an enabled Interrupt, or a reset. For example, the processor could put itself to sleep and be programmed to wake when a serial port received a new byte of data, or an external button was pressed by a user.

Deep sleep mode: Here the main oscillator (Fig. 15.4) and PLLs are powered down. The internal R−C oscillator can continue running, and can run the WDT, which can cause a wake-up. The 32-kHz RTC continues, and can generate an interrupt. The SRAM and processor keep their contents, and the flash memory is in stand-by, ready for a quick wake-up. Wake-up is by reset, by the RTC, or by any other interrupt which can function without a clock. Clock dividers and PLLs must be reconfigured, as needed, on wake-up.

Power down mode: This is similar to deep sleep, but the internal R−C oscillator and flash memory are turned off. The RTC can continue running and can be used for wake-up. Wake-up time is thus a little longer. Clock dividers and PLLs must be reconfigured, as needed, on wake-up.

Deep power down mode: In this mode all power is switched off, except to the RTC. Wake-up is only through external reset, or from the RTC.

15.5.5 Applying Operating System Sleep Modes

The OS support for low power was introduced in Section 15.4.3. We can test it here, with Program Example 15.6. The program moves through a series of clearly identifiable phases, and a current measurement can be made during each. It starts immediately with 5 s of sleep, invoking the already familiar **Thisthread::sleep_for()**. This is placed here because the OS guide claims that deep sleep is always enabled as **main()** starts, but may subsequently be disabled by mechanisms not always obvious to the programmer. Following a text introduction to screen, the program flashes the test LED 10 times, while using **wait_us()** for delays; thus the CPU is kept "busy." A test for deep-sleep permission follows, and sleep is re-entered. Deep sleep is then disabled, and a further 5 s of sleep follows.

```
/*Program Example 15.6
Testing Sleep modes using OS functions
Runs on either LPC1768 or F401RE without adjustment
*/

#include "mbed.h"
DigitalOut testled(LED1);
//int i;

int main(){
  ThisThread::sleep_for(5000); //An initial sleep, at start of program
  printf("TESTING CURRENT CONSUMPTION\r\n");
  printf("______________________________\r\n");
  //Normal operation for 5s
  printf("Normal continuous operation\r\n");
  for(int i = 10; i > 0; i-){
    testled = 1;
    wait_us(100); //Keep the CPU busy
    testled = 0;
    wait_us(499900);
  }

  // Deep sleep for 5 seconds
  printf("Deep sleep allowed: %i\r\n", sleep_manager_can_deep_sleep());
  //Disable input from STDIN, will release deep sleep lock (Ref 15.2)
  mbed_file_handle(STDIN_FILENO)->enable_input(false);
  ThisThread::sleep_for(5000);

  //Lock deep sleep
  printf("Locking deep sleep\r\n");
  sleep_manager_lock_deep_sleep_internal();

  // Sleep for 5 seconds
  printf("Deep sleep allowed: %i\r\n", sleep_manager_can_deep_sleep());
  ThisThread::sleep_for(5000);
}
```

Program Example 15.6: Programming with OS sleep modes

Note from Reference 2.13 the possible impact on sleep modes that the chosen build profile (develop, debug, or release) makes. Then compile and run Program Example 15.6 on an Mbed LPC1768, applying the circuit of Fig. 15.10. (See Exercise 15.8 for guidance on running on the F401RE.) For correct current readings, the USB cable must be disconnected, so we cannot benefit from the messages to screen within the program—we use these later. Changes in current are superimposed on the high overall consumption of the Mbed board, so they may not appear to be of great significance. Can you find any correlation with information provided in Table 15.8?

15.6 Reviewing the Nucleo F401RE Clock and Power Features

An understanding of the full details of the F401RE clock and power supply structures demands a very careful reading of References 2.6 and 2.7; we give just an overview here. Broadly speaking, the F401RE and LPC1768 microcontrollers have many similarities, although of course there are numerous differences in detail.

Like the LPC1768, the F401RE can accommodate two external crystals, as main and RTC clock sources. It also has a high-speed (16-MHz) internal RC oscillator, and a low-speed (32 kHz) internal RC oscillator; this latter can drive the WDT and/or RTC. A PLL is available to multiply the main clock frequency.

On the Nucleo board the F401RE microcontroller has its own 32-kHz RTC crystal (labeled X2 on the board), but takes its 8 MHz main oscillator from the interface microcontroller crystal (X1). There is a location for a dedicated crystal for the main microcontroller (X3), but it would need to be fitted by the user. The internal system clock runs at 84 MHz.

A selection of current consumption figures for the Nucleo board is given in Table 15.9. While these are measured under different operating conditions, so they cannot be compared directly with the LPC1768, they display similar patterns. As expected, current consumption is highly dependent on clock frequency, as well as the number of peripherals in use.

Three low-power modes are available, with example current consumption figures appearing in the table. They are summarized here:

Sleep mode: The CPU is stopped; all peripherals continue to operate. Exit is through interrupt or reset.

Stop mode: Clocks (except for RTC) and PLL are stopped. The contents of SRAM and registers are retained. Exit can be by external interrupt or RTC wakeup.

Standby mode: Clocks (except for RTC) are switched off; SRAM and register contents are lost. Exit can be by reset or RTC wakeup. This gives the lowest power consumption.

The Mbed OS maps its sleep mode to the F401RE sleep mode, and deep sleep to stop.

Table 15.9: Overview: F401RE current consumption characteristics.

Condition	Clock Frequency (MHz)	Supply Current
External clock, all peripherals enabled, code running from flash memory	84	23.2 mA
External clock, all peripherals enabled, code running from flash memory	40	10.8 mA
External clock, all peripherals disabled. code running from flash memory	84	12.3 mA
External clock, all peripherals disabled. code running from flash memory	40	6.0 mA
Sleep mode: External clock, all peripherals enabled	84	16.6 mA
Sleep mode: External clock, all peripherals disabled	84	5.3 mA
Stop mode: All oscillators off, main regulator usage	—	111 uA
Standby mode: All oscillators (inc. RTC) off	—	1.8 uA

Taken from Tables 20–30 of Reference 2.6.

Exercise 15.8

Compile and run Program Example 15.6 on a Nucleo F401RE board. Recall that we can very conveniently measure current flow to the microcontroller itself: remove the jumper labeled IDD (shown as JP6 in Fig. 2.5), which lies between the microcontroller and the two pushbutton switches, and insert an ammeter. Does the current consumption in general correlate with information provided in Table 15.9 and the accompanying text?

■

15.7 Getting Serious about Low Power: The M0/M0+ Cores and the Zero Gecko

We have seen that it is possible to manipulate current consumption in a microcontroller and its surrounding circuit, given a knowledge of which variables have an influence. But where do we turn if a really low-power design is needed?

As outlined in Chapter 1, there are a number of versions of the Cortex core, as summarized in Table 1.1. The simplest are the M0 and M0+ cores. These have a highly optimized gate count and are specifically designed for small-scale, low-power, and low-cost applications. This combination leads to many advantages in the embedded world, for example, in intelligent or networked sensors or any small-scale or portable device. Full details on these cores can be found in Reference 15.4.

The EFM Zero Gecko starter kit is a development board based around the Silicon Labs Zero Gecko microcontroller. Based on an M0+ Cortex core, it forms a useful example of a truly low-power Mbed-enabled design. It is designed for extreme low power, and—of great interest to us—has an on-board *energy profiler*, which can measure current consumption from 0.1 μA to 50 mA. The board layout is shown in Fig. 15.11.

The Silicon Labs Simplicity Studio development environment can be downloaded free from the company website (Reference 15.5). This has a number of demo programs available on it for the kit, which can be downloaded directly to the device. A user's guide to the board is available, Reference 15.6.

Once a program is running, the energy profiler gives a power consumption reading. Fig. 15.12 shows the energy profiler at work, running the "analog and digital clock example" from the range of demos available. A quick glance at this shows the diagnostic power available to the developer. Supply current (left vertical axis) and supply voltage (right vertical axis) appear. It is interesting to see even the current peaks sitting well below 1 mA. Compare this with the current measurements earlier in this chapter! In addition to this, the data block on the right shows average current and power. Most important, perhaps, it shows total energy transferred; this is the ultimate piece of information we need when running from a battery. This simple example does not implement *code correlation*, an option whereby power consumption can be displayed against the actual line of code being executed. However, this important capability can be implemented.

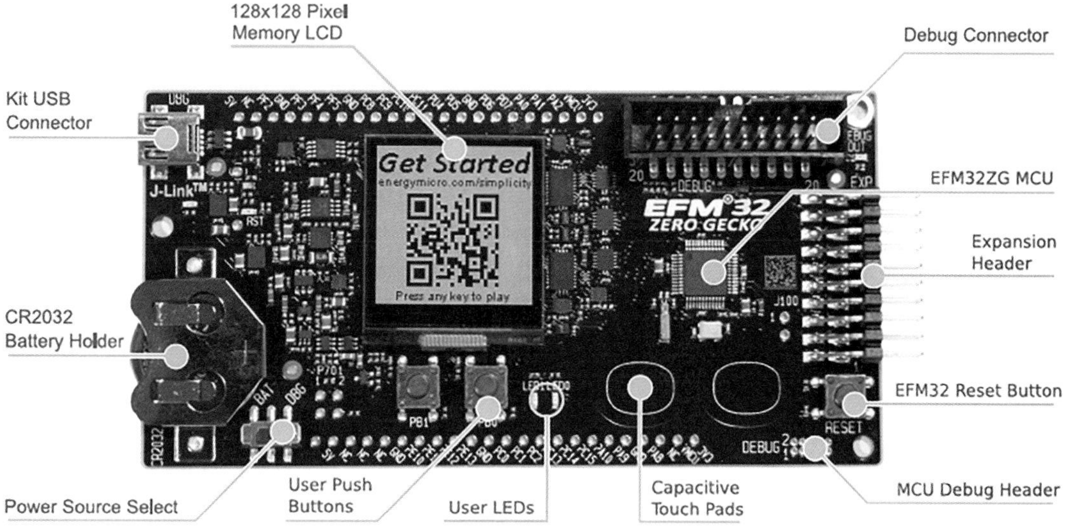

Figure 15.11
The Mbed-enabled EFM32 Zero Gecko Starter Kit. *Image courtesy of Silicon Labs.*

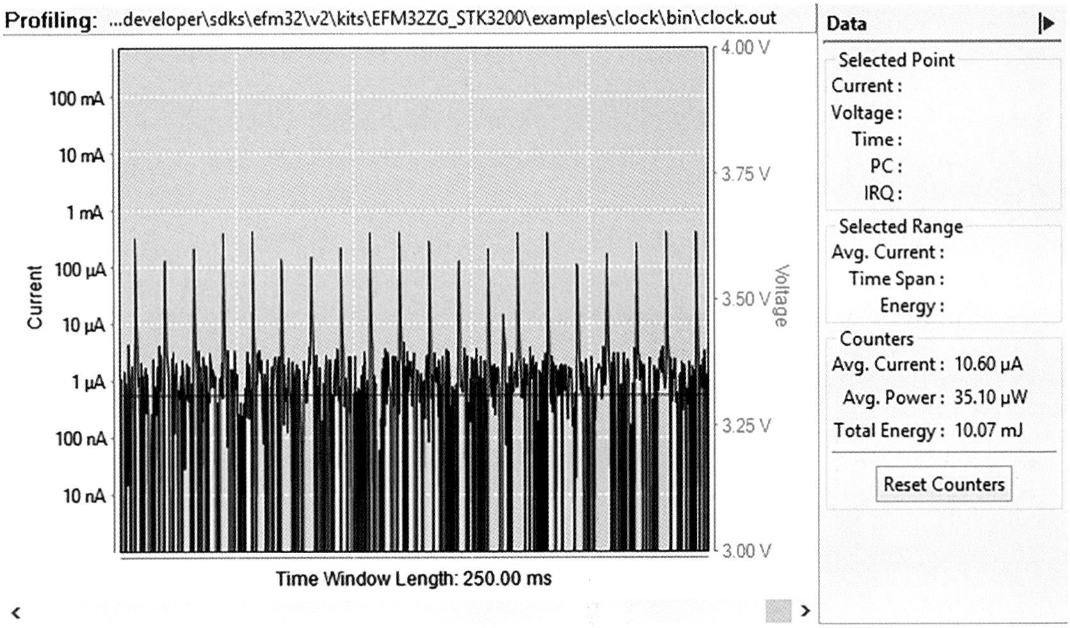

Figure 15.12
The EFM32 Zero Gecko energy profiler.

If you are working with this development board, you have the choice of doing so with the Simplicity Studio environment, or with one of the Mbed development environments. It provides an excellent pathway for study in genuinely low-power applications.

Chapter Review

- Distribution of power supplies and clocks in microcontrollers can be understood from studying data sheets and user manuals; this leads to valuable insights for the programmer and circuit designer.
- Power consumption mechanisms in digital electronics are well understood. Microcontroller consumption data is generally available in good detail, including for both the LPC1768 and F401RE microcontrollers.
- Sophisticated techniques, both hardware and software, exist to minimize power consumption. In the Mbed OS, software techniques center around use of *sleep* and *deep-sleep* modes.
- Power consumption in the Mbed LPC1768 can be optimized, but this development board is not ideal for low-power applications.
- Power consumption in the Nucleo F401RE can be monitored and optimized, with the possibility of energy efficiency, which is of interest for low-power applications.

- For competitive low-power design, circuits must be designed specifically and rigorously for that purpose. The EFM32 Zero Gecko, applying the M0+ Cortex core, is one example where this has been achieved.

Quiz

1. Name two advantages and two disadvantages of R–C and quartz oscillators.
2. Following reset, the LPC1768 always starts running from the internal R–C oscillator. Why is this?
3. In a certain application, the main oscillator in an LPC1768 application is running at 18.000 MHz, the PLL multiplies by 8, and the lower 7 bits of register **CCLKCFG** are set to 5. What is the frequency of **cclk**?
4. A certain logic circuit is powered from 3.0 V. It has a quiescent current of 120 nA, and an "equivalent capacitance" in the circuit of 56 pF. Applying Eq. 15.1, what is its current consumption when the clock frequency is 1 kHz, when it is 1 MHz, and when it is not clocked at all?
5. An Mbed LPC1768 is found to draw 140 mA, when powered from 4 AAA cells in series, each of capacity 1175 mAh. Approximately how long will the cells last if they run continuously?
6. Access these answers from the LPC1768 User Manual:
   a. Explain why only the main oscillator may be used as clock source for the USB.
   b. There is a required range of *output* frequency for PLL0. What is it?
7. Propose situations where each of the LPC1768 low-power modes (sleep, deep sleep, and so on) can be used effectively.
8. Manufacturer's data shows that the F401RE microcontroller has, among others, clock sources labeled LSE and LSI. What does each of these stand for, and what is each used for?
9. This chapter mentions three clock oscillators in the LPC1768, but four in the STMF401RE. Explain the differences, and outline any benefits of one or the other arrangement.
10. The power consumption of a digital circuit is being carefully monitored. It is found that connecting a long cable to a digital interface marginally increases the power consumption, even though nothing is connected at the far end of the cable. Why is this?
11. Identify the main characteristics of the M0 and the M0+ processor cores, and the differences between them.
12. A designer wishes to estimate power consumption characteristics of a new product based on the LPC1769, and applies the current consumption data of Table 15.8. A 3.3 V battery is available, with capacity 1200 mAh. She anticipates behavior where the processor will need to wake once per minute to perform a task, made up of two parts. In the first part, the processor must run with a clock speed of 120 MHz, PLL

enabled, for 200 ms. It then runs at 12 MHz, PLL disabled, for 400 ms. For the rest of the time it is in power-down mode. For the purposes of this question, the current taken by the rest of the circuit can be assumed to be negligible.

a. Estimate the average current drawn from the battery.
b. Estimate the battery life.

References

15.1. Hugo Zijlmans, PLL0 config script, 2010. https://os.Mbed.com/users/hugozijlmans/notebook/pll0-config-script/.

15.2. Power management (sleep). https://os.Mbed.com/docs/Mbed-os/v6.15/apis/power-management-sleep.html.

153. *Using the LPC1700 power modes, AN10915, Rev. 01-25*. NXP; February 2010.

154. Joseph Yiu. *The Definitive Guide to ARM Cortex-M0 and Cortex-M0+ Processors*. 2nd ed. Elsevier; 2015.

155. Silicon Labs web site. http://www.silabs.com/.

156. *EFM 32 Starter Kit, EFM32ZG-STK3200 User Manual*. Silicon Labs; 2013.

APPENDIX A

Some Number Systems

A.1 Binary, Decimal, and Hexadecimal

The number system we are most familiar with is decimal, which makes use of 10 different symbols to represent numbers: 0, 1, 2, 3, 4, 5, 6, 7, 8, and 9. Each of these symbols represents a number, and we make larger numbers by using groups of symbols. In this case, the digit farthest to the right represents units, the next represents tens, the next hundreds, and so on. For example, the number 249, shown in Fig. A.1, is evaluated by adding the values in each position:

$$2\ hundreds + 4\ tens + 9\ units = 249$$

or

$$2x10^2 + 4x10^1 + 9x10^0 = 249.$$

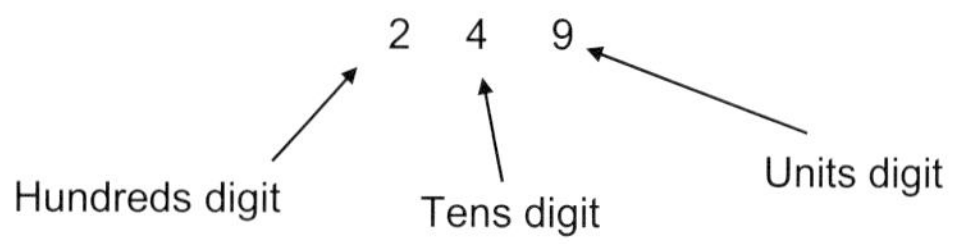

Figure A.1
The decimal number 249

The *base* or *radix* of the decimal system just described is 10. We almost certainly count in the decimal system due to the accident of having 10 fingers and thumbs on our hands. There is nothing intrinsically correct or superior about it. It is quite possible to count in other bases, and the world of digital computing almost forces us to do this.

The binary counting system has a base or radix of 2. Therefore, it uses just two symbols, normally 0 and 1. These are called binary digits, or bits. Numbers are made up of groups of digits. Again, the value each digit represents depends upon its position in the number. Therefore, the 4-bit number 1101, shown in Fig. A.2, is interpreted as:

$$1x8 + 1x4 + 0x2 + 1x1 = 8 + 4 + 1 = 13.$$

Similarly, the value 0110 binary = 6 decimal. Note that we refer to the units digit as "bit 0," or the least-significant bit (LSB or lsb). The two's digit is called "bit 1" and so on, up to the most-significant bit (MSB or msb).

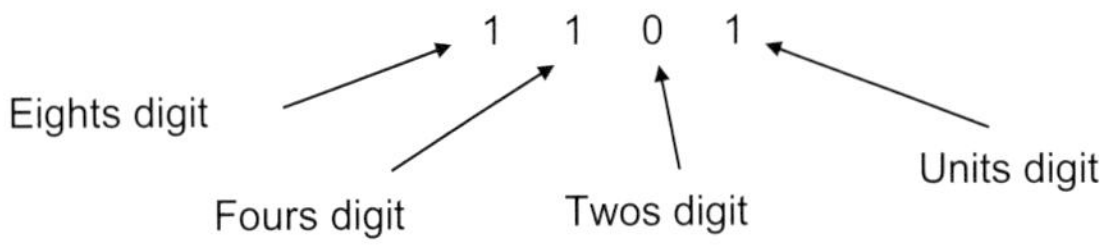

Figure A.2
The binary number 1101

We are obviously interested in binary representation of numbers, because that is how digital machines perform mathematical operations. But sometimes there are just too many 0s and 1s to keep track of. We often group bits in fours, or multiples of fours. A single *byte* is made up of 2 groups of 4, which is 8 bits. Each bit in the byte represents a power of 2, as shown in Fig. A.2, but going up from 8 to 16, 32, 64, and 128. With 1 byte we can count up to 255 in decimal; for example:

$$0000\ 0000\ binary = 0\ decimal$$
$$1001\ 1100\ binary = 128 + 16 + 8 + 4 = 156\ decimal$$
$$1111\ 1111\ binary = 128 + 64 + 32 + 16 + 8 + 4 + 2 + 1 = 255\ decimal$$

The range from 0 to 255 offered by a single byte is not very great, however. To perform mathematical calculations to a high accuracy, we need to work with larger numbers; for example: 16-, 24-, or 32-bit. With 16 bits we can count up to 65,535; for example:

$$0111\ 0111\ 0111\ 0110\ binary = 30582\ decimal$$
$$1111\ 1111\ 1111\ 1111\ binary = 65535\ decimal.$$

Working with large binary numbers is not easy for a human; we just cannot absorb all those 1s and 0s. A very convenient alternative is to use hexadecimal, which works to base 16. Each binary group of 4 bits is now directly represented by one hexadecimal digit. Now, we need to represent 16 numbers with just a single digit. The decimal digits for numbers 0 to 9 are used as in decimal, but to represent numbers 10 to 15 we use letters: "A" or "a" is adopted to represent decimal 10. Similarly, "B" = 11, "C" = 12, "D" = 13, "E" = 14, "F" = 15. By convention, "0x" can be placed before a number to indicate that it is hexadecimal. For example, 0xE5 = (14 × 16) + 5 = 229. Table A.1 shows the equivalent 4-bit values in binary, hexadecimal, and decimal.

Using hexadecimal means that one digit can represent any number from 0 to 15, while the next digit represents 16s, and the next (16 x 16)s or 256s. This is illustrated with the hexadecimal number 0x371 in Fig. A.3. Here we see that 371 in hexadecimal is (3x256) + (7x16) + 1 = 881 in decimal.

Because we can now represent 4-bit numbers with a single digit, an 8-bit number is represented with 2 hexadecimal digits, and a 16-bit number with 4 digits. You can then see that by individually looking at groups of 4 bits, we can easily generate the hexadecimal equivalent, as with the following examples:

255 decimal = 1111 1111 = 0xFF
156 decimal = 1001 1100 = 0x9C
65535 decimal = 1111 1111 1111 1111 = 0xFFFF
30582 decimal = 0111 0111 0111 0110 = 0x7776

Table A.1: 4-bit values in binary, hexadecimal, and decimal

4-bit Binary Number	Hexadecimal Equivalent	Decimal Equivalent
0 0 0 0	0x0	0
0 0 0 1	0x1	1
0 0 1 0	0x2	2
0 0 1 1	0x3	3
0 1 0 0	0x4	4
0 1 0 1	0x5	5
0 1 1 0	0x6	6
0 1 1 1	0x7	7
1 0 0 0	0x8	8
1 0 0 1	0x9	9
1 0 1 0	0xA	10
1 0 1 1	0xB	11
1 1 0 0	0xC	12
1 1 0 1	0xD	13
1 1 1 0	0xE	14
1 1 1 1	0xF	15

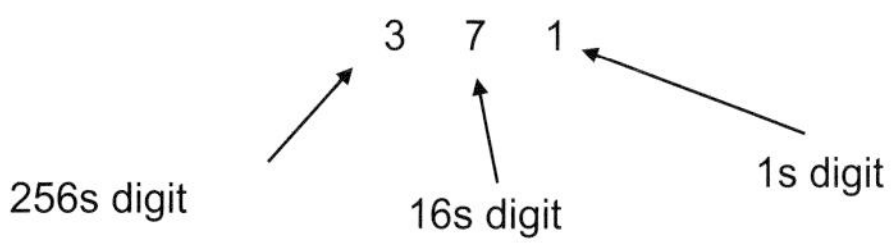

Figure A.3
Hexadecimal number system

While we correctly use the two values of 0 and 1 in all binary numbers above, it's worth noting that different terminology is sometimes used when we apply electronic circuits to represent these numbers. This is done particularly by those who are thinking more in terms of the circuit than of the numbers. This terminology is shown in Table A.2.

Table A.2: Some terminology for logic values

Logic 0	Logic 1
0	1
Off	On
Low	High
Clear	Set
Open	Closed

A.2 Representation of Negative Numbers—Two's Complement

Simple binary numbers allow only the representation of unsigned numbers, which under normal circumstances are considered to be positive. Yet we must have a way of representing negative numbers as well. A simple way of doing this is by offsetting the available range of numbers. We do this by coding the largest anticipated negative number as 0 and counting up from there. In the 8-bit range, with symmetrical offset, we can represent −128 as 00000000; 1000000 then represents 0, and 11111111 represents +127. This method of coding is called *Offset Binary* and is illustrated in the right-hand column of Table A.3. It is used on occasion (e.g., in analog-to-digital converter outputs), but its usefulness is limited, as it is not easy to do arithmetic with.

Let us consider an alternative approach. Suppose we took an 8-bit binary down counter and clocked it from any value down to, and then below, 0. We would get this sequence of numbers:

Binary	Decimal
0000 0101	5
0000 0100	4
0000 0011	3
0000 0010	2
0000 0001	1
0000 0000	0
1111 1111	−1?
1111 1110	−2?
1111 1101	−3?
1111 1100	−4?
1111 1011	−5?

Table A.3: Two's complement and offset binary

Two's Complement	Decimal	Offset Binary
0111 1111	+127	1111 1111
0111 1110	+126	1111 1110
	:	
	:	
0000 0001	+1	1000 0001
0000 0000	0	1000 0000
1111 1111	−1	0111 1111
1111 1110	−2	0111 1110
	:	
	:	
1000 0010	−126	0000 0010
1000 0001	−127	0000 0001
1000 0000	−128	0000 0000

This gives a possible means of representing negative numbers—effectively, we subtract the magnitude of the negative number from 0, within the limits of the 8-bit number, or whatever other size is in use. This representation is called *two's complement*. It can be shown that using two's complement leads to correct results when simple binary addition and subtraction are applied. Two's complement notation can be applied to binary words of any size.

The two's complement of an n-bit number is found by subtracting it from 2^n, leading to its two's complement negative value. For example, the two's complement 4-bit value for 3, or 0011 in binary, is $2^4 - 3 = 13$ in decimal, or 1101 in binary.

Rather than doing this error-prone subtraction, an easier way of reaching the same result is to complement (i.e., change 1 to 0, and 0 to 1) all the bits of the positive number, and then add 1. Hence to find − 5 we follow the procedure:

original number		**complement all**		**add 1**
0000 0101 (+5)	→	1111 1010	→	1111 1011 (−5 in 2's comp)

To convert back, simply subtract 1 and complement again. Importantly, the most significant bit of a two's complement number acts as a "sign bit," 1 for negative, and 0 for positive. The 8-bit binary range, shown both for two's complement and offset binary, appears in Table A.3.

Table A.4: Number ranges for differing word sizes.

Number of Bits	Unsigned Binary	Two's Complement
8	0 to 255	−128 to +127
12	0 to 4095	−2048 to + 2047
16	0 to 65,535	−32,768 to + 32,767
24	0 to 16,777,215	−8,388,608 to + 8,388,607
32	0 to 4,294,967,295	−2,147,483,648 to + 2,147,483,647

In general, the range of an n-bit two's complement number is from $-2^{(n-1)}$ to $+\{2^{(n-1)} - 1\}$. Table A.4 summarizes the ranges available for some commonly used values of n.

A.3 Floating Point Number Representation

Numbers described so far appear to be all integers—we haven't been able to represent any sort of fractional number—and their range is limited by the size of the binary word representing them. Suppose we need to represent really large or small fractional numbers? Another way of expressing a number, which greatly widens the range, is called *floating point* representation.

In general, a number can be represented as a x r^e, where a is the *mantissa*, r is the radix, and e is the exponent. This is sometimes called scientific notation. For example, the decimal number 12.3 can be represented as

$$1.23 \times 10^1 \textit{ or}$$
$$.123 \times 10^2 \textit{ or}$$
$$12.3 \times 10^0 \textit{ or}$$
$$123 \times 10^{-1} \textit{ or}$$
$$1230 \times 10^{-2}$$

Floating point notation adapts and applies scientific notation to the computer world. The name is derived from the way the binary point can be allowed to float by adjusting the value of the exponent to make best use of the bits available in the mantissa. Standard formats exist for representing numbers by their sign, mantissa, and exponent, and a host of hardware and software techniques exist to process numbers represented in this way. Their disadvantage lies in their greater complexity, and usually lower processing speed and higher cost. However, for flexible use of numbers in the computing world they are essential.

The most widely recognized and used format is the *IEEE Standard for Floating-Point Arithmetic* (known as IEEE 754). In single precision form this makes use of 32-bit

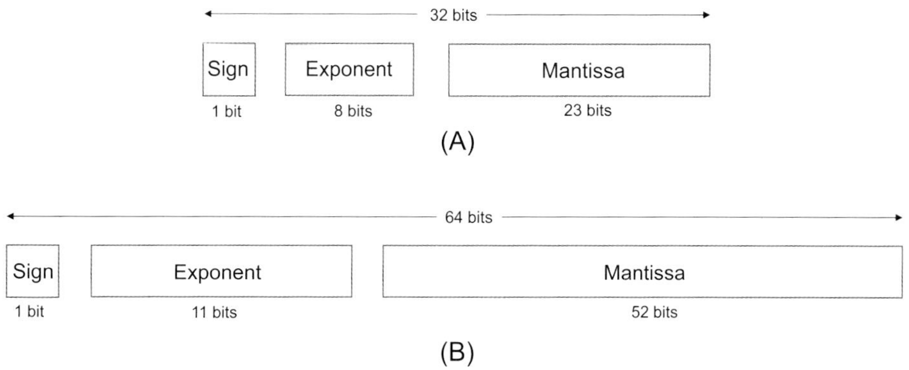

Figure A.4
IEEE 754 floating point format: (A) 32-bit single precision; (B) 64-bit double precision

representation for a number, with 23 bits for the mantissa, 8 bits for the exponent, and a sign bit, as represented in Fig. A.4(a). The binary point is assumed to be just to the left of the MSB of the mantissa. A further bit, always 1 for a non-zero number, is added to the mantissa, making it effectively a 24-bit number. Zero is represented by 4 zero bytes. The number 127 is subtracted from the exponent, leading to an effective range of exponents from − 126 to + 127. Exponent 255 (leading to 128, when 127 is subtracted) is reserved to represent infinity. The value of a number represented in this format is then:

$$(-1)^{\text{sign}} \text{ x } 2^{(\text{exponent}-127)} \text{ x } 1.\text{mantissa}.$$

This allows number representation in the range:

$$\pm 1.175494 \text{ x } 10^{-38} \text{ to } \pm 3.402823 \text{ x } 10^{+38}.$$

A 64-bit double precision representation is now also widely used, as shown in Fig. A.4(b).

APPENDIX B

Some C Essentials, with a Little C++

B.1 A word about C

This appendix aims to summarize the main features of the C language as used in this book, and should be adequate for the purpose of supporting the book. If you are a C novice, however, it's worth having another reference source available, such as Reference B.1, particularly as you consider the more advanced features. The appendix ends with a small section introducing some essential C++ features.

As you progress through the book, you will find yourself jumping around within the material of this appendix. Don't feel you need to read it sequentially. Instead, read the different sections as they're referenced from the book.

B.2 Elements of a C program

B.2.1 Keywords

C has a number of keywords whose use is defined. A programmer cannot use a keyword for any other purpose, for example, as a variable name. Keywords are summarized in Tables B.1, B.2, and B.3.

B.2.2 Program features and layout

Simply speaking, a C program is made up of the following:

Declarations

All variables in C must be declared before they can be defined and used, giving—at a minimum—the variable name and its data type. A declaration is terminated with a semicolon. In simple programs, declarations appear as one of the first things in the program. They can also occur within the program, with significance attached to the location of the declaration.

For example:

```
float exchange_rate;
int new_value;
```

Table B.1: C keywords associated with data type and structure definition

Word	Summary Meaning	Word	Summary Meaning
char	A single character, usually 8-bit	**signed**	A qualifier applied to **char** or **int** (default for **char** and **int** is signed)
const	Data that will not be modified	**sizeof**	Returns the size in bytes of a specified item, which may be variable, expression or array
double	A 'double precision' floating-point number	**struct**	Allows definition of a data structure
enum	Defines variables that can take named integer values	**typedef**	Creates new name for existing data type
float	A 'single precision' floating-point number	**union**	A memory block shared by 2 or more variables, of any data type
int	An integer value, usually 32-bit	**unsigned**	A qualifier applied to **char** or **int** (default for **char** and **int** is signed)
long	An extended integer value, usually 32- or 64-bit	**void**	No value or type
short	A short integer value, usually 16-bit	**volatile**	A variable which can be changed by factors other than the program code

Table B.2: C keywords associated with program flow

Word	Summary Meaning	Word	Summary Meaning
break	Causes exit from a loop	**for**	Defines a repeated loop – loop is executed as long as condition associated with **for** remains true
case	Identifies options for selection within a **switch** expression	**goto**	Program execution moves to labelled statement
continue	Allows a program to skip to the end of a **for**, **while** or **do** statement	**if**	Starts conditional statement; if condition is true, associated statement or code block is executed
default	Identifies default option in a **switch** expression, if no matches found	**return**	Returns program execution to calling routine, causing also return of any data value specified by function
do	Used with **while** to create loop, in which statement or code block following **do** is executed once, and then repeated as long as **while** condition is true	**switch**	Used with **case** to allow selection of a number of alternatives; **switch** has an associated expression which is tested against a number of **case** options
else	Used with **if**, and precedes alternative statement or code block to be executed if **if** condition is not true	**while**	Defines a repeated loop – loop is executed as long as condition associated with **while** remains true

Table B.3: C keywords associated with data storage class

Word	Summary Meaning	Word	Summary Meaning
auto	Variable exists only within block within which it is defined. This is the default class, and so rarely used	**register**	Variable to be stored in a CPU register; thus, address operator (&) has no effect
extern	Declares data defined elsewhere	**static**	Declares variable which exists throughout program execution; the location of its declaration affects in what part of the program it can be referenced

declares a variable called **exchange_rate** as a floating point number, and another variable called **new_value** as a signed integer. The data types are keywords seen in the preceding tables.

Statements

Statements are where the action of the program takes place. They perform mathematical or logical operations, and establish program flow. Every statement which is not a block (see below) ends with a semicolon. Statements are executed in the sequence in which they appear in the program, except where program branches take place.

For example, this line is a statement:

```
counter = counter + 1;
```

Space and layout

There is not a strict layout format to which C programs must adhere. The way the program is laid out and the use of space are both used to enhance clarity. Blank lines and indents in lines, for example, are ignored by the compiler, but used by the programmer to optimize the program layout.

As an example, the program that the Mbed compiler always starts up with, shown as Program Example 2.1, *could* be written as shown here. However, it wouldn't be easy to read. It's the semicolons, at the end of each statement, and the brackets which in reality define much of the program structure.

```
int main(){DigitalOut led(LED1);while(true){led=!led;ThisThread::sleep_for(500ms);}}
```

Comments

Two ways of commenting are used. One is to place the comment between the markers /* and */. This is useful for a block of text information running over several lines. Alternatively, when 2 forward slash symbols (//) are used, the compiler ignores any text which follows on that line only, which can then be used for comment.

For example:

```
/*A program which flashes Mbed LED1 on and off,
Demonstrating use of digital output and wait functions. */
#include "mbed.h" //include the mbed header file as part of this program
```

Code blocks

Declarations and statements can be grouped together into *blocks*. A block is contained within braces, such as { and }. Blocks can be and are written within other blocks, each within its own pair of braces. Keeping track of these pairs of braces is an important pastime in C programming, as in a complex piece of software there can be numerous pairs nested within each other. It's generally a good idea to avoid nesting blocks too deeply, as it makes the code quite hard to follow. Instead, splitting a deeply nested function into multiple functions (see Section B.4) can significantly enhance readability and reduce coding errors.

B.2.3 Compiler directives

Compiler directives are messages to the compiler and do not directly lead to program code. Compiler directives all start with a hash (`#`). Two examples follow.

#include

The **#include** directive directly inserts another file into the file that invokes the directive. This provides a feature for combining a number of files as if they were 1 large file. Angled brackets (<>) are used to enclose files held in a directory different from the current working directory, and hence often for library files not written by the current author. Quotation marks are used to contain a file located within the current working directory, and hence often user defined.

For example:

```
#include "mbed.h"
```

#define

The **#define** directive allows use of names for specific constants. For example, to use the number $\pi = 3.141592$ in the program, we could create a `#define` for the name "PI" and assign that number to it, as shown:

```
#define PI 3.141592
```

The name 'PI' is then used in the code whenever the number is needed. When compiling, the compiler replaces the name in the **#define** with the value that has been specified.

B.3 Variables and data

B.3.1 Declaring, naming, and initializing

Variables must be named, and their data type declared, before they can be used in a program. Keywords from Table B.1 are used for this. For example,

```
int MyVariable;
```

declares "MyVariable" as a data type int (signed integer).

It is possible to initialize (or define) the variable at the same time as declaration; for example,

```
int MyVariable = 25;
```

declares **MyVariable** as an integer and sets it to an initial value of 25.

It is possible to give variables meaningful names, while still avoiding excessive length, for example, "Height," "InputFile," "Area." Variable names must start with a letter or underscore; no other punctuation marks are allowed. Variable names are case sensitive.

B.3.2 Data types

When a variable declaration is made, the compiler reserves a section of memory for it whose size depends on the type invoked. Examples of the link between data type, number range, and memory size are shown in Table B.4. It is interesting to compare these with information on number types given in Appendix A. Note that the actual memory size applied to data types can vary between compilers; for example, in the case of the **long** data type, if the compiler produces code for a 64-bit platform, **long** will be 64 bits (or 8 bytes) wide. If the compiler produces code for a 32-bit platform, **long** will be 32 bits (or 4 bytes) wide.

Note that further data types, such uint8_t or int16_t, are available through the C standard library, as mentioned in Section B.9.1. These are widely used in the example programs in the book.

B.3.3 Working with data

When numbers are written in a C program, the default radix (number base) for integers is decimal, with no leading 0 (zero). We can also work with numbers in floating point decimal, octal, or hexadecimal format, depending on what is most convenient, and what number type and range are required. In general, it's easiest to work in decimal, but if a variable represents a register bit field or a port address, then it can work better to manipulate the data in hexadecimal. For time-critical applications

Table B.4: Example C data types, as implemented by the Mbed compiler

Data Type	Description	Length (bytes)	Range
char	Character	1	0 to 255
signed char	Character	1	−128 to +127
unsigned char	Character	1	0 to 255
short	Integer	2	−32768 to +32767
unsigned short	Integer	2	0 to 65535
int	Integer	4	−2147483648 to + 2147483647
unsigned int	Integer	4	0 to 4294967295
long	Integer	4 or 8	−2147483648 to + 2147483647 or −9223372036854775808 to 9223372036854775807
unsigned long	Integer	4 or 8	0 to 4294967295 or 0 to 18446744073709551615
float	Floating point, single precision	4	$1.17549435 \times 10^{-38}$ to $3.40282347 \times 10^{+38}$
double	Floating point, double precision	8	$2.22507385850720138 \times 10^{-308}$ to $1.79769313486231571 \times 10^{+308}$

it is important to remember that floating point calculations can take much longer than fixed point. Octal numbers are identified with a leading 0. Hexadecimal numbers are prefixed with 0x.

For example, if a variable **MyVariable** is of type **char**, we can perform the following examples to assign a number to that variable. The value for each of the following is the same:

```
MyVariable = 14; //a decimal example
MyVariable = 0x0E; //a hexadecimal example
MyVariable = 016; //an octal example
```

B.3.4 Changing data type: casting

Data can be changed from one data type to another by *type casting*. This is done using the cast operator, seen at the bottom of Table B.5. For example, in the line that follows, **size** has been declared as **char,** and **sum** as **int.** However, their division will result in an integer division being performed, and therefore losing precision. To coerce the compiler into making a floating point division, the **sum** variable is cast to a floating point:

```
mean = (float)sum / size;
```

Table B.5: C operators

Precedence and Order	Operation	Symbol	Precedence and Order	Operation	Symbol
Parentheses and array access operators					
1, L to R	Function calls	( )	1, L to R	Point at member	X->Y
1, L to R	Subscript	[ ]	1, L to R	Select member	X.Y
Arithmetic operators					
4, L to R	Add	X+Y	3, L to R	Multiply	X*Y
4, L to R	Subtract	X−Y	3, L to R	Divide	X/Y
2, R to L	Unary plus	+X	3, L to R	Modulus	%
2, R to L	Unary minus	−X			
Relational operators					
6, L to R	Greater than	X>Y	6, L to R	Less than or equal to	X<=Y
6, L to R	Greater than or equal to	X>=Y	7, L to R	Equal to	X==Y
6, L to R	Less than	X<Y	7, L to R	Not equal to	X!=Y
Logical operators					
11, L to R	AND (1 if both X and Y are not 0)	X&&Y	2, R to L	NOT (1 if X=0)	!X
12, L to R	OR (1 if either or both X or Y are not 0)	X||Y			
Bitwise operators					
8, L to R	Bitwise AND	X&Y	2, L to R	Ones complement (bitwise NOT)	~X
10, L to R	Bitwise OR	X|Y	5, L to R	Right shift. X is shifted right Y times	X >>Y
9, L to R	Bitwise XOR	X^Y	5, L to R	Left shift. X is shifted left Y times	X<<Y
Assignment operators					
14, R to L	Assignment	X=Y	14, R to L	Bitwise AND assign	X&=Y
14, R to L	Add assign	X+=Y	14, R to L	Bitwise inclusive OR assign	X|=Y
14, R to L	Subtract assign	X−=Y	14, R to L	Bitwise exclusive OR assign	X^=Y
14, R to L	Multiply assign	X*=Y	14, R to L	Right shift assign	X>>=Y
14, R to L	Divide assign	X/=Y	14, R to L	Left shift assign	X<<=Y
14, R to L	Remainder assign	X%=Y			
Increment and decrement operators					
2, R to L	Preincrement	++X	2, R to L	Postincrement	X++
2, R to L	Predecrement	− −X	2, R to L	Postdecrement	X− −

Continued

Table B.5: C operators—cont'd

Precedence and Order	Operation	Symbol	Precedence and Order	Operation	Symbol
Conditional operators					
13, R to L	Evaluate *either* X (if Z ≠ 0) *or* Y (if Z=0)	Z?X:Y	15, L to R	Evaluate X first, followed by Y	X,Y
'Data interpretation' operators					
2, R to L	The object or function pointed to by X	*X	2, R to L	The address of X	&X
2, R to L	Cast – the value of X, with (scalar) type specified	(*type*) X	2, R to L	The size of X, in bytes	sizeof(X)

Some type conversions may be done by the compiler implicitly. It is better programming practice not to depend on this, but to do it explicitly; however, it's important to be aware of such implicit conversions, in case they cause unexpected behavior.

B.4 Functions

A function is a section of code which can be called from another part of the program. So if a particular piece of code is to be used or duplicated many times, we can write it once as a function, and then call that function whenever the specific operation is required. Using functions saves coding time and improves readability by making the code neater.

Data can be passed to functions and returned from them. Such data elements, called *parameters*, must be of a type which is declared in advance. Only 1 return value is allowed, whose type must also be declared. The data elements passed to the variable when the function is invoked, called the *argument(s),* are a *copy* of the original. Therefore, the function does not itself modify the value of the variable named. Thus, the impact of the function should be predictable and controlled.

A function is defined in a program by a block of code having particular characteristics. Its first line forms the function header, with the format:

```
return_type function_name (variable_type_1 variable_name_1, variable_type_2
                                                        variable_name_2,...)
```

An example is shown in Fig. B.1. The return type is given first. In this example, the keyword **float** is used. After the function name, in brackets, 1 or more data types may be listed, which identify the arguments which must be passed *to* the function. In this case, 2 arguments are sent, 1 of type **char** and 1 of type **float.** Following the function header, a pair of braces enclose the code which makes up the function itself. This could be anything from a single line to many pages. The final statement of a function with a return type that

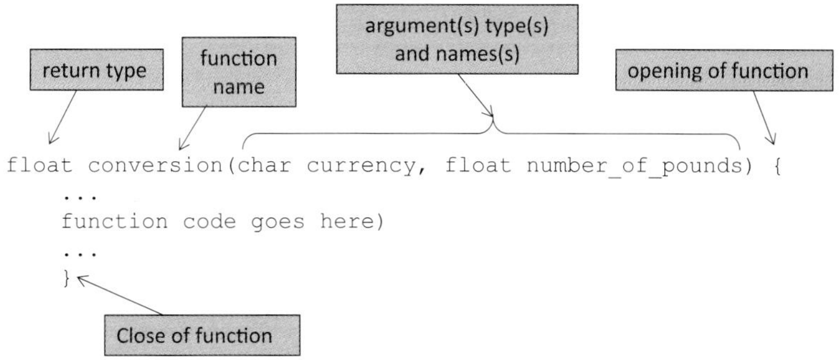

Figure B.1
Function definition example

is not void must be a **return** statement, which will specify the value returned to the calling function.

B.4.1 The main *function*

The core code of any C program is contained within its "main" function. Other functions may be written outside **main( )**, and called from within it. Program execution starts at the beginning of **main( ).** It must follow the structure just described, and must be defined as follows:

```
int main(void) {
    ...
}
```

This indicates that the **main( )** function takes no arguments, and returns an integer. The return value from main should normally be 0, but can be non-zero to indicate an error condition.

B.4.2 Function prototypes

Just like variables, functions must be declared at the start of a program, before the main function. The declaration statements for functions are called prototypes. Each function in the code must have an associated prototype for it to be located by the compiler, unless it is defined in full before its first use. The format is the same as for the function header.

For example, the following function prototype applies to the function header seen above:

```
float conversion(char currency, float number_of_pounds);
```

This describes a function that takes inputs of a character value for the selected currency and a floating point (decimal) value for the number of pounds to be converted. The function returns the decimal monetary value in the specified currency.

B.4.3 Function definitions

The actual function code is called the *function definition*. For example:

```
float conversion(char currency, float number_of_pounds) {
  float exchange_rate;
  switch(currency) {
    case 'U': exchange_rate = 1.22;         // US Dollars
      break;
    case 'E': exchange_rate = 1.12;         // Euros
      break;
    case 'Y': exchange_rate = 157.42;       // Japan Yen
      break;
    default: exchange_rate = 1;
      break;
  }
  exchange_value = number_of_pounds * exchange_rate;
  return exchange_value;
}
```

This function can be called any number of times from within the main C program, or from another function, as in this statement:

```
ten_pounds_in_yen = conversion('Y', 10.45);
```

The structure of this function is explained in Section B.6.2.

B.4.4 Using the static storage class with functions

The static storage class specifier is useful for defining variables within functions, where the data inside the function must be remembered between function calls. For example, if a function within a real-time system is used to calculate a digital filter output, the function should always remember its previous data values. In this case, data values inside the function should be defined as static, for example, as shown below:

```
float movingaveragefilter(float data_in) {
  static float data_array[10];        // define static float data array
  for (int i = 8; i >= 0; i--) {
    data_array[i+1] = data_array[i]; —// shift each data value along
  }                                   // (the oldest data value is discarded)
  data_array[0] = data_in;            —// place new data at index 0
  float sum = 0;
  for (int i = 0; i < 10; i++) {
    sum = sum + data_array[i];        —// calculate sum of data array
  }
  return sum / 10;                     // return average value of array
}
```

If **static** is used in the declaration of a global variable (i.e., a variable that is not declared inside a function), then that variable cannot be accessed from outside the file it is defined in.

B.5 Operators

C has a wide set of operators, shown in Table B.5. The symbols used are familiar, but their application is *not* always the same as in conventional algebra. For example, a single "equals" symbol, "=," is used to assign a value to a variable. A double equals sign, "==," is used to represent the conventional "equal to."

Operators have a certain order of precedence, shown in the table. The compiler applies this order when it evaluates a statement. If more than 1 operator at the same level of precedence occurs in a statement, then those operators are evaluated in turn, either left to right or right to left, as shown in the table. For example, the line

```
counter = counter + 1;
```

contains 2 operators. Table B.5 shows that the addition operator has precedence level 4, while all assign operators have precedence 14. Therefore, the addition is evaluated first, followed by the assign. In words, we could say that the new value of the variable **counter** has been assigned the previous value of **counter**, plus 1.

B.6 Flow control: conditional branching

Flow control covers the different forms of branching and looping available in C. As branching and looping can lead to programming errors, C provides clear structures to improve programming reliability.

B.6.1 if *and* else

"If" statements always start with the **if** keyword, followed by a logical condition. If the condition is satisfied, then the code block which follows is executed. If the condition is not satisfied, then the code is not executed. There may or may not also be following **else** or **else if** statements.

Syntax:

```
if (Condition1) {
   ...C statements here
} else if (Condition2) {
   ...C statements here
} else if (Condition3) {
   ...C statements here
} else {
   ...C statements here
}
```

The **if** and **else** statements are evaluated in sequence:

> **else if** statements are only evaluated if all previous **if** and **else** conditions have failed;
> **else** statements are only executed if all previous conditions have failed.

For example, in the above example, the `else if (Condition2)` will only be executed if Condition1 has failed.

Example:

```
if (data > 10) {
  data += 5;                 // If we reach this point, data must be > 10
} else if (data > 5) {       // If we reach this point, data must be <= 10
  data -= 3;
} else {                     // If we reach this point, data must be <= 5
  nVal = 0;
}
```

B.6.2 switch *statements and using* break

The **switch** statement allows a selection to be made of one out of several actions, based on the value of a variable or expression given in the statement. An example of this structure has already appeared, in the example function in Section B.4.3. The structure uses no less than 4 C keywords. Selection is made from a list of **case** statements, each with an associated label—note that a colon following a text word defines it as a label. If the label equals the **switch** expression, then the action associated with that **case** is executed. The **default** action (which is optional) occurs if none of the **case** statements are satisfied. The **break** keyword terminates each **case** condition, causing program execution to continue after the **switch** code block.

B.7 Flow control: program loops

B.7.1 while *loops*

A **while** loop is a simple mechanism for repeating a section of code until a certain condition is satisfied. The condition is stated in brackets after the word **while**, with the conditional code block following. For example:

```
int i = 1;
while (i < 10) {
  ... C statements here
  i++;           // increment i
}
```

Here the value of **i** is defined outside the loop; it is then updated within the loop. Eventually **i** increments to 10, at which point the loop terminates. The condition associated with the **while** statement is evaluated at the start of each loop iteration; the loop then only runs if the condition is found to be true.

B.7.2 for loops

The **for** loop allows a different form of looping, in that the dependent variable is updated automatically every time the loop is repeated. It defines an initialized variable, a condition for looping, and an update statement. Note that the update takes place at the end of each loop iteration. If the updated variable is no longer true for the loop condition, the loop stops and program flow continues For example:

```
for (j = 0; j < 10; j++) {
     ... C statements here
}
```

Here the initial condition is `j=0` and the update value is `j++`, that is, **j** is incremented. This means that **j** increments with each loop. When **j** becomes 10 (i.e., after 10 loops), the condition `j<10` is no longer satisfied, so the loop does not continue any further.

B.7.3 Infinite loops

We often require a program to loop forever, particularly in a super-loop program structure. An infinite loop can be implemented by either of the following loops:

```
while (1) {           //while(true) may also be used
  ... continuously called C statements here
}
```

Or

```
for (;;) {
   ... continuously called C statements here
 }
```

B.7.4 Exiting loops with break

The **break** keyword can also be used to exit from a **for** or **while** loop, at any time within the loop. For example:

```
while (i > 5) {
    ... C statements here
    if (fred == 1)
```

```
        break;
      ... C statements here
  }                               //end of while
  //execution continues here on loop completion, or on break
```

B.8 Derived data types

In addition to the fundamental data types, there are further data types which can be derived from them. Example types that we use are described in this section.

B.8.1 Arrays and strings

An array is a series of data elements, each of which has the same type. Any data type can be used. Array elements are stored in consecutive memory locations. An array is declared with its name and the data type of its elements; it is recognized by the use of the square brackets which follow the name. The number of elements and their value can also be specified. For example, the declaration

```
char message1[8];
```

defines an array called **message1**, containing 8 characters. Alternatively, it can be left to the compiler to deduce the array length, as seen in the 2 examples here:

```
char item1[] = "Apple";
int nTemp[] = {5,15,20,25};
```

In each of these the array is initialized as it is declared.

Elements within an array can be accessed with an index, starting with value 0. Therefore, for the first example above, **message1[0]** selects the first element and **message1[7]** the last. An access to **message1[8]** would be outside the boundary of the array, and would give invalid data. The index can be replaced by any variable which represents the required value.

Importantly, the name of an array is set equal to the address of the initial element. Therefore, when an array name is passed in a function, what is passed is this address.

A string is a special array of type **char** that is ended by the NULL (\0) character. The null character allows code to search for the end of a string. The size of the string array must therefore be 1 byte greater than the string itself, to contain this character. For example, a 20-character string could be declared:

```
char MyString[21];            // 20 characters plus null
```

B.8.2 Pointers

Instead of specifying a variable by name, we can specify its address. In C terminology such an address is called a *pointer.* A pointer can be loaded with the address of a variable by using the unary operator "&," like this:

```
my_pointer = &fred;
```

This loads the variable **my_pointer** with the *address* of the variable **fred**; **my_pointer** is then said to *point* to **fred**.

Doing things the other way round, the value of the variable pointed to by a pointer can be specified by prefixing the pointer with the '*' operator. For example, ***my_pointer** can be read as "the value pointed to by **my_pointer**." The * operator, used in this way, is sometimes called the *dereferencing* or *indirection* operator. The indirect value of a pointer, for example ***my_pointer**, can be used in an expression just like any other variable.

A pointer is declared by the data type it points to. Thus,

```
int *my_pointer;
```

indicates that **my_pointer** points to a variable of type **int**.

We can also use pointers with arrays, because an array is really just a number of data values stored at consecutive memory locations. So if the following is defined,

```
int dataarray[]={3,4,6,2,8,9,1,4,6};  // define an array of arbitrary values
int *ptr;                             // define a pointer
ptr = &dataarray[0];                  // assign pointer to the address of
                                      // the first element of the data array
```

given the previous declarations, the following statements will therefore be true:

```
*ptr == 3;            // the first element of the array pointed to
*(ptr+1) == 4;        // the second element of the array pointed to
*(ptr+2) == 6;        // the third element of the array pointed to
```

So array searching can be done by moving the pointer value to the correct array offset. Pointers are required for a number of reasons, but one simple reason is that the C standard does not allow us to pass arrays of data to and from functions, so we must use pointers instead to get around this.

B.8.3 Structures and unions

Structures and *unions* are both sets of related variables, defined through the C keywords **struct** and **union**. In a way they are like arrays, but in both cases they can be of data elements of *different* types, and rather than the data elements having an index, they have names.

Structure elements, called *members*, are arranged sequentially, with the members occupying successive locations in memory. Sometimes the compiler might insert spacing between elements, called padding. A structure is declared by invoking the **struct** keyword, followed by an optional name (called the structure *tag*), followed by a list of the structure members, each of these itself forming a declaration. For example,

```
struct resistor {int val; char pow; char tol;};
```

declares a structure with tag **resistor**, which holds the value (**val**), power rating (**pow**), and tolerance (**tol**) of a resistor.

Structure elements are identified by specifying the name of the variable and the name of the member, separated by a full stop (period). Like a structure, a union can hold different types of data. Unlike the structure, union elements all begin at the same address—they overlap. Hence the union can represent only one of its members at any one time, and the size of the union is the size of the largest element. It is up to the programmer to track which type is currently stored. Unions are declared in a format similar to that of structures.

Unions, structures, and arrays can occur within each other.

B.9 Libraries and standard functions

B.9.1 Libraries and the C standard library

Because C is a simple language, much of its functionality derives from standard functions and macros which are available in the libraries accompanying any compiler. A C library is a set of precompiled functions which can be linked into the application. These may be supplied with a compiler, be available in company, or be public domain. Notably, there is a *Standard Library,* defined in the C ANSI standard. There are a number of standard header files, used for different groups of functions within the standard library. For example, the **math.h** header file is used for a range of mathematical functions including all trigonometric functions, while **stdio.h** contains the standard input and output functions, including for example **printf( )**.

As examples, Program Example 4.3 uses the standard **sin( )** function, linked through the **math.h** header file; Program Example 5.4 (and many others) uses the standard **printf( )** function, linked through **stdio.h;** Program Example 6.7 uses the **rand( )** function, linked through **stdlib.h**; Program Example 7.4 and others use the **stdint.h** library to define further data types, for example **int16_t**, or **uint8_t** (with versions for 8, 16, 32, or 64 bits available for each). Both **math.h** and **stdio.h** are, incidentally, included automatically by **Mbed.h**.

B.9.2 Using printf

This versatile function provides formatted output, typically for displaying information on a screen or over a serial port connection. Text, data, formatting, and control formatting can be specified. Only summary information is provided here; a full statement can be found in Reference B.1.

The string of characters or data to be printed is contained within inverted commas; control and specification codes or characters can be contained within this sequence, identified by their start character \ or %. Data name(s) whose value(s) is/are to be printed appear after the closing of the inverted commas.

Control characters

These are single characters, preceded by a back slash, which mostly control layout of the printed text. Examples are given in Table B.6.

Data formatting options

As soon as you want to print data values to screen, you need to understand the formatting options available. The general format is shown in Fig. B.2.

Flags set the justification of the output; the default is right justification. Common options are:

Table B.6: Control characters for printf()

Character	Action
\a	alert (bell/beep)
\b	backspace
\n	new line
\r	carriage return
\t	horizontal tab
\v	vertical tab

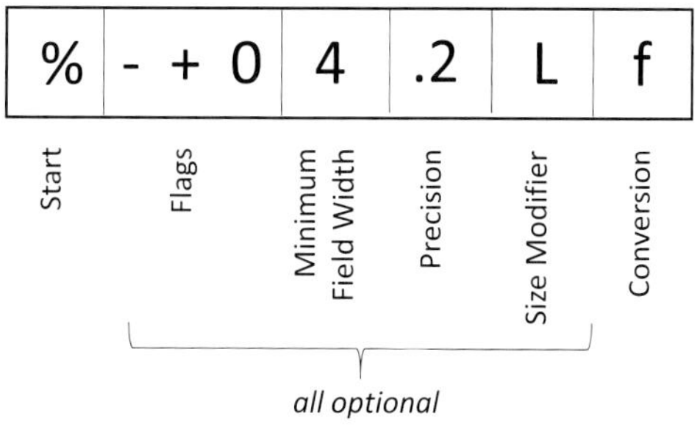

Figure B.2
Formatting printf( ) output

+ the sign to be included, whether positive or negative,
– left justify,
(blank) indicate only negative sign, otherwise a blank,
0 fill field with leading zeros.

Width sets the minimum width, in number of characters.

Precision gives the maximum number of characters to print after a decimal point.

Size Modifier sets the input size, often not used. L is for long double, l is long, h is short or char.

Conversion indicates the type of the displayed data. Examples are given in Table B.7.

There must be 1 variable name provided for each use of %, listed after the output string, whose data type must match the type indicated in the preceding format specifier.

Examples:

```
printf("ADC Data Values")
```

The message "ADC Data Values" is printed. The cursor remains at the end of the text.

```
printf("ADC Data Values\n\r");
```

The message "ADC Data Values" is printed. The cursor is moved by \n and \r to force a new line and carriage return.

```
printf("%1.3f",ADCdata);
```

This prints the value of the floating point variable **ADCdata.** A *conversion specifier*, initiated by the % character, defines the format. Within this the "f" specifies floating point, and the .3 causes output to 3 decimal places.

```
printf("%1.3f \n\r",ADCdata);
```

As above, but includes \n and \r to force a new line and carriage return.

```
printf("random number is %d\n\r", r_delay);
```

Table B.7: Conversion specifiers for printf( )

Character	Description	Argument Type
c	character	char
d, i	a signed integer	int, short, long
f	float in fixed point notation	double, float
s	string, terminated by '\0'	string
u	unsigned integer	int, short, long
x, X	Unsigned integer in hexadecimal; x as lower case, X as upper case	int, short, long

This prints a text message, followed by the value of **int** variable **r_delay.**

```
printf("Time taken was %f seconds\n", t_meas);
```

This prints a text message, followed by the value of **t_meas,** which is of type **float.**

```
printf("x = %+1.2fg\t y = %+1.2fg\t z = %+1.2fg\n\r", x, y, z);
```

This prints the **float** values of x, y, and z, tabulating each. For example:

```
x = -0.16g      y = +0.92g      z = +0.12g
```

The **printf( )** function returns an integer value giving the number of bytes printed.

B.10 File access operations

In C we can open files, read and write data, and also scan through files to specific locations, even searching for particular types of data. The commands for input and output operations are all defined by the C **stdio** library**.**

The stdio library uses the concept of *streams* to manage the data flow, where a stream is a sequence of bytes. All streams have similar properties, even though the actual implementation of their use is varied, depending on things like the source and destination of their flow. Streams are represented in the stdio library as pointers to FILE objects, which uniquely identify the stream. We can store data in files (as chars) or we can store words and strings (as character arrays). A summary of useful stdio file access library functions is given in Table B.8.

B.10.1 Opening and closing files

A file can be opened with the following command:

```
FILE* pFile = fopen("datafile.txt", "w");
```

This assigns a pointer with name `pFile` to the file at the specific location given in the `fopen` statement. In this example, the *access mode* is specified with a "w." A number of other file open access modes and their specific meanings are shown in Table B.9. When "w" is the access mode (meaning *write access*), if the file doesn't already exist then the `fopen` command will automatically create it in the specified location.

When we have finished using a file for reading or writing, it should be closed, for example, with

```
fclose(pFile);
```

Table B.8: Useful stdio library functions

Function	Format	Summary Action
fclose	`int fclose (FILE * stream)`	closes a file
fgetc	`int fgetc (FILE * stream)`	gets a character from a stream
fgets	`char * fgets (char * str, int num, FILE * stream)`	gets a string from a stream
fopen	`FILE * fopen (const char * filename, const char * mode)`	opens the file of type FILE and name filename
fprintf	`int fprintf (FILE * stream, const char * format, ... )`	writes formatted data to a file
fputc	`int fputc (int character, FILE * stream)`	writes a character to a stream
fputs	`int fputs (const char * str, FILE * stream)`	writes a string to a stream
fseek	`int fseek (FILE * stream, long int offset, int origin)`	moves the file pointer to the specified location

str: An array containing the null-terminated sequence of characters to be written.
stream: Pointer to a FILE object that identifies the stream where the string is to be written.

Table B.9: Access modes for fopen( )

Access Mode	Action
"r"	Open an existing file for reading.
"w"	Create a new empty file for writing. If a file of the same name already exists, it will be deleted and replaced with a blank file.
“a”	Append to a file. Write operations result in data being appended to the end of the file. If the file does not exist, a new blank file will be created.
"r+"	Open an existing file for both reading and writing.
"w+"	Create a new empty file for both reading and writing. If a file of the same name already exists, it will be deleted and replaced with a blank file.
"a+"	Open a file for appending or reading. Write operations result in data being appended to the end of the file. If the file does not exist, a new blank file will be created.

B.10.2 Writing and reading file data

If the intention is to store numerical data, this can be done in a simple way by storing individual 8-bit data values. The **fputc** command allows this, as follows:

```
char write_var = 0x0F;
fputc(write_var, pFile);
```

This stores the 8-bit variable **write_var** to the data file. The data can also be read from a file to a variable as follows:

```
read_var = fgetc(pFile);
```

Using the **stdio** commands, it is also possible to read and write words and strings with **fgets( ), fputs( )** and to write formatted data with **fprintf( )**.

It is also possible to search or move through files looking for particular data elements. When data are read from a file, the file pointer can be moved with the **fseek( )** function. For example, the following command will reposition the file pointer to the 8th byte in the text file:

```
fseek (pFile, 8, SEEK_SET);     // move file pointer to 8 bytes from the start
```

The first term of the **fseek( )** function is the stream (the file pointer name), the second term is the pointer offset value, and the third term is the origin for where the offset should be applied. There are 3 possible values for the origin, as shown in Table B.10. The origin values have predefined names as shown.

Further details on the **stdio** commands and syntax can be found in Reference B.1.

B.11 Using multiple files

B.11.1 Header files

All but the simplest of C programs are made up of more than one file. Generally, there are many files, which are combined together in the process of compiling—for example, original source files combining with standard library files. To aid this process, a key section of any library file is detached and created as a separate *header* file.

Header file names end in **.h**; the file typically includes declarations of constants and function prototypes, and links on to other library files. The function definitions themselves stay in the associated **.c** or **.cpp** files. In order to use the features of the header file and the file(s) it invokes, it must be included within any program accessing it, using **#include**. We see **Mbed.h** being included in almost every program in the book. Note that the **.c** file where the function declarations appear must also include the header file. The functions of the C standard library, outlined in Section B.9, are linked via their header files.

Table B.10: fseek( ) origin values

Origin Value	Description
SEEK_SET	Beginning of file
SEEK_CUR	Current position of the file pointer
SEEK_END	End of file

B.11.2 The compilation process

A very simplified version of the compilation process is shown in Figure B.3. In summary, first a *pre-processor* looks at a particular source file and implements any defined pre-processor directives. The pre-processor also identifies and prepares the header files that will be used by the C++ program files, and prepares the final output that will be sent to the compiler. The compiler then takes each pre-processed C/C++ file and compiles to generate a number of *object files* for the program. In doing so, the compiler ensures that the source files do not contain any syntax errors and that the object and library files are formatted correctly for the linker. The *linker* manages the allocation of memory for the microprocessor and ensures that all object and library files are linked to each other correctly. In undertaking the task, the linker may uncover programming faults associated with memory allocation and capacity. The linker generates a single executable binary (**.bin**) file, which can be downloaded to the microprocessor memory.

B.11.3 Using #ifndef and #endif directives

With multiple related files, it is possible for the program to contain more than one definition of a variable, function or object, which is not allowed by the compiler. The

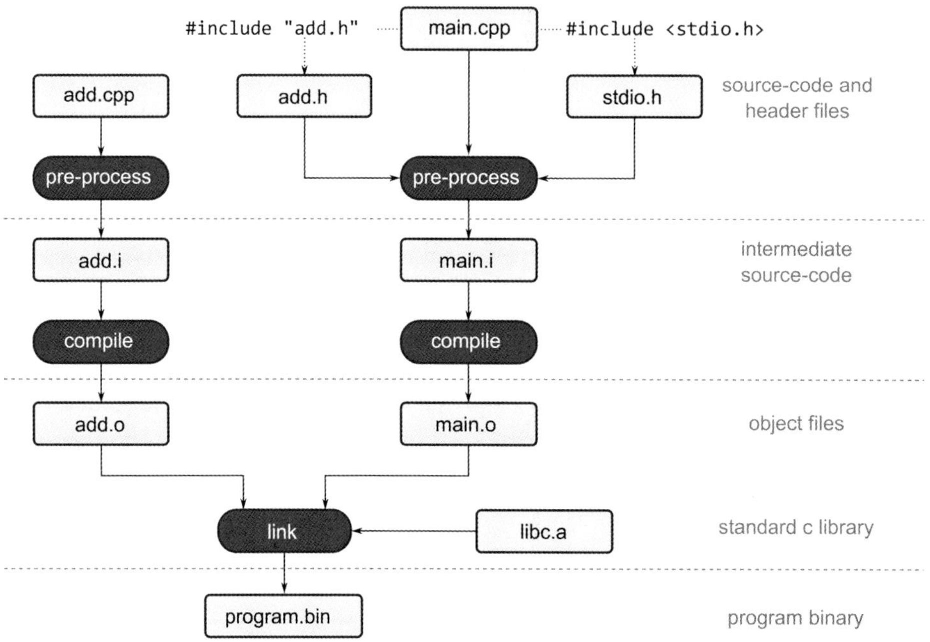

Figure B.3
C/C++ program compile and link process

#ifndef directive ("if not defined") can be used to provide a solution to the problem. In using header files it is possible (and indeed good practice) to use a conditional statement to define variables and function prototypes only if they have not previously been defined. The **#ifndef** directive allows a conditional statement based on the existence of a **#define** value. If the **#define** value has not previously been defined then that value and all of the header file's variables and prototypes are defined. If the **#define** value has previously been declared then the header file's contents are not implemented by the pre-processor (as they must certainly have already been implemented). This ensures that all header file declarations are only added once to the project. The example code below represents a template header file structure using the **#ifndef** condition.

```
#ifndef VARIABLE_H     // if VARIABLE_H has not previously been defined
#define VARIABLE_H     // define it now

// header declarations here...

#endif                 // end of the if directive
```

This programming pattern is called an *include guard*. An alternative solution to this problem is to use the **#pragma once** directive at the beginning of the header file, which instructs the pre-processor to only include the contents of this file once. It is a non-standard directive, but it is supported by all major compilers. It also solves the problem of 2 separate header files using the same **#ifndef/#define** to protect their contents, which would interfere with each other.

B.11.4 Using Mbed objects globally

All Mbed objects must be defined in an "owner" source file. However, we may wish to use those objects from within other source files in the project, in other words, "globally." This can be done by also declaring the Mbed object in the associated owner's header file. When an Mbed object is declared for global use, it cannot be declared static. The **extern** specifier should be used to declare the object in each source file that wants to access it. For example, a file called **my_functions.cpp** may define a **DigitalOut** object called "**RedLed**" as follows:

```
DigitalOut RedLed(p5);
```

If any other source files need to manipulate **RedLed**, the object must be declared in the **functions.h** header file using the **extern** specifier, as follows:

```
extern DigitalOut RedLed;
```

Note that the specific Mbed pins don't need to be redefined in the header file, as these will have already been specified in the object declaration in **my_functions.cpp**.

B.12 Toward professional practice

The readability of a C program is much enhanced by good layout on the page or screen. This helps to produce good and error-free code and enhances the ability to understand, maintain, and upgrade it. Many companies impose a "house style" on C code written for them; such guides can be found on-line.

The very simple style guide we have adopted in this book should be exemplified in the programs reproduced. It includes the following features:

- monospace font applied
- opening header text block, giving overview of program action
- blank lines used to separate major code sections
- extensive commenting
- opening brace of any code block placed on line which initiates it
- code within any code block indented 2 spaces compared with code immediately outside block (many use indentation of 4 spaces, but this can push code off the edge of a book page)
- closing brace stands alone, indented to align with line initiating code block

A useful and interesting set of guidelines which relate to the development of the Mbed OS is given in Reference B.2.

Version control is essential practice in any professional environment. This can be managed by the IDE, and/or by the programmer(s) through entries in the program header text block. To save space, we have not displayed version control information within the programs as reproduced in the book, although this was done in the development process. Minimal version control within the source code would include date of origin, name of original author, name of person making most recent revision, and revision date.

B.13 Some C++ snippets

C++ is a superset of C, and is now a huge and complex language. This section gives the briefest of glimpses of some of those features of C++ needed to get going in this book. These add to the C features already covered. References B.3, B.4, and B.5 are all very useful for further reference and reading.

C++ was first commercially released in 1985, following development of "C with Classes," and became an ISO standard in 1998. In 2011, the ISO C++11 version was launched, with many more features and an enlarged standard library. Versions C++14, C++17, and C++20 were issued in 2014, 2017, and 2020, each offering significant further features and improvements.

B.13.1 Classes and objects

Earlier, we introduced the concept of a **struct** data type as one that is a container for one or more type. We now must introduce the concept of *visibility* to explain how **classes** relate to **structs**. Visibility dictates which parts of the code can access the members defined within, and visibility labels are used to specify this. In C++, a **class** is exactly the same as a **struct**, except the members of the type have a different *default* visibility. There are 3 different visibility levels:

- **public** – the member is accessible to any code outside the class or struct
- **protected** – the member is accessible only within the class or struct, and any subclasses
- **private** – the member is accessible only within the class or struct

Structs, by default, have **public** visibility, which means any member is accessible outside the **struct**. Consider the following snippet:

```
struct my_struct {
  int member1;
  int member2;
private:
  int member3;
};
```

This declares a struct type called my_struct that has 2 public integer members, **member1** and **member2**, and a private integer member, **member3**. Any code can access the public members freely For example,

```
my_struct m;
m.member1 = 5;
m.member2 = 10; // COMPILES OK
```

Here, a variable called **m** is declared to be of type **my_struct**. Then, each member of the **struct** is initialized with a value. But the private member is inaccessible:

```
my_struct m;
m.member3 = 5; // COMPILER ERROR: 'int my_struct::member3 is private within this
context'
```

In contrast to structs, classes by default have **private** visibility, which means members can only be accessed by functions defined inside the class. Members can be made public by placing them after a **public** label:

```
class my_class {
  int private_member;
public:
  int public_member;
};
```

C++ extends C to allow functions to be defined within a **struct**, and therefore also a **class**. Functions are also subject to visibility rules—public functions can be called from outside the class. Classes and structs can also inherit from other classes and structs, therefore taking on the members and functions of those. If class A inherits from class B, A is said to be a subclass of B, and B is the superclass of A.

An *object* is an *instance* of a class. Think of a **class** or **struct** as a specification (or description) of the layout of data. An **object** is a particular realization of this specification. To create an **object**, you *instantiate* a **class**.

```
class my_class {    // declare a class called my_class
  int tag;
public:
  void set_tag(int val) { tag = val; }
  int get_tag() const { return tag; }
};

my_class obja;  // define an instance of my_class called obja
my_class objb;  // define an instance of my_class called objb
```

The above example shows the declaration of a class called **my_class**. The class declares a single private field called **tag** and 2 public functions that update or retrieve the value of **tag**. Then 2 instances of the class are constructed.

The Mbed OS APIs depend extensively on the use of classes and member functions. The user invokes, for example, a constructor to create an instance of a driver, like **DigitalIn**, with certain initialized features, like pin names or other settings. Member functions of **DigitalIn** are then available to the user. In Program Example 7.4, in the lines below, an instance of the class **SPI** is instantiated, with the resulting object called **acc.** The member function of **SPI, format( )**, is later invoked. Different classes could also have functions called **format( )**, but they would remain distinct from each other due to this mechanism.

```
SPI acc(D11,D12,D13);
...
acc.format(8,3);
```

B.13.2 More keywords and operators

C++ introduces numerous additional keywords and operators. A few examples of such keywords are given in Table B.11.

While there are a number of new operators in C++, we only introduce here the scope resolution operator: **a::b**. This relates item **b** to the scope in which it was declared, **a.** For example, we have used the line below in numerous program examples. It is invoking the function **sleep_for( )**, in namespace **ThisThread.**

```
ThisThread::sleep_for(500ms);
```

Table B.11: Example keywords C++ adds to C

Example Keyword	Meaning
and, bitand, bitor...	A set of keywords which replicate certain logical operators.
bool	Boolean data type.
char8_t, char16_t, char32_t	Integer data types, of 8, 16 or 32 bits width (giving greater precision of identifying data type).
namespace	A mechanism indicating that certain declarations or features belong together, and should be identified as such.
public	See Section B.13.1.
private	See Section B.13.1.
using	Brings names from a namespace into the current namespace.

B.13.3 The C++ standard library

Like C, C++ has a standard library. This includes the C library (with very minor adjustments), but goes very much further; in fact, it makes up around two-thirds of the C++ definition. As with C, library features should be included in the program with **#include.** The standard library applies the **namespace** named **std** (see example below).

B.13.4 Time features and chrono

The **chrono** standard library provides very extensive time-related facilities. It was introduced to bring standardization to the diverse hardware and software timing techniques which were in use. It deals with clocks, time points and durations. Chrono elements are defined under **std::chrono namespace**. Three clock sources are defined: **system_clock, steady_clock,** and **high_resolution_clock.** Chrono recognizes the standard units of **h, min, s, ms, us, ns.**

The function **duration_cast( )** converts a duration into a standard unit of time. For example, Program Example 6.3 invokes the **chrono** library in the first line shown below. It later prints a time duration, derived from a reading of **timer_2,** where **duration_cast( )** converts the timer reading into milliseconds. Similar usage occurs in Program Example 10.4.

```
using namespace std::chrono;
...
printf("The time taken was %llu milliseconds\n",
duration_cast<milliseconds>(timer_2.elapsed_time()).count());
```

References

B.1. Peter Prinz and Ulla Kirch-Prinz. *C Pocket Reference*. O'Reilly; 2002.
B.2. Mbed Style Guide. https://os.Mbed.com/docs/Mbed-os/v6.15/contributing/style.html.
B.3. Stroustrup Bjarne. *A Tour of C++*. 3rd ed. Addison Wesley; 2022
B.4. Loudon Kyle. *C++ Pocket Reference*. O'Reilly; 2003
B.5. The C++ Resources Network. www.cplusplus.com.

APPENDIX C

Mbed LPC1768 Technical Data

Note: This short Appendix outlines electrical details of the Mbed LPC1768. Similar data is published for the STM32F401RE microcontroller and the Nucleo F401RE board; for example, in Reference 2.6.

C.1 Summary Technical Details of the Mbed LPC1768

The Mbed LPC1768 development board is built around an LPC1768 microcontroller, made by NXP Semiconductors. The LPC1768 is in turn designed around an ARM Cortex-M3 core, with 512 KB FLASH memory, 64 KB RAM, and a wide range of interfaces, including Ethernet, USB Device, CAN, SPI, I^2C, and other I/O. It runs at 96 MHz.

Summary characteristics and operating conditions are given below.

Package

- 40-pin DIP package, with 0.1-inch spacing between pins, and 0.9 inches between the 2 rows.
- Overall size: 2 inches x 1 inch, 53 mm × 26 mm.

Power

- Powered through the USB connection, or 4.5−9.0 V applied to VIN (see Fig. 2.1).
- Power consumption < 200 mA (~100 mA with Ethernet disabled).
- Real-time clock battery backup input VB; 1.8−3.3 V at this input keeps the real-time clock running. This requires 27 μA, which can be supplied by a coin cell.
- 3.3 V regulated output on VOUT available to power peripherals.
- 5.0 V from USB available on VU, available only when USB is connected.
- Total supply is current-limited to 500 mA.
- Digital IO pins are 3.3 V, 40 mA each, 400 mA maximum total.

Reset

- nR - Active-low reset pin with identical action to the reset button. Pull-up resistor is on the board.

C.2 LPC1768 Electrical Characteristics

To interface successfully to any microcontroller, we have to satisfy certain electrical conditions. Many digital components are designed to be compatible with each other, so in many situations we don't need to worry about this. However, once it comes to applying nonstandard devices, it is very important to gain an understanding of interfacing requirements. Input signals have to lie within certain thresholds in order to be interpreted correctly as logic levels. We also need to understand the ability of an output to drive external loads which may be connected to it.

These operating conditions are referred to in Chapter 3 and from time to time onward. They are specified precisely, and in very great detail, by the manufacturer of any electronic component or integrated circuit in the relevant data sheet. It is part of the skill of a professional design engineer to know where to access these details, to know how to interpret them, and to be able to design to meet the criteria specified. The hobbyist can engage creatively or intuitively in some trial and error and hope for the best. The professional needs to be able to predict the performance of a design analytically.

When interfacing with the Mbed LPC1768 data lines (i.e., lines excluding power supply or Ethernet), we are interfacing directly with the LPC1768. We therefore turn to the LPC1768 data sheet, Reference 2.3. For a novice reader this is, however, a very complex document. We have therefore extracted a very limited amount of data which contains some of the main details that are needed.

C.2.1 Port Pin Interface Characteristics

Table C.1 is drawn mainly from Table 8 of Reference 2.3, and defines port pin operating characteristics. The format of Reference 2.3 is retained as far as possible, but for simplicity a few of the finer details of operation are removed. Each parameter that appears in the table is then described below.

<u>Supply voltage (3.3 V), $V_{DD(3V3)}$</u>

This is the power supply voltage connected to the $V_{DD(3V3)}$ pin of the microcontroller. Although nominally 3.3 V, the minimum acceptable supply voltage for this pin is seen to be 2.4 V, and the maximum 3.6 V. This pin is one of several supply inputs, and supplies the port pins.

<u>LOW-level input current, I_{IL}</u>

This is the current flowing into a port pin, with pull-up resistor disabled, when the input voltage is at 0 V. This very low current implies a very high input impedance.

Table C.1: Selected port pin characteristics

Symbol	Parameter	Conditions	Minimum	Typical	Maximum	Unit
$V_{DD(3V3)}$	supply voltage (3.3 V)	external rail	2.4	3.3	3.6	V
I_{IL}	LOW-level input current	$V_I = 0$ V; on-chip pull-up resistor disabled	—	0.5	10	nA
I_{IH}	HIGH-level input current	$V_I = V_{DD(3V3)}$; on-chip pull-down resistor disabled	—	0.5	10	nA
V_I	input voltage	pin configured to provide a digital function	0	—	5	V
V_O	output voltage	output active	0	—	$V_{DD(3V3)}$	V
V_{IH}	HIGH-level input voltage		$0.7V_{DD(3V3)}$	—	—	V
V_{IL}	LOW-level input voltage		—	—	$0.3V_{DD(3V3)}$	V
V_{OH}	HIGH-level output voltage	$I_{OH} = -4$ mA	$V_{DD(3V3)} - 0.4$	—	—	V
V_{OL}	LOW-level output voltage	$I_{OL} = 4$ mA	—	—	0.4	V
I_{OH}	HIGH-level output current	$V_{OH} = V_{DD(3V3)} - 0.4$ V	−4	—	—	mA
I_{OL}	LOW-level output current	$V_{OL} = 0.4$ V	4	—	—	mA
I_{OHS}	HIGH-level short-circuit output current	$V_{OH} = 0$ V	—	—	−45	mA
I_{OLS}	LOW-level short-circuit output current	$V_{OL} = V_{DD(3V3)}$	—	—	50	mA
I_{pd}	pull-down current	$V_I = 5$ V	10	50	150	uA
I_{pu}	pull-up current	$V_I = 0$ V	−15	−50	−85	uA

HIGH-level input current, I_{IH}

This is the current flowing into a port pin, with pull-down resistor disabled, when the input voltage is equal to $V_{DD(3V3)}$. This very low current implies a very high input impedance.

Input voltage, V_I

This indicates the acceptable range of input voltages to any port pin. Unsurprisingly, the minimum is 0 V. Interestingly, the maximum is 5 V, showing that the pin input can actually exceed the supply voltage. This is a very useful feature, and allows interfacing to a system supplied from 5 V.

Output voltage, V_O

This indicates the range of output voltages that a port pin can source. The limits are 0 V and the supply voltage.

HIGH-level input voltage, V_{IH}

This parameter defines the range of voltages for an input to be recognzed as a Logic 1. Any input voltage exceeding $0.7V_{DD(3V3)}$ is interpreted as logic 1. In this context there is no need for a typical or maximum value, although the maximum is contained within the definition for V_I. The concept of V_{IH} is illustrated in Fig. 3.3.

LOW-level input voltage, V_{IL}

This parameter defines the range of voltages for an input to be recognized as a Logic 0. Any input voltage less than $0.3V_{DD(3V3)}$ is interpreted as Logic 0. In this context there is no need for a typical or minimum value, although the minimum is contained within the definition for V_I. The concept of V_{IL} is illustrated in Fig. 3.3.

HIGH-level output voltage, V_{OH}

This indicates the output voltage for a Logic 1 output. The value is defined for a load current of 0.4mA, where the convention is applied that a current flowing out of a logic gate terminal is negative. Consider that a circuit such as Fig. 3.4b applies. For this output current the minimum output voltage is ($V_{DD(3V3)}$ - 0.4) V. With no output current the output voltage will be equal to $V_{DD(3V3)}$. The values quoted imply an approximate output resistance of 100 Ω, under these operating conditions.

LOW-level output voltage, V_{OL}

This indicates the output voltage for a Logic 0 output. The value is defined for a load current of 0.4 mA, where the convention is applied that a current flowing out of a logic gate terminal is positive. Consider that a circuit such as Fig. 3.4c applies. For this output current the maximum output voltage is 0.4 V. With no output current the output voltage will be equal to 0 V. The values quoted imply an approximate output resistance of 100 Ω, under these operating conditions.

HIGH-level output current, I_{OH}

This gives the same information as the HIGH-level output voltage, V_{OH}.

LOW-level output current, I_{OL}

This gives the same information as the LOW-level output voltage, V_{OL}.

Table C.2: Selected limiting values

Symbol	Parameter	Conditions	Minimum	Maximum	Unit
$V_{DD(3V3)}$	supply voltage (3.3 V)	external rail	−0.5	+4.6	V
V_I	input voltage	5 V tolerant I/O pins; only valid when the $V_{DD(3V3)}$ supply voltage is present	−0.5	+5.5	V
V_{IA}	analog input voltage	on ADC related pins	−0.5	+5.1	V
I_{DD}	supply current	per supply pin	—	100	mA
I_{SS}	ground current	per ground pin	—	100	mA
$P_{tot(pack)}$	total power dissipation (per package)	based on package heat transfer, not device power consumption	—	1.5	W

HIGH-level short-circuit output current, I_{OHS}

This gives the maximum output current if the output is at Logic 1, but is short-circuited to ground.

LOW-level short-circuit output current, I_{OLS}

This gives the maximum output current if the output is a Logic 0 but is connected to the supply, $V_{DD(3V3)}$.

Pull-down current, I_{pd}

This is the current which flows due to the internal pull-down resistor, when enabled, when $V_I = 5$ V.

Pull-up current, I_{pu}

This is the current which flows due to the internal pull-up resistor, when enabled, when $V_I = 0$ V.

C.2.2 Limiting Values

The values in the previous section showed the limits to which operating conditions can be taken, while maintaining normal operation. Limiting values, also called absolute maximum values, define the limits which must always be observed; otherwise device damage occurs. For example, although a single port pin can supply up to 45 mA when short-circuited, one must also take note of the limiting value for supply and ground pin currents. Check Fig. 15.2 for number of supply and ground pins. Table C.2 is taken from Table 6 of Reference 2.3, and shows some of the key limiting values of the LPC1768.

APPENDIX D

Parts List

To get started with the practical activities in this book you will of course need either the Mbed LPC1768 or the Nucleo F401RE development boards. There is some flexibility in almost any other component or subsystem after that. The items below are valid at the time of writing; in most cases, it is easy to substitute other suppliers, and sometimes equivalent parts. As is the way with electronics, parts do change or become obsolete, and you may have to seek an alternative. You must check that each code is correct for the part you wish to order, before placing the order. Note also that some codes link to multiples of the component identified, for example, for resistors.

Table D.1: Components and devices used in the book—new parts introduced in each chapter

Description	Supplier	Supplier Code
Chapter 2		
NXP Mbed prototyping board LPC1768	uk.farnell.com/	1761179
Nucleo board: STM32F401RET6 MCU - NUCLEO-F401RE	uk.farnell.com/	2394223
Prototyping breadboard	www.rapidonline.com	34-1330 OR 34-1331
Jumper wire kit	www.rapidonline.com	34-0495 OR 34-0677
Chapter 3		
Mbed Application Board	www.rs-online.com	769-4182
LED, 5 V, red	www.rapidonline.com	56-1500
LED, 5 V, green	www.rapidonline.com	56-1505
Switch PCB mount SPDT	www.rapidonline.com	76-0200
Photointerrupter/slotted optoswitch	www.rapidonline.com	58-0944
Kingbright 7-segment display	www.rapidonline.com	57-0294
Switching transistor, ZVN4206 ***OR***	www.rapidonline.com	47-0162
BC107	www.rapidonline.com	81-0010

Continued

Table D.1: Components and devices used in the book—new parts introduced in each chapter—cont'd

Description	Supplier	Supplier Code
220-Ω resistor	www.rapidonline.com	62-3434
10k resistor	www.rapidonline.com	62-3474
Motor, 6 V, DC	www.rapidonline.com	37-0445
Battery box, 4xAA	www.rapidonline.com	18-2913 OR 18-2909
Diode, IN4001	www.rapidonline.com	47-3420
Chapter 4		
Servo Hitec HS-422	https:/uk.robotshop.com/	HS-422
Piezo transducer ("Buzzer")	www.rapidonline.com	35-0200
Chapter 5		
10k linear potentiometer	www.rapidonline.com	65-0715
NORPS12 light-dependent resistor	www.rs-online.com	914-6714
10k resistor	www.rapidonline.com	62-3474
LM35 temperature sensor	www.rapidonline.com	82-0240
Chapter 6		
ICL7611 op amp	www.rapidonline.com	82-0782
Chapter 7		
(2 Mbed targets and 2 breadboards required for some builds)		
Triple axis accelerometer breakout, ADXL345	www.sparkfun.com	SEN-09836
4k7 resistor	www.rapidonline.com	62-3577
Pushbutton switch, red	www.rapidonline.com	78-0160
Pushbutton switch, blue	www.rapidonline.com	78-0170
Digital temperature sensor breakout, TMP102	www.sparkfun.com	SEN-13314
Ultrasonic range finder, SRF08	https://uk.robotshop.com/	SRF08
Optional: IoT Discovery kit, as listed in Chapter 12		

Table D.1: Components and devices used in the book—new parts introduced in each chapter—cont'd

Description	Supplier	Supplier Code
Chapter 8		
PC1602 alphanumeric LCD display	www.rs-online.com	294-8695
uLCD-144-G2 GFX color LCD module	www.sparkfun.com	LCD-11377
Chapter 9		
(no new items)		
Chapter 10		
Breakout board for MicroSD transflash	www.sparkfun.com	BOB-00544
Kingston MicroSD card (with standard SD adaptor)	www.rs-online.com	126-4363
Chapter 11		
B-L475E-IOT01A discovery kit for IoT node		
B-L475E-IOT01A1 (915 MHz – US) *or*	uk.farnell.com/	2708777
B-L475E-IOT01A2 (868 MHz – EU)	uk.farnell.com/	2708778
XBee 2-mW module with PCB antenna (Series ZB) (2 off)	www.rs-online.com	122-5775
XBee Explorer USB	www.coolcomponents.co.uk	WRL-11812
Breakout board for XBee module (2 off, 1 if Mbed app board used) with 2-mm 10-pin XBee Socket (4 off)	www.coolcomponents.co.uk	BOB-08276 PRT-08272
Chapters 12–14		
(no new items)		
Chapter 15		
Optional: EFM32 Zero Gecko Starter Kit	http://uk.farnell.com/	2409172

APPENDIX E

Using a Host Terminal Emulator

E.1 Introducing host terminal applications

A terminal emulator allows a host computer to send or receive data from another device or computer through a variety of links. Terminal emulators take the form of a software package running on the host computer which creates a screen image on the computer screen, through which settings can be selected. They provide a context for data transfer using the host computer keyboard for data input. Such a terminal is particularly useful for displaying messages from a development board to a computer screen. It can become a very useful tool for user interfacing and software debugging. For example, status or error messages can be embedded in a program running on the target system and transferred to the terminal emulator when certain points in the program are reached.

A number of free and open source terminal emulators are available, some optimized for different operating systems or host computer types. They are similar, and we don't recommend one type over another, but mention these three here:

Tera Term: For download visit Reference E.1.
CoolTerm: Favored by users of Apple Mac. For download visit Reference E.2.
PuTTY: For download visit Reference E.3.

The Mbed development environments Mbed Studio and Keil Studio Cloud also have inbuilt serial terminals. You may find this adequate for your needs, or you may prefer to have one of these emulators as an alternative or backup.

We outline the installation steps for Tera Term as an example. The others will have similar stages and requirements.

E.2 Setting up Tera Term

Download Tera Term from Reference E.1 and install on your computer.

Open the Tera Term application. You will need to perform the following configuration:

- Select File -> New Connection (or just press Alt+N).
- Select the *Serial* radio button and select the *Mbed Serial Port* from the dropdown menu, as seen in Fig. E.1. Don't worry which COM port it appears on.
- Click *OK.*

Figure E.1
Setting up the Tera Term connection.

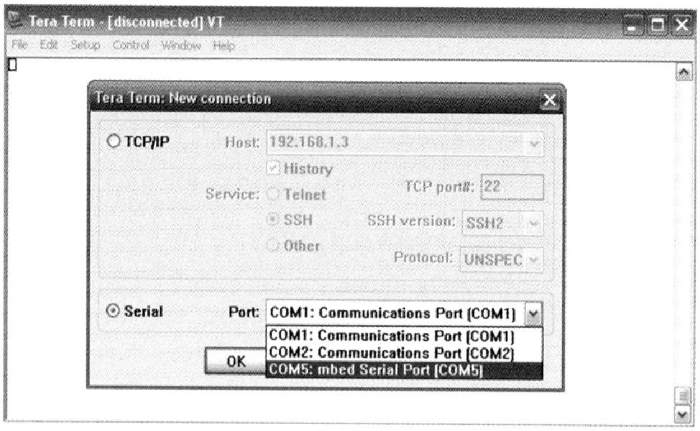

If *Mbed Serial Port* is not in the dropdown menu, a Windows serial driver may be missing, or there may be a connection problem.

E.3 Testing the terminal emulator

To test a terminal running with an Mbed target, create a new project of suitable name in the IDE, and enter the code of Program Example E.1. This code simply greets the world, and then reads keyboard input as it comes and displays it to the terminal. To show that the character value is genuinely being seen by the Mbed, the program adds one to the ASCII code, meaning that every number is incremented by 1, and every letter is echoed back to the terminal as the letter following in the alphabet, b instead of a, c instead of b, and so on. See Section 7.8 for an underlying explanation of the program.

```
/* Program Example E.1: Print to the PC, then read keyboard characters and pass
back (slightly modified!)
Works on both LPC1768 and F401RE, with no external connections on either. */

#include "mbed.h"
BufferedSerial serial_port(USBTX, USBRX);    // define transmitter and receiver
DigitalOut led (LED1);
char buf[14]="Hello World\n\r";          //declare buffer and load text message

int main() {
  serial_port.write(buf,13);
  while(true) {
    if (serial_port.readable()==1){     //is there a character to be read?
      serial_port.read(buf,1);       //if yes, then read it
      buf[0]=buf[0]+1;               //add 1, to show the program is active
      serial_port.write(buf,1);
      led=!led;                      //diagnostic
    }
  }
}
```

Program Example E.1 Transferring keyboard characters to screen

References

E.1. Tera Term home page. http://www.teraterm.org/.
E.2. CoolTerm download page. http://freeware.the-meiers.org.
E.3. PuTTY home page. https://putty.org/.

Index

Note: Page numbers followed by f indicate figures, t indicate tables and b indicate boxes.

H

I

J

K

L

Q

R

S